*MATHEMATICAL
METHODS IN SCIENCE
AND ENGINEERING*

MATHEMATICAL METHODS IN SCIENCE AND ENGINEERING

Ş. SELÇUK BAYIN

Middle East Technical University
Ankara, Turkey

A JOHN WILEY & SONS, INC., PUBLICATION

Published by John Wiley & Sons, Inc., Hoboken, New Jersey.
Published simultaneously in Canada.

For general information on our other products and services or for technical support, please contact our
Customer Care Department within the United States at (800) 762-2974, outside the United States at
(317) 572-3993 or fax (317) 572-4002.

Wiley also publishes its books in a variety of electronic formats. Some content that appears in print may
not be available in electronic format. For information about Wiley products, visit our web site at
www.wiley.com.

Library of Congress Cataloging-in-Publication Data is available.

　ISBN-13 978-0-470-04142-0
　ISBN-10 0-470-04142-0

Printed in the United States of America.

10 9 8 7 6 5 4 3 2 1

Contents

Preface

Courses on mathematical methods of physics are among the essential courses for graduate programs in physics, which are also offered by most engineering departments. Considering that the audience in these courses comes from all subdisciplines of physics and engineering, the content and the level of mathematical formalism has to be chosen very carefully. Recently the growing interest in interdisciplinary studies has brought scientists together from physics, chemistry, biology, economy, and finance and has increased the demand for these courses in which upper-level mathematical techniques are taught. It is for this reason that the mathematics departments, who once overlooked these courses, are now themselves designing and offering them.

Most of the available books for these courses are written with theoretical physicists in mind and thus are somewhat insensitive to the needs of this new multidisciplinary audience. Besides, these books should not only be tuned to the existing practical needs of this multidisciplinary audience but should also play a lead role in the development of new interdisciplinary science by introducing new techniques to students and researchers.

About the Book

We give a coherent treatment of the selected topics with a style that makes advanced mathematical tools accessible to a multidisciplinary audience. The book is written in a modular way so that each chapter is actually a review of its subject and can be read independently. This makes the book very useful as a reference for scientists. We emphasize physical motivation and the multidisciplinary nature of the methods discussed.

The entire book contains enough material for a three-semester course meeting three hours a week. However, the modular structure of the book gives enough flexibility to adopt the book for several different advanced undergraduate and graduate-level courses. Chapter 1 is a philosophical prelude about physics, mathematics, and mind for the interested reader. It is not a part of the curriculum for courses on mathematical methods of physics. Chapters 2-8, 12, 13 and 15-19 have been used for a two-semester compulsory graduate course meeting three hours a week. Chapters 16-20 can be used for an introductory graduate course on Green's functions. For an upper-level undergraduate course on special functions, colleagues have used Chapters 1-8. Chapter 14 on fractional calculus can be expanded into a one-term elective course supported by projects given to students. Chapters 2-11 can be used in an introductory graduate course, with emphasis given to Chapters 8-11 on Sturm-Liouville theory, factorization method, coordinate transformations, general tensors, continuous groups, Lie algebras, and representations.

Students are expected to be familiar with the topics generally covered during the first three years of the science and engineering undergraduate curriculum. These basically comprise the contents of the books *Advanced Calculus* by Kaplan, *Introductory Complex Analysis* by Brown and Churchill, and *Differential Equations* by Ross, or the contents of books like *Mathematical Methods in Physical Sciences* by Boas, *Mathematical Methods: for Students of Physics and Related Fields* by Hassani, and *Essential Mathematical Methods for Physicists* by Arfken and Weber. Chapters (10 and 11) on coordinates, tensors, and groups assume that the student has already seen orthogonal transformations and various coordinate systems. These are usually covered during the third year of the undergraduate physics curriculum at the level of *Classical Mechanics* by Marion or *Theoretical Mechanics* by Bradbury. For the sections on special relativity (in Chapter 10) we assume that the student is familiar with basic special relativity, which is usually covered during the third year of undergraduate curriculum in modern physics courses with text books like *Concepts of Modern Physics* by Beiser.

Three very interesting chapters on the method of factorization, fractional calculus, and path integrals are included for the first time in a text book on mathematical methods. These three chapters are also extensive reviews of these subjects for beginning researchers and advanced graduate students.

Summary of the Book

In Chapter 1 we start with a philosophical prelude about physics, mathematics, and mind.

In Chapters 2-6 we present a detailed discussion of the most frequently encountered special functions in science and engineering. This is also very timely, because during the first year of graduate programs these functions are used extensively. We emphasize the fact that certain second-order partial differential equations are encountered in many different areas of science, thus allowing one to use similar techniques. First we approach these partial differential equations by the method of separation of variables and reduce them to a set of ordinary differential equations. They are then solved by the method of series, and the special functions are constructed by imposing appropriate boundary conditions. Each chapter is devoted to a particular special function, where it is discussed in detail. Chapter 7 introduces hypergeometric equation and its solutions. They are very useful in parametric representations of the commonly encountered second-order differential equations and their solutions. Finally our discussion of special functions climaxes with Chapter 8, where a systematic treatment of their common properties is given in terms of the Sturm-Liouville theory. The subject is now approached as an eigenvalue problem for second-order linear differential operators.

Chapter 9 is one of the special chapters of the book. It is a natural extension of the chapter on Sturm-Liouville theory and approaches second-order differential equations of physics and engineering from the viewpoint of the theory of factorization. After a detailed analysis of the basic theory we discuss specific cases. Spherical harmonics, Laguerre polynomials, Hermite polynomials, Gegenbauer polynomials, and Bessel functions are revisited and studied in detail with the factorization method. This method is not only an interesting approach to solving Sturm-Liouville systems, but also has deep connections with the symmetries of the system.

Chapter 10 presents an extensive treatment of coordinates, their transformations, and tensors. We start with the Cartesian coordinates, their transformations, and Cartesian tensors. The discussion is then extended to general coordinate transformations and general tensors. We also discuss Minkowski spacetime, coordinate transformations in spacetime, and four-tensors in detail. We also write Maxwell's equations and Newton's dynamical theory in covariant form and discuss their transformation properties in spacetime.

In Chapter 11 we discuss continuous groups, Lie algebras, and group representations. Applications to the rotation group, special unitary group, and homogeneous Lorentz group are discussed in detail. An advanced treatment of spherical harmonics is given in terms of the rotation group and its representations. We also discuss symmetry of differential equations and extension (prolongation) of generators.

Chapters 12 and 13 deal with complex analysis. We discuss the theory of analytic functions, mappings, and conformal and Schwarz-Christoffel transformations with interesting examples like the fringe effects of a parallel plate capacitor and fluid flow around an obstacle. We also discuss complex integrals, series, and analytic continuation along with the methods of evaluating some definite integrals.

Chapter 14 introduces the basics of fractional calculus. After introducing the experimental motivation for why we need fractional derivatives and integrals, we give a unified representation of the derivative and integral and extend it to fractional orders. Equivalency of different definitions, examples, properties, and techniques with fractional derivatives are discussed. We conclude with examples from Brownian motion and the Fokker-Planck equation. This is an emerging field with enormous potential and with applications to physics, chemistry, biology, engineering, and finance. For beginning researchers and instructors who want to add something new and interesting to their course, this self-contained chapter is an excellent place to start.

Chapter 15 contains a comprehensive discussion of infinite series: tests of convergence, properties, power series, and uniform convergence along with the methods of evaluating sums of infinite series. An interesting section on divergent series in physics is added with a discussion of the Casimir effect.

Chapter 16 treats integral transforms. We start with the general definition, and then the two most commonly used integral transforms, Fourier and Laplace transforms, are discussed in detail with their various applications and techniques.

Chapter 17 is on variational analysis. Cases with different numbers of dependent and independent variables are discussed. Problems with constraints, variational techniques in eigenvalue problems, and the Rayleigh-Ritz method are among other interesting topics covered.

In Chapter 18 we introduce integral equations. We start with their classification and their relation to differential equations and vice versa. We continue with the methods of solving integral equations and conclude with the eigenvalue problem for integral operators, that is, the Hilbert-Schmidt theory.

In Chapter 19 (and 20) we present Green's functions, and this is the second climax of this book, where everything discussed so far is used and their connections seen. We start with the time-independent Green's functions in one dimension and continue with three-dimensional Green's functions. We discuss their applications to electromagnetic theory and the Schrödinger equation. Next we discuss first-order time-dependent Green's functions with applications to diffusion problems and the time-dependent Schrödinger equation. We introduce the propagator interpretation and the compounding of propagators. We conclude this section with second-order time-dependent Green's functions, and their application to the wave equation and discuss advanced and retarded solutions.

Chapter 20 is an extensive discussion of path integrals and their relation to Green's functions. During the past decade or so path integrals have found wide range of applications among many different fields ranging from physics to finance. We start with the Brownian motion, which is considered a prototype of many different processes in physics, chemistry, biology, finance etc. We discuss the Wiener path integral approach to Brownian motion. After the Feynman-Kac formula is introduced, the perturbative solution of the Bloch equation is given. Next an interpretation of $V(x)$ in the Bloch equation is given, and we continue with the methods of evaluating path integrals. We also discuss the Feynman path integral formulation of quantum mechanics along with the phase space approach to Feynman path integrals.

Story of the Book

Since 1989, I have been teaching the graduate level 'Methods of Mathematical Physics I & II' courses at the Middle East Technical University in Ankara. Chapters 2-8 with 12 and 13 have been used for the first part and Chapters 15-19 for the second part of this course, which meets three hours a week. Whenever possible I prefer to introduce mathematical techniques through physical applications. Examples are often used to extend discussions of specific techniques rather than as mere exercises. Topics are introduced in a logical sequence and discussed thoroughly. Each sequence climaxes with a part where the material of the previous chapters is unified in terms of a general theory, as in Chapter 8 (and 9) on the Sturm-Liouville theory, or with a part that utilizes the gains of the previous chapters, as in Chapter 19 (and 20) on Green's functions. Chapter 9 is on factorization method, which is a natural extension of our discussion on the Sturm-Liouville theory. It also presents a different and advanced treatment of special functions. Similarly, Chapter 20 on path integrals is a natural extension of our chapter on Green's functions. Chapters 10 and 11 on coordinates, tensors, and continuous groups have been located after Chapter 9 on the Sturm-Liouville theory and the factorization method. Chapters 12 and 13 are on complex techniques, and they are self-contained. Chapter 14 on fractional calculus can either be integrated into the curriculum of the mathematical methods of physics courses or used independently.

During my lectures and first reading of the book I recommend that readers view equations as statements and concentrate on the logical structure of the discussions. Later, when they go through the derivations, technical details become understood, alternate approaches appear, and some of the questions are answered. Sufficient numbers of problems are given at the back of each chapter. They are carefully selected and should be considered an integral part of the learning process.

In a vast area like mathematical methods in science and engineering, there is always room for new approaches, new applications, and new topics. In fact, the number of books, old and new, written on this subject shows how dynamic this field is. Naturally this book carries an imprint of my style and lectures. Because the main aim of this book is pedagogy, occasionally I have followed other books when their approaches made perfect sense to me. Sometimes I indicated this in the text itself, but a complete list is given at the back. Readers of this book will hopefully be well prepared for advanced graduate studies in many areas of physics. In particular, as we use the same terminology and style, they should be ready for full-term graduate courses based on the books: *The Fractional Calculus* by Oldham and Spanier and *Path Inte-*

grals in Physics, Volumes I and II by Chaichian and Demichev, or they could jump into the advanced sections of these books, which have become standard references in their fields.

I recommend that students familiarize themselves with the existing literature. Except for an isolated number of instances I have avoided giving references within the text. The references at the end should be a good first step in the process of meeting the literature. In addition to the references at the back, there are also three websites that are invaluable to students and researchers: For original research, http://lanl.arxiv.org/ and the two online encyclopedias: http://en.wikipedia.org and http://scienceworld.wolfram.com/ are very useful. For our chapters on special functions these online encyclopedias are extremely helpful with graphs and additional information.

A precursor of this book (Chapters 1-8, 12, 13, and 15-19) was published in Turkish in 2000. With the addition of two new chapters on fractional calculus and path integrals, the revised and expanded version appeared in 2004 as 440 pages and became a widely used text among the Turkish universities. The positive feedback from the Turkish versions helped me to prepare this book with a minimum number of errors and glitches. For news and communications about the book we will use the website http://www.physics.metu.edu.tr/~ bayin, which will also contain some relevant links of interest to readers.

S. BAYIN

ODTÜ
Ankara/TURKEY
April 2006

Acknowledgments

I would like to pay tribute to all the scientists and mathematicians whose works contributed to the subjects discussed in this book. I would also like to compliment the authors of the existing books on mathematical methods of physics. I appreciate the time and dedication that went into writing them. Most of them existed even before I was a graduate student. I have benefitted from them greatly. I am indebted to Prof. K.T. Hecht of the University of Michigan, whose excellent lectures and clear style had a great influence on me. I am grateful to Prof. P.G.L. Leach for sharing his wisdom with me and for meticulously reading Chapters 1 and 9 with 14 and 20. I also thank Prof. N. K. Pak for many interesting and stimulating discussions, encouragement, and critical reading of the chapter on path integrals. I thank Wiley for the support by a grant during the preparation of the camera ready copy. My special thanks go to my editors at Wiley, Steve Quigley, Susanne Steitz, and Danielle Lacourciere for sharing my excitement and their utmost care in bringing this book into existence.

I finally thank my wife, Adalet, and daughter, Sumru, for their endless support during the long and strenuous period of writing, which spanned over several years.

Ş.S.B.

1

NATURE and MATHEMATICS

The most incomprehensible thing about this universe is that it is comprehensible
— Albert Einstein

When man first opens his eyes into this universe, he encounters an endless variety of events and shivers as he wonders how he will ever survive in this enormously complex system. However, as he contemplates he begins to realize that the universe is not hostile and there is some order among all this diversity.

As he wanders around, he inadvertently kicks stones on his path. As the stones tumble away, he notices that the smaller stones not only do not hurt his feet, but also go further. Of course, he quickly learns to avoid the bigger ones. The sun, to which he did not pay too much attention at first, slowly begins to disappear; eventually leaving him in cold and dark. At first this scares him a lot. However, what a joy it must be to witness the sun slowly reappearing in the horizon. As he continues to explore, he realizes that the order in this universe is also dependable. Small stones, which did not hurt him, do not hurt him another day in another place. Even though the sun eventually disappears, leaving him in cold and dark, he is now confident that it will reappear. In time he learns to live in communities and develops languages to communicate with his fellow human beings. Eventually the quality and the number of observations he makes increase. In fact, he even begins to undertake projects that require careful recording and interpretation of data that span over several generations. As in Stonehenge he even builds an agricultural computer to find the crop times. A similar version of this story is actually repeated with every newborn.

For man to understand nature and his place in it has always been an instinctive desire. Along this endeavour he eventually realizes that the everyday language developed to communicate with his fellow human beings is not sufficient. For further understanding of the law and order in the universe, a new language, richer and more in tune with the inner logic of the universe, is needed. At this point physics and mathematics begin to get acquainted. With the discovery of coordinate systems, which is one of the greatest constructions of the free human mind, foundations of this relation become ready. Once a coordinate system is defined, it is possible to reduce all the events in the universe to numbers. Physical processes and the law and order that exists among these events can now be searched among these numbers and could be expressed in terms of mathematical constructs much more efficiently and economically. From the motion of a stone to the motions of planets and stars, it can now be understood and expressed in terms of the dynamical theory of Newton:

$$\overrightarrow{F} = m\frac{d^2\overrightarrow{r}}{dt^2} \tag{1.1}$$

and his law of gravitation

$$\overrightarrow{F} = G\frac{m_1 m_2}{r^2}\widehat{e}_r. \tag{1.2}$$

Newton's theory is full of the success stories that very few theories will ever have for years to come. Among the most dramatic is the discovery of Neptune. At the time small deviations from the calculated orbit of Uranus were observed. At first the neighboring planets, Saturn and Jupiter, were thought to be the cause. However, even after the effects of these planets were subtracted, a small unexplained difference remained. Some scientists questioned even the validity of Newton's theory. However, astronomers, putting their trust in Newton's theory, postulated the existence of another planet as the source of these deviations. From the amount of the deviations they calculated the orbit and the mass of this proposed planet. They even gave a name to it: Neptune. Now the time had come to observe this planet. When the telescopes were turned into the calculated coordinates: Hello! Neptune was there. In the nineteenth century, when Newton's theory was joined by Maxwell's theory of electromagnetism, there was a time when even the greatest minds like Bertrand Russell began to think that physics might have come to an end, that is, the existing laws of nature could in principle explain all physical phenomena.

Actually, neither Newton's equations nor Maxwell's equations are laws in the strict sense. They are based on some assumptions. Thus it is probably more appropriate to call them theories or models. We frequently make assumptions in science. Sometimes in order to concentrate on a special but frequently encountered case, we keep some of the parameters constant to avoid unnecessary complications. At other times, because of the complexity of the problem, we restrict our treatment to certain domains like small velocities, high temperatures, weak fields, etc. However, the most important of all are the assumptions that sneak into our theories without our awareness. Such assumptions are actually manifestations of our prejudices about nature. They come so naturally to us that we usually do not notice their presence. In fact, it sometimes takes generations before they are recognized as assumptions. Once they are identified and theories are reformulated, dramatic changes take place in our understanding of nature.

Toward the beginning of the twentieth century the foundations of Newton's dynamical theory are shaken by the introduction of new concepts like the wave-particle duality and the principle of uncertainty. It eventually gives way to quantum mechanics. Similarly, Galilean relativity gives way to the special theory of relativity when it is realized that there is an upper limit to velocities in nature, which is the speed of light. Newton's theory of gravitation also gives way to Einstein's theory of gravitation when it is realized that absolute space and flat geometry are assumptions valid only for slowly moving systems and near small masses.

However, the development of science does not take place by leaving the successful theories of the past in desolation either. Yes, the wave-particle duality, the principle of uncertainty, and a new type of determinism are all essential elements of quantum mechanics, which are all new to Newton's theory. However, it is also true that in the classical limit, $\hbar \rightarrow 0$, quantum mechanics reduces to Newton's theory and for many important physical and astronomical phenomena, quantum mechanical treatment is not practical. In such cases Newton's theory is still the economical theory to use. Similarly, even though there is an upper limit to velocity, for many practical problems speed of light can be taken as infinity, thus making Galilean relativity still useful. Even though Newton's theory of gravitation has been replaced by Einstein's theory, for a large class of astronomical problems the curvature of spacetime can be neglected. For these problems Newton's theory still remains an excellent working theory.

1.1 MATHEMATICS AND NATURE

As time goes on, the mathematical techniques and concepts used to understand nature develop and increase in number. Today we have been rather successful in representing physical processes in terms of mathematics, but one thing has never changed. Mathematics is a world of numbers, and, if we have to understand nature by mathematics, we have to transform it into numbers first. However, aside from integers all the other numbers are constructs of the free human mind. Besides, mathematics has a certain logical structure to it, thus implying a closed or complete system. Considering that our knowledge of the universe is far too limited to be understood by logic, one naturally wonders why mathematics is so successful as a language. What is the secret of this mysterious relation between physics and mathematics?

In 1920 Hilbert suggested that mathematics be formulated on a solid and complete logical foundation such that all mathematics can be derived from a finite and consistent system of axioms. This philosophy of mathematics is usually called **formalism**. In 1931 Gödel shattered the foundations of the formal approach to mathematics with his famous **incompleteness theorem**. This theorem not only showed that Hilbert's goal is impossible but also proved to be only the first in a series of deep and counterintuitive statements about rigor and provability of mathematics. Could Gödel's incompleteness theorem be the source of this mysterious relation between mathematics and nature?

It is true that certain mathematical models have been rather successful in expressing the law and order in the universe. However, this does not mean that all possible mathematical models and concepts will somehow find a place in science. If we could have extended our understanding of nature by logical extensions of the existing theories, physics would have been rather easy. Sometimes physicists are almost hypnotized by the mathematical beauty and the sophistication of their theories, so that they begin to lose contact with nature. We should not get upset if it happens that nature has not preferred our way.

1.2 LAWS OF NATURE

At first we had only the dynamical theory of Newton and his theory of gravitation. Then came Maxwell's theory of electromagnetism. After the discovery of quantum mechanics in the early twentieth century, there was a brief period when it was thought that everything in nature could in principle be explained in terms of the three elementary particles; electron, proton, and neutron, and the electromagnetic and gravitational interactions between them. Not for long; The discovery of strong and weak interactions along with a proliferation of new particles complicated the picture. Introduction of quarks as the new elementary constituents of these particles did not help, either. Today string theorists are trying to build a theory of everything in which all known interactions are unified.

What is a true law of nature? In my opinion genuine laws of nature are relatively simple and in general expressed as inequalities like the uncertainty principle:

$$\Delta x \Delta p \geq \hbar \tag{1.3}$$

and the second law of entropy:

$$\delta S(\text{ total entropy of the universe}) \geq 0. \tag{1.4}$$

Others, which are expressed in terms of equalities, are theories or models based on certain assumptions and subject to change in time.

1.3 MATHEMATICS AND MIND

Almost everywhere mathematics is a very useful and powerful language in expressing the law and order in the universe. However, mathematics is also a world of ideas, and these ideas occur as a result of some physical processes at the cellular and molecular level in our brain. Today not just our physical properties like the eye and hair colors but also the human psyche is thought to be linked to our genes. We have taken important strides in identifying parts of our genes that are responsible for certain properties. Research is ongoing in developing technologies that will allow to us to remove or replace parts of our genes that may represent a potential hazard to our health. Scientists are working on mechanisms to silence or turn off bad genes in a cell. This mechanism will eventually lead to the development of new medicines for protecting cells from hostile genes and treating diseases. Even though we still have a long way to go, we have covered important distance in understanding and controlling our genetic code.

To understand and codify ideas in terms of some basic physical processes naturally requires a significantly deeper level of understanding of our brain and its processes. If ideas could be linked to certain physical processes at the molecular and cellular level, then there could also exist a finite upper limit to the number of ideas, no matter how absurd they may be, that we could ever devise. This limit basically implies that one's brain has a finite phase space, which allows only a finite number of configurations corresponding to ideas. This also means that there is an upper limit to all the mathematical statements, theorems, concepts, etc. that we could ever imagine. We simply cannot think of anything that requires a process that either violates some of the fundamental laws of nature or requires a brain with a larger phase space. A quick way to improve this limit is to have a bigger brain. In fact, to some extent nature has already utilized this alternative. It is evidenced in fossils that, as humans evolved, brain size increased dramatically. The average brain size of *Homo habilis*, who lived approximately 2 million years ago, was approximately 750 cc. *Homo erectus*, who lived 1.7−1 million years ago, averaged 900 cc in brain size. The modern human skull holds a brain of around 1400 cc.

However, brain size and intelligence are only correlated loosely. A much more stringent limit to our mental capacity naturally comes from the inner efficiency of our brain. Research on subjects like brain stimulators, hard wiring of our brain, and mind reading machines are all aiming at a faster and much more efficient use of our brain. A better understanding of our brain may also bring a more efficient way of using our creativity, much needed at times of crisis or impasses, the working of which is now left to chance. The possibility of tracing ideas to their origins in terms of physical processes at the molecular and cellular level and also the possibility of codifying them with respect to some finite, probably small, number of key processes implies that the relation between mathematics and nature may actually work both ways.

1.4   IS MATHEMATICS THE ONLY LANGUAGE FOR NATURE?

We have been extremely successful with mathematics in understanding and expressing the law and order in the universe. However, can there be other languages? Can the universe itself serve as its own language? It is known that intrinsically different phenomena occasionally satisfy similar mathematical equations. For example, in two-dimensional electrostatic problems the potential satisfies

$$\overrightarrow{\nabla}^2 \Phi(x,y) = -\frac{\rho(x,y)}{\varepsilon_0}, \tag{1.5}$$

where Φ is the electrostatic potential, ρ is the charge density, and ε_0 is the constant permittivity of vacuum. Now consider an elastic sheet stretched over a cylindrical frame like a drum head with uniform tension τ. If we push this sheet by small amounts, its displacement from its equilibrium position, $u(x,y)$, satisfies the equation

$$\overrightarrow{\nabla}^2 u(x,y) = -\frac{f(x,y)}{\tau}, \tag{1.6}$$

where $f(x,y)$ is the applied force. If we make the identification

$$\Phi(x,y) \rightarrow u(x,y) \tag{1.7}$$

$$\frac{\rho(x,y)}{\varepsilon_0} \rightarrow \frac{f(x,y)}{\tau}, \tag{1.8}$$

all the electrostatics problems with infinite charged sheets, long parallel wires, or charged cylinders have a representation in terms of a stretched membrane. In fact, this method has been used to solve complex electrical problems. By pushing rods and bars at various heights against a membrane corresponding to the potentials of a set of electrodes, we can obtain the electric potential by simply reading the displacement of the membrane. The analogy can even be carried further. If we put little balls on the membrane, their motion is approximately the corresponding motion of electrons in the corresponding electric field. This method has actually been used to obtain the complicated geometry of many photomultipliers.

The limitation of this method is that Equation (1.6) is valid only for small displacements of the membrane. Also, the difficulty in preparing a membrane with uniform tension restricts the accuracy. However, the beauty of the method is that we can find the solution of a complex boundary value problem without actually solving a partial differential equation. Note that even though we have not solved the boundary value problem explicitly, we have still used mathematics to link the two phenomena.

Recently scientists have been intrigued by the uncanny similarity between the propagation of light in curved spacetime and the propagation of sound in uneven flow. Scientists are now trying to exploit these similarities to gain insight into the microscopic structure of spacetime. Even black holes have acoustic counterparts. Acoustic analogs of the Casimir effect, which is usually introduced as a purely quantum mechanical phenomenon, are now being investigated with technological applications in mind. The development of fast computers has slowed the development of this approach. However, the fact that nature could also be its own language is something to keep in mind.

1.5 NATURE AND MANKIND

What is our place in this universe? What is our role? Why does this universe exist? Man has probably asked questions like these since the beginning of time. Are we any closer to the answers? If we discover the theory of everything, will at least some of these questions be answered? Scientist or not, everybody has wondered about these issues.

Let us now imagine a civilization the entire universe of which is all the existing novels. Members of this civilization are amazed by the events depicted in these novels and wonder about the reason behind all the drama and the intricate relations among the characters. One day, one of their scientists comes up with a model, claims that all these novels are composed of a finite number of words, and prepares a dictionary. They all get excited, and the experimentalists begin to search every sentence and every paragraph that they can find. In time a few additions and subtractions are made to this dictionary, but one thing does not change: Their universe is made up of a finite number of words. As they are happy with this theory, a new scientist comes along and claims that all these words in the dictionary and the novels themselves are actually made up of a small number of letters, numbers, and punctuation marks. After intense testing, this theory also finds enormous support and its author is hailed with their greatest honors. Naturally this story goes on and as the quality of their observations increases, they begin to discover grammatical rules. The rules of grammar are actually the laws of nature in this universe. It is clear that grammar rules alone cannot tell us why a novel is written, but it is not possible to understand a novel properly without knowing the grammar rules, either.

As we said, scientist or not, everybody has wondered why this universe exists and what our place in this magnificent system is. Even though no simple answers exist, it is incredible that almost everybody has somehow come to a peaceful coexistence with such questions. What we should realize is that such questions do not have a single answer. With analogies like the one we just gave, one may only get a glimpse of one of the many facets of truth. Somebody else may come up with another analogy that may be as intriguing as this one. Starting from the success of simulation experiments it has been argued that the universe acts like a giant computer, where matter is its hardware and the laws of nature are its software. Now the question to be answered becomes: Who built this computer, and for what is it being used?

2

LEGENDRE EQUATION and POLYNOMIALS

Many of the second-order partial differential equations of physics and engineering can be written as

$$\overrightarrow{\nabla}^2 \Psi(x, y, z) + k^2 \Psi(x, y, z) = F(x, y, z), \qquad (2.1)$$

where k in general is a function of coordinates. Some examples for these equations are:

1. If k and $F(x, y, z)$ are zero, Equation (2.1) becomes the Laplace equation

$$\overrightarrow{\nabla}^2 \Psi(x, y, z) = 0, \qquad (2.2)$$

   which is encountered in many different areas of science like electrostatics, magnetostatics, laminar (irrotational) flow, surface waves, heat transfer, and gravitation.

2. When the right-hand side of the Laplace equation is different from zero, we have the Poisson equation:

$$\overrightarrow{\nabla}^2 \Psi = F(x, y, z), \qquad (2.3)$$

   where $F(x, y, z)$ represents sources in the system.

3. The Helmholtz wave equation is given as

$$\overrightarrow{\nabla}^2 \Psi(x, y, z) \pm k_0^2 \Psi(x, y, z) = 0, \qquad (2.4)$$

where k_0 is a constant.

4. Another important example is the time-independent Schrödinger equation

$$-\frac{\hbar^2}{2m}\overrightarrow{\nabla}^2\Psi(x,y,z) + V\Psi(x,y,z) = E\Psi(x,y,z),\qquad(2.5)$$

where $F(x,y,z)$ in Equation (2.1) is zero and k is now given as

$$k^2 = \frac{2m}{\hbar^2}[E - V(x,y,z)].\qquad(2.6)$$

All these equations are linear and second-order partial differential equations. Separation of variables, Green's functions, and integral transforms are among the most frequently used techniques for obtaining analytic solutions. In addition to these there are also numerical techniques like Runge-Kutta. Appearance of similar differential equations in different areas of science allows one to adopt techniques developed in one area into another. Of course, the variables and interpretation of the solutions will be very different. Also, one has to be aware of the fact that boundary conditions used in one area may not be appropriate for another. For example, in electrostatics charged particles can only move perpendicular to the conducting surfaces, whereas in laminar (irrotational) flow fluid elements follow the contours of the surfaces; thus even though the Laplace equation is to be solved in both cases, solutions obtained in electrostatics may not always have meaningful counterparts in laminar flow.

2.1   LEGENDRE EQUATION

We now solve Equation (2.1) in spherical polar coordinates by using the method of separation of variables. We consider cases where k is only a function of the radial coordinate, and also we take F as zero. The time-independent Schrödinger Equation (2.5) written for central force problems, where

$$V(x,y,z) = V(r)\qquad(2.7)$$

is an important example for such cases. We first separate r and the (θ,ϕ) variables and write the solution, $\Psi(r,\theta,\phi)$, as

$$\Psi(r,\theta,\phi) = R(r)Y(\theta,\phi).\qquad(2.8)$$

This basically assumes that the radial dependence of the solution is independent of the (θ,ϕ) dependence and vice versa. Substituting this in Equation

(2.1) we get

$$\frac{1}{r^2}\frac{\partial}{\partial r}\left[r^2\frac{\partial}{\partial r}R\left(r\right)Y\left(\theta,\phi\right)\right] + \frac{1}{r^2\sin\theta}\frac{\partial}{\partial\theta}\left[\sin\theta\frac{\partial}{\partial\theta}R\left(r\right)Y\left(\theta,\phi\right)\right]$$

$$+\frac{1}{r^2\sin^2\theta}\frac{\partial^2}{\partial\phi^2}R\left(r\right)Y\left(\theta,\psi\right) + k^2\left(r\right)R\left(r\right)Y\left(\theta,\psi\right) = 0. \tag{2.9}$$

After multiplying the above equation by

$$\frac{r^2}{R\left(r\right)Y\left(\theta,\phi\right)} \tag{2.10}$$

and collecting the (θ,ϕ) dependence on the right-hand side we obtain

$$\frac{1}{R\left(r\right)}\frac{\partial}{\partial r}\left[r^2\frac{\partial}{\partial r}R\left(r\right)\right] + k^2\left(r\right)r^2 =$$

$$-\frac{1}{\sin\theta}\frac{1}{Y\left(\theta,\phi\right)}\frac{\partial}{\partial\theta}\left[\sin\theta\frac{\partial}{\partial\theta}Y\left(\theta,\phi\right)\right] - \frac{1}{\sin^2\theta Y\left(\theta,\phi\right)}\frac{\partial^2 Y\left(\theta,\phi\right)}{\partial\phi^2}. \tag{2.11}$$

Since r and (θ,ϕ) are independent variables, this equation can be satisfied for all r and (θ,ϕ) only when both sides of the equation are equal to the same constant. We show this constant with λ, which is also called the separation constant. Now Equation (2.11) reduces to the following two equations:

$$\frac{d}{dr}\left(r^2\frac{dR\left(r\right)}{dr}\right) + r^2k^2\left(r\right)R\left(r\right) - \lambda R\left(r\right) = 0 \tag{2.12}$$

and

$$\frac{1}{\sin\theta}\frac{\partial}{\partial\theta}\left[\sin\theta\frac{\partial Y\left(\theta,\phi\right)}{\partial\theta}\right] + \frac{1}{\sin^2\theta}\frac{\partial^2 Y\left(\theta,\phi\right)}{\partial\phi^2} + \lambda Y\left(\theta,\phi\right) = 0. \tag{2.13}$$

Equation (2.12) for $R(r)$ is now an ordinary differential equation. We also separate the θ and the ϕ variables in $Y\left(\theta,\phi\right)$ as

$$Y\left(\theta,\phi\right) = \Theta\left(\theta\right)\Phi\left(\phi\right) \tag{2.14}$$

and call the new separation constant m^2, and write

$$\frac{\sin\theta}{\Theta\left(\theta\right)}\frac{d}{d\theta}\left[\sin\theta\frac{d\Theta}{d\theta}\right] + \lambda\sin^2\theta = -\frac{1}{\Phi\left(\phi\right)}\frac{d^2\Phi\left(\phi\right)}{d\phi^2} = m^2. \tag{2.15}$$

We now obtain the differential equations to be solved for $\Theta\left(\theta\right)$ and $\Phi\left(\phi\right)$ as

$$\sin^2\theta\frac{d^2\Theta\left(\theta\right)}{d\theta^2} + \cos\theta\sin\theta\frac{d\Theta\left(\theta\right)}{d\theta} + \left[\lambda\sin^2\theta - m^2\right]\Theta\left(\theta\right) = 0 \tag{2.16}$$

and

$$\frac{d^2\Phi(\phi)}{d\phi^2} + m^2\Phi(\phi) = 0. \tag{2.17}$$

In summary, using the method of separation of variables we have reduced the partial differential Equation (2.9) to three ordinary differential Equations, (2.12), (2.16), and (2.17). During this process two constant parameters, λ and m, called the separation constants have entered into our equations, which so far have no restrictions on them.

2.1.1 Method of Separation of Variables

In the above discussion the fact that we are able to separate the solution is closely related to our use of the spherical polar coordinates, which reflect the symmetry of the central potential best. If we had used the Cartesian coordinates, the potential would be given as $V(x, y, z)$ and the solution would not be separable, that is

$$\Psi(x, y, z) \neq X(x)Y(y)Z(z).$$

Whether a given partial differential equation is separable or not is closely related to the symmetries of the physical system. Even though a proper discussion of this point is beyond the scope of this book, we refer the reader to Stephani (p. 193) and suffice by saying that if a partial differential equation is not separable in a given coordinate system it is possible to check the existence of a coordinate system in which it would be separable, and if such a coordinate system exists it is possible to construct it with the generators of the symmetries.

Among the three ordinary differential Equations (2.12), (2.16), and (2.17), Equation (2.17) can be solved immediately with the general solution

$$\Phi(\phi) = Ae^{im\phi} + Be^{-im\phi}, \tag{2.18}$$

where m is still unrestricted. Using the periodic boundary condition

$$\Phi(\phi + 2\pi) = \Phi(\phi), \tag{2.19}$$

it is seen that m could only take integer values: $0, \pm 1, \pm 2, \dots$. Note that in anticipation of applications to quantum mechanics we have taken the two linearly independent solutions as $e^{\pm im\phi}$. For other problems $\sin m\phi$ and $\cos m\phi$ is preferred.

For the differential equation to be solved for $\Theta(\theta)$ we define a new independent variable

$$x = \cos\theta, \qquad (\theta \in [0, \pi], \ x \in [-1, 1]) \tag{2.20}$$

and write

$$(1 - x^2) \frac{d^2 Z(x)}{dx^2} - 2x \frac{dZ(x)}{dx} + \left[\lambda - \frac{m^2}{(1 - x^2)} \right] Z(x) = 0. \qquad (2.21)$$

For $m = 0$ this equation is called the **Legendre equation,** and for $m \neq 0$ it is known as the **associated Legendre equation.**

2.2   SERIES SOLUTION OF THE LEGENDRE EQUATION

Starting with the

$$m = 0$$

case we write the Legendre equation as

$$(1 - x^2) \frac{d^2 Z(x)}{dx^2} - 2x \frac{dZ(x)}{dx} + \lambda Z(x) = 0, \ x \in [-1, 1]. \qquad (2.22)$$

This has two regular singular points at $x = \pm 1$. Since these points are at the end points of our interval, using the Frobenius method we can try a series solution about

$$x = 0 \qquad (2.23)$$

as

$$Z(x) = \sum_{k=0}^{\infty} a_k x^{k+\alpha}. \qquad (2.24)$$

Substituting this into Equation (2.22) we get

$$\sum_{k=0}^{\infty} a_k (k + \alpha) (k + \alpha - 1) x^{k+\alpha-2}$$

$$- \sum_{k=0}^{\infty} x^{k+\alpha} [(k + \alpha)(k + \alpha - 1) + 2(k + \alpha) - \lambda] a_k = 0. \qquad (2.25)$$

We write the first two terms of first series in the above equation explicitly as

$$a_0 \alpha (\alpha - 1) x^{\alpha-2} + a_1 (\alpha + 1) \alpha x^{\alpha-1} + \sum_{k'=2}^{\infty} a_{k'} (k' + \alpha) (k' + \alpha - 1) x^{k'+\alpha-2}$$

$$(2.26)$$

and make the variable change

$$k' = k + 2, \qquad (2.27)$$

to write Equation (2.25) as

$$a_0 \alpha (\alpha - 1) x^{\alpha - 2} + a_1 (\alpha + 1) \alpha x^{\alpha - 1}$$

$$+ \sum_{k=0}^{\infty} x^{k+\alpha} \left\{ a_{k+2} (k + 2 + \alpha) (k + 1 + \alpha) - a_k [(k + \alpha) (k + \alpha + 1) - \lambda] \right\} = 0.$$

$$(2.28)$$

From the uniqueness of power series this equation cannot be satisfied for all x, unless the coefficients of all the powers of x vanish simultaneously, which gives us the following relations among the coefficients:

$$a_0 \alpha (\alpha - 1) = 0, \quad a_0 \neq 0, \tag{2.29}$$

$$a_1 (\alpha + 1) \alpha = 0, \tag{2.30}$$

$$\frac{a_{k+2}}{a_k} = \frac{[(k + \alpha) (k + \alpha + 1) - \lambda]}{(k + 1 + \alpha) (k + \alpha + 2)}, \quad k = 0, 1, 2... . \tag{2.31}$$

Equation (2.29), obtained by setting the coefficient of the lowest power of x to zero, is called the **indicial equation.** Assuming $a_0 \neq 0$, the two roots of the indicial equation give the values of α as

$$\alpha = 0 \quad \text{and} \quad \alpha = 1. \tag{2.32}$$

The remaining Equations (2.30) and (2.31) give us the recursion relation among the remaining coefficients. Starting with $\alpha = 1$ we obtain

$$a_{k+2} = a_k \frac{(k + 1) (k + 2) - \lambda}{(k + 2) (k + 3)}, \quad k = 0, 1, 2... . \tag{2.33}$$

For $\alpha = 1$ Equation (2.30) implies

$$a_1 = 0, \tag{2.34}$$

hence all the remaining nonzero coefficients are obtained as

$$a_2 = a_0 \frac{(2 - \lambda)}{6}, \tag{2.35}$$

$$a_3 = a_1 \frac{(6 - \lambda)}{12} = 0, \tag{2.36}$$

$$a_4 = a_2 \frac{(12 - \lambda)}{20}, \tag{2.37}$$

$$\vdots$$

This gives the series solution for $\alpha = 1$ as

$$Z_1(x) = a_0 \left[x + \frac{(2-\lambda)}{6} x^3 + \frac{(2-\lambda)(12-\lambda)}{120} x^5 + \cdots \right]. \qquad (2.38)$$

Similarly for the $\alpha = 0$ value, Equations (2.29) and (2.30) give us

$$a_0 \neq 0 \quad \text{and} \quad a_1 \neq 0. \qquad (2.39)$$

Now the recursion relation becomes

$$a_{k+2} = a_k \frac{k(k+1) - \lambda}{(k+1)(k+2)}, \qquad k = 0, 1, 2, ..., \qquad (2.40)$$

which gives the remaining nonzero coefficients as

$$a_2 = a_0 \left(-\frac{\lambda}{2} \right),$$

$$a_3 = a_1 \left(\frac{2-\lambda}{6} \right),$$

$$a_4 = a_2 \left(\frac{6-\lambda}{12} \right),$$

$$a_5 = a_3 \left(\frac{12-\lambda}{20} \right), \qquad (2.41)$$

$$a_6 = a_4 \left(\frac{20-\lambda}{30} \right),$$

$$\vdots$$

Now the series solution for the $\alpha = 0$ value is obtained as

$$Z_2(x) = a_0 \left[1 - \frac{\lambda}{2} x^2 - \frac{\lambda(6-\lambda)}{2 \cdot 12} x^4 + \cdots \right]$$

$$+ a_1 \left[x + \frac{(2-\lambda)}{6} x^3 + \frac{(2-\lambda)(12-\lambda)}{120} x^5 + \cdots \right]. \qquad (2.42)$$

The Legendre equation is a second-order linear ordinary differential equation, and in general it will have two linearly independent solutions. Since a_0 and a_1 take arbitrary values, the solution for the $\alpha = 0$ root also contains the solution for the $\alpha = 1$ root; hence the general solution can be written as

$$Z(x) = C_0 \left[1 - \left(\frac{\lambda}{2} \right) x^2 - \left(\frac{\lambda}{2} \right) \left(\frac{6-\lambda}{12} \right) x^4 + \cdots \right]$$

$$+ C_1 \left[x + \frac{(2-\lambda)}{6} x^3 + \frac{(2-\lambda)(12-\lambda)}{120} x^5 + \cdots \right], \qquad (2.43)$$

where C_0 and C_1 are two integration constants to be determined from the boundary conditions. These series are called the **Legendre series.**

2.2.1 Frobenius Method

We have used the Frobenius method to find the Legendre series. A second-order linear homogeneous ordinary differential equation with two linearly independent solutions may be put in the form

$$\frac{d^2y}{dx^2} + P(x)\frac{dy}{dx} + Q(x) = 0. \tag{2.44}$$

If x_0 is no worse then a regular singular point, that is, if

$$\lim_{x \to x_0} (x - x_0)P(x) \to \text{finite} \tag{2.45}$$

and

$$\lim_{x \to x_0} (x - x_0)^2 Q(x) \to \text{finite}, \tag{2.46}$$

then we can seek a series solution of the form

$$y(x) = \sum_{k=0}^{\infty} a_k(x - x_0)^{k+\alpha}, \quad a_0 \neq 0. \tag{2.47}$$

Substituting this series into the above differential equation and setting the coefficient of the lowest power of $(x - x_0)$ with $a_0 \neq 0$ gives us a quadratic equation for α called the indicial equation. For almost all the physically interesting cases the indicial equation has two real roots. This gives us the following possibilities for the two linearly independent solutions of the differential equation (Ross):

1. If the two roots $(\alpha_1 > \alpha_2)$ differ by a noninteger, then the two linearly independent solutions are given as

$$y_1(x) = |x - x_0|^{\alpha_1} \sum_{k=0}^{\infty} a_k(x - x_0)^k, \quad a_0 \neq 0$$

and $\hspace{8cm}$ (2.48)

$$y_2(x) = |x - x_0|^{\alpha_2} \sum_{k=0}^{\infty} b_k(x - x_0)^k, \quad b_0 \neq 0.$$

2. If $(\alpha_1 - \alpha_2) = N$, where $\alpha_1 > \alpha_2$ and N is a positive integer, then the two linearly independent solutions are given as

$$y_1(x) = |x - x_0|^{\alpha_1} \sum_{k=0}^{\infty} a_k(x - x_0)^k, \quad a_0 \neq 0, \tag{2.49}$$

and

$$y_2(x) = |x - x_0|^{\alpha_2} \sum_{k=0}^{\infty} b_k(x - x_0)^k + Cy_1(x) \ln|x - x_0|, \quad b_0 \neq 0. \tag{2.50}$$

The second solution contains a logarithmic singularity, where C is a constant that may or may not be zero. Sometimes α_2 will contain both solutions; hence it is advisable to start with the smaller root with the hopes that it might provide the general solution.

3. If the indicial equation has a double root, that is, $\alpha_1 = \alpha_2$, then the Frobenius method yields only one series solution. In this case the two linearly independent solutions can be taken as

$$y(x, \alpha_1) \quad \text{and} \quad \left. \frac{\partial y(x, \alpha)}{\partial \alpha} \right|_{\alpha = \alpha_1}, \tag{2.51}$$

where the second solution diverges logarithmically as $x \to x_0$. In the presence of a double root the Frobenius method is usually modified by taking the two linearly independent solutions as

$$y_1(x) = |x - x_0|^{\alpha_1} \sum_{k=0}^{\infty} a_k (x - x_0)^k, \quad a_0 \neq 0$$

and $\tag{2.52}$

$$y_2(x) = |x - x_0|^{\alpha_1 + 1} \sum_{k=0}^{\infty} b_k (x - x_0)^k + y_1(x) \ln |x - x_0|.$$

In all these cases the general solution is written as

$$y(x) = A_1 y_1(x) + A_2 y_2(x). \tag{2.53}$$

2.3   LEGENDRE POLYNOMIALS

Legendre series are convergent in the interval $(-1, 1)$. This can easily be checked by the ratio test. To see how they behave at the end points, $x = \pm 1$, we take the $k \to \infty$ limit of the recursion relation, Equation (2.40), to obtain

$$\frac{a_{k+2}}{a_k} \to 1. \tag{2.54}$$

For sufficiently large k values this means that both series behave as

$$Z(x) = \cdots + a_k x^k \left(1 + x^2 + x^4 + \cdots \right). \tag{2.55}$$

The series inside the parentheses is nothing but the geometric series:

$$\left(1 + x^2 + x^4 + \cdots \right) = \frac{1}{1 - x^2}. \tag{2.56}$$

Hence both of the Legendre series diverge at the end points as $1/(1 - x^2)$. However, the end points correspond to the north and the south poles of a sphere. Because the problem is spherically symmetric, there is nothing special about these points. Any two diametrically opposite points can be chosen to serve as the end points. Hence we conclude that the physical solution should be finite everywhere on a sphere. To avoid the divergence at the end points, we terminate the Legendre series after a finite number of terms. We accomplish this by restricting the separation constant λ to integer values given by

$$\lambda = l(l+1), \qquad l = 0, 1, 2, ... \ . \tag{2.57}$$

With this restriction on λ, one of the Legendre series in Equation (2.43) terminates after a finite number of terms while the other one still diverges at the end points. Choosing the coefficient of the divergent series in the general solution as zero, we obtain the polynomial solutions of the Legendre equation as

$$Z(x) = P_l(x), \quad l = 0, 1, 2, ... \ .$$

Legendre Polynomials

$$P_0(x) = 1$$
$$P_1(x) = x$$
$$P_2(x) = \left(\frac{1}{2}\right)[3x^2 - 1]$$
$$P_3(x) = \left(\frac{1}{2}\right)[5x^3 - 3x] \tag{2.58}$$
$$P_4(x) = \left(\frac{1}{8}\right)[35x^4 - 30x^2 + 3]$$
$$P_5(x) = \left(\frac{1}{8}\right)[63x^5 - 70x^3 + 15x].$$
$$P_6(x) = \left(\frac{1}{16}\right)[231x^6 - 315x^4 + 105x^2 - 5].$$

These polynomials are called the **Legendre polynomials,** and they are finite everywhere on a sphere. They are defined so that their value at $x = 1$ is one. In general they can be expressed as

$$P_l(x) = \sum_{n=0}^{\left[\frac{l}{2}\right]} \frac{(-1)^n (2l - 2n)!}{2^l (l - 2n)! (l - n)! n!} x^{l-2n}, \tag{2.59}$$

where $\left[\frac{l}{2}\right]$ means the greatest integer in the interval ($\frac{l}{2}$, $\frac{l}{2}-1$]. Restriction of λ to certain integer values for finite solutions everywhere is a physical (boundary) condition and has very significant physical consequences. In quantum mechanics it means that magnitude of the angular momentum is quantized. In wave mechanics, like the standing waves on a string fixed at both ends, it means that waves on a sphere can only have certain wavelengths.

2.3.1   Rodriguez Formula

Another definition of the Legendre polynomials is given as

$$P_l(x) = \frac{1}{2^l l!} \frac{d^l}{dx^l} \left(x^2 - 1\right)^l, \tag{2.60}$$

which is called the Rodriguez formula. To show that this is equivalent to Equation (2.59) we use the binomial formula (Dwight)

$$(x + y)^m = \sum_{n=0}^{\infty} \frac{m!}{n! \, (m - n)!} x^n y^{m-n} \tag{2.61}$$

to write Equation (2.60) as

$$P_l(x) = \frac{1}{2^l l!} \frac{d^l}{dx^l} \sum_{n=0}^{l} \frac{l! \, (-1)^n}{n! \, (l - n)!} x^{2l-2n}. \tag{2.62}$$

We now use the formula

$$\frac{d^l x^m}{dx^l} = \frac{m!}{(m - l)!} x^{m-l} \tag{2.63}$$

to obtain

$$P_l(x) = \sum_{n=0}^{\left[\frac{l}{2}\right]} \frac{(-1)^n}{2^l} \frac{(2l - 2n)!}{n! \, (l - n)! \, (l - 2n)!} x^{l-2n}, \tag{2.64}$$

thus proving the equivalence of Equations (2.60) and (2.59).

2.3.2   Generating Function

Another way to define the Legendre polynomials is by using a generating function, $T(x,t)$, which is defined as

$$T(x,t) = \frac{1}{\sqrt{1 - 2xt + t^2}} = \sum_{l=0}^{\infty} P_l(x) t^l, \qquad |t| < 1. \tag{2.65}$$

To show that $T(x,t)$ generates the Legendre polynomials we write it as

$$T(x,t) = \frac{1}{[1 - t(2x - t)]^{\frac{1}{2}}} \tag{2.66}$$

and use the binomial expansion

$$(1 - x)^{-\frac{1}{2}} = \sum_{l=0}^{\infty} \frac{\left(-\frac{1}{2}\right)! \, (-1)^l \, x^l}{l! \left(-\frac{1}{2} - l\right)!}. \tag{2.67}$$

We derive the useful relation:

$$
\begin{cases}
\dfrac{\left(-\dfrac{1}{2}\right)!}{\left(-\dfrac{1}{2}-l\right)!} = \\[4mm]
\dfrac{\left(-\dfrac{1}{2}\right)\left(-\dfrac{1}{2}-1\right)\left(\dfrac{1}{2}-2\right)\cdots}{\left(-\dfrac{1}{2}-l\right)\left(-\dfrac{1}{2}-l-1\right)\cdots} \\[4mm]
= \dfrac{(-1)^l\left[\left(\dfrac{1}{2}\right)\left(\dfrac{1}{2}+1\right)\left(\dfrac{1}{2}+2\right)\cdots\left(-\dfrac{1}{2}-l\right)\left(-\dfrac{1}{2}-l-1\right)\cdots\right]}{\left[\left(-\dfrac{1}{2}-l\right)\left(-\dfrac{1}{2}-l-1\right)\cdots\right]} \\[4mm]
= (-1)^l\left[\left(\dfrac{1}{2}\right)\left(\dfrac{1}{2}+1\right)\left(\dfrac{1}{2}+2\right)\cdots\left(\dfrac{1}{2}+l-1\right)\right] \\[4mm]
= (-1)^l\,\dfrac{1\cdot 3\cdot 5\cdots(2l-1)}{2^l} \\[4mm]
= (-1)^l\,\dfrac{(2l)!}{2^{2l}l!},
\end{cases}
\tag{2.68}
$$

to write Equation (2.67) as

$$
(1-x)^{-\frac{1}{2}} = \sum_{l=0}^{\infty} \frac{(2l)!\,(-1)^{2l}}{2^{2l}\,(l!)^2}\,x^l .
\tag{2.69}
$$

We use this in Equation (2.66) to write

$$
\frac{1}{(1-t\,(2x-t))^{\frac{1}{2}}} = \sum_{l=0}^{\infty} \frac{(2l)!\,(-1)^{2l}\,t^l}{2^{2l}\,(l!)^2}\,(2x-t)^l .
\tag{2.70}
$$

Using the binomial formula again, we expand the factor

$$
(2x-t)^l
\tag{2.71}
$$

to write

$$
\sum_{l=0}^{\infty} \frac{(2l)!\,(-1)^{2l}\,t^l}{2^{2l}\,(l!)^2} \sum_{k=0}^{l} \frac{l!}{k!\,(l-k)!}\,(2x)^{l-k}\,(-t)^k
\tag{2.72}
$$

$$
= \sum_{l=0}^{\infty} \sum_{k=0}^{l} \frac{(2l)!\,(-1)^k\,(2x)^{l-k}\,t^{k+l}}{2^{2l}\,l!\,k!\,(l-k)!} .
\tag{2.73}
$$

We now rearrange the double sum by the substitutions

$$k \to n \quad \text{and} \quad l \to l - n \tag{2.74}$$

to write

$$T(x,t) = \sum_{l=0}^{\infty} \left[\sum_{n=0}^{[\frac{l}{2}]} \frac{(-1)^n (2l - 2n)!}{2^l (l - n)! n! (l - 2n)!} x^{l-2n} \right] t^l. \tag{2.75}$$

Comparing this with the right-hand side of Equation (2.65), which is

$$\sum_{l=0}^{\infty} P_l(x) t^l, \tag{2.76}$$

we obtain

$$P_l(x) = \sum_{n=0}^{[\frac{l}{2}]} \frac{(-1)^n (2l - 2n)!}{2^l (l - n)! n! (l - 2n)!} x^{l-2n}. \tag{2.77}$$

2.3.3 Recursion Relations

We differentiate the generating function for the Legendre polynomials with respect to t to get

$$\frac{\partial}{\partial t} T(x,t) = -\frac{-2(x-t)}{2(1 - 2xt + t^2)^{\frac{3}{2}}} = \sum_{l=1}^{\infty} P_l(x) l\, t^{l-1}. \tag{2.78}$$

We rewrite this as

$$(x - t) \sum_{l=0}^{\infty} P_l(x) t^l = \sum_{l=1}^{\infty} P_l(x) l\, t^{l-1} (1 - 2xt + t^2) \tag{2.79}$$

and expand in powers of t to get

$$\sum_{l=0}^{\infty} t^l (2l + 1) x P_l(x) = \sum_{l'=1}^{\infty} P_{l'} l' t^{l'-1} + \sum_{l''=0}^{\infty} t^{l''+1} (l'' + 1) P_{l''}(x). \tag{2.80}$$

We now make the substitutions

$$l' = l + 1 \quad \text{and} \quad l'' = l - 1 \tag{2.81}$$

and collect equal powers of t^l to write

$$\sum_{l=0}^{\infty} [(2l + 1) x P_l(x) - P_{l+1}(x)(l + 1) - l P_{l-1}(x)] t^l = 0. \tag{2.82}$$

This equation can only be satisfied for all values of t if the expression inside the square brackets is zero for all l, thus giving the recursion relation

$$(2l + 1) x P_l (x) = (l + 1) P_{l+1} (x) + l P_{l-1} (x).$$ (2.83)

Another recursion relation is obtained by differentiating $T (x, t)$ with respect to x and following similar steps as

$$P_l (x) = P'_{l+1} (x) + P'_{l-1} (x) - 2x P'_l (x).$$ (2.84)

It is possible to find other recursion relations, which are very useful in manipulations with the Legendre polynomials.

2.3.4   Special Values

In various applications one needs special values of the Legendre polynomials at the points $x = \pm 1$ and $x = 0$. If we write $x = \pm 1$ in the generating function Equation (2.65) we find

$$\frac{1}{(1 \mp t)} = \sum_{l=0}^{\infty} P_l (1) t^l (\pm 1)^l.$$ (2.85)

Expanding the left-hand side by using the binomial formula and comparing equal powers of t, we obtain

$$P_l (1) = 1$$
$$\text{and}$$
$$P_l (-1) = (-1)^l.$$ (2.86)

Similarly, we write $x = 0$ in the generating function to get

$$\frac{1}{\sqrt{1 + t^2}} = \sum_{l=0}^{\infty} P_l (0) t^l$$ (2.87)

$$= \sum_{t=0}^{\infty} (-1)^l \frac{(2l)!}{2^{2l} (l!)^2} t^{2l}.$$ (2.88)

This leads us to the special values:

$$P_{2s+1} (0) = 0$$ (2.89)

and

$$P_{2l} (0) = \frac{(-1)^l (2l)!}{2^{2l} (l!)^2}.$$ (2.90)

2.3.5   Special Integrals

1. In applications we frequently encounter the integral

$$\int_0^1 dx\, P_l(x). \tag{2.91}$$

Using the recursion relation Equation (2.84) we can write this integral as

$$\int_0^1 dx\, P_l(x) = \int_0^1 dx\, \left[P'_{l+1}(x) + P'_{l-1}(x) - 2x P'_l(x) \right]. \tag{2.92}$$

The right hand side can be integrated as

$$\int_0^1 dx\, P_l(x) = P_{l+1}(1) + P_{l-1}(1) - P_{l+1}(0) - P_{l-1}(0) - 2x P_l(x)\big|_0^1$$

$$+2 \int_0^1 dx\, P_l(x). \tag{2.93}$$

This is simplified by using the special values and leads to

$$\int_0^1 dx\, P_l(x) = P_{l+1}(0) + P_{l-1}(0) \tag{2.94}$$

and

$$\int_0^1 dx\, P_l(x) = \begin{cases} 0 & l \geq 2 \text{ and even} \\[2mm] 1 & l = 0 \\[2mm] \dfrac{1}{2(s+1)} P_{2s}(0), & l = 2s+1,\ s = 0, 1, \ldots \ . \end{cases} \tag{2.95}$$

2. Another integral useful in dipole calculations is

$$\int_{-1}^1 dx\, x P_l(x) P_k(x). \tag{2.96}$$

Using the recursion relation, Equation (2.83), we can write this as

$$\int_{-1}^1 dx\, x P_l(x) P_k(x) = \int_{-1}^1 dx \frac{P_l(x)}{(2k+1)} \left[(k+1) P_{k+1}(x) + k P_{k-1}(x) \right], \tag{2.97}$$

which leads to

$$\int_{-1}^{1} dx \; x P_l(x) P_k(x) = \begin{cases} 0, & k \neq l \pm 1, \\[2mm] \dfrac{l}{(2l-1)} \dfrac{2}{(2l+1)}, & k = l-1, \\[2mm] \dfrac{l+1}{(2l+3)} \dfrac{2}{(2l+1)}, & k = l+1. \end{cases} \tag{2.98}$$

In general one can show the integral:

$$\int_{-1}^{1} dx \; x^l P_n(x) = \frac{2^{n+1} l! \left(\dfrac{l+n}{2}\right)!}{(l+n+1)! \left(\dfrac{l-n}{2}\right)!}, \qquad l-n = |\text{even integer}|. \tag{2.99}$$

2.3.6 Orthogonality and Completeness

We can also write the Legendre equation [Eq. (2.22)] as

$$\frac{d}{dx}\left[(1-x^2)\frac{dP_l(x)}{dx}\right] + l(l+1)P_l(x) = 0. \tag{2.100}$$

Multiplying this equation with $P_{l'}(x)$ and integrating over x in the interval $[-1,1]$, we get

$$\int_{-1}^{1} P_{l'}(x) \left\{ \frac{d}{dx}\left[(1-x^2)\frac{dP_l(x)}{dx}\right] + l(l+1)P_l(x) \right\} dx = 0. \tag{2.101}$$

Using integration by parts this can be written as

$$\int_{-1}^{1} \left[(x^2-1)\frac{dP_l(x)}{dx}\frac{dP_{l'}(x)}{dx} + l(l+1)P_{l'}(x)P_l(x)\right] dx = 0. \tag{2.102}$$

Interchanging l and l' in Equation (2.102) and subtracting the result from Equation (2.102) we get

$$[l(l+1) - l'(l'+1)]\int_{-1}^{1} P_{l'}(x)P_l(x)\,dx = 0. \tag{2.103}$$

For $l \neq l'$ this equation gives

$$\int_{-1}^{1} P_{l'}(x)P_l(x)\,dx = 0 \tag{2.104}$$

and for $l = l'$ it gives

$$\int_{-1}^{1} P_{l'}(x) P_l(x) \, dx = N_l, \qquad (2.105)$$

where N_l is a finite normalization constant. We can evaluate the value of N_l by using the Rodriguez formula [Eq. (2.60)]. We write

$$N_l = \int_{-1}^{1} P_l^2(x) \, dx \qquad (2.106)$$

$$= \frac{1}{2^{2l} (l!)^2} \int_{-1}^{1} \frac{d^l}{dx^l} (x^2 - 1)^l \frac{d^l}{dx^l} (x^2 - 1)^l \, dx, \qquad (2.107)$$

and after l-fold integration by parts we obtain

$$N_l = \frac{(-1)^l}{2^{2l} (l!)^2} \int_{-1}^{1} (x^2 - 1)^l \frac{d^{2l}}{dx^{2l}} (x^2 - 1)^l \, dx. \qquad (2.108)$$

Using the Leibniz formula

$$\frac{d^m}{dx^m} A(x) B(x) = \sum_{s=0}^{m} \frac{m!}{s! (m-s)!} \frac{d^s A}{dx^s} \frac{d^{m-s} B}{dx^{m-s}}, \qquad (2.109)$$

we evaluate the $2l$-fold derivative of $(x^2 - 1)^l$ as $(2l)!$. Thus Equation (2.108) becomes

$$N_l = \frac{(2l)!}{2^{2l} (l!)^2} \int_{-1}^{1} (1 - x^2)^l \, dx. \qquad (2.110)$$

We now write $(1 - x^2)^l$ as

$$(1 - x^2)^l = (1 - x^2)(1 - x^2)^{l-1} = (1 - x^2)^{l-1} + \frac{x}{2l} \frac{d}{dx} (1 - x^2)^l \qquad (2.111)$$

to obtain

$$N_l = \frac{(2l - 1)}{2l} N_{l-1} + \frac{(2l - 1)!}{2^{2l} (l!)^2} \int_{-1}^{1} x d \left[(1 - x^2)^l \right]. \qquad (2.112)$$

This gives

$$N_l = \frac{(2l - 1)}{2l} N_{l-1} - \frac{1}{2l} N_l \qquad (2.113)$$

or

$$(2l + 1) N_l = (2l - 1) N_{l-1}, \qquad (2.114)$$

which means that the value of

$$(2l+1) N_l \tag{2.115}$$

is a constant independent of l. Evaluating Equation (2.115) for $l = 0$ gives 2, which determines the normalization constant as

$$N_l = \frac{2}{(2l+1)}. \tag{2.116}$$

Using N_l we can now define set of polynomials $\{U_l(x)\}$ as

$$U_l(x) = \sqrt{\frac{2l+1}{2}} P_l(x), \tag{2.117}$$

which satisfies the orthogonality relation

$$\int_{-1}^{1} U_{l'}(x) U_l(x) \, dx = \delta_{ll'}. \tag{2.118}$$

At this point we suffice by saying that this set is also complete, that is in terms of this set any sufficiently well-behaved and at least piecewise continuous function $\Psi(x)$ can be expressed as an infinite series in the interval $[-1, 1]$ as

$$\Psi(x) = \sum_{l=0}^{\infty} C_l U_l(x). \tag{2.119}$$

We will be more specific about what is meant by sufficiently well-behaved when we discuss the Sturm-Liouville theory in Chapter 8. To evaluate the constants C_l we multiply both sides by $U_{l'}(x)$ and integrate over $[-1, 1]$:

$$\int_{-1}^{1} U_{l'}(x) \Psi(x) \, dx = \sum_{l=0}^{\infty} C_l \int U_{l'}(x) U_l(x) \, dx. \tag{2.120}$$

Using the orthogonality relation [Eq. (2.118)] we can free the constants C_l under the summation sign and obtain

$$C_l = \int_{-1}^{1} U_l(x) \Psi(x) \, dx. \tag{2.121}$$

Orthogonality and the completeness of the Legendre polynomials are very useful in applications.

Example 2.1. *Legendre polynomials and electrostatics problems:* .

    To find the electric potential Ψ in vacuum one has to solve the Laplace equation

$$\vec{\nabla}^2 \Psi(\vec{r}) = 0. \tag{2.122}$$

For problems with azimuthal symmetry in spherical polar coordinates potential does not have any ϕ dependence, hence in the ϕ-dependent part of the solution [Eq. (2.18)] we set

$$m = 0. \tag{2.123}$$

The differential equation to be solved for the r-dependent part is now found by taking

$$k^2 = 0 \tag{2.124}$$

in Equation (2.12) as

$$\frac{d^2 R}{dr^2} + \frac{2}{r}\frac{dR}{dr} - \frac{l(l+1)}{r^2} R(r) = 0. \tag{2.125}$$

Linearly independent solutions of this equation are easily found as r^l and $\dfrac{1}{r^{l+1}}$, thus giving the general solution of Equation (2.122) as

$$\Psi(r,\theta) = \sum_{l=0}^{\infty} \left[A_l r^l + \frac{B_l}{r^{l+1}} \right] P_l \left(x = \cos\theta \right). \tag{2.126}$$

We now calculate the electric potential outside a spherical conductor with radius a, where the upper hemisphere is held at potential V_0 and the lower hemisphere is held at potential $-V_0$ and that are connected by an insulator at the center. Since the potential cannot diverge at infinity, we set the coefficients A_l to zero and write the potential for the outside as

$$\Psi(r,\theta) = \sum_{l=0}^{\infty} \frac{B_l}{r^{l+1}} P_l(x), \quad r \geq a. \tag{2.127}$$

To find the coefficients B_l, we use the boundary conditions at $r = a$ as

$$\Psi(a,\theta) = \sum_{l=0}^{\infty} \frac{B_l}{a^{l+1}} P_l(x) = \begin{cases} V_0 & 0 < x \leq 1 \\ -V_0 & 0 > x \geq -1 \end{cases}. \tag{2.128}$$

We multiply both sides by $P_{l'}(x)$ and integrate over x and use the orthogonality relation to get

$$\int_{-1}^{1} \Psi(a,x) P_l(x)\, dx = \frac{B_l}{a^{l+1}} \frac{2}{(2l+1)}, \tag{2.129}$$

$$V_0 \int_{0}^{1} dx P_l(x) - V_0 \int_{-1}^{0} dx P_l(x) = \frac{2B_l}{(2l+1)\, a^{l+1}}, \tag{2.130}$$

$$B_l = \frac{(2l+1)\,a^{l+1}}{2} V_0 \int_0^1 \left[1 - (-1)^l\right] P_l(x)\,dx. \qquad (2.131)$$

For the even values of l the expansion coefficients are zero. For the odd values of l we use the result Equation (2.95) to write

$$B_{2s+1} = \frac{(4s+3)}{2} \frac{P_{2s}(0)}{(2s+2)} a^{2s+2}(2V_0), \quad s = 0, 1, 2, \dots. \qquad (2.132)$$

Substituting these in Equation (2.127) we finally obtain the potential outside the sphere as

$$\Psi(r,\theta) = V_0 \sum_{s=0}^{\infty} (4s+3) \frac{P_{2s}(0)}{(2s+2)} \frac{a^{2s+2}}{r^{2s+2}} P_{2s+1}(\cos\theta). \qquad (2.133)$$

Potential inside can be found similarly.

2.4   ASSOCIATED LEGENDRE EQUATION AND ITS SOLUTIONS

We now consider the associated Legendre Equation (2.21) for the

$$m \neq 0 \qquad (2.134)$$

values and try a series solution around $x = 0$ of the form

$$Z(x) = \sum_{k=0}^{\infty} a_k x^k. \qquad (2.135)$$

Now the recursion relation becomes

$$(k+4)(k+3)\,a_{k+4} + \left[(\lambda - m^2) - 2(k+2)^2\right] a_{k+2} + \left[k(k+1) - \lambda\right] a_k = 0. \qquad (2.136)$$

Compared with the recursion relation for Legendre Equation (2.33) this has three terms, which is not very practical to use. In such situations, in order to get a two-term recursion relation we study the behavior of the differential equation near the end points. For the points near $x = 1$ we introduce a new variable

$$y = (1 - x). \qquad (2.137)$$

Now Equation (2.21) becomes

$$(2-y)\,y\,\frac{d^2 Z(y)}{dy^2} + 2(1-y)\frac{dZ(y)}{dy} + \left[\lambda - \frac{m^2}{y(2-y)}\right] Z(y) = 0. \qquad (2.138)$$

In the limit as $y \to 0$ this equation can be approximated by

$$2y \frac{d^2 Z(y)}{dy^2} + 2 \frac{dZ(y)}{dy} - m^2 \frac{Z(y)}{2y} = 0. \tag{2.139}$$

To find the solution, we try a power dependence of the form $Z(y) = y^n$ to determine n as $\pm m$. Hence the two linearly independent solutions are

$$y^{m/2} \quad \text{and} \quad y^{-m/2}. \tag{2.140}$$

For $m \geq 0$ the solution that remains finite as $y \to 0$ is $y^{\frac{m}{2}}$. Similarly for points near $x = -1$, we use the substitution

$$y = (1 + x) \tag{2.141}$$

and obtain the finite solution in the limit $y \to 0$ as $y^{m/2}$. We now substitute in the associated Legendre Equation (2.21) a solution of the form

$$Z(x) = (1 + x)^{m/2} (1 - x)^{m/2} f(x) \tag{2.142}$$
$$= \left(1 - x^2\right)^{m/2} f(x),$$

which gives the differential equation to be solved for $f(x)$ as

$$\left(1 - x^2\right) \frac{d^2 f}{dx^2} - 2x(m+1) \frac{df(x)}{dx} + [\lambda - m(m+1)] f(x) = 0. \tag{2.143}$$

Note that this equation is valid for both the positive and the negative values of m. We now try a series solution in this equation:

$$f(x) = \sum_k a_k x^{k+\alpha} \tag{2.144}$$

and obtain a two-term recursion relation

$$a_{k+2} = a_k \frac{[(k+m)(k+m+1) - \lambda]}{(k+2)(k+1)}. \tag{2.145}$$

Since in the limit as k goes to infinity the ratio of two successive terms, $\dfrac{a_{k+2}}{a_k}$, goes to one, this series also diverges at the end points; thus to get a finite solution we restrict the separation constant λ to the values

$$(k+m)[(k+m)+1] - \lambda = 0, \tag{2.146}$$

$$\lambda = (k+m)[(k+m)+1]. \tag{2.147}$$

Defining a new integer

$$l = k + m \tag{2.148}$$

we obtain

$$\lambda = l(l+1) \tag{2.149}$$

and

$$k = l - m . \tag{2.150}$$

Since k takes only positive integer values, m can only take the values

$$m = -l, ..., 0, ..., l . \tag{2.151}$$

2.4.1   Associated Legendre Polynomials

To obtain the associated Legendre polynomials we write the equation that the Legendre polynomials satisfy as

$$\left(1 - x^2\right) \frac{d^2 P_l(x)}{dx^2} - 2x \frac{dP_l(x)}{dx} + l(l+1) P_l(x) = 0. \tag{2.152}$$

Using the Leibniz formula

$$\frac{d^m}{dx^m} [A(x) B(x)] = \sum_{s=0}^{m} \frac{m!}{s!(m-s)!} \left[\frac{d^s A}{dx^s}\right] \left[\frac{d^{m-s} B}{dx^{m-s}}\right], \tag{2.153}$$

we differentiate Equation (2.152) m times to obtain

$$\left(1 - x^2\right) P_l^{(m+2)}(x) - 2xm P_l^{(m+1)}(x) - \frac{2m(m-1)}{2} P_l^{(m)}(x)$$
$$= 2x P_l^{(m+1)}(x) + 2m P_l^{(m)}(x) - l(l+1) P_l^{(m)}(x). \tag{2.154}$$

After simplification this becomes

$$\left(1 - x^2\right) P_l^{(m+2)}(x) - 2x(m+1) P_l^{(m+1)}(x)$$

$$+ [l(l+1) - m(m+1)] P_l^{(m)}(x) = 0, \tag{2.155}$$

where

$$P_l^{(m)}(x) = \frac{d^m}{dx^m} P_l(x). \tag{2.156}$$

Comparing Equation (2.155) with Equation (2.143) we obtain $f(x)$ as

$$f(x) = \frac{d^m}{dx^m} P_l(x). \tag{2.157}$$

Using Equation (2.142), we can now write the finite solutions of the **associated Legendre equation** (2.21) as

$$Z(x) = P_l^m(x) = \left(1 - x^2\right)^{\frac{m}{2}} \frac{d^m}{dx^m} P_l(x), \quad m \geq 0. \tag{2.158}$$

These polynomials, $P_l^m(x)$, are called the **associated Legendre polynomials**.

For the negative values of m, associated Legendre polynomials are defined as

$$P_l^{-m}(x) = (-1)^m \frac{(l-m)!}{(l+m)!} P_l^m(x), \quad m \geq 0. \tag{2.159}$$

2.4.2 Orthogonality of the Associated Legendre Polynomials

To derive the orthogonality relation of the associated Legendre polynomials we use the Rodriguez formula [Eq. (2.60)] for the Legendre polynomials to write

$$\int_{-1}^1 P_l^m(x) P_{l'}^m(x)\, dx = \frac{(-1)^m}{2^{l+l'} l! l'!} \int_{-1}^1 X^m \frac{d^{l+m}}{dx^{l+m}} X^l \frac{d^{l'+m}}{dx^{l'+m}} X^{l'}\, dx, \tag{2.160}$$

where

$$X = x^2 - 1, \tag{2.161}$$

$$P_l^m(x) = \left(1 - x^2\right)^{m/2} \frac{d^m}{dx^m} P_l(x) \tag{2.162}$$

and

$$P_l(x) = \frac{1}{2^l l!} \frac{d^l}{dx^l} \left(x^2 - 1\right)^l. \tag{2.163}$$

The integral in Equation (2.160) after $(l' + m)$-fold integration by parts becomes

$$(-1)^{l'+m} \int_{-1}^1 \frac{d^{l'+m}}{dx^{l'+m}} \left[X^m \frac{d^{l+m}}{dx^{l+m}} X^l \right] X^{l'}\, dx. \tag{2.164}$$

Using the Leibniz formula, Equation (2.153), we get

$$(-1)^{l'+m} \int_{-1}^1 X^{l'} \sum_\lambda \binom{l'+m}{\lambda} \left[\frac{d^{l'+m-\lambda}}{dx^{l'+m-\lambda}} X^m \right] \left[\frac{d^{l+m+\lambda}}{dx^{l+m+\lambda}} X^l dx \right]. \tag{2.165}$$

Since the highest power in X^m is x^{2m} and the highest power in X^l is x^{2l}, the summation is empty unless the inequalities

$$l' + m - \lambda \leq 2m \tag{2.166}$$

and

$$l + m + \lambda \leq 2l \tag{2.167}$$

are simultaneously satisfied. The first inequality gives

$$\lambda \geq l' - m, \tag{2.168}$$

while the second one gives

$$\lambda \leq l - m. \tag{2.169}$$

For $m \geq 0$, if we assume $l < l'$ the summation [Eq. (2.165)] does not contain any term that is different from zero; hence the integral is zero. Since the expression in Equation (2.160) is symmetric with respect to l' and l, this result is also valid for $l > l'$. When $l = l'$ these inequalities can be satisfied only for the single value of $\lambda = l - m$. Now the summation contains only one term, and Equation (2.165) becomes

$$(-1)^{l+m} \int_{-1}^{1} X^l \binom{l+m}{l-m} \frac{d^{2m}}{dx^{2m}} X^m \frac{d^{2l}}{dx^{2l}} X^l dx$$

$$= (-1)^{l+m} \binom{l+m}{l-m} (2l)! \, (2m)! \int_{-1}^{1} X^l dx. \tag{2.170}$$

This integral can be evaluated as

$$\int_{-1}^{1} X^l dx = \int_{-1}^{1} \left(x^2 - 1 \right)^l dx = 2 \left(-1 \right)^l \int_{0}^{\frac{\pi}{2}} (\sin \theta)^{2l+1} d\theta$$

$$= \frac{(-1)^l \, 2^{l+1} l!}{3.5 \ldots (2l+1)}$$

$$= \frac{(-1)^l \, 2^{2l+1} \, (l!)^2}{(2l+1)!}. \tag{2.171}$$

Since the binomial coefficients are given as

$$\binom{l+m}{l-m} = \frac{(l+m)!}{(l-m)! \, (2m)!}, \tag{2.172}$$

the orthogonality relation of the associated Legendre polynomials is obtained as

$$\int_{-1}^{1} P_l^m (x) \, P_{l'}^m (x) \, dx = \frac{(-1)^m}{2^{2l} \, (l!)^2} \frac{(l+m)! \, (-1)^{l+m}}{(l-m)! \, (2m)!} (2l)! \, (2m)! \frac{(-1)^l \, 2^{2l+1} \, (l!)^2}{(2l+1)!} \delta_{ll'},$$

$$= \frac{(l+m)!}{(l-m)!} \left[\frac{2}{(2l+1)} \right] \delta_{ll'}. \tag{2.173}$$

Associated Legendre Polynomials

$$P_0^0(x) = 1$$

$$P_1^1(x) = (1 - x^2)^{1/2} = \sin\theta$$

$$P_2^1(x) = 3x(1 - x^2)^{1/2} = 3\cos\theta\sin\theta$$

$$P_2^2(x) = 3(1 - x^2) = 3\sin^2\theta$$

$$P_3^1(x) = \frac{3}{2}(5x^2 - 1)(1 - x^2)^{1/2} = \frac{3}{2}(5\cos^2\theta - 1)\sin\theta$$

$$P_3^2(x) = 15x(1 - x^2) = 15\cos\theta\sin^2\theta$$

$$P_3^3(x) = 15(1 - x^2)^{3/2} = 15\sin^3\theta.$$

2.5   SPHERICAL HARMONICS

We have seen that the solution of Equation (2.17) with respect to the independent variable ϕ is given as

$$\Phi(\phi) = Ae^{im\phi} + Be^{-im\phi}. \tag{2.174}$$

Imposing the periodic boundary condition $\Phi_m(\phi + 2\pi) = \Phi_m(\phi)$, it is seen that the separation constant m has to take $\pm$integer values. However, in Section 2.4 we have also seen that m must be restricted further to the integer values $-l, ..., 0, ..., l$. We can now define another complete and orthonormal set as

$$\left\{ \Phi_m(\phi) = \frac{1}{\sqrt{2\pi}} e^{im\phi} \right\}, \quad m = -l, ..., 0, ..., l. \tag{2.175}$$

This set satisfies the orthogonality relation

$$\int_0^{2\pi} d\phi \Phi_{m'}(\phi)\Phi_m^*(\phi) = \delta_{mm'}. \tag{2.176}$$

We now combine the two sets $\{\Phi_m(\phi)\}$ and $\{P_l^m(\theta)\}$ to define a new complete and orthonormal set called the **spherical harmonics** as

$$Y_l^m(\theta, \phi) = (-1)^m \sqrt{\frac{2l+1}{4\pi} \frac{(l-m)!}{(l+m)!}} e^{im\phi} P_l^m(\cos\theta), \quad m \geq 0. \tag{2.177}$$

In conformity with applications to quantum mechanics and atomic spectroscopy, we have introduced the factor $(-1)^m$. It is also called the **Condon-Shortley phase**. The definition of spherical harmonics can easily be extended to the negative m values as

$$Y^{-m}(\theta, \phi) = (-1)^m Y_l^{m*}(\theta, \phi), \quad m \geq 0. \tag{2.178}$$

The orthogonality relation of $Y_l^m(\theta, \phi)$ is given as

$$\int_0^{2\pi} d\varphi \int_0^\pi d\theta \sin \theta Y_{l'}^{m'*}(\theta, \phi) Y_l^m(\theta, \phi) = \delta_m^{m'} \delta_l^{l'}. \tag{2.179}$$

Since they also form a complete set, any sufficiently well-behaved and at least piecewise continuous function $g(\theta, \phi)$ can be expressed in terms of $Y_l^m(\theta, \phi)$ as

$$g(\theta, \phi) = \sum_{l=0}^\infty \sum_{m=-l}^{m=l} A_m^l Y_l^m(\theta, \phi), \tag{2.180}$$

where the expansion coefficients A_m^l are given as

$$A_m^l = \int \int d\phi d\theta \sin \theta g(\theta, \phi) Y_l^{m*}(\theta, \phi). \tag{2.181}$$

Looking back at Equation (2.13), we see that the spherical harmonics satisfy the differential equation

$$\frac{1}{\sin \theta} \frac{\partial}{\partial \theta} \left[\sin \theta \frac{\partial Y_l^m(\theta, \phi)}{\partial \theta} \right] + \frac{1}{\sin^2 \theta} \frac{\partial^2 Y_l^m(\theta, \phi)}{\partial \phi^2} + l(l+1) Y_l^m(\theta, \phi) = 0. \tag{2.182}$$

If we rewrite this equation as

$$\left[\frac{1}{\sin \theta} \frac{\partial}{\partial \theta} \left[\sin \theta \frac{\partial}{\partial \theta} \right] + \frac{1}{\sin^2 \theta} \frac{\partial^2}{\partial \phi^2} \right] Y_l^m(\theta, \phi) = -l(l+1) Y_l^m(\theta, \phi), \tag{2.183}$$

the left-hand side is nothing but the square of the angular momentum operator (aside from a factor of $\hbar$) in quantum mechanics, which is given as

$$\vec{L}^2 = (\vec{r} \times \vec{p})^2 = \left(\vec{r} \times \frac{\hbar}{i} \vec{\nabla} \right)^2$$

$$= -\hbar^2 \left[\frac{1}{\sin \theta} \frac{\partial}{\partial \theta} \left[\sin \theta \frac{\partial}{\partial \theta} \right] + \frac{1}{\sin^2 \theta} \frac{\partial^2}{\partial \phi^2} \right]. \tag{2.184}$$

In quantum mechanics the fact that the separation constant λ is restricted to integer values means that the magnitude of the angular momentum is quantized. From Equation (2.183) it is seen that the spherical harmonics are also the eigenfunctions of the $\overrightarrow{L}^2$ operator.

Spherical Harmonics $Y_l^m(\theta, \phi)$

$$l = 0 \quad \left\{ Y_0^0 = \frac{1}{\sqrt{4\pi}} \right.$$

$$l = 1 \quad \left\{ \begin{array}{l} Y_1^1 = -\sqrt{\dfrac{3}{8\pi}} \sin\theta e^{i\phi} \\[3mm] Y_1^0 = +\sqrt{\dfrac{3}{4\pi}} \cos\theta \\[3mm] Y_1^{-1} = +\sqrt{\dfrac{3}{8\pi}} \sin\theta e^{-i\phi} \end{array} \right.$$

$$l = 2 \quad \left\{ \begin{array}{l} Y_2^2 = +\dfrac{1}{4}\sqrt{\dfrac{15}{2\pi}} \sin^2\theta e^{2i\phi} \\[3mm] Y_2^1 = -\sqrt{\dfrac{15}{8\pi}} \sin\theta \cos\theta e^{i\phi} \\[3mm] Y_2^0 = +\sqrt{\dfrac{5}{4\pi}}(\dfrac{3}{2}\cos^2\theta - \dfrac{1}{2}) \\[3mm] Y_2^{-1} = +\sqrt{\dfrac{15}{8\pi}} \sin\theta \cos\theta e^{-i\phi} \\[3mm] Y_2^{-2} = +\dfrac{1}{4}\sqrt{\dfrac{15}{2\pi}} \sin^2\theta e^{-2i\phi} \end{array} \right.$$

$$l = 3 \begin{cases} Y_3^3 = -\frac{1}{4}\sqrt{\frac{35}{4\pi}} \sin^3 \theta e^{3i\phi} \\[2ex] Y_3^2 = +\frac{1}{4}\sqrt{\frac{105}{2\pi}} \sin^2 \theta \cos \theta e^{2i\phi} \\[2ex] Y_3^1 = -\frac{1}{4}\sqrt{\frac{21}{4\pi}} \sin \theta (5\cos^2 \theta - 1) e^{i\phi} \\[2ex] Y_3^0 = +\sqrt{\frac{7}{4\pi}}(\frac{5}{2}\cos^3 \theta - \frac{3}{2}\cos \theta) \\[2ex] Y_3^{-1} = +\frac{1}{4}\sqrt{\frac{21}{4\pi}} \sin \theta (5\cos^2 \theta - 1) e^{-i\phi} \\[2ex] Y_3^{-2} = +\frac{1}{4}\sqrt{\frac{105}{2\pi}} \sin^2 \theta \cos \theta e^{-2i\phi} \\[2ex] Y_3^{-3} = +\frac{1}{4}\sqrt{\frac{35}{4\pi}} \sin^3 \theta e^{-3i\phi} \end{cases}$$

Problems

2.1   Locate and classify the singular points of each of the following differential equations:

i) Laguerre equation:

$$x\frac{d^2 y_n}{dx^2} + (1 - x)\frac{dy_n}{dx} + n y_n = 0$$

ii) Harmonic oscillator equation:

$$\frac{d^2 \Psi_\varepsilon(x)}{dx^2} + \left(\varepsilon - x^2\right)\Psi_\varepsilon(x) = 0$$

iii) Bessel equation:

$$x^2 J_m''(x) + x J_m'(x) + (x^2 - m^2)J_m(x) = 0$$

iv)

$$(x^2 - 4x)\frac{d^2 y}{dx^2} + (x + 8)\frac{dy}{dx} + 2y = 0$$

v)

$$(x^4 - 2x^3 + x^2)\frac{d^2 y}{dx^2} + (x - 1)\frac{dy}{dx} + 2x^2 y = 0$$

vi) Chebyshev equation:

$$(1 - x^2)\frac{d^2y}{dx^2} - x\frac{dy}{dx} + n^2 y = 0$$

vii) Gegenbauer equation:

$$(1 - x^2)\frac{d^2 C_n^\lambda(x)}{dx^2} - (2\lambda + 1)x\frac{dC_n^\lambda(x)}{dx} + n(n + 2\lambda)C_n^\lambda(x) = 0$$

viii) Hypergeometric equation:

$$x(1 - x)\frac{d^2y(x)}{dx^2} + [c - (a + b + 1)x]\frac{dy(x)}{dx} - aby(x) = 0$$

ix) Confluent Hypergeometric equation:

$$z\frac{d^2y(z)}{dz^2} + [c - z]\frac{dy(z)}{dz} - ay(z) = 0$$

2.2 For the following differential equations use the Frobenius method to find solutions about $x = 0$:

i)

$$2x^3\frac{d^2y}{dx^2} + 5x^2\frac{dy}{dx} + x^3 y = 0$$

ii)

$$x^3\frac{d^2y}{dx^2} + 3x^2\frac{dy}{dx} + (x^3 + \frac{8}{9}x)y = 0$$

iii)

$$x^3\frac{d^2y}{dx^2} + 3x^2\frac{dy}{dx} + (x^3 + \frac{3}{4}x)y = 0$$

iv)

$$x^2\frac{d^2y}{dx^2} + 3x\frac{dy}{dx} + (2x + 1)y = 0$$

v)

$$x^3\frac{d^2y}{dx^2} + x^2\frac{dy}{dx} + (8x^3 - 9x)y = 0$$

vi)

$$x^2\frac{d^2y}{dx^2} + x\frac{dy}{dx} + x^2 y = 0$$

vii)

$$x\frac{d^2y}{dx^2} + (1-x)\frac{dy}{dx} + 4y = 0$$

viii)

$$2x^3\frac{d^2y}{dx^2} + 5x^2\frac{dy}{dx} + (x^3 - 2x)y = 0$$

2.3 Find finite solutions of the equation

$$(1-x^2)\frac{d^2y}{dx^2} - x\frac{dy}{dx} + n^2y = 0$$

in the interval $x \in [-1, 1]$ for $n =$ integer.

2.4 Consider a spherical conductor with radius a, with the upper hemisphere held at potential V_0 and the lower hemisphere held at potential $-V_0$, which are connected by an insulator at the center. Show that the electric potential inside the sphere is given as

$$\Psi(r,\theta) = V_0 \sum_{l=0}^{\infty}(-1)^l(\frac{r}{a})^{2l+1}\frac{(2l)!}{(2^l l!)^2}\frac{4l+3}{2l+2}P_{2l+1}(\cos\theta).$$

2.5 Using the Frobenius method, show that the two linearly independent solutions of

$$\frac{d^2R}{dr^2} + \frac{2}{r}\frac{dR}{dr} - \frac{l(l+1)}{r^2}R = 0,$$

are given as

$$r^l \text{ and } r^{-(l+1)}.$$

2.6 The amplitude of a scattered wave is given as

$$f(\theta) = \gamma\sum_{l=0}^{\infty}(2l+1)(e^{i\delta_l}\sin\delta_l)P_l(\cos\theta),$$

where θ is the scattering angle, l is the angular momentum, and δ_l is the phase shift caused by the central potential causing the scattering. If the total scattering cross section is

$$\sigma_{total} = \int_0^{2\pi}\int_0^{\pi}d\phi d\theta\sin\theta\,|f(\theta)|^2,$$

show that

$$\sigma_{total} = 4\pi\gamma^2 \sum_{l=0}^{\infty} (2l+1)\sin^2 \delta_l.$$

2.7   Prove the following recursion relations:

$$P_l(x) = P'_{l+1}(x) + P'_{l-1}(x) - 2xP'_l(x)$$

$$P'_{l+1}(x) - P'_{l-1}(x) = (2l+1)P_l(x)$$

$$P'_{l+1}(x) - xP'_l(x) = (l+1)P_l(x)$$

2.8   Use the Rodriguez formula to prove

$$P'_l(x) = xP'_{l-1}(x) + lP_{l-1}(x)$$

and

$$P_l(x) = xP_{l-1}(x) + \frac{x^2-1}{l}P'_{l-1}(x),$$

where $l = 1, 2, ...$.

2.9   Show that the Legendre polynomials satisfy the following relations:
i)

$$\frac{d}{dx}\left[(1-x^2)P'_l(x)\right] + l(l+1)P_l(x) = 0$$

ii)

$$P_{l+1}(x) = \frac{(2l+1)xP_l(x) - lP_{l-1}(x)}{l+1}, \quad l \geq 1$$

2.10   Derive the normalization constant, N_l, in the orthogonality relation

$$\int_{-1}^{1} P_{l'}(x)P_l(x)\,dx = N_l\delta_{ll'}$$

of the Legendre polynomials by using the generating function.

2.11   Show the integral

$$\int_{-1}^{1} dx\, x^l P_n(x) = \frac{2^{n+1}l!\left(\dfrac{l+n}{2}\right)!}{(l+n+1)!\left(\dfrac{l-n}{2}\right)!},$$

where

$$(l - n) = |\text{even integer}| .$$

2.12 Show that the associated Legendre polynomials with negative m values are given as

$$P_l^{-m}(x) = (-1)^m \frac{(l - m)!}{(l + m)!} P_l^m(x), \quad m \geq 0.$$

2.13 Expand the Dirac delta function in a series of Legendre polynomials in the interval $[-1, 1]$.

2.14 A metal sphere is cut into sections that are separated by a very thin insulating material. One section extending from $\theta = 0$ to $\theta = \theta_0$ at potential V_0 and the second section extending from $\theta = \theta_0$ to $\theta = \pi$ is grounded. Find the electrostatic potential outside the sphere.

2.15 The equation for the surface of a liquid drop (nucleus) is given by

$$r^2 = a^2 (1 + \varepsilon_2 \frac{Z^2}{r^2} + \varepsilon_4 \frac{Z^4}{r^4}),$$

where Z, ε_2, and ε_4 are given constants. Express this in terms of the Legendre polynomials as

$$r^2 = a^2 \sum_l C_l P_l(\cos\theta).$$

2.16 Show that the inverse distance between two points in three dimensions can be expressed in terms of the Legendre polynomials as

$$\frac{1}{|\vec{x} - \vec{x}'|} = \frac{1}{\sqrt{r^2 + r'^2 - 2rr' \cos\theta}}$$

$$= \sum_{l=0}^{\infty} \frac{r_<^l}{r_>^{l+1}} P_l(\cos\theta),$$

where $r_<$ and $r_>$ denote the lesser and the greater of r and r', respectively.

2.17 Evaluate the sum

$$S = \sum_{l=0}^{\infty} \frac{x^{l+1}}{l + 1} P_l(x).$$

Hint: Try using the generating function for the Legendre polynomials.

2.18 If two solutions $y_1(x)$ and $y_2(x)$ are linearly dependent, then their Wronskian

$$W[y_1(x), y_2(x)] = y_1(x)y_2'(x) - y_1'(x)y_2(x)$$

vanishes identically. What is the Wronskian of two solutions of the Legendre equation?

2.19   The Jacobi polynomials $P_n^{(a,b)}(\cos\theta)$, where $n =$ positive integer and a, b are arbitrary real numbers, are defined by the Rodriguez formula

$$P_n^{(a,b)}(x) = \frac{(-)^n}{2^n n!(1-x)^a(1+x)^b} \frac{d^n}{dx^n}\left[(1-x)^{n+a}(1+x)^{n+b}\right], \quad |x| < 1.$$

Show that the polynomial can be expanded as

$$P_n^{(a,b)}(\cos\theta) = \sum_{k=0}^{n} A(n, a, b, k)\left(\sin\frac{\theta}{2}\right)^{2n-2k}\left(\cos\frac{\theta}{2}\right)^{2k}.$$

Determine the coefficients $A(n, a, b, k)$ for the special case, where a and b are both integers.

2.20   Find solutions of the differential equation

$$2x(x-1)\frac{d^2y}{dx^2} + (10x - 3)\frac{dy}{dx} + \left[8 + \frac{1}{x} - 2\lambda\right]y(x) = 0,$$

satisfying the condition

$$y(x) = \text{finite}$$

in the entire interval $x \in [0, 1]$. Write the solution explicitly for the third lowest value of λ.

3

LAGUERRE
POLYNOMIALS

For the central force problems, solutions of the time-independent Schrödinger equation can be separated as

$$R(r)Y_l^m(\theta, \phi),\tag{3.1}$$

where the angular part is the spherical harmonics and the radial part comes from the solutions of the differential equation

$$\frac{d}{dr}\left(r^2\frac{dR_l(r)}{dr}\right) + r^2k^2(r)R_l(r) - l(l+1)R_l(r) = 0,\tag{3.2}$$

where

$$k^2 = \frac{2m}{\hbar^2}[E - V(r)].$$

If we substitute

$$R_l(r) = \frac{u_{E,l}(r)}{r},\tag{3.3}$$

the differential equation to be solved for $u_{E,l}(r)$ becomes

$$-\frac{\hbar^2}{2m}\frac{d^2u_{E,l}}{dr^2} + \left[\frac{\hbar^2l(l+1)}{2mr^2} + V(r)\right]u_{E,l}(r) = Eu_{E,l}(r).\tag{3.4}$$

To indicate that the solutions depend on the energy and the angular momentum values, we have written

$$u_{E,l}(r).$$

For single-electron atom models potential energy is given as Coulomb's law:

$$V(r) = -\frac{Ze^2}{r},$$ (3.5)

where Z is the atomic number and e is the electron's charge. A series solution in Equation (3.4) yields a three-term recursion relation. To get a two-term recursion relation we investigate the behavior of the differential equation near the end points, 0 and ∞, which suggests that we try a solution of the form

$$u_{E,l}(\rho) = \rho^{l+1} e^{-\rho} w(\rho),$$ (3.6)

where we have defined a dimensionless variable $\rho = r\sqrt{2m\,|E|\,/\hbar^2}$. Because electrons in an atom are bounded, their energy values are negative. We can simplify the differential equation for $w(\rho)$ further by the definitions

$$\rho_0 = \sqrt{\frac{2m}{|E|}}\frac{Ze^2}{\hbar}$$ (3.7)

and

$$\frac{E}{V} = \frac{\rho}{\rho_0},$$ (3.8)

to write

$$\rho\frac{d^2w}{d\rho^2} + 2(l+1-\rho)\frac{dw}{d\rho} + [\rho_0 - 2(l+1)]w(\rho) = 0.$$ (3.9)

We now try a series solution

$$w(\rho) = \sum_{k=0}^{\infty} a_k \rho^k,$$ (3.10)

which gives us a two-term recursion relation

$$\frac{a_{k+1}}{a_k} = \frac{2(k+l+1) - \rho_0}{(k+1)(k+2l+2)}.$$ (3.11)

In the limit as $k \to \infty$ the ratio of two successive terms, $\dfrac{a_{k+1}}{a_k}$, goes as $\dfrac{2}{k}$; hence the infinite series in Equation (3.10) diverges as $e^{2\rho}$, which also implies that $R_{E,l}(r)$ diverges as $r^l e^r \sqrt{2m|E|/\hbar^2}$. Since

$$|R_{E,l}(r)Y_l^m(\theta,\varphi)|^2$$ (3.12)

represents the probability density of the electron, for physically acceptable solutions $R_{E,l}(r)$ must be finite everywhere. In particular, as $r \to \infty$ it should go to zero. Hence for a finite solution in the interval $[0, \infty]$, we terminate the series [Eq. (3.10)] by restricting ρ_0 (energy) to the values

$$\rho_0 = 2(N + l + 1), \quad N = 0, 1, 2, \dots \,. \tag{3.13}$$

Since l takes integer values, we introduce a new quantum number, n, and write the energy levels of a single-electron atom as

$$E_n = -\frac{Z^2 m e^4}{2\hbar^2 n^2}, \quad n = 1, 2, \dots \,, \tag{3.14}$$

which are nothing but the Bohr energy levels.

Substituting Equation (3.13) in Equation (3.9) we obtain the differential equation to be solved for $w(\rho)$ as

$$\frac{\rho}{2}\frac{d^2 w}{d\rho^2} + (l + 1 - \rho)\frac{dw}{d\rho} + Nw(\rho) = 0, \tag{3.15}$$

solutions of which can be expressed in term of the **associated Laguerre polynomials** .

3.1   LAGUERRE EQUATION AND POLYNOMIALS

The Laguerre equation is defined as

$$x\frac{d^2 y}{dx^2} + (1 - x)\frac{dy}{dx} + ny = 0, \tag{3.16}$$

where n is a constant. Using the Frobenius method, we substitute a series solution about the regular singular point $x = 0$ as

$$y(x, s) = \sum_{r=0}^{\infty} a_r x^{s+r} \tag{3.17}$$

and obtain a two-term recursion relation

$$a_{r+1} = a_r \frac{(s + r - n)}{(s + r + 1)^2}. \tag{3.18}$$

In this case the indicial equation has a double root,

$$s = 0, \tag{3.19}$$

where the two linearly independent solutions are given as

$$y(x, 0) \quad \text{and} \quad \left.\frac{\partial y(x, s)}{\partial s}\right|_{s=0}. \tag{3.20}$$

The second solution diverges logarithmically as $x \to 0$. Hence for finite solutions everywhere we keep only the first solution, $y(x, 0)$, which has the recursion relation

$$a_{r+1} = -a_r \frac{(n-r)}{(r+1)^2}.$$ (3.21)

This gives us the infinite series solution as

$$y(x) = a_0 \left\{ 1 - \frac{nx}{1^2} + \frac{n(n-1)}{(2!)^2} x^2 + \cdots + \frac{(-1)^r n(n-1) \ldots (n-r+1)}{(r!)^2} x^r + \cdots \right\}.$$ (3.22)

From the recursion relation [Eq. (3.21)], it is seen that in the limit as $r \to \infty$ the ratio of two successive terms has the limit $a_{r+1}/a_r \to 1/r$; hence this series diverges as e^x for large x. We now restrict n to integer values to obtain finite polynomial solutions as

$$y(x) = a_0 \sum_{r=0}^{n} (-1)^r \frac{n(n-1) \cdots (n-r+1)}{(r!)^2} x^r$$

$$= a_0 \sum_{r=0}^{n} (-1)^r \frac{n! x^r}{(n-r)! (r!)^2}.$$ (3.23)

Laguerre polynomials are defined by setting $a_0 = 1$ in Equation (3.23) as

$$L_n(x) = \sum_{r=0}^{n} (-1)^r \frac{n! x^r}{(n-r)! (r!)^2}.$$ (3.24)

3.2   OTHER DEFINITIONS OF LAGUERRE POLYNOMIALS

3.2.1   Generating Function of Laguerre Polynomials

The generating function of the Laguerre polynomials is defined as

$$T(x, t) = \frac{e^{\frac{-xt}{(1-t)}}}{(1-t)} = \sum_{n=0}^{\infty} L_n(x) t^n, \quad |t| < 1.$$ (3.25)

To see that this gives the same polynomials as Equation (3.24), we expand the left-hand side as power series:

$$\frac{1}{(1-t)} e^{\frac{-xt}{(1-t)}} = \frac{1}{(1-t)} \sum_{r=0}^{\infty} \frac{1}{r!} \left[-\frac{xt}{1-t} \right]^r$$

$$= \sum_{r=0}^{\infty} \frac{(-1)^r}{r!} \frac{x^r t^r}{(1-t)^{r+1}}.$$ (3.26)

Using the binomial formula

$$\frac{1}{(1-t)^{r+1}} = 1 + (r+1)t + \frac{(r+1)(r+2)}{2!}t^2 + \cdots$$

$$= \sum_{s=0}^{\infty} \frac{(r+s)!}{r!s!}t^s, \tag{3.27}$$

Equation (3.26) becomes

$$\frac{1}{(1-t)}\exp\left\{-\frac{xt}{(1-t)}\right\} = \sum_{r,s=0}^{\infty}(-1)^r\frac{(r+s)!}{(r!)^2 s!}x^r t^{r+s}. \tag{3.28}$$

Defining a new dummy variable as

$$n = r + s, \tag{3.29}$$

we now write

$$\sum_{n=0}^{\infty}\left[\sum_{r=0}^{\infty}(-1)^r\frac{n!}{(r!)^2(n-r)!}x^r\right]t^n = \sum_{n=0}^{\infty}L_n(x)t^n \tag{3.30}$$

and compare equal powers of t. Since

$$s = n - r \ge 0, \tag{3.31}$$

$r \le n$; thus we obtain the Laguerre polynomials $L_n(x)$ as

$$L_n(x) = \sum_{r=0}^{n}(-1)^r\frac{n!}{(r!)^2(n-r)!}x^r. \tag{3.32}$$

3.2.2 Rodriguez Formula for the Laguerre Polynomials

Another definition of the Laguerre polynomials is given in terms of the Rodriguez formula:

$$L_n(x) = \frac{e^x}{n!}\frac{d^n}{dx^n}\left(x^n e^{-x}\right). \tag{3.33}$$

To show the equivalence of this formula with the other definitions we use the Leibniz formula

$$\frac{d^n}{dx^n}(fg) = \sum_{r=0}^{n}\frac{n!}{(n-r)!r!}\frac{d^{n-r}f}{dx^{n-r}}\frac{d^r g}{dx^r} \tag{3.34}$$

to write

$$\frac{e^x}{n!}\frac{d^n}{dx^n}\left(x^n e^{-x}\right) = \frac{e^x}{n!}\sum_{r=0}^{n}\frac{n!}{(n-r)!r!}\frac{d^{n-r}x^n}{dx^{n-r}}\frac{d^r e^{-x}}{dx^r}. \tag{3.35}$$

We now use

$$\frac{d^p x^q}{dx^p} = q\,(q-1)\cdots(q-p+1)\,x^{q-p}$$

$$= \frac{q!}{(q-p)!}x^{q-p} \tag{3.36}$$

to obtain

$$\frac{e^x}{n!}\frac{d^n}{dx^n}\left(x^n e^{-x}\right) = \frac{e^x}{n!}\sum_{r=0}^{n}\frac{n!}{(n-r)!r!}\frac{n!}{r!}x^r\,(-1)^r\,e^{-x}$$

$$= \sum_{r=0}^{n}(-1)^r\,\frac{n!x^r}{(r!)^2\,(n-r)!} \tag{3.37}$$

$$= L_n\,(x)\,.$$

3.3   ORTHOGONALITY OF LAGUERRE POLYNOMIALS

To show that the Laguerre polynomials form an orthogonal set, we evaluate the integral

$$I_{nm} = \int_0^{\infty} e^{-x} L_n\,(x)\,L_m\,(x)\,dx. \tag{3.38}$$

Using the generating function definition of the Laguerre polynomials we write

$$\frac{1}{1-t}\exp\left\{-\frac{xt}{(1-t)}\right\} = \sum_{n=0}^{\infty} L_n\,(x)\,t^n \tag{3.39}$$

and

$$\frac{1}{1-s}\exp\left\{-\frac{xs}{(1-s)}\right\} = \sum_{n=0}^{\infty} L_m\,(x)\,s^m. \tag{3.40}$$

We first multiply Equations (3.39) and (3.40) and then the result with e^{-x} to write

$$\sum_{n,m=0}^{\infty} e^{-x} L_n\,(x)\,L_m\,(x)\,t^n s^m = \frac{e^{-x}\exp\left\{-\dfrac{xt}{(1-t)}\right\}\exp\left\{-\dfrac{xs}{(1-s)}\right\}}{(1-t)\qquad\qquad(1-s)}.$$

$$\tag{3.41}$$

Interchanging the integral and the summation signs and integrating with respect to x gives us

$$\sum_{n,m=0}^{\infty} \left[\int_0^\infty e^{-x} L_n(x) L_m(x) \, dx \right] t^n s^m$$
$$= \int_0^\infty \frac{e^{-x} \exp\left\{-\dfrac{xt}{(1-t)}\right\} \exp\left\{-\dfrac{xs}{(1-s)}\right\}}{(1-t)\,(1-s)} \, dx. \qquad (3.42)$$

It is now seen that the value of the integral in Equation (3.38) can be obtained by expanding

$$I = \int_0^\infty \frac{e^{-x} \exp\left\{-\dfrac{xt}{(1-t)}\right\} \exp\left\{-\dfrac{xs}{(1-s)}\right\}}{(1-t)\,(1-s)} \, dx \qquad (3.43)$$

in powers of t and s and then by comparing the equal powers of $t^n s^m$ with the left-hand side of Equation (3.42). If we write I as

$$I = \frac{1}{(1-t)\,(1-s)} \int_0^\infty \exp\left\{-x\left(1 + \frac{t}{1-t} + \frac{s}{1-s}\right)\right\} dx, \qquad (3.44)$$

the integral can be taken to yield

$$I = \frac{1}{(1-t)\,(1-s)} \left[\frac{-\exp\left\{-x\left(1 + \frac{t}{1-t} + \frac{s}{1-s}\right)\right\}}{1 + \left[\frac{t}{(1-t)}\right] + \left[\frac{s}{(1-s)}\right]} \right]_0^\infty \qquad (3.45)$$

$$= \frac{1}{(1-t)\,(1-s)} \left[\frac{1}{1 + \left[\frac{t}{(1-t)}\right] + \left[\frac{s}{(1-s)}\right]} \right] \qquad (3.46)$$

$$= \frac{1}{1-st} \qquad (3.47)$$

$$= \sum_{n=0}^{\infty} s^n t^n. \qquad (3.48)$$

This leads us to the orthogonality relation for the Laguerre polynomials as

$$\int_0^\infty e^{-x} L_n(x) L_m(x) \, dx = \delta_{nm}.$$

Compared with the Legendre polynomials, we say that the Laguerre polynomials are orthogonal with respect to the weight function e^{-x}.

3.4 OTHER PROPERTIES OF LAGUERRE POLYNOMIALS

3.4.1 Recursion Relations

Using the method we have used for the Legendre polynomials, we can obtain two recursion relations for the Laguerre polynomials. We first differentiate the generating function with respect to t to obtain

$$(n+1) L_{n+1}(x) = (2n+1-x) L_n(x) - n L_{n-1}(x). \qquad (3.49)$$

Differentiating the generating function with respect to x gives us the second recursion relation

$$x L'_n(x) = n L_n(x) - n L_{n-1}(x). \qquad (3.50)$$

Another useful recursion relation is given as

$$L'_n(x) = -\sum_{r=0}^{n-1} L_r(x). \qquad (3.51)$$

Laguerre Polynomials

$$
\begin{aligned}
L_0(x) &= & 1 \\
L_1(x) &= & -x + 1 \\
L_2(x) &= & (1/2!)\left(x^2 - 4x + 2\right) \\
L_3(x) &= & (1/3!)\left(-x^3 + 9x^2 - 18x + 6\right) \\
L_4(x) &= & (1/4!)\left(x^4 - 16x^3 + 72x^2 - 96x + 24\right) \\
L_5(x) &= & (1/5!)\left(-x^5 + 25x^4 - 200x^3 + 600x^2 - 600x + 120\right)
\end{aligned}
\qquad (3.52)
$$

3.4.2 Special Values of Laguerre Polynomials

Taking $x = 0$ in the generating function we find

$$\sum_{n=0}^{\infty} L_n(0) t^n = \frac{1}{1-t} \qquad (3.53)$$

$$= \sum_{n=0}^{\infty} t^n. \qquad (3.54)$$

This gives us the special value

$$L_n(0) = 1. \qquad (3.55)$$

Another special value is obtained by writing the Laguerre equation at $x = 0$ as

$$\left[x \frac{d^2 L_n(x)}{dx^2} + (1-x) \frac{d}{dx} L_n(x) + n L_n(x) \right]_{x=0} = 0, \qquad (3.56)$$

which gives

$$L'_n(0) = -n. \tag{3.57}$$

3.5 ASSOCIATED LAGUERRE EQUATION AND POLYNOMIALS

The associated Laguerre equation is given as

$$x\frac{d^2y}{dx^2} + (k + 1 - x)\frac{dy}{dx} + ny = 0, \tag{3.58}$$

which reduces to the Laguerre equation for $k = 0$. Solution of Equation (3.58) can be found by the following theorem:

Theorem: Let $Z(x)$ be a solution of the Laguerre equation of order $(n + k)$, then $\dfrac{d^k Z(x)}{dx^k}$ satisfies the associated Laguerre equation.

Proof: We write the Laguerre equation of order $(n + k)$ as

$$x\frac{d^2 Z}{dx^2} + (1 - x)\frac{dZ}{dx} + (n + k)Z(x) = 0. \tag{3.59}$$

Using the Leibniz formula [Eq. (3.34)], k-fold differentiation of Equation (3.59) gives

$$x\frac{d^{k+2}Z}{dx^{k+2}} + k\frac{d^{k+1}Z}{dx^{k+1}} + (1 - x)\frac{d^{k+1}Z}{dx^{k+1}} + k(-1)\frac{d^k Z}{dx^k} + (n + k)\frac{d^k Z}{dx^k} = 0. \tag{3.60}$$

Rearranging this, we obtain the desired result as

$$x\frac{d^2}{dx^2}\left[\frac{d^k Z}{dx^k}\right] + (k + 1 - x)\frac{d}{dx}\left[\frac{d^k Z}{dx^k}\right] + n\left[\frac{d^k Z}{dx^k}\right] = 0. \tag{3.61}$$

Using the definition of the Laguerre polynomials [Eq. (3.32)], we can now write the associated Laguerre polynomials as

$$L_n^k(x) = (-1)^k \frac{d^k}{dx^k} \sum_{r=0}^{n+k} (-1)^r \frac{(n + k)! x^r}{(n + k - r)!(r!)^2}. \tag{3.62}$$

Since k-fold differentiation of x^r is going to give zeroes for the $r < k$ values, we can write

$$L_n^k(x) = (-1)^k \frac{d^k}{dx^k} \sum_{r=k}^{n+k} (-1)^r \frac{(n + k)! x^r}{(n + k - r)!(r!)^2}, \tag{3.63}$$

$$L_n^k(x) = (-1)^k \sum_{r=k}^{n+k} (-1)^r \frac{(n+k)!}{(n+k-r)!\,(r!)^2\,(r-k)!} x^{r-k}. \tag{3.64}$$

Defining a new dummy variable s as

$$s = r - k, \tag{3.65}$$

we find the final form of the associated Laguerre polynomials as

$$L_n^k(x) = \sum_{s=0}^{n} (-1)^s \frac{(n+k)!x^s}{(n-s)!\,(k+s)!s!}. \tag{3.66}$$

3.6   PROPERTIES OF ASSOCIATED LAGUERRE POLYNOMIALS

3.6.1   Generating Function

The generating function of the associated Laguerre polynomials is defined as

$$T(x,t) = \frac{\exp\left[-\dfrac{xt}{(1-t)}\right]}{(1-t)^{k+1}} = \sum_{n=0}^{\infty} L_n^k(x)\, t^n, \quad |t| < 1. \tag{3.67}$$

To prove this we write the generating function of the Laguerre polynomials as

$$\frac{\exp\left[-\dfrac{xt}{(1-t)}\right]}{(1-t)^{k+1}} = \sum_{n=0}^{\infty} L_n(x)\, t^n, \tag{3.68}$$

which gives us

$$\frac{d^k}{dx^k}\left[\frac{\exp\left[-\dfrac{xt}{(1-t)}\right]}{(1-t)}\right] = \frac{d^k}{dx^k}\sum_{n=k}^{\infty} L_n(x)\, t^n. \tag{3.69}$$

This can also be written as

$$\left[\frac{-t}{(1-t)}\right]^k \frac{\exp\left[-\dfrac{xt}{(1-t)}\right]}{(1-t)} = \sum_{n=0}^{\infty} \frac{d^k}{dx^k} L_{n+k}(x)\, t^{n+k}. \tag{3.70}$$

We now use the relation

$$L_n^k(x) = (-1)^k \frac{d^k}{dx^k} L_{n+k}(x) \tag{3.71}$$

to write

$$(-1)^k \frac{t^k}{(1-t)^{k+1}} \exp\left[-\frac{xt}{(1-t)}\right] = \sum_{n=0}^{\infty} (-1)^k L_n^k(x) t^{n+k}, \qquad (3.72)$$

which leads us to the desired result

$$\frac{\exp\left[-\frac{xt}{(1-t)}\right]}{(1-t)^{k+1}} = \sum_{n=0}^{\infty} L_n^k(x) t^n. \qquad (3.73)$$

3.6.2 Rodriguez Formula and Orthogonality

The Rodriguez formula for the associated Laguerre polynomials is given as

$$L_n^k(x) = \frac{e^x x^{-k}}{n!} \frac{d^n}{dx^n}\left[e^{-x} x^{n+k}\right]. \qquad (3.74)$$

Their orthogonality relation is:

$$\int_0^\infty e^{-x} x^k L_n^k(x) L_m^k(x)\, dx = \frac{(n+k)!}{n!} \delta_{nm}, \qquad (3.75)$$

where the weight function is given as

$$(e^{-x} x^k). \qquad (3.76)$$

3.6.3 Recursion Relations

Some frequently used recursion relations of the associated Laguerre polynomials are given as

$$(n+1) L_{n+1}^k(x) = (2n+k+1-x) L_n^k(x) - (n+k) L_{n-1}^k(x) \qquad (3.77)$$

$$x\frac{d}{dx} L_n^k(x) = n L_n^k(x) - (n+k) L_{n-1}^k(x) \qquad (3.78)$$

$$L_{n-1}^k(x) + L_n^{k-1}(x) = L_n^k(x). \qquad (3.79)$$

Problems

3.1 We have seen that the Schrödinger equation for a single-electron atom is written as

$$-\frac{\hbar^2}{2m}\frac{d^2 u_{E,l}}{dr^2} + \left[\frac{\hbar^2 l(l+1)}{2mr^2} - \frac{Ze^2}{r}\right] u_{E,l}(r) = E u_{E,l}(r).$$

i) Without any substitutions, convince yourself that the above equation gives a three-term recursion relation and then derive the substitution

$$u_{E,l}(\rho) = \rho^{l+1} e^{-\rho} w(\rho),$$

which leads to a differential equation with a two-term recursion relation for $w(\rho)$. We have defined a dimensionless variable $\rho = r\sqrt{2m|E|/\hbar^2}$. Hint: Study the asymptotic forms and the solutions of the differential equation at the end points of the interval $[0, \infty]$.

ii) Show that the differential equation for $w(\rho)$ has the recursion relation

$$\frac{a_{k+1}}{a_k} = \frac{2(k+l+1) - \rho_0}{(k+1)(k+2l+2)},$$

where

$$\rho_0 = \sqrt{\frac{2m}{|E|}} \frac{Ze^2}{\hbar}.$$

3.2   Derive the recursion relations

$$(n+1) L_{n+1}(x) = (2n+1-x) L_n(x) - n L_{n-1}(x),$$

$$x L'_n(x) = n L_n(x) - n L_{n-1}(x),$$

and

$$L'_n(x) = -\sum_{r=0}^{n-1} L_r(x).$$

3.3   Show that the associated Laguerre polynomials satisfy the orthogonality relation

$$\int_0^\infty e^{-x} x^k L_n^k(x) L_m^k(x) \, dx = \frac{(n+k)!}{n!} \delta_{nm}.$$

3.4   Write the normalized wave function of the hydrogen atom in terms of the spherical harmonics and the associated Laguerre polynomials.

3.5   Using the generating function

$$\frac{\exp\left[-\dfrac{xt}{(1-t)}\right]}{(1-t)^{k+1}} = \sum_{n=0}^\infty L_n^k(x) t^n,$$

derive the Rodriguez formula for $L_n^k(x)$.

3.6   Find the expansion of $\exp(-kx)$ in terms of the associated Laguerre polynomials.

3.7   Show the special value

$$L_n^k(0) = \frac{(n+k)!}{n!k!}$$

for the associated Laguerre polynomials.

3.8
i) Using the Frobenius method, find a series solution about $x = 0$ to the differential equation

$$x\frac{d^2C}{dx^2} + \frac{dC}{dx} + (\lambda - \frac{x}{4})C = 0, \qquad x \in [0, \infty].$$

ii) Show that solutions regular in the entire interval $[0, \infty]$ must be of the form

$$C_n(x) = e^{-x/2}\overline{L}_n(x),$$

with $\lambda = n+1/2$, $n = 0, 1, 2, ...$, where $\overline{L}_n(x)$ satisfies the differential equation

$$x\frac{d^2\overline{L}_n}{dx^2} + (1-x)\frac{d\overline{L}_n}{dx} + n\overline{L}_n = 0.$$

iii) With the integration constant

$$a_n = (-1)^n,$$

find the general expression for the coefficients a_{n-j} of $\overline{L}_n(x)$.
iv) Show that this polynomial can also be defined by the generating function

$$T(x,t) = \frac{\exp\left[-\dfrac{xt}{(1-t)}\right]}{(1-t)} = \sum_{n=0}^{\infty} \frac{\overline{L}_n(x)}{n!}t^n$$

or the Rodriguez formula

$$\overline{L}_n(x) = e^x\frac{d^n}{dx^n}\left(x^n e^{-x}\right).$$

v) Derive two recursion relations connecting

$$\overline{L}_{n+1}, \overline{L}_n \text{ and } \overline{L}_{n-1}$$

and

$$\overline{L}_n' \text{ with } \overline{L}_n, \overline{L}_{n-1}.$$

vi) Show that $C_n(x)$ form an orthogonal set, that is,

$$\int_0^\infty dx e^{-x} \overline{L}_n(x) \overline{L}_m(x) = 0 \quad \text{for} \quad n \neq m$$

and calculate the integral

$$\int_0^\infty dx e^{-x} \left[\overline{L}_n(x)\right]^2.$$

Note: Some books use $\overline{L}_n$ for their definition of Laguerre polynomials.

3.9   Starting with the generating function definition

$$\frac{e^{\dfrac{-xt}{(1-t)}}}{(1-t)} = \sum_{n=0}^\infty L_n(x) t^n,$$

derive the Rodriguez formula

$$L_n(x) = \frac{e^x}{n!} \frac{d^n}{dx^n} \left(x^n e^{-x}\right)$$

for the Laguerre polynomials.

3.10   Using the series definition of the Laguerre polynomials, show that

$$L_n'(0) = -n,$$
$$L_n''(0) = \frac{1}{2}n(n-1).$$

3.11   In quantum mechanics the radial part of Schrödinger's equation for the three-dimensional harmonic oscillator is given as

$$\frac{d^2 R(x)}{dx^2} + \frac{2}{x}\frac{dR(x)}{dx} + \left(\epsilon - x^2 - \frac{l(l+1)}{x^2}\right) R(x) = 0,$$

where x and ϵ are defined in terms of the radial distance r and the energy E as

$$x = \frac{r}{\sqrt{\dfrac{\hbar}{m\omega}}} \quad \text{and} \quad \epsilon = \frac{E}{\hbar\omega/2}.$$

l takes the integer values $l = 0, 1, 2....$. Show that the solutions of this equation can be expressed in terms of the associated Laguerre polynomials of argument x^2.

4

HERMITE POLYNOMIALS

The operator form of the time-independent Schrödinger equation is given as

$$\mathbf{H}\Psi\left(\overrightarrow{x}\right) = E\Psi\left(\overrightarrow{x}\right), \tag{4.1}$$

where $\Psi\left(\overrightarrow{x}\right)$ is the wave function, $\mathbf{H}$ is the Hamiltonian operator, and E stands for the energy eigenvalues. $\mathbf{H}$ is usually obtained from the classical Hamiltonian by replacing $\overrightarrow{x}$ (position) and $\overrightarrow{p}$ (momentum) with their operator counterparts:

$$\overrightarrow{x} \to \overrightarrow{x}, \tag{4.2}$$

$$\overrightarrow{p} \to \frac{\hbar}{i}\overrightarrow{\nabla}.$$

For the one-dimensional harmonic oscillator, the Hamiltonian operator is obtained from the classical Hamiltonian

$$H = \frac{p^2}{2m} + \frac{m\omega^2 x^2}{2} \tag{4.3}$$

as

$$\mathbf{H}(x) = -\frac{\hbar^2}{2m}\frac{d^2}{dx^2} + \frac{m\omega^2 x^2}{2}. \tag{4.4}$$

This leads us to the following Schrödinger equation:

$$\frac{d^2\Psi\left(x\right)}{dx^2} + \frac{2m}{\hbar^2}\left(E - \frac{m\omega^2 x^2}{2}\right)\Psi\left(x\right) = 0. \tag{4.5}$$

Defining two dimensionless variables

$$x' = \frac{x}{\sqrt{\hbar/m\omega}} \quad \text{and} \quad \varepsilon = \frac{E}{(\hbar\omega/2)}, \qquad (4.6)$$

and dropping the prime in x', we obtain the differential equation to be solved for the wave function as

$$\frac{d^2\Psi(x)}{dx^2} + \left(\varepsilon - x^2\right)\Psi(x) = 0, \quad x \in [0, \infty], \qquad (4.7)$$

which is closely related to the **Hermite equation,** and its solutions are given in terms of the **Hermite polynomials**.

4.1 HERMITE EQUATION AND POLYNOMIALS

We need to find a finite solution to differential Equation (4.7) in the entire interval $[0, \infty]$. However, direct application of the Frobenius method gives us a three-term recursion relation. To get a two-term recursion relation we again look at the behavior of the solution near the singularity at infinity.

First we make the substitution

$$x = \frac{1}{\xi}, \qquad (4.8)$$

which transforms the differential equation into the form

$$\frac{d^2\Psi(\xi)}{d\xi^2} + \frac{2}{\xi}\frac{d\Psi(\xi)}{d\xi} + \frac{1}{\xi^4}\left[\varepsilon - \frac{1}{\xi^2}\right]\Psi(\xi) = 0. \qquad (4.9)$$

It is clear that the singularity at infinity is essential. Because it is at the end point of our interval it does not pose any difficulty in finding a series solution about the origin. We now consider differential Equation (4.7) in the limit as $x \to \infty$, where it behaves as

$$\frac{d^2\Psi(x)}{dx^2} - x^2\Psi(x) = 0. \qquad (4.10)$$

This has two solutions, $\exp(-\frac{x^2}{2})$ and $\exp(\frac{x^2}{2})$. Since $\exp(\frac{x^2}{2})$ blows up at infinity, we use the first solution and substitute into Equation (4.7) a solution of the form

$$\Psi(x) = h(x)\exp(-\frac{x^2}{2}), \qquad (4.11)$$

which leads to the following differential equation for $h(x)$:

$$\frac{d^2h}{dx^2} - 2x\frac{dh}{dx} + (\varepsilon - 1)h(x) = 0. \qquad (4.12)$$

We now try a series solution of the form

$$h(x) = \sum_{k=0}^{\infty} a_k x^k, \tag{4.13}$$

which gives a two-term recursion relation:

$$a_{k+2} = a_k \frac{(2k+1-\varepsilon)}{(k+2)(k+1)}. \tag{4.14}$$

Since the ratio of two successive terms has the limit $\lim_{k\to\infty} \frac{a_{k+2}}{a_k} \to 2k$, this series asymptotically behaves as e^{x^2}. Thus the wave function diverges as

$$\lim_{x\to\infty} \Psi(x) \to e^{x^2/2}. \tag{4.15}$$

A physically meaningful solution must be finite in the entire interval $[0, \infty]$; hence we terminate the series after a finite number of terms. This is accomplished by restricting the energy of the system to certain integer values as

$$\varepsilon - 1 = 2n \ , \ n = 0, 1, 2... \ . \tag{4.16}$$

Now the recursion relation [Eq. (4.14)] becomes

$$a_{k+2} = a_k \frac{(2k-2n)}{(k+2)(k+1)}. \tag{4.17}$$

Thus we obtain the polynomial solutions of Equation (4.12) as

$$\begin{array}{lll} n = 0 & h_0(x) = & a_0 \\ n = 1 & h_1(x) = & a_1 x \\ n = 2 & h_2(x) = & a_0 \left(1 - 2x^2\right) \ . \\ \vdots & \vdots & \vdots \end{array} \tag{4.18}$$

From the recursion relation [Eq. (4.17)] we can write the coefficients of the decreasing powers of x for the nth-order polynomial as

$$a_{n-2j} = (-1)^j \frac{n(n-1)(n-2)(n-3)\cdots(n-2j+1)}{2^j 2.4...(2j)} a_n, \tag{4.19}$$

$$a_{n-2j} = \frac{(-1)^j n!}{(n-2j)!} \frac{1}{2^j 2^j j!} a_n. \tag{4.20}$$

When we take a_n as $a_n = 2^n$ we obtain the Hermite polynomials

$$H_n(x) = \sum_{j=0}^{\left[\frac{n}{2}\right]} \frac{(-1)^j 2^{n-2j} n!}{(n-2j)!j!} x^{n-2j}, \tag{4.21}$$

which satisfy the differential equation

$$H_n''(x) - 2xH_n'(x) + 2nH_n(x) = 0. \tag{4.22}$$

Hermite Polynomials

$$H_0(x) = 1$$
$$H_1(x) = 2x$$
$$H_2(x) = -2 + 4x^2$$
$$H_3(x) = -12x + 8x^3$$
$$H_4(x) = 12 - 48x^2 + 16x^4$$
$$H_5(x) = 120x - 160x^3 + 32x^5.$$

Going back to the energy parameter E, we find

$$E = \frac{\hbar\omega}{2}\varepsilon, \tag{4.23}$$

$$E = \hbar\omega\left(n + \frac{1}{2}\right), \quad n = 0, 1, 2, \dots . \tag{4.24}$$

This means that in quantum mechanics a one-dimensional harmonic oscillator can only oscillate with the energy values given above.

4.2   OTHER DEFINITIONS OF HERMITE POLYNOMIALS

4.2.1   Generating Function

The generating function for the Hermite polynomials is given as

$$T(t, x) = e^{-t^2 + 2xt} = \sum_{n=0}^{\infty} \frac{H_n(x)}{n!} t^n. \tag{4.25}$$

To show that this is equivalent to our former definition, Equation (4.21), we write the left-hand side as

$$e^{t(2x-t)} = \sum_{n=0}^{\infty} \frac{t^n (2x - t)^n}{n!} \tag{4.26}$$

$$= \sum_{n=0}^{\infty} \sum_{m=0}^{\infty} (-1)^m \frac{n! 2^{n-m}}{m! (n-m)! n!} x^{n-m} t^{n+m}. \tag{4.27}$$

Making the replacement

$$n + m \rightarrow n' \tag{4.28}$$

and dropping primes, we obtain

$$e^{t(2x-t)} = \sum_{n=0}^{\infty} \sum_{m=0}^{\left[\frac{n}{2}\right]} (-1)^m \frac{2^{n-2m} x^{n-2m}}{m! (n-2m)!} t^n. \tag{4.29}$$

Comparing this with the right-hand side of Equation (4.25), which is

$$\sum_{n=0}^{\infty} \frac{H_n(x)}{n!} t^n, \tag{4.30}$$

we see that $H_n(x)$ is the same as given in Equation (4.21).

4.2.2 Rodriguez Formula

Another definition for the Hermite polynomials is given by the Rodriguez formula

$$H_n(x) = (-1)^n e^{x^2} \frac{d^n}{dx^n} \left[e^{-x^2} \right]. \tag{4.31}$$

To see that this is equivalent to the generating function [Eq. (4.25)] we write the Taylor series expansion of an arbitrary function $F(t)$ as

$$F(t) = \sum_{n=0}^{\infty} \frac{d^n F(t)}{dt^n} \bigg|_{t=0} \frac{t^n}{n!}. \tag{4.32}$$

Comparing this with Equation (4.25) we obtain

$$H_n(x) = \left[\frac{\partial^n}{\partial t^n} e^{2tx - t^2} \right]_{t=0} \tag{4.33}$$

$$= \frac{\partial^n}{\partial t^n} e^{x^2 - (x-t)^2} \bigg|_{t=0} \tag{4.34}$$

$$= e^{x^2} \frac{\partial^n}{\partial t^n} e^{-(x-t)^2} \bigg|_{t=0}. \tag{4.35}$$

For an arbitrary differentiable function we can write

$$\frac{\partial}{\partial t} f(x-t) = -\frac{\partial}{\partial x} f(x-t), \tag{4.36}$$

hence

$$\frac{\partial^n}{\partial t^n} f(x-t) = (-1)^n \frac{\partial^n}{\partial x^n} f(x-t). \tag{4.37}$$

Applying this to Equation (4.35), we obtain the Rodriguez formula as

$$H_n(x) = (-1)^n e^{x^2} \frac{\partial^n}{\partial x^n} e^{-(x-t)^2} \bigg|_{t=0} = (-1)^n e^{x^2} \frac{d^n}{dx^n} e^{-x^2}. \tag{4.38}$$

4.3 RECURSION RELATIONS AND ORTHOGONALITY

Differentiating the generating function of the Hermite polynomials, first with respect to x and then with respect to t, we obtain two recursion relations:

$$H_{n+1}(x) = 2xH_n(x) - 2nH_{n-1}(x), \ (n \geqslant 1), \ H_1(x) = 2xH_0(x) \qquad (4.39)$$

and

$$H_n'(x) = 2nH_{n-1}(x), \ (n \geqslant 1), \ H_0'(x) = 0. \qquad (4.40)$$

To show the orthogonality of the Hermite polynomials we evaluate the integral

$$I_{nm} = \int_{-\infty}^{\infty} dx e^{-x^2} H_n(x) H_m(x) \ , \qquad (n \geq m). \qquad (4.41)$$

Using the Rodriguez formula, we write Equation (4.41) as

$$I_{nm} = (-1)^n \int_{-\infty}^{\infty} dx e^{-x^2} e^{x^2} \left[\frac{d^n}{dx^n} e^{-x^2} \right] H_m(x). \qquad (4.42)$$

After n-fold integration by parts and since $n > m$, we obtain

$$I_{nm} = (-1)^n \frac{d^{n-1}}{dx^{n-1}} \left(e^{-x^2} \right) H_m(x) \Big|_{-\infty}^{\infty} + (-1)^{n+1} \int_{-\infty}^{\infty} dx \left[\frac{d^{n-1}}{dx^{n-1}} e^{-x^2} \right] H_m'(x)$$

$$\vdots \qquad\qquad (4.43)$$

$$= (-1)^{2n} \int_{-\infty}^{\infty} dx e^{-x^2} \frac{d^n}{dx^n} H_m.$$

Since the x dependence of the mth-order Hermite polynomial goes as

$$H_m(x) = 2^m x^m + a_{m-2} x^{m-2} + \cdots , \qquad (4.44)$$

we obtain

$$I_{nm} = \begin{cases} 0 & , \ n > m \\ 2^n n! \sqrt{\pi} & , \ n = m \end{cases} , \qquad (4.45)$$

where we have used

$$\frac{d^n H_n(x)}{dx^n} = 2^n n! \text{ and } 2 \int_0^{\infty} dx e^{-x^2} = \sqrt{\pi}.$$

We now write the orthogonality relation as

$$\int_{-\infty}^{\infty} dx e^{-x^2} H_n(x) H_m(x) = 2^n n! \sqrt{\pi} \delta_{nm}. \qquad (4.46)$$

Using Equation (4.46) we can define a set of polynomials, $\{\phi_n(x)\}$, where $\phi_n(x)$ are defined as

$$\phi_n(x) = \frac{1}{\sqrt{2^n n! \sqrt{\pi}}} e^{-\frac{x^2}{2}} H_n(x), \quad n = 0, 1, 2, \dots \qquad (4.47)$$

and which satisfies the orthogonality relation

$$\int_{-\infty}^{\infty} dx \phi_n(x) \phi_m(x) = \delta_{nm}.$$

Since this set is also complete, any sufficiently well-behaved function in the interval $[-\infty, \infty]$ can be expanded in terms of $\{\phi_n(x)\}$ as

$$f(x) = \sum_{n=0}^{\infty} C_n \phi_n(x), \qquad (4.48)$$

where the coefficients C_n are given as

$$C_n = \int_{-\infty}^{\infty} dx' f(x') \phi_n(x'). \qquad (4.49)$$

Example 4.1. *Gaussian and the Hermite polynomials:* In quantum mechanics the wave function of a particle localized around x_0 can be given as a Gaussian:

$$f(x) = A e^{-\frac{1}{2}(x-x_0)^2}, \qquad (4.50)$$

where A is the normalization constant, which is determined by requiring the area under $f(x)$ to be unity. Let us find the expansion of this function in terms of the Hermite polynomials as

$$f(x) = \sum_{n=0}^{\infty} C_n \frac{e^{-x^2/2} H_n(x)}{\sqrt{2^n n! \sqrt{\pi}}}. \qquad (4.51)$$

This expansion corresponds to the representation of the wave function of a particle under the influence of a harmonic oscillator potential in terms of the harmonic oscillator energy eigenfunctions. Expansion coefficients C_n are determined from the integral

$$C_n = \frac{A}{\sqrt{2^n n! \sqrt{\pi}}} \int_{-\infty}^{\infty} d\xi \exp\left[-\frac{(\xi-x_0)^2}{2} - \frac{\xi^2}{2}\right] H_n(\xi). \qquad (4.52)$$

Writing this as

$$C_n = \frac{A}{\sqrt{2^n n! \sqrt{\pi}}} \int_{-\infty}^{\infty} d\xi e^{-\xi^2} \exp\left[2\xi \left(\frac{x_0}{2}\right) - \left(\frac{x_0}{2}\right)^2\right] H_n(\xi) e^{-x_0^2/4} \tag{4.53}$$

and defining a new parameter

$$\frac{x_0}{2} = t, \tag{4.54}$$

and using the generating function [Eq. (4.25)], we obtain

$$C_n = \frac{A}{\sqrt{2^n n! \sqrt{\pi}}} e^{-x_0^2/4} \int_{-\infty}^{\infty} d\xi e^{-\xi^2} \left[\sum_{m=0}^{\infty} \frac{H_m(\xi)}{m!} \left(\frac{x_0}{2}\right)^m\right] H_n(\xi). \tag{4.55}$$

We now use the orthogonality relation [Eq. (4.46)] of the Hermite polynomials to obtain

$$C_n = \frac{A}{\sqrt{2^n n! \sqrt{\pi}}} e^{-x_0^2/4} \left(\frac{x_0}{2}\right)^n \frac{1}{n!} \int d\xi e^{-\xi^2} \left[H_n(\xi)\right]^2, \tag{4.56}$$

$$C_n = \frac{A}{\sqrt{2^n n! \sqrt{\pi}}} e^{-x_0^2/4} \left(\frac{x_0}{2}\right)^n \frac{1}{n!} 2^n n! \sqrt{\pi}, \tag{4.57}$$

$$C_n = A e^{-x_0^2/4} \left(\frac{x_0}{2}\right)^n \frac{\sqrt{2^n}}{\sqrt{n!}} \pi^{\frac{1}{4}}. \tag{4.58}$$

Probability of finding a particle in the nth energy eigenstate is given as

$$|C_n|^2.$$

Example 4.2. *Dipole calculations in quantum mechanics:* In quantum mechanics and in electric dipole calculations we encounter integrals like

$$I_{nm} = e \int_{-\infty}^{\infty} dx \phi_n(x) \phi_m(x) x, \tag{4.59}$$

where e is the electric charge. Let us write this as

$$I_{nm} = \frac{e}{\sqrt{2^n n! \sqrt{\pi}} \sqrt{2^m m! \sqrt{\pi}}} \left[\int_{-\infty}^{\infty} dx e^{-x^2} H_n(x) H_m(x) x\right]. \tag{4.60}$$

We now use the generating function definition of the Hermite polynomials to write

$$\int_{-\infty}^{\infty} dx T(t, x) S(s, x) e^{-x^2} x$$

$$= \sum_{n=0}^{\infty} \sum_{m=0}^{\infty} \frac{1}{n! m!} \left[\int_{-\infty}^{\infty} dx e^{-x^2} H_n(x) H_m(x) x\right] t^n s^m. \tag{4.61}$$

If we show the expression inside the square brackets on the right-hand side as J_{nm} , integral I_{nm} will be given as

$$I_{nm} = \frac{e}{\sqrt{2^n n! \sqrt{\pi}} \sqrt{2^m m! \sqrt{\pi}}} J_{nm}. \tag{4.62}$$

Writing the left-hand side of Equation (4.61) explicitly we get

$$\int_{-\infty}^{\infty} dx e^{-t^2 + 2tx} e^{-s^2 + 2sx} e^{-x^2} x$$

$$= \int_{-\infty}^{\infty} dx e^{-(x-(s+t))^2} e^{2st} \{[x - (s+t)] + (s+t)\}$$

$$= e^{2st} \left\{ \int_{-\infty}^{\infty} u e^{-u^2} du + (s+t) \int_{-\infty}^{\infty} du e^{-u^2} \right\}$$

$$= e^{2st} (s+t) \sqrt{\pi}, \tag{4.63}$$

where we have defined

$$u = x - (s+t). \tag{4.64}$$

Expanding this in power series of t and s gives us

$$\int_{-\infty}^{\infty} dx e^{-t^2 + 2tx} e^{-s^2 + 2sx} e^{-x^2} x = \sqrt{\pi} \sum_{k=0}^{\infty} \frac{2^k s^k t^{k+1}}{k!} + \sqrt{\pi} \sum_{k=0}^{\infty} \frac{2^k s^{k+1} t^k}{k!}. \tag{4.65}$$

Finally, by comparing with

$$\sum_{n=0}^{\infty} \sum_{m=0}^{\infty} \frac{1}{n! m!} [J_{nm}] t^n s^m, \tag{4.66}$$

we obtain the desired result as

$$J_{nm} = 0 \qquad \Rightarrow \quad I_{nm} = 0 \qquad \qquad \text{for} \quad m \neq n \mp 1$$

$$J_{n,n+1} = \sqrt{\pi} 2^n (n+1)! \quad \Rightarrow \quad I_{n,n+1} = e\left[(n+1)/2\right]^{1/2} \quad \text{for} \quad m = n+1$$

$$J_{n,n-1} = \sqrt{\pi} 2^{n-1} n! \quad \Rightarrow \quad I_{n,n-1} = e\sqrt{n/2} \qquad \qquad \text{for} \quad m = n-1.$$

We can also write this result as

$$J_{nm} = \sqrt{\pi} 2^{n-1} n! \delta_{n-1,m} + \sqrt{\pi} 2^n (n+1)! \delta_{n+1,m}.$$

Problems

4.1   For the Hermite polynomials given the recursion relation

$$a_{k+2} = a_k \frac{(2k - 2n)}{(k + 2)(k + 1)},$$

show that one can write the coefficients of the decreasing powers of x for the nth-order polynomial as

$$a_{n-2j} = (-1)^j \frac{n(n-1)(n-2)(n-3)\cdots(n-2j+1)}{2^j 24 \cdots 2j} a_n$$

or

$$a_{n-2j} = \frac{(-1)^j n!}{(n-2j)!} \frac{1}{2^j 2^j j!} a_n.$$

4.2   For a three-dimensional harmonic oscillator the Schrödinger equation is given as

$$-\frac{\hbar^2}{2m} \vec{\nabla}^2 \Psi(\vec{r}) + \frac{1}{2} m\omega^2 r^2 \Psi(\vec{r}) = E\Psi(\vec{r}).$$

Using the separation of variables technique find the ordinary differential equations to be solved for r, θ, and ϕ.

4.3   Quantum mechanics of the-three dimensional harmonic oscillator leads to the following differential equation for the radial part of the wave function:

$$\frac{d^2 R(x)}{dx^2} + \frac{2}{x} \frac{dR(x)}{dx} + \left[\epsilon - x^2 - \frac{l(l+1)}{x^2} \right] R(x) = 0,$$

where x and ϵ are defined in terms of the radial distance r and the energy E as

$$x = \frac{r}{\sqrt{\dfrac{\hbar}{m\omega}}} \quad \text{and} \quad \epsilon = \frac{E}{\hbar\omega/2}$$

and l takes integer values $l = 0, 1, 2...$.
i) Examine the nature of the singular point at $x = \infty$.
ii) Show that in the limit as $x \to \infty$, the solution goes as

$$R \to e^{-x^2/2}.$$

iii) Using the Frobenius method, find an infinite series solution about $x = 0$ in the interval $[0, \infty]$. Check the convergence of your solution. Should your solution be finite everywhere, including the end points of your interval? why?

iv) For finite solutions everywhere in the interval $[0, \infty]$, what restrictions do you have to impose on the physical parameters of the system. v) For $l = 0, 1,$ and 2 find explicitly the solutions corresponding to the three smallest values of ϵ.

4.4   Show the integral

$$\int_{-\infty}^{\infty} x^2 e^{-x^2} H_n(x) H_n(x) dx = \pi^{1/2} 2^n n! \left(n + \frac{1}{2}\right).$$

4.5   Prove the orthogonality relation

$$\int_{-\infty}^{\infty} e^{-x^2} H_m(x) H_n(x) dx = 2^n n! \sqrt{\pi} \delta_{mn}$$

by using the generating function definition of $H_n(x)$.

4.6   Expand x^{2k} and x^{2k+1} in terms of the Hermite polynomials, where $k = 0, 1, 2...,$ to establish the results

$$x^{2k} = \frac{(2k)!}{2^{2k}} \sum_{n=0}^{k} \frac{H_{2n}(x)}{(2n)!(k-n)!}$$

and

$$x^{2k+1} = \frac{(2k+1)!}{2^{2k+1}} \sum_{n=0}^{k} \frac{H_{2n+1}(x)}{(2n+1)!(k-n)!}.$$

4.7   Show the following integrals:

$$\int_{-\infty}^{\infty} x e^{-x^2/2} H_n(x) dx = \left\{ \begin{array}{c} 0 \\ \dfrac{2\pi(n+1)!}{[(n+1)/2]!} \end{array} \right\} \text{for} \left\{ \begin{array}{c} n \text{ even} \\ n \text{ odd} \end{array} \right\},$$

$$\int_{-\infty}^{\infty} e^{-x^2/2} H_n(x) dx = \left\{ \begin{array}{c} 2\pi n!/(n/2)! \\ 0 \end{array} \right\} \text{for} \left\{ \begin{array}{c} n \text{ even} \\ n \text{ odd} \end{array} \right\}.$$

4.8   Show that

$$H_n(0) dx = \left\{ \begin{array}{c} 0 \\ (-1)^m \dfrac{(2m)!}{m!} \end{array} \right\} \text{for} \left\{ \begin{array}{c} n \text{ odd} \\ n = 2m \end{array} \right\}.$$

4.9   For positive integers $k, m,$ and n, show that

$$\int_{-\infty}^{\infty} x^k e^{-x^2} H_m(x) H_{m+n}(x) dx$$

$$= \left\{ \begin{array}{c} 0 \\ \sqrt{\pi} 2^m (m+k)! \end{array} \right\} \text{for} \left\{ \begin{array}{c} n > k \\ n = k \end{array} \right\}.$$

4.10   Prove that

$$\int_{-\infty}^{\infty} e^{-a^2 x^2} H_{2n}(x) dx = \frac{(2n)!}{n!} \frac{\sqrt{\pi}}{a} \left[\frac{1-a^2}{a^2} \right]^n,$$

where

$$\text{Re}\, a^2 > 0 \text{ and } n = 0, 1, 2, \dots .$$

4.11   Prove the expansions

$$e^{t^2} \cos 2xt = \sum_{n=0}^{\infty} \frac{(-1)^n H_{2n}(x)}{(2n)!} t^{2n}, \quad |t| < \infty$$

and

$$e^{t^2} \sin 2xt = \sum_{n=0}^{\infty} \frac{(-1)^n H_{2n+1}(x)}{(2n+1)!} t^{2n+1}, \quad |t| < \infty.$$

Note that these can be regarded as the generating functions for the even and the odd Hermite polynomials.

4.12   Show that the integral

$$\int_{-\infty}^{\infty} x^m e^{-x^2} H_n(x) dx = 0$$

is true for m integer and

$$0 \leqslant m \leqslant n - 1.$$

4.13   The hypergeometric equation is given as

$$x(1-x)\frac{d^2 y}{dx^2} + [\gamma - (\alpha + \beta + 1)x]\frac{dy}{dx} - \alpha\beta y(x) = 0,$$

where α, β, and γ are arbitrary constants, ($\gamma \neq$ integer and $\gamma \neq 0$).
i) Show that it has the general solution

$$y(x) = C_0 F(\alpha, \beta, \gamma; x) + C_1 F(\alpha - \gamma + 1, \beta - \gamma + 1, 2 - \gamma; x),$$

valid for the region $|x| < 1$ and C_0 and C_1 are arbitrary integration constants, and the hypergeometric function is defined by

$$F(\alpha, \beta, \gamma; x) = \sum_{k=0}^{\infty} \frac{(\alpha)_k (\beta)_k}{(\gamma)_k} \frac{x^k}{k!}$$

with $(\alpha)_k = \alpha(\alpha + 1)(\alpha + 2) \cdots (\alpha + k - 1)$.
ii) If a regular series solution is required for the entire interval $[-1, 1]$, the

above series will not serve as the solution. What conditions do you have to impose on α, β to ensure a regular solution in this case?
iii) Show that Legendre polynomials can be expressed as

$$P_l(x) = F(-l, l+1, 1; \frac{1-x}{2}).$$

4.14   Establish the following connections between the Hermite and the Laguerre polynomials:

$$L_n^{-1/2}(x) = \frac{(-1)^n}{2^{2n}n!} H_{2n}(\sqrt{x})$$

$$L_n^{1/2}(x) = \frac{(-1)^n}{2^{2n+1}n!} \frac{H_{2n+1}(\sqrt{x})}{\sqrt{x}}.$$

4.15   Derive the following recursion relations:

$$H_{n+1}(x) = 2xH_n(x) - 2nH_{n-1}(x)$$

and

$$H_n'(x) = 2nH_{n-1}(x).$$

5

GEGENBAUER and CHEBYSHEV POLYNOMIALS

In the study of oscillations and waves, sine and cosine functions play a central role. They come from the solutions of the wave (Helmholtz) equation in Cartesian coordinates with the appropriate boundary conditions. They also form a basis for representing general waves and oscillations of various types, shapes, and sizes. Solutions of the angular part of the Helmholtz equation in spherical polar coordinates are the spherical harmonics. Analogous to the oscillations of a piece of string, spherical harmonics correspond to the oscillations of a two-sphere. Spherical harmonics also form a complete set of orthonormal functions; hence they are very important in many theoretical and practical applications. To represent the oscillations of a three-sphere (hypersphere), along with the spherical harmonics we need the Gegenbauer polynomials. They are very useful in cosmology and quantum field theory in curved backgrounds. Both the spherical harmonics and the Gegenbauer polynomials are combinations of sines and cosines. Chebyshev polynomials form another complete and orthonormal set of functions, which are closely related to the Gegenbauer polynomials.

5.1 COSMOLOGY AND GEGENBAUER POLYNOMIALS

Standard models in cosmology are generally accepted as accurately describing the global properties of the universe like homogeneity, isotropy, and expansion. Among the standard models, closed universes correspond to the surface of a hypersphere (three-sphere), where the line element is given as

$$ds^2 = dt^2 - R_0(t)^2[d\chi^2 + \sin^2\chi d\theta^2 + \sin^2\chi \sin^2\theta d\phi^2], \quad \text{(we set c=1)}. \quad (5.1)$$

Angular coordinates χ, θ, and ϕ have the ranges

$$\chi \in [0, \pi], \; \theta \in [0, \pi], \; \phi \in [0, 2\pi], \quad (5.2)$$

t is the universal time, and $R_0(t)$ is the radius of the hypersphere. We now consider the wave equation for the massless conformal scalar field in a closed static universe,

$$\Box \Phi(t, \chi, \theta, \phi) + \frac{1}{R_0^2}\Phi(t, \chi, \theta, \phi) = 0, \quad (5.3)$$

where $\Box$ is the d'Alembert (wave) operator,

$$\Box = g_{\mu\nu}\partial^\mu\partial_\nu, \quad (5.4)$$

with ∂_ν standing for the covariant derivative. Explicit evaluation of Equation (5.3) is beyond the scope of this chapter (see Chapter 10); hence we suffice by saying that a separable solution of the form

$$\Phi(t, \chi, \theta, \phi) = T(t)X(\chi)Y(\theta, \phi) \quad (5.5)$$

reduces Equation (5.3) to

$$\left[\frac{1}{T(t)}\frac{d^2T(t)}{dt^2}\right]$$
$$-\frac{1}{R_0^2 X(\chi)}\left(\frac{d^2X(\chi)}{d\chi^2} + \frac{2\cos\chi}{\sin\chi}\frac{dX(\chi)}{d\chi} - 1\right) \quad (5.6)$$
$$-\frac{1}{R_0^2\sin^2\chi}\left[\frac{1}{Y(\theta,\phi)}\left(\frac{\partial^2 Y(\theta,\phi)}{\partial\theta^2} + \frac{\cos\theta}{\sin\theta}\frac{\partial Y(\theta,\phi)}{\partial\theta} + \frac{1}{\sin^2\theta}\frac{\partial^2 Y(\theta,\phi)}{\partial\phi^2}\right)\right] = 0.$$

Since t, χ, θ, and ϕ are independent coordinates, differential equations to be solved for T, X and Y are easily found as

$$\frac{1}{T(t)}\frac{d^2T(t)}{dt^2} = -\omega^2, \quad (5.7)$$

$$\frac{1}{Y(\theta,\phi)}\left(\frac{\partial^2 Y(\theta,\phi)}{\partial\theta^2} + \frac{\cos\theta}{\sin\theta}\frac{\partial Y(\theta,\phi)}{\partial\theta} + \frac{1}{\sin^2\theta}\frac{\partial^2 Y(\theta,\phi)}{\partial\phi^2}\right) = \lambda, \quad (5.8)$$

$$\sin^2\chi\frac{d^2X(\chi)}{d\chi^2} + 2\sin\chi\cos\chi\frac{dX(\chi)}{d\chi} + \left(\omega^2 - \frac{1}{R_0^2}\right)R_0^2\sin^2\chi X(\chi) = -\lambda X(\chi).$$

$$(5.9)$$

ω and λ are two separation constants. For wave problems ω corresponds to the angular frequency.

Two linearly independent solutions of Equation (5.7) can be immediately written as

$$T(t) = e^{i\omega t} \text{ and } e^{-i\omega t}, \tag{5.10}$$

while the Second Equation (5.8) is nothing but the differential equation [Eq. (2.182)] that the spherical harmonics satisfy with λ and m given as

$$\lambda = -l(l+1), \quad l = 0, 1, 2, ..., \text{ and } m = 0, \pm 1, ..., \pm l. \tag{5.11}$$

Before we try a series solution in Equation (5.9) we make the substitution

$$X(\chi) = C_0 \sin^l \chi C(\cos\chi), \tag{5.12}$$

where

$$x = \cos\chi, \ x \in [-1, 1] \tag{5.13}$$

and obtain the following differential equation for $C(x)$:

$$\left(1 - x^2\right)\frac{d^2C(x)}{dx^2} - (2l+3)x\frac{dC(x)}{dx} + \left[-l(l+2) + (\omega^2 - \frac{1}{R_0^2})R_0^2\right]C(x) = 0. \tag{5.14}$$

Substitution (5.12) is needed to ensure a two-term recursion relation with the Frobenius method. This equation has two regular singular points at the end points $x = \pm 1$. We now try a series solution of the form

$$C(x) = \sum_{k=0}^{\infty} a_k x^{k+\alpha} \tag{5.15}$$

to get

$$a_0\alpha(\alpha - 1)x^{\alpha-2} + a_1\alpha(\alpha + 1)x^{\alpha-1}$$

$$+ \sum_{k=0}^{\infty}\{a_{k+2}(k + \alpha + 2)(k + \alpha + 1)$$

$$-a_k[(k + \alpha)(k + \alpha - 1) + (2l+3)(k + \alpha) - A]\}x^{k+\alpha} = 0. \tag{5.16}$$

In this equation A is defined as

$$A = -l(l+2) + (\omega^2 - \frac{1}{R_0^2})R_0^2. \tag{5.17}$$

Equation (5.16) cannot be satisfied for all x unless the coefficients of all the powers of x are zero, that is

$$a_0\alpha(\alpha - 1) = 0, \ \ a_0 \neq 0, \tag{5.18}$$

$$a_1\alpha(\alpha + 1) = 0, \tag{5.19}$$

$$a_{k+2} = a_k \left[\frac{(k + \alpha)(k + \alpha - 1) + (2l + 3)(k + \alpha) - A}{(k + \alpha + 2)(k + \alpha + 1)} \right], \tag{5.20}$$

$$k = 0, 1, 2, \dots \ . \tag{5.21}$$

The indicial Equation (5.18) has two roots, 0 and 1. Starting with the smaller root, $\alpha = 0$, we obtain the general solution as

$$C(x) = a_0 \left[1 - \frac{A}{2}x^2 - \left(\frac{2 + 2(2l + 3) - A}{3.4} \right) \frac{A}{2}x^4 + \cdots \right]$$

$$+ a_1 \left[x + \frac{(2l + 3) - A}{2.3}x^3 + \cdots \right], \tag{5.22}$$

where a_0 and a_1 are two integration constants and the recursion relation for the coefficients is given as

$$a_{k+2} = a_k \left[\frac{k(k - 1) + (2l + 3)k - A}{(k + 2)(k + 1)} \right], \ \ k = 0, 1, 2, \dots \ . \tag{5.23}$$

From the limit

$$\lim_{k \to \infty} \frac{a_{k+2}}{a_k} \to 1, \tag{5.24}$$

we see that both of these series diverge at the end points, $x = \pm 1$, as $\dfrac{1}{1 - x^2}$: To avoid this divergence we terminate the series by restricting ωR_0 to integer values given by

$$\omega_N = \frac{(N + 1)}{R_0}, \ \ N = 0, 1, 2, \dots \ . \tag{5.25}$$

Polynomial solutions obtained in this way can be expressed in terms of the Gegenbauer polynomials. Note that these frequencies mean that one can only fit integer multiples of full wavelengths around the circumference, $2\pi R_0$, of the universe, that is,

$$(1 + N)\lambda_N = 2\pi R_0, \ \ N = 0, 1, 2, \dots \ . \tag{5.26}$$

Using the relation

$$\omega_N = 2\pi / \lambda_N$$

we easily obtain the frequencies of Equation (5.25).

5.2   GEGENBAUER EQUATION AND ITS SOLUTIONS

The Gegenbauer equation is in general written as

$$(1 - x^2)\frac{d^2 C_n^\lambda(x)}{dx^2} - (2\lambda + 1)x\frac{dC_n^\lambda(x)}{dx} + n(n + 2\lambda)C_n^\lambda(x) = 0. \qquad (5.27)$$

For $\lambda = 1/2$, this equation reduces to the Legendre equation. For the integer values of n, its solutions reduce to the Gegenbauer or the Legendre polynomials as:

$$C_n^\lambda(x) = \sum_{r=0}^{[n/2]} (-1)^r \frac{\Gamma(n - r + \lambda)}{\Gamma(\lambda)r!(n - 2r)!}(2x)^{n-2r}. \qquad (5.28)$$

5.2.1   Orthogonality and the Generating Function

The orthogonality relation of the Gegenbauer polynomials is given as

$$\int_{-1}^{1}(1 - x^2)^{\lambda - \frac{1}{2}}C_n^\lambda(x)C_m^\lambda(x)dx = 2^{1-2\lambda}\frac{\pi\Gamma(n + 2\lambda)}{(n + \lambda)\Gamma^2(\lambda)\Gamma(n + 1)}\delta_{nm}. \qquad (5.29)$$

The generating function of the Gegenbauer polynomials is defined as

$$\frac{1}{(1 - 2xt + t^2)^\lambda} = \sum_{n=0}^{\infty} C_n^\lambda(x)t^n, \quad |t| < 1, \ |x| \leq 1, \ \lambda > -1/2. \qquad (5.30)$$

We can now write the solution of Equation (5.14) in terms of the Gegenbauer polynomials as $C_{N-l}^{l+1}(x)$, and the complete solution for the wave Equation (5.3) becomes

$$\Phi(t, \chi, \theta, \phi) = (c_1 e^{i\omega_N t} + c_2 e^{-i\omega_N t})(\sin^l \chi)C_{N-l}^{l+1}(\cos \chi)Y_l^m(\theta, \phi) \qquad (5.31)$$

5.3   CHEBYSHEV EQUATION AND POLYNOMIALS

5.3.1   Chebyshev Polynomials of the First Kind

Polynomials defined as

$$T_n(\cos \chi) = \cos(n\chi), \quad n = 0, 1, 2... \qquad (5.32)$$

are called the Chebyshev polynomials of first kind, and they satisfy the Chebyshev equation

$$(1 - x^2)\frac{d^2 T_n(x)}{dx^2} - x\frac{dT_n(x)}{dx} + n^2 T_n(x) = 0, \qquad (5.33)$$

where we have defined

$$x = \cos \chi. \qquad (5.34)$$

5.3.2   Relation of Chebyshev and Gegenbauer Polynomials

The Chebyshev equation after $(l + 1)$-fold differentiation yields

$$(1 - x^2)\frac{d^{l+3}(\cos n\chi)}{dx^{l+3}} - (2l + 3)x\frac{d^{l+2}(\cos n\chi)}{dx^{l+2}} \tag{5.35}$$

$$+ \left[-l^2 - 2l - 1 + n^2\right]\frac{d^{l+1}(\cos n\chi)}{dx^{l+1}} = 0,$$

where $n = 1, 2, \ldots$. We now rearrange this as

$$\left\{(1 - x^2)\frac{d^2}{dx^2} - (2l + 3)x\frac{d}{dx} + \left[-l(l + 2) + n^2 - 1\right]\right\}\left[\frac{d^{l+1}(\cos n\chi)}{dx^{l+1}}\right] = 0,$$

$$\left\{(1 - x^2)\frac{d^2}{dx^2} - [2(l + 1) + 1]\,x\frac{d}{dx} + (n - l - 1)\,[(n - l - 1) + 2(l + 1)]\right\}$$

$$\times \left[\frac{d^{l+1}(\cos n\chi)}{dx^{l+1}}\right] = 0 \tag{5.36}$$

and compare with Equation (5.27) to obtain the following relation between the Gegenbauer and the Chebyshev polynomials of the first kind:

$$C_{n-l-1}^{l+1}(x) = \frac{d^{l+1}(\cos n\chi)}{dx^{l+1}} \tag{5.37}$$

$$= \frac{d^{l+1}T_n(x)}{dx^{l+1}}, \quad n = 1, 2, \ldots . \tag{5.38}$$

5.3.3   Chebyshev Polynomials of the Second Kind

Chebyshev polynomials of the second kind are defined as

$$U_n(x) = \sin(n\chi), \quad n = 0, 1, 2\ldots \ , \tag{5.39}$$

where $x = \cos\chi$. Chebyshev polynomials of the first and second kinds are linearly independent, and they both satisfy the Chebyshev Equation (5.33).

In terms of x the Chebyshev polynomials are written as

$$T_n(x) = \sum_{r=0}^{\left[\frac{n}{2}\right]} (-1)^r \frac{n!}{(2r)!(n - 2r)!}(1 - x^2)^r x^{n-2r} \tag{5.40}$$

and

$$U_n(x) = \sum_{r=0}^{[(n-1)/2]} (-1)^r \frac{n!}{(2r + 1)!(n - 2r - 1)!}(1 - x^2)^{r+\frac{1}{2}}x^{n-2r-1}. \tag{5.41}$$

For some n values Chebyshev polynomials are given as

Chebyshev Polynomials of the First Kind

$$T_0 - 1$$

$$T_1(x) = x$$

$$T_2(x) = 2x^2 - 1$$

$$T_3(x) = 4x^3 - 3x \qquad (5.42)$$

$$T_4(x) = 8x^4 - 8x^2 + 1$$

$$T_5(x) = 16x^5 - 20x^3 + 5x$$

Chebyshev Polynomials of the Second Kind

$$U_0 = 0$$

$$U_1(x) = \sqrt{(1 - x^2)}$$

$$U_2(x) = \sqrt{(1 - x^2)}(2x)$$

$$U_3(x) = \sqrt{(1 - x^2)}(4x^2 - 1) \qquad (5.43)$$

$$U_4(x) = \sqrt{(1 - x^2)}(8x^3 - 4x)$$

$$U_5(x) = \sqrt{(1 - x^2)}(16x^4 - 12x^2 + 1)$$

5.3.4   Orthogonality and the Generating Function of Chebyshev Polynomials

The generating functions of the Chebyshev polynomials are given as

$$\frac{1-t^2}{1-2tx+t^2} = T_0(x) + 2\sum_{n=1}^{\infty} T_n(x)t^n \ , \ |t| < 1, \ |x| \le 1 \tag{5.44}$$

and

$$\frac{\sqrt{1-x^2}}{1-2tx+t^2} = \sum_{n=0}^{\infty} U_{n+1}(x)t^n. \tag{5.45}$$

Their orthogonality relations are

$$\int_{-1}^{1} \frac{T_m(x)T_n(x)}{\sqrt{1-x^2}} dx = \left\{ \begin{array}{cc} 0 & m \ne n \\ \pi/2 & m = n \ne 0 \\ \pi & m = n = 0 \end{array} \right\} \tag{5.46}$$

and

$$\int_{-1}^{1} \frac{U_m(x)U_n(x)}{\sqrt{1-x^2}} dx = \left\{ \begin{array}{cc} 0 & m \ne n \\ \pi/2 & m = n \ne 0 \\ 0 & m = n = 0 \end{array} \right\}. \tag{5.47}$$

5.3.5   Another Definition for the Chebyshev Polynomials of the Second Kind

Sometimes the polynomials defined as

$$\overline{U}_0(x) = 1$$

$$\overline{U}_1(x) = 2x$$

$$\overline{U}_2(x) = 4x^2 - 1 \tag{5.48}$$

$$\overline{U}_3(x) = 8x^3 - 4x$$

$$\overline{U}_4(x) = 16x^4 - 12x^2 + 1$$

$$\vdots$$

are also referred to as the Chebyshev polynomials of the second kind. They are related to $U_n(x)$ by

$$\sqrt{1-x^2}\overline{U}_n(x) = U_{n+1}(x), \quad n = 0, 1, 2, \dots . \tag{5.49}$$

$\overline{U}_n(x)$ satisfy the differential equation

$$(1-x^2)\frac{d^2\overline{U}_n(x)}{dx^2} - 3x\frac{d\overline{U}_n(x)}{dx} + n(n+2)\overline{U}_n(x) = 0, \qquad (5.50)$$

and their orthogonality relation is given as

$$\int_{-1}^{1} dx\sqrt{1-x^2}\overline{U}_m(x)\overline{U}_n(x) = \frac{\pi}{2}\delta_{mn}. \qquad (5.51)$$

Note that even though $\overline{U}_m(x)$ are polynomials, $U_m(x)$ are not. The generating function for $\overline{U}_m(x)$ is given as

$$\frac{1}{(1-2xt+t^2)} = \sum_{m=0}^{\infty} \overline{U}_m(x)t^m, \quad |t| < 1, \ |x| < 1. \qquad (5.52)$$

$T_n(x)$ and $\overline{U}_n(x)$ satisfy the recursion relations

$$(1-x^2)T'_n(x) = -nxT_n(x) + nT_{n-1}(x) \qquad (5.53)$$

and

$$(1-x^2)\overline{U}'_n(x) = -nx\overline{U}_n(x) + (n+1)\overline{U}_{n-1}(x). \qquad (5.54)$$

Special Values of the Chebyshev Polynomials

$$T_n(1) = 1,$$
$$T_n(-1) = (-1)^n,$$
$$T_{2n}(0) = (-1)^n,$$
$$T_{2n+1}(0) = 0,$$
$$U_n(1) = 0,$$
$$U_n(-1) = 0,$$
$$U_{2n}(0) = 0,$$
$$U_{2n+1}(0 = (-1)^n.$$

Problems

5.1 Observe that the equation

$$\sin^2\chi\frac{d^2X(\chi)}{d\chi^2} + 2\sin\chi\cos\chi\frac{dX(\chi)}{d\chi} + \left(\omega^2 - \frac{1}{R_0^2}\right)R_0^2\sin^2\chi X(\chi) = -\lambda X(\chi)$$

gives a three-term recursion relation and then drive the transformation

$$X(\chi) = C_0 \sin^l \chi C(\cos \chi),$$

which gives a differential equation for $C(\cos \chi)$ with a two-term recursion relation.

5.2   Using the line element

$$ds^2 = c^2 dt^2 - R_0(t)^2 [d\chi^2 + \sin^2 \chi d\theta^2 + \sin^2 \chi \sin^2 \theta d\phi^2],$$

find the spatial volume of a closed universe. What is the circumference?

5.3   Show that the solutions of

$$\left(1 - x^2\right) \frac{d^2 C(x)}{dx^2} - (2l + 3)x \frac{dC(x)}{dx} + \left[-l(l + 2) + (\omega_N^2 - \frac{1}{R_0^2})R_0^2\right] C(x) = 0$$

can be expressed in terms of Gegenbauer polynomials as

$$C_{N-l}^{l+1}(x),$$

where

$$\omega_N = \frac{(N + 1)}{R_0}, \quad N = 0, 1, 2, \dots .$$

5.4   Show the orthogonality relation of the Gegenbauer polynomials:

$$\int_{-1}^{1} (1 - x^2)^{\lambda - \frac{1}{2}} C_n^\lambda(x) C_m^\lambda(x) dx = 2^{1-2\lambda} \frac{\pi \Gamma(n + 2\lambda)}{(n + \lambda)\Gamma^2(\lambda)\Gamma(n + 1)} \delta_{nm}.$$

5.5   Show that the generating function

$$\frac{1}{(1 - 2xt + t^2)^\lambda} = \sum_{n=0}^{\infty} C_n^\lambda(x) t^n, \quad |t| < 1, \ |x| \le 1, \ \lambda > -1/2$$

can be used to define the Gegenbauer polynomials.

5.6   Using the Frobenius method, find a series solution to the Chebyshev equation

$$(1 - x^2) \frac{d^2 y(x)}{dx^2} - x \frac{dy(x)}{dx} + n^2 y(x) = 0, \quad x \in [-1, 1].$$

For finite solutions in the entire interval $[-1, 1]$ do you have to restrict n to integer values?

5.7   Show the following special values:

$$T_n(1) = 1$$
$$T_n(-1) = (-1)^n$$
$$T_{2n}(0) = (-1)^n$$
$$T_{2n+1}(0) = 0$$

and

$$\overline{U}_n(1) = n + 1$$
$$\overline{U}_n(-1) = (-1)^n(n+1)$$
$$\overline{U}_{2n}(0) = (-1)^n$$
$$\overline{U}_{2n+1}(0) = 0.$$

5.8   Show the relations

$$T_n(-x) = (-1)^n T_n(x),$$
$$\overline{U}_n(-x) = (-1)^n \overline{U}_n(x).$$

5.9   Using the generating function

$$\frac{1}{(1 - 2xt + t^2)} = \sum_{m=0}^{\infty} \overline{U}_m(x)t^m, \quad |t| < 1, \ |x| < 1,$$

show that

$$\overline{U}_m(x) = \sum_{k=0}^{[m/2]} (-1)^k \frac{(m-k)!}{k!(m-2k)!} (2x)^{m-2k}.$$

5.10   Show that $T_n(x)$ and $U_n(x)$ satisfy the recursion relations

$$(1 - x^2)T_n'(x) = -nxT_n(x) + nT_{n-1}(x)$$

and

$$(1 - x^2)U_n'(x) = -nxU_n(x) + nU_{n-1}(x).$$

5.11   Using the generating function

$$\frac{1}{(1 - 2xt + t^2)^\lambda} = \sum_{n=0}^{\infty} C_n^\lambda(x)t^n, \quad |t| < 1, \ |x| \le 1, \ \lambda > -1/2,$$

show

$$C_n^\lambda(x) = \sum_{r=0}^{[n/2]} (-1)^r \frac{\Gamma(n-r+\lambda)}{\Gamma(\lambda)r!(n-2r)!}(2x)^{n-2r}.$$

5.12   Let $x = \cos\chi$ and find a series expansion of

$$C(x) = \frac{d^{l+1}(\cos n\chi)}{d(\cos\chi)^{l+1}}$$

in terms of x.

5.13   Using

$$C_n^\lambda(x) = \sum_{r=0}^{[n/2]} (-1)^r \frac{\Gamma(n-r+\lambda)}{\Gamma(\lambda)r!(n-2r)!}(2x)^{n-2r},$$

show that for $\lambda = 1/2$ Gegenbauer polynomials reduce to the Legendre polynomials, that is

$$C_n^{1/2}(x) = P_n(x).$$

5.14   Prove the recursion relations

$$T_{n+1}(x) - 2xT_n(x) + T_{n-1}(x) = 0$$

and

$$U_{n+1}(x) - 2xU_n(x) + U_{n-1}(x) = 0.$$

5.15   Chebyshev polynomials $T_n(x)$ and $U_n(x)$ can be related to each other. Show the relations

$$(1 - x^2)^{1/2}T_n(x) = U_{n+1}(x) - xU_n(x)$$

and

$$(1 - x^2)^{1/2}U_n(x) = xT_n(x) - T_{n+1}(x).$$

5.16   Obtain the Chebyshev expansion

$$(1 - x^2)^{1/2} = \frac{2}{\pi}\left[1 - 2\sum_{s=1}^{\infty}(4s^2 - 1)^{-1}T_{2s}(x)\right].$$

6

BESSEL FUNCTIONS

The important role that trigonometric and hyperbolic functions play in the study of oscillations is well known. The equation of motion of a uniform rigid rod of length $2l$ suspended from one end and oscillating freely in a plane is given as

$$I\ddot{\theta} = -mgl\sin\theta. \tag{6.1}$$

In this equation I is the moment of inertia, m is the mass of the rod, g is the acceleration of gravity, and θ is the angular displacement of the rod from its equilibrium position. For small oscillations we can approximate $\sin\theta$ with θ; thus the general solution is given in terms of trigonometric functions as

$$\theta(t) = A\cos\omega_0 t + B\sin\omega_0 t, \quad \left(\omega_0^2 = mgl/I\right). \tag{6.2}$$

Suppose the rod is oscillating inside a viscous fluid exerting a drag force proportional to $\dot{\theta}$. Now the equation of motion will be given as

$$I\ddot{\theta} = -k\dot{\theta} - mgl\theta, \tag{6.3}$$

where k is the drag coefficient. For low viscosity the general solution is still expressed in terms of trigonometric functions albeit an exponentially decaying amplitude. However, for high viscosity, $(k/2I)^2 > \omega_0^2$, we need the hyperbolic functions, where the general solution is now given as

$$\theta(t) = e^{-(k/2I)t}\left[A\cosh q_0 t + B\sinh q_0 t\right], \quad \left(q_0^2 = (k/2I)^2 - \omega_0^2\right). \tag{6.4}$$

83

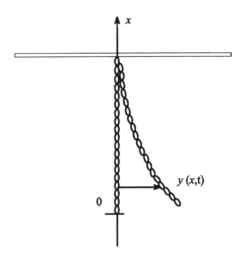

Fig. 6.1   Flexible chain

We now consider small oscillations of a flexible chain with uniform density $\rho_0(g/cm)$ and length l. We assume that the loops are very small compared to the length of the chain. We show the distance measured upwards from the free end of the chain with x and use $y(x,t)$ to represent the displacement of the chain from its equilibrium position (Fig. 6.1). For small oscillations we assume that the change in y with x is small; hence $\partial y/\partial x \ll 1$. We can write the y-component of the tension along the chain as $T_y(x) = \rho_0 gx(\partial y/\partial x)$. This gives the restoring force on a mass element of length Δx as

$$T_y(x + \Delta x) - T_y(x) = \frac{\partial}{\partial x}\left(\rho_0 gx\frac{\partial y}{\partial x}\right)\Delta x.$$

We can now write the equation of motion of a mass element of length Δx as

$$(\rho_0\Delta x)\frac{\partial^2 y(x,t)}{\partial t^2} = \frac{\partial}{\partial x}\left[\rho_0 gx\frac{\partial y(x,t)}{\partial x}\right]\Delta x,$$

$$\frac{\partial^2 y(x,t)}{\partial t^2}\Delta x = g\frac{\partial}{\partial x}\left[x\frac{\partial y(x,t)}{\partial x}\right]\Delta x. \tag{6.5}$$

Since Δx is small but finite, we obtain the differential equation to be solved for $y(x,t)$ as

$$\frac{\partial^2 y(x,t)}{\partial t^2} = g\frac{\partial}{\partial x}\left[x\frac{\partial y(x,t)}{\partial x}\right]. \tag{6.6}$$

We separate the variables as

$$y(x,t) = u(x)v(t), \tag{6.7}$$

to obtain

$$\frac{\ddot{v}(t)}{v(t)} = \frac{g}{u(x)} \left[\frac{du}{dx} + x\frac{d^2u}{dx^2} \right] = -\omega^2, \tag{6.8}$$

where ω is a separation constant. Solution for $v(t)$ can be written immediately as

$$v(t) = c_0 \cos(\omega t - \delta), \tag{6.9}$$

while $u(x)$ satisfies the differential equation

$$x\frac{d^2u}{dx^2} + \frac{du}{dx} + \frac{\omega^2}{g}u(x) = 0. \tag{6.10}$$

After defining a new independent variable

$$z = 2\sqrt{\frac{x}{g}}, \tag{6.11}$$

the differential equation to be solved for $u(z)$ becomes

$$\frac{d^2u}{dz^2} + \frac{1}{z}\frac{du}{dz} + \omega^2 u(z) = 0. \tag{6.12}$$

To express the solutions of this equation we need a new type of function called the Bessel function. This problem was first studied by Bernoulli in 1732, however, he did not recognize the general nature of these functions. As we shall see, this equation is a special case of Bessel's equation.

6.1 BESSEL'S EQUATION

If we write the Laplace equation in cylindrical coordinates as

$$\frac{\partial^2 \Psi}{\partial \rho^2} + \frac{1}{\rho}\frac{\partial \Psi}{\partial \rho} + \frac{1}{\rho^2}\frac{\partial^2 \Psi}{\partial \phi^2} + \frac{\partial^2 \Psi}{\partial z^2} = 0 \tag{6.13}$$

and try a separable solution of the form

$$\Psi(\rho, \phi, z) = R(\rho)\Phi(\phi)Z(z), \tag{6.14}$$

we obtain three ordinary differential equations to be solved for $R(\rho)$, $\Phi(\phi)$, and $Z(z)$:

$$\frac{d^2 Z(z)}{dz^2} - k^2 Z(z) = 0, \tag{6.15}$$

$$\frac{d^2 \Phi(\phi)}{d\phi^2} + m^2 \Phi(\phi) = 0, \tag{6.16}$$

$$\frac{d^2 R(\rho)}{d\rho^2} + \frac{1}{\rho}\frac{dR(\rho)}{d\rho} + (k^2 - \frac{m^2}{\rho^2})R(\rho) = 0. \tag{6.17}$$

Solutions of the first two equations can be written immediately as

$$Z(z) = \bar{c}_1 e^{kz} + \bar{c}_2 e^{-kz} \tag{6.18}$$

and

$$\Phi(\phi) = c_1 e^{im\phi} + c_2 e^{-im\phi}. \tag{6.19}$$

The remaining Equation (6.17) is known as the Bessel equation and, with the definitions

$$x = k\rho \text{ and } R(\rho) = J_m(x), \tag{6.20}$$

can be written as

$$J_m''(x) + \frac{1}{x}J_m'(x) + (1 - \frac{m^2}{x^2})J_m(x) = 0. \tag{6.21}$$

Solutions of this equation are called the Bessel functions of order m, and they are shown as $J_m(x)$.

6.2   SOLUTIONS OF BESSEL'S EQUATION

6.2.1   Bessel Functions $J_{\pm m}(x)$, $N_m(x)$, and $H_m^{(1,2)}(x)$

Series solution of Bessel's equation is given as

$$J_m(x) = \sum_{r=0}^{\infty} \frac{(-1)^r}{r!\Gamma(m+r+1)}(\frac{x}{2})^{m+2r}, \tag{6.22}$$

which is called the Bessel function of the first kind of order m. A second solution can be written as

$$J_{-m}(x) = \sum_{r=0}^{\infty} \frac{(-1)^r}{r!\Gamma(-m+r+1)}(\frac{x}{2})^{-m+2r}. \tag{6.23}$$

However, the second solution is independent of the first solution only for the noninteger values of m. For the integer values of m the two solutions are related by

$$J_{-m}(x) = (-1)^m J_m(x). \tag{6.24}$$

When m takes integer values, the second and linearly independent solution can be taken as

$$N_m(x) = \frac{\cos m\pi J_m(x) - J_{-m}(x)}{\sin m\pi}, \tag{6.25}$$

which is called the Neumann function or the Bessel function of the second kind. Note that $N_m(x)$ and $J_m(x)$ are linearly independent even for the integer values of m. Hence it is common practice to take $N_m(x)$ and $J_m(x)$ as the two linearly independent solutions for all n.

Other linearly independent solutions of Bessel's equation are given as the Hankel functions:

$$H_m^{(1)}(x) = J_m(x) + iN_m(x), \tag{6.26}$$

$$H_m^{(2)}(x) = J_m(x) - iN_m(x). \tag{6.27}$$

They are also called the Bessel functions of the third kind.

In the limit as $x \to 0$ Bessel function $J_m(x)$ is finite for $m \geq 0$ and behaves as

$$\lim_{x \to 0} J_m(x) \to \frac{1}{\Gamma(m+1)} (\frac{x}{2})^m. \tag{6.28}$$

All the other functions diverge as

$$\lim_{x \to 0} N_m(x) \to \left\{ \begin{array}{ll} \frac{2}{\pi} \left[\ln(\frac{x}{2}) + \gamma \right], & m = 0, \\ \\ -\frac{\Gamma(m)}{\pi} (\frac{2}{x})^m, & m \neq 0, \end{array} \right\} \tag{6.29}$$

where $\gamma = 0.5772...$. In the limit as $x \to \infty$ functions $J_m(x)$, $N_m(x)$, $H_m^{(1)}(x)$, and $H_m^{(2)}(x)$ behave as

$$J_m(x) \underset{x \to \infty}{\to} \sqrt{\frac{2}{\pi x}} \cos(x - \frac{m\pi}{2} - \frac{\pi}{4}), \tag{6.30}$$

$$N_m(x) \underset{x \to \infty}{\to} \sqrt{\frac{2}{\pi x}} \sin(x - \frac{m\pi}{2} - \frac{\pi}{4}). \tag{6.31}$$

$$H_m^{(1)}(x) \underset{x \to \infty}{\to} \sqrt{\frac{2}{\pi x}} \exp\left[i(x - \frac{m\pi}{2} - \frac{\pi}{4}) \right], \tag{6.32}$$

$$H_m^{(2)}(x) \underset{x \to \infty}{\to} \sqrt{\frac{2}{\pi x}} \exp\left[-i(x - \frac{m\pi}{2} - \frac{\pi}{4}) \right]. \tag{6.33}$$

6.2.2   Modified Bessel Functions $I_m(x)$ and $K_m(x)$

If we take the argument of the Bessel functions $J_m(x)$ and $H_m^{(1)}(x)$ as imaginary we obtain the modified Bessel functions

$$I_m(x) = \frac{J_m(ix)}{i^m} \tag{6.34}$$

and

$$K_m(x) = \frac{\pi i}{2}(i)^m H_m^{(1)}(ix). \tag{6.35}$$

These functions are linearly independent solutions of the differential equation

$$\frac{d^2 R(x)}{dx^2} + \frac{1}{x}\frac{dR(x)}{dx} - (1 + \frac{m^2}{x^2})R(x) = 0. \tag{6.36}$$

Their $x \to 0$ and $x \to \infty$ limits are given as (real $m \geq 0$)

$$\lim_{x \to 0} I_m(x) \to \frac{x^m}{2^m \Gamma(m+1)}, \tag{6.37}$$

$$\lim_{x \to 0} K_m(x) \to \left\{ \begin{array}{ll} -\left[\ln(\frac{x}{2}) + \gamma\right], & m = 0, \\ \\ \frac{\Gamma(m)}{2}(\frac{2}{x})^m, & m \neq 0 \end{array} \right\} \tag{6.38}$$

and

$$\lim_{x \to \infty} I_m(x) \to \frac{1}{\sqrt{2\pi x}}e^x\left[1 + 0(\frac{1}{x})\right], \tag{6.39}$$

$$\lim_{x \to \infty} K_m(x) \to \sqrt{\frac{\pi}{2x}}e^{-x}\left[1 + 0(\frac{1}{x})\right]. \tag{6.40}$$

6.2.3   Spherical Bessel Functions $j_l(x)$, $n_l(x)$, and $h_l^{(1,2)}(x)$

Spherical Bessel functions $j_l(x)$, $n_l(x)$, and $h_l^{(1,2)}(x)$ are defined as

$$j_l(x) = \sqrt{\frac{\pi}{2x}}J_{l+\frac{1}{2}}(x),$$

$$n_l(x) = \sqrt{\frac{\pi}{2x}}N_{l+\frac{1}{2}}(x), \tag{6.41}$$

$$h_l^{(1,2)}(x) = (\frac{\pi}{2x})^{1/2}\left[J_{l+\frac{1}{2}}(x) \pm iN_{l+\frac{1}{2}}(x)\right].$$

Bessel functions with half integer indices, $J_{l+\frac{1}{2}}(x)$ and $N_{l+\frac{1}{2}}(x)$, satisfy the differential equation

$$\frac{d^2y(x)}{dx^2} + \frac{1}{x}\frac{dy(x)}{dx} + \left[1 - \frac{(l+\frac{1}{2})^2}{x^2}\right]y(x) = 0, \qquad (6.42)$$

while the spherical Bessel functions, $j_l(x)$, $n_l(x)$, and $h_l^{(1,2)}(x)$ satisfy

$$\frac{d^2y(x)}{dx^2} + \frac{2}{x}\frac{dy(x)}{dx} + \left[1 - \frac{l(l+1)}{x^2}\right]y(x) = 0, \qquad (6.43)$$

Spherical Bessel functions can also be defined as

$$j_l(x) = (-x)^l \left(\frac{1}{x}\frac{d}{dx}\right)^l \frac{\sin x}{x}, \qquad (6.44)$$

$$n_l(x) = (-x)^l \left(\frac{1}{x}\frac{d}{dx}\right)^l \left(-\frac{\cos x}{x}\right). \qquad (6.45)$$

Asymptotic forms of the spherical Bessel functions are given as

$$\begin{aligned}
j_l(x) &\to \frac{x^l}{(2l+1)!!}\left(1 - \frac{x^2}{2(2l+1)} + \cdots\right), \quad x \ll 1, \\
n_l(x) &\to -\frac{(2l-1)!!}{x^{l+1}}\left(1 - \frac{x^2}{2(1-2l)} + \cdots\right), \quad x \ll 1, \\
j_l(x) &= \frac{1}{x}\sin(x - \frac{l\pi}{x}), \quad x \gg 1, \\
n_l(x) &= -\frac{1}{x}\cos(x - \frac{l\pi}{x}), \quad x \gg 1,
\end{aligned} \qquad (6.46)$$

where $(2l+1)!! = (2l+1)(2l-1)(2l-3)\cdots 5\cdot 3\cdot 1$.

6.3   OTHER DEFINITIONS OF THE BESSEL FUNCTIONS

6.3.1   Generating Function

Bessel function $J_n(x)$ can be defined by a generating function $T(x,t)$ as

$$T(x,t) = \exp\left[\frac{1}{2}x\left(t - \frac{1}{t}\right)\right]$$

$$= \sum_{n=-\infty}^{\infty} t^n J_n(x). \qquad (6.47)$$

6.3.2 Integral Definitions

Bessel function $J_n(x)$ also has the following integral definitions:

$$J_n(x) = \frac{1}{\pi} \int_0^\pi \cos\left[n\varphi - x\sin\varphi\right] d\varphi, \qquad n = 0, 1, 2, \ldots \tag{6.48}$$

and

$$J_n(x) = \frac{(\frac{1}{2}x)^n}{\sqrt{\pi}\,\Gamma(n + \frac{1}{2})} \int_{-1}^1 \left(1 - t^2\right)^{n-\frac{1}{2}} \cos xt\, dt, \qquad \left(n > -\frac{1}{2}\right). \tag{6.49}$$

6.4 RECURSION RELATIONS OF THE BESSEL FUNCTIONS

Using the series definitions of the Bessel functions we can obtain the following recursion relations

$$J_{m-1}(x) + J_{m+1}(x) = \frac{2m}{x} J_m(x) \tag{6.50}$$

and

$$J_{m-1}(x) - J_{m+1}(x) = 2J_m'(x). \tag{6.51}$$

First by adding and then by subtracting these equations we also obtain the relations

$$J_{m-1}(x) = \frac{m}{x} J_m(x) + J_m'(x) \tag{6.52}$$

and

$$J_{m+1}(x) = \frac{m}{x} J_m(x) - J_m'(x). \tag{6.53}$$

Other Bessel functions, N_n, $H_n^{(1)}$, and $H_n^{(2)}$, satisfy the same recursion relations.

6.5 ORTHOGONALITY AND THE ROOTS OF THE BESSEL FUNCTIONS

From the asymptotic form [Eq. (6.30)] of the Bessel function it is clear that it has infinitely many roots:

$$J_n(x_{nl}) = 0, \qquad l = 1, 2, 3, \ldots . \tag{6.54}$$

x_{nl} stands for the lth root of the nth order Bessel function. When n takes integer values the first three roots are given as

$$
\begin{array}{llllll}
n = 0 & x_{0l} = 2.405 & 5.520 & 8.654 & \cdots & \\
n = 1 & x_{1l} = 3.832 & 7.016 & 10.173 & \cdots & (6.55) \\
n = 2 & x_{2l} = 5.136 & 8.417 & 11.620 & \cdots &
\end{array}
$$

Higher-order roots are approximately given by the formula

$$
x_{nl} \simeq l\pi + \left(n - \frac{1}{2}\right)\frac{\pi}{2}. \tag{6.56}
$$

The Bessel functions' orthogonality relation in the interval $[0, a]$ is given as

$$
\int_0^a \rho J_n\left(x_{nl}\frac{\rho}{a}\right)J_n\left(x_{nl'}\frac{\rho}{a}\right)d\rho = \frac{a^2}{2}\left[J_{n+1}(x_{nl})\right]^2 \delta_{ll'}, \; n \geq -1. \tag{6.57}
$$

Since Bessel functions also form a complete set, any sufficiently smooth function, $f(\rho)$, in the interval

$$
\rho \in [0, a] \tag{6.58}
$$

can be expanded as

$$
f(\rho) = \sum_{l=1}^{\infty} A_{nl}J_n\left(x_{nl}\frac{\rho}{a}\right), \; n \geq -1, \tag{6.59}
$$

where the expansion coefficients A_{nl} are found from

$$
A_{nl} = \frac{2}{a^2 J_{n+1}^2(x_{nl})}\int_0^a \rho f(\rho)J_n\left(x_{nl}\frac{\rho}{a}\right)d\rho. \tag{6.60}
$$

6.6 BOUNDARY CONDITIONS FOR THE BESSEL FUNCTIONS

For the roots given in Equation (6.55) we have used the Dirichlet boundary condition, that is,

$$
R(a) = 0. \tag{6.61}
$$

In terms of the Bessel functions this condition implies

$$
J_n(ka) = 0 \tag{6.62}
$$

and gives us the infinitely many roots [Eq. (6.55)] shown as

$$
x_{nl}. \tag{6.63}
$$

Now the functions

$$\{J_n(x_{nl}\frac{\rho}{a})\}, \quad n \geq 0 \tag{6.64}$$

form a complete and orthogonal set with respect to the index l. The same conclusion holds for the Neumann boundary condition

$$\frac{dR(\rho)}{d\rho}\bigg|_{\rho=a} = 0 \tag{6.65}$$

and the general boundary condition

$$\left[A_0\frac{dR(\rho)}{d\rho} + B_0 R(\rho)\right]_{\rho=a} = 0. \tag{6.66}$$

In terms of the Bessel function $J_n(kr)$, Neumann and general boundary conditions are written as

$$k\frac{dJ_n(x)}{dx}\bigg|_{x=ka} = 0 \tag{6.67}$$

and

$$\left[A_0 J_n(x) + B_0 k\frac{dJ_n(x)}{dx}\right]_{x=ka} = 0, \tag{6.68}$$

respectively. For the Neumann boundary condition [Eq. (6.67)] there exist infinitely many roots, which can be found from tables. However, for the general boundary condition roots depend on the values that A_0 and B_0 take; thus each case must be handled separately by numerical analysis. From all three types of boundary conditions we obtain a complete and orthogonal set as

$$\{J_n(x_{nl}\frac{r}{a})\}, \quad l = 1, 2, 3, \dots . \tag{6.69}$$

Example 6.1. Flexible chain problem: We now return to the flexible chain problem, where the equation of motion was written as

$$\frac{d^2u}{dz^2} + \frac{1}{z}\frac{du}{dz} + \omega^2 u = 0. \tag{6.70}$$

General solution of this equation is given as

$$u(z) = a_0 J_0(\omega z) + a_1 N_0(\omega z), \quad (z = 2\sqrt{x/g}). \tag{6.71}$$

Since $N_0(\omega z)$ diverges at the origin, we choose a_1 as zero and obtain the displacement of the chain from its equilibrium position as (Fig. 6.2)

$$y(x, t) = a_0 J_0(2\omega\sqrt{x/g})\cos(\omega t - \delta). \tag{6.72}$$

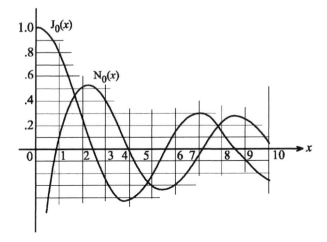

Fig. 6.2 J_0 and N_0 functions

If we impose the condition

$$y(l, t) = J_0(2\omega_n \sqrt{l/g}) = 0,$$

we find the normal modes of the chain as

$$2\omega_n \sqrt{l/g} = 2.405, \ 5.520, \dots \ , \quad (n = 1, 2, \dots). \tag{6.73}$$

If the shape of the chain at $t = 0$ is given as $f(x)$, we can write the solution as

$$y(x, t) = \sum_{n=1}^{\infty} A_n J_0(2\omega_n \sqrt{x/g}) \cos(\omega_n t - \delta), \tag{6.74}$$

where the expansion coefficients are given as

$$A_n = \frac{2}{J_1^2(\omega_n)} \int_0^1 z f(\frac{g}{4} z^2) J_0(\omega_n z) dz.$$

Example 6.2. *Tsunamis and wave motion in a channel:* The equation of motion for one dimensional waves in a channel with breadth $b(x)$ and depth $h(x)$ is given as

$$\frac{\partial^2 \eta}{\partial t^2} = \frac{g}{b} \frac{\partial}{\partial x} \left(hb \frac{\partial \eta}{\partial x} \right), \tag{6.75}$$

where $\eta(x, t)$ is the displacement of the water surface from its equilibrium position and g is the acceleration of gravity. If the depth of the channel varies uniformly from the end of the channel, $x = 0$, to the mouth $(x = a)$ as $h(x) = h_0 x/a$ we can try a separable solution of the form

$$\eta(x, t) = A(x) \cos(\omega t + \alpha) \tag{6.76}$$

to find the differential equation that $A(x)$ satisfies as

$$\frac{d}{dx}\left(x \frac{dA}{dx}\right) + kA = 0, \tag{6.77}$$

where $k = \omega^2 a/gh_0$. Solution that is finite at $x = 0$ can easily be obtained as

$$A(x) = A_0\left(1 - \frac{kx}{1^2} + \frac{k^2 x^2}{1^2 \cdot 2^2} - \cdots\right) \tag{6.78}$$

or as

$$A(x) = A_0 J_0(2k^{1/2} x^{1/2}). \tag{6.79}$$

After evaluating the constant A_0 we write the final solution as

$$\eta(x, t) = C \frac{J_0(2k^{1/2} x^{1/2})}{J_0(2k^{1/2} a^{1/2})} \cos(\omega t + \alpha). \tag{6.80}$$

With an appropriate normalization a snapshot of this wave is shown in Fig. 6.3. Note how the amplitude increases and the wavelength decreases as shallow waters is reached. If hb is constant or at least a slow varying function of position, we can take it outside the brackets in Equation (6.75), thus obtaining the wave velocity as $\sqrt{hg}$. This is characteristic of tsunamis, which are wave trains caused by sudden displacement of large amounts of water by earthquakes, volcanos, meteors, etc. Tsunamis have wavelengths in excess of 100 km and their period is around one hour. In the Pacific Ocean, where typical water depth is 4000 m, tsunamis travel with velocities over 700 km/h. Since the energy loss of a wave is inversely proportional to its wavelength, tsunamis could travel transoceanic distances with little energy loss. Because of their huge wavelengths they are imperceptible in deep waters; however, in reaching shallow waters they compress and slow down. Thus to conserve energy their amplitude increases to several or tens of meters in height as they reach the shore.

When both the breadth and the depth vary as $b(x) = b_0 x/a$ and $h(x) = h_0 x/a$, respectively, the differential equation to be solved for $A(x)$ becomes

$$x \frac{d^2 A}{dx^2} + 2\frac{dA}{dx} + kA = 0, \tag{6.81}$$

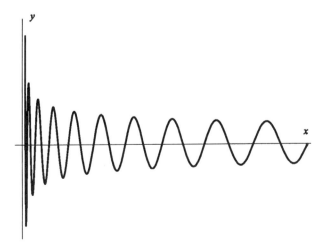

Fig. 6.3 Channel waves

where $k = \omega^2 a/gh_0$ as before. The solution is now obtained as

$$\eta(x,t) = A_0\left(1 - \frac{kx}{(1 \cdot 2)} + \frac{k^2 x^2}{(1 \cdot 2) \cdot (2 \cdot 4)} - \cdots\right)\cos(\omega t + \alpha), \quad (6.82)$$

which is

$$\eta(x,t) = A_0 \frac{J_1(2k^{1/2}x^{1/2})}{k^{1/2}x^{1/2}}\cos(\omega t + \alpha). \quad (6.83)$$

6.7 WRONSKIANS OF PAIRS OF SOLUTIONS

The Wronskian of a pair of solutions of a second-order linear differential equation is defined by the determinant

$$W[u_1(x), u_2(x)] = \begin{vmatrix} u_1(x) & u_2(x) \\ u_1'(x) & u_2'(x) \end{vmatrix} \quad (6.84)$$

$$= u_1 u_2' - u_2 u_1'.$$

The two solutions are linearly independent if and only if their Wronskian does not vanish identically. We now calculate the Wronskian of a pair of solutions of Bessel's equation

$$u''(x) + \frac{1}{x}u'(x) + \left(1 - \frac{m^2}{x^2}\right)u(x) = 0, \quad (6.85)$$

thus obtaining a number of formulas that are very helpful in various calculations. For two solutions u_1 and u_2 we write

$$\frac{d}{dx}(xu_1') + (1 - \frac{m^2}{x^2})u_1(x) = 0, \tag{6.86}$$

$$\frac{d}{dx}(xu_2') + (1 - \frac{m^2}{x^2})u_2(x) = 0. \tag{6.87}$$

We multiply the second equation by u_1 and subtract the result from the first equation multiplied by u_2 to get

$$\frac{d}{dx}\{xW[u_1(x), u_2(x)]\} = 0. \tag{6.88}$$

This means

$$W[u_1(x), u_2(x)] = \frac{C}{x}, \tag{6.89}$$

where C is a constant independent of x but depends on the pair of functions whose Wronskian is calculated. For example,

$$W[J_m(x), N_m(x)] = \frac{2}{\pi x}, \tag{6.90}$$

$$W\left[J_m(x), H_m^{(2)}(x)\right] = -\frac{2i}{\pi x}, \tag{6.91}$$

$$W\left[H_m^{(1)}(x), H_m^{(2)}(x)\right] = -\frac{4i}{\pi x}. \tag{6.92}$$

Since C is independent of x it can be calculated by using the asymptotic forms of these functions in the limit $x \to 0$ as

$$C = \lim_{x \to 0} xW[u_1(x), u_2(x)]. \tag{6.93}$$

Problems

6.1   Drive the following recursion relations:

$$J_{m-1}(x) + J_{m+1}(x) = \frac{2m}{x} J_m(x), \quad m = 1, 2, \ldots$$

and

$$J_{m-1}(x) - J_{m+1}(x) = 2J_m'(x), \quad m = 1, 2, \ldots$$

Use the first equation to express a Bessel function of arbitrary order ($m = 0, 1, 2, \ldots$) in terms of J_0 and J_1. Show that for $m = 0$ the second equation is replaced by

$$J_0'(x) = -J_1(x).$$

6.2   Write the wave equation

$$\nabla^2 \Psi(\overrightarrow{r}, t) - \frac{1}{c^2} \frac{\partial^2}{\partial t^2} \Psi(\overrightarrow{r}, t) = 0$$

in spherical polar coordinates. Using the method of separation of variables show that the solutions for the radial part are given in terms of the spherical Bessel functions.

6.3   Use the result in Problem 6.2 to find the solutions for a spherically split antenna. On the surface, $r = a$, take the solution as

$$\Psi(\overrightarrow{r}, t)\big|_{r=a} = \begin{cases} V_0 e^{-iw_0 t} & 0 < \theta < \pi/2 \\ \\ -V_0 e^{-iw_0 t} & \frac{\pi}{2} < \theta < \pi \end{cases}$$

and assume that in the limit as $r \to \infty$ solution behaves as

$$\psi \approx \frac{1}{r} e^{i(k_0 r - w_0 t)}, \quad k = k_0 = w_0/c.$$

6.4   Solve the wave equation

$$\overrightarrow{\nabla}^2 \Psi - \frac{1}{v^2} \frac{\partial^2 \Psi}{\partial t^2} = 0, \quad k = \frac{\omega}{c} = \text{wave number},$$

for the oscillations of a circular membrane with radius a and clamped at the boundary. What boundary conditions did you use? What are the lowest three modes?

6.5   Verify the following Wronskians:

$$W\left[J_m(x), N_m(x)\right] = +\frac{2}{\pi x},$$

$$W\left[J_m(x), H_m^{(2)}(x)\right] = -\frac{2i}{\pi x},$$

$$W\left[H_m^{(1)}(x), H_m^{(2)}(x)\right] = -\frac{4i}{\pi x}.$$

6.6   Find the constant C in the Wronskian

$$W\left[I_m(x), K_m(x)\right] = \frac{C}{x}.$$

6.7   Show that the stationary distribution of temperature, $T(\rho, z)$, in a cylinder of length l and radius a with one end held at temperature T_0 while the rest of the cylinder is held at zero is given as

$$T(\rho, z) = 2T_0 \sum_{n=1}^{\infty} \frac{J_0(x_n \frac{\rho}{a}) \sinh\left(x_n \frac{l-z}{a}\right)}{x_n J_1(x_n) \sinh\left(x_n \frac{l}{a}\right)}.$$

Hint: Use cylindrical coordinates and solve the Laplace equation,

$$\vec{\nabla}^2 T(\rho, z) = 0,$$

by the method of separation of variables.

6.8   Consider the cooling of an infinitely long cylinder heated to an initial temperature $f(\rho)$. Solve the heat transfer equation

$$c\rho_0 \frac{\partial T(\rho, t)}{\partial t} = k\vec{\nabla}^2 T(\rho, t),$$

with the boundary condition

$$\left.\frac{\partial T}{\partial \rho} + hT\right|_{\rho=a} = 0$$

and the initial condition

$$T(\rho, 0) = f(\rho) \quad \text{(finite)}.$$

$T(\rho, t)$ is the temperature distribution in the cylinder and the physical parameters of the problem are defined as

$$k - \text{thermal conductivity}$$
$$c - \text{ heat capacity}$$
$$\rho_0 - \text{ density}$$
$$\lambda - \text{ emissivity}$$

and $h = \lambda/k$. Hint: Use the method of separation of variables and show that the solution can be expressed as

$$T(\rho, t) = \sum_{n=1}^{\infty} C_n J_0(x_n \frac{\rho}{a}) e^{-x_n^2 t/a^2 b}, \qquad (b = c\rho_0/k),$$

then find C_n so that the initial condition $T(\rho, 0) = f(\rho)$ is satisfied. Where does x_n come from?

7

HYPERGEOMETRIC FUNCTIONS

The majority of the second-order linear ordinary differential equations of science and engineering can be conveniently expressed in terms of the three parameters (a, b, c) of the hypergeometric equation:

$$x(1-x)\frac{d^2y(x)}{dx^2} + [c - (a+b+1)x]\frac{dy(x)}{dx} - aby(x) = 0. \qquad (7.1)$$

Solutions of the hypergeometric equation are called the (Gauss's) hypergeometric functions, and they are shown as $F(a, b, c; x)$.

7.1 HYPERGEOMETRIC SERIES

The hypergeometric equation has three regular singular points at $x = 0, 1$ and ∞; hence we can find a series solution about the origin by using the Frobenius method. Substituting the series

$$y = \sum_{r=0}^{\infty} a_r x^{s+r}, \quad a_0 \neq 0, \qquad (7.2)$$

into Equation (7.1) gives us

$$x\left(1-x\right)\sum_{r=0}^{\infty}a_r\left(s+r\right)\left(s+r-1\right)x^{s+r-2}$$

$$+\left\{c-\left(a+b+1\right)x\right\}\sum_{r=0}^{\infty}a_r\left(s+r\right)x^{s+r-1}$$

$$-ab\sum_{r=0}^{\infty}a_r x^{s+r}=0,\tag{7.3}$$

which we write as

$$\sum_{r=0}^{\infty}a_r\left(s+r\right)\left(s+r-1\right)x^{s+r-1}$$

$$-\sum_{r=0}^{\infty}a_r\left(s+r\right)\left(s+r-1\right)x^{s+r}$$

$$+c\sum_{r=0}^{\infty}a_r\left(s+r\right)x^{s+r-1}$$

$$-\left(a+b+1\right)\sum_{r=0}^{\infty}a_r\left(s+r\right)x^{s+r}-ab\sum_{r=0}^{\infty}a_r x^{s+r}=0.\tag{7.4}$$

After rearranging we obtain

$$\sum_{r=0}^{\infty}\left[\left(s+r\right)\left(s+r-1\right)+c\left(s+r\right)\right]a_r x^{s+r-1}$$

$$-\sum_{r=1}^{\infty}\left[\left(s+r-1\right)\left(s+r-2\right)+\left(a+b+1\right)\left(s+r-1\right)+ab\right]a_{r-1}x^{s+r-1}=0.$$

$$\tag{7.5}$$

Writing the first term explicitly this becomes

$$\left[s\left(s-1\right)+sc\right]a_0 x^{s-1}$$

$$+\sum_{r=1}^{\infty}\{[(s+r)(s+r-1)+c(s+r)]a_r$$

$$-a_{r-1}\left[\left(s+r-1\right)\left(s+r-2\right)+ab+\left(a+b+1\right)\left(s+r-1\right)\right]\}x^{s+r-1}=0.$$

$$\tag{7.6}$$

Setting the coefficients of all the equal powers of x to zero gives us the indicial equation

$$\left[s\left(s-1\right)+sc\right]a_0=0,\quad a_0\neq0\tag{7.7}$$

and the recursion relation

$$a_r = \frac{(s+r-1+a)(s+r-1+b)}{(s+r)(s+r-1+c)} a_{r-1}, \quad r \geq 1. \tag{7.8}$$

Roots of the indicial equation are

$$s = 0 \quad \text{and} \quad s = 1 - c. \tag{7.9}$$

For $s = 0$ we write the recursion relation as

$$a_r = \frac{(r-1+a)(r-1+b)}{(r-1+c)r} a_{r-1}, \quad r \geq 1 \tag{7.10}$$

and obtain the following coefficients for the series:

$$a_1 = \frac{ab}{c} a_0,$$

$$a_2 = \frac{(a+1)(b+1)}{(c+1)2} a_1,$$

$$a_3 = \frac{(a+2)(b+2)}{(c+2)3} a_2,$$

$$a_4 = \frac{(a+3)(b+3)}{(c+3)4} a_3,$$

$$\vdots$$

where the general term is

$$a_k = a_0 \frac{a(a+1)(a+2)\cdots(a+k-1)b(b+1)\cdots(b+k-1)}{c(c+1)\cdots(c+k-1)1\cdot2\cdot3\cdots k}. \tag{7.11}$$

Now the series solution can be written explicitly as

$$y_1(x) = a_0 \left[1 + \frac{ab}{c} \frac{x}{1!} + \frac{a(a+1)b(b+1)}{c(c+1)2!} \frac{x^2}{2!} + \cdots \right],$$

$$= a_0 \frac{\Gamma(c)}{\Gamma(a)\Gamma(b)} \sum_{k=0}^{\infty} \frac{\Gamma(a+k)\Gamma(b+k)}{\Gamma(c+k)} \frac{x^k}{k!}, \tag{7.12}$$

where

$$c \neq 0, -1, -2, \ldots. \tag{7.13}$$

Similarly for the other root,

$$s = 1 - c, \tag{7.14}$$

the recursion relation becomes

$$a_r = a_{r-1} \frac{(r + a - c)(r + b - c)}{r(1 - c + r)}, \quad r \geq 1, \tag{7.15}$$

which gives the following coefficients for the second series:

$$a_1 = a_0 \frac{(a - c + 1)(b - c + 1)}{(2 - c)},$$

$$a_2 = a_1 \frac{(a - c + 2)(b - c + 2)}{2(3 - c)},$$

$$a_3 = a_2 \frac{(a - c + 3)(b - c + 3)}{3(4 - c)},$$

$$\vdots$$

where the general term can be written as

$$a_k = a_0 \left[\frac{(a - c + 1)(a - c + 2) \cdots (a - c + k)(b - c + 1) \cdots (b - c + k)}{(2 - c)(3 - c) \cdots (k + 1 - c) k!} \right]. \tag{7.16}$$

Now the second series solution becomes

$$y_2(x) = a_0 x^{1-c} \sum_{k=0}^{\infty} a_k x^k, \tag{7.17}$$

$$y_2(x) = a_0 x^{1-c} \left[1 + \frac{(a + 1 - c)(1 + b - c)}{(2 - c)} \frac{x}{1!} + \cdots \right], \tag{7.18}$$

where $c \neq 2, 3, 4\ldots$. If we set a_0 to 1, $y_1(x)$ becomes the hypergeometric function (or series):

$$y_1(x) = F(a, b, c; x). \tag{7.19}$$

The hypergeometric function is convergent in the interval $|x| < 1$. For convergence at the end point $x = 1$ one needs $c > a + b$, and for convergence at $x = -1$ one needs $c > a + b - 1$. Similarly the second solution, $y_2(x)$, can be expressed in term of the hypergeometric function as

$$y_2(x) = x^{1-c} F(a - c + 1, b - c + 1, 2 - c; x), \quad c \neq 2, 3, 4, \ldots . \tag{7.20}$$

Thus the general solution of the hypergeometric equation is

$$y(x) = AF(a, b, c; x) + Bx^{1-c} F(a - c + 1, b - c + 1, 2 - c; x). \tag{7.21}$$

Hypergeometric functions are also written as $_2F_1(a, b, c; x)$.

Similarly, one can find series solutions about the regular singular point $x = 1$ as

$$y_3(x) = F(a, b, a + b - c; 1 - x), \tag{7.22}$$

$$y_4(x) = (1 - x)^{c-a-b} F(c - a, c - b, c - a - b + 1; 1 - x). \tag{7.23}$$

The interval of convergence of these series is $0 < x < 2$. Series solutions appropriate for the singular point at infinity are given as

$$y_5(x) = x^{-a} F(a, a - c + 1, a - b + 1; x^{-1}), \tag{7.24}$$

$$y_6(x) = x^{-b} F(b, b - c + 1, b - a + 1; x^{-1}), \tag{7.25}$$

which converge for $|x| > 1$. These constitute the six solutions found by Kummer. Since the hypergeometric equation can only have two linearly independent solutions, there are linear relations among them like;

$$\begin{aligned} F(a, b, c; x) &= \frac{\Gamma(c)\Gamma(c - a - b)}{\Gamma(c - a)\Gamma(c - b)} F(a, b, a + b - c + 1; 1 - x) \\ &+ \frac{\Gamma(c)\Gamma(a + b - c)}{\Gamma(a)\Gamma(b)} (1 - x)^{c-a-b} F(c - a, c - b, c - a - b + 1; 1 - x). \end{aligned} \tag{7.26}$$

The basic integral representation of hypergeometric functions is:

$$F(a, b, c; x) = \frac{\Gamma(c)}{\Gamma(b)\Gamma(c - b)} \int_0^1 \frac{t^{b-1}(1 - t)^{c-b-1} dt}{(1 - tx)^a}, \quad (\text{real } c > \text{real } b > 0). \tag{7.27}$$

This integral, which can be proven by expanding $(1 - tx)^{-a}$ in binomial series and integrating term by term, transforms into an integral of the same type by Euler's hypergeometric transformations:

$$\begin{aligned} t &\rightarrow t, \\ t &\rightarrow 1 - t, \\ t &\rightarrow t/(1 - x + tx), \\ t &\rightarrow (1 - t)/(1 - tx). \end{aligned} \tag{7.28}$$

Applications of the 4 Euler transformations to the 6 Kummer solutions give all the possible 24 forms of the solutions of hypergeometric equation. These solutions and a list of 20 relations among them can be found in Erdelyi et.al.

7.2 HYPERGEOMETRIC REPRESENTATIONS OF SPECIAL FUNCTIONS

The majority of the special functions can be represented in terms of hypergeometric functions. If we change the independent variable in Equation (7.1) to

$$x = \frac{(1-\xi)}{2}, \tag{7.29}$$

the hypergeometric equation becomes

$$\left(1-\xi^2\right)\frac{d^2y}{d\xi^2} + [(a+b+1-2c)-(a+b+1)\xi]\frac{dy}{d\xi} - aby = 0. \tag{7.30}$$

Choosing the parameters a, b, and c as

$$a = -\nu, \ b = \nu + 1, \ \text{and} \ c = 1, \tag{7.31}$$

Equation (7.30) becomes

$$\left(1-\xi^2\right)\frac{d^2y}{d\xi^2} - 2\xi\frac{dy}{d\xi} + \nu\left(\nu+1\right)y = 0. \tag{7.32}$$

This is nothing but the Legendre equation, finite solutions of which are given as the Legendre polynomials:

$$y_\nu\left(\xi\right) = P_\nu\left(\xi\right). \tag{7.33}$$

Thus the Legendre polynomials can be expressed in terms of the hypergeometric functions as

$$P_\nu\left(\xi\right) = F\left(-\nu, \nu+1, 1; \frac{1-\xi}{2}\right), \qquad \nu = 0, 1, 2, \ldots . \tag{7.34}$$

Similarly, we can write the associated Legendre polynomials as

$$P_n^m\left(x\right) = \frac{(n+m)!}{(n-m)!}\frac{\left(1-x^2\right)^{m/2}}{2^m m!}F\left(m-n, m+n+1, m+1; \frac{1-x}{2}\right) \tag{7.35}$$

and the Gegenbauer polynomials as

$$C_n^\lambda\left(x\right) = \frac{\Gamma\left(n+2\lambda\right)}{n!\Gamma\left(2\lambda\right)}F\left(-n, n+2\lambda, \lambda+\frac{1}{2}; \frac{1-x}{2}\right). \tag{7.36}$$

The main reason for our interest in hypergeometric functions is that so many of the second-order linear ordinary differential equations encountered in physics and engineering can be expressed in terms of $F(a, b, c; x)$.

7.3   CONFLUENT HYPERGEOMETRIC EQUATION

The hypergeometric equation:

$$z(1-z)\frac{d^2y(z)}{dz^2} + [c-(a+b+1)z]\frac{dy(z)}{dz} - aby(z) = 0, \tag{7.37}$$

has three regular singular points at $z = 0, 1$, and ∞. By setting $z = x/b$ and taking the limit as $b \to \infty$ we can merge the singularities at b and infinity. This gives us the confluent hypergeometric equation as

$$x\frac{d^2y}{dx^2} + (c - x)\frac{dy}{dx} - ay = 0, \tag{7.38}$$

solutions of which are the confluent hypergeometric functions, which are shown as $M(a, c; x)$. The confluent hypergeometric equation has a regular singular point at $x = 0$ and an essential singularity at infinity. Bessel functions, $J_n(x)$, and the Laguerre polynomials, $L_n(x)$, can be written in terms of the solutions of the confluent hypergeometric equation as

$$J_n(x) = \frac{e^{-ix}}{n!}\left(\frac{x}{2}\right)^n M\left(n + \frac{1}{2}, 2n + 1; 2ix\right), \tag{7.39}$$

$$L_n(x) = M(-n, 1; x). \tag{7.40}$$

Linearly independent solutions of Equation (7.38) are given as

$$y_1(x) = M(a, c, x) = 1 + \frac{a}{c}\frac{x}{1!} + \frac{a(a + 1)}{c(c + 1)}\frac{x^2}{2!} + \frac{a(a + 1)(a + 2)}{c(c + 1)(c + 2)}\frac{x^3}{3!} + \cdots, \tag{7.41}$$

$$c \neq 0, -1, -2, \ldots$$

and

$$y_2(x) = x^{1-c}M(a + 1 - c, 2 - c; x), \qquad c \neq 2, 3, 4, \ldots \tag{7.42}$$

Integral representation of the confluent hypergeometric functions, which are also shown as $_1F_1(a, b; x)$, can be given as

$$M(a, c; x) = \frac{\Gamma(c)}{\Gamma(a)\Gamma(c - a)}\int_0^1 dt\, e^{xt}t^{a-1}(1 - t)^{c-a-1}, \quad (\text{real}\, c > \text{real}\, a > 0). \tag{7.43}$$

Problems

7.1 Show that the Hermite polynomials can be expressed as

$$H_{2n}(x) = (-1)^n\frac{(2n)!}{n!}M(-n, \frac{1}{2}; x^2),$$

$$H_{2n+1}(x) = (-1)^n\frac{2(2n + 1)!}{n!}xM(-n, \frac{3}{2}; x^2).$$

7.2   Show that associated Legendre polynomials can be written as

$$P_n^m(x) = \frac{(n+m)!}{(n-m)!} \frac{(1-x^2)^{\frac{m}{2}}}{2^m m!} F\left(m-n, m+n+1, m+1; \frac{1-x}{2}\right).$$

7.3   Derive the Kummer formula

$$M(a, c, x) = e^x M(c - a, c; -x).$$

7.4   Show that the associated Laguerre polynomials could be written as

$$L_n^k(x) = \frac{(n+k)!}{n!k!} M(-n, k+1; x).$$

7.5   Show that the modified Bessel functions can be expressed as

$$I_n(x) = \frac{e^{-x}}{n!} \left(\frac{x}{2}\right)^n M\left(n + \frac{1}{2}, 2n + 1; 2x\right).$$

7.6   Write the Chebyshev polynomials in terms of the hypergeometric functions.

7.7   Show that Gegenbauer polynomials can be expressed as

$$C_n^\lambda(x) = \frac{\Gamma(n+2\lambda)}{n!\Gamma(2\lambda)} F\left(-n, n+2\lambda, \lambda + \frac{1}{2}; \frac{1-x}{2}\right).$$

7.8   Express the solutions of

$$t(1-t^2)\frac{d^2y}{dt^2} + 2[\gamma - \frac{1}{2} - (\alpha + \beta + \frac{1}{2})t^2]\frac{dy}{dt} - 4\alpha\beta t y(t) = 0$$

in terms of the hypergeometric functions. Hint: Try the substitution $x = t^2$.

7.9   Show the following relations:

(a) $(1-x)^{-\alpha} = F(\alpha, \beta, \beta; x)$ (b) $\ln(1-x) = -xF(1, 1, 2; x)$

(c) $\sin^{-1} x = xF(\frac{1}{2}, \frac{1}{2}, \frac{3}{2}; x^2)$ (d) $e^x = M(\alpha, \alpha; x)$.

7.10   Derive the following integral representation of the hypergeometric function:

$$F(a, b, c; x) = \frac{\Gamma(c)}{\Gamma(b)\Gamma(c-b)} \int_0^1 t^{b-1}(1-t)^{c-b-1}(1-tx)^{-a} dt.$$

Use the relation between the beta and the gamma functions:

$$B(p, q) = \frac{\Gamma(p)\Gamma(q)}{\Gamma(p+q)},$$

where the beta function is defined as $B(p, q) = \int_0^1 t^{p-1}(1-t)^{q-1} dt$, $p > 0$, $q > 0$.

8

STURM-LIOUVILLE THEORY

The majority of the frequently encountered partial differential equations in physics and engineering can be solved by the method of separation of variables. This method helps us to reduce a second-order partial differential equation into a set of ordinary differential equations, which includes some new parameters called the separation constants. We have seen that solutions of these equations with the appropriate boundary conditions have properties reminiscent of an eigenvalue problem. In this chapter we study these properties systematically in terms of the Sturm-Liouville theory.

8.1 SELF-ADJOINT DIFFERENTIAL OPERATORS

We define a second-order linear differential operator in the interval $[a, b]$ as

$$\pounds = P_0(x) \frac{d^2}{dx^2} + P_1(x) \frac{d}{dx} + P_2(x), \qquad (8.1)$$

where $P_i(x)$ are real functions with the first $(2 - i)$ derivatives continuous. Also, in the open interval (a, b), $P_0(x)$ does not vanish even though it could have zeroes at the end points. We now define the adjoint operator $\bar{\pounds}$ as

$$\bar{\pounds}u(x) = \frac{d^2}{dx^2}[P_0(x)u(x)] - \frac{d}{dx}[P_1(x)u(x)] + P_2(x)u(x) \qquad (8.2)$$

$$= [P_0 \frac{d^2}{dx^2} + (2P_0' - P_1) \frac{d}{dx} + P_0''(x) - P_1'(x) + P_2(x)]u(x). \qquad (8.3)$$

The sufficient and necessary condition for an operator $\mathcal{L}$ to be self-adjoint, that is,

$$\mathcal{L} = \bar{\mathcal{L}} \tag{8.4}$$

is,

$$P_0'(x) = P_1(x). \tag{8.5}$$

A self-adjoint operator can also be written in the form

$$\bar{\mathcal{L}}u(x) = \mathcal{L}u(x) = \frac{d}{dx}\left[p(x)\frac{du(x)}{dx}\right] + q(x)u(x), \tag{8.6}$$

where

$$p(x) = P_0(x), \quad q(x) = P_2(x). \tag{8.7}$$

This is also called the **first canonical form**. A non-self-adjoint operator can always be put into self-adjoint form by multiplying it with

$$\frac{1}{P_0(x)}\exp\left[\int^x \frac{P_1(x)}{P_0(x)}dx\right]. \tag{8.8}$$

Among the equations we have seen, the Legendre equation is self adjoint, whereas the Hermite and the Laguerre equations are not.

8.2   STURM-LIOUVILLE SYSTEMS

The operator $\mathcal{L}$ defined in Equation (8.6) is called the **Sturm-Liouville operator**. Using this operator we can define a differential equation as

$$\mathcal{L}u(x) = -\lambda\omega(x)u(x), \tag{8.9}$$

which is called the **Sturm-Liouville equation**. This equation defines an eigenvalue problem for the operator $\mathcal{L}$, with the eigenvalue $-\lambda$, eigenfunction $u(x)$, and $\omega(x)$ as the weight function. The weight function satisfies the condition $\omega(x) > 0$ except for a finite number of isolated points, where it could have zeroes.

It is clear that a differential equation alone can not be a complete description of a physical problem. One also needs the boundary conditions to determine the integration constants. We now supplement the above differential equation with the following boundary conditions:

$$v\left(x\right)p(x)\frac{du\left(x\right)}{dx}\bigg|_{x=a} = 0 \, , \tag{8.10}$$

$$v\left(x\right)p(x)\frac{du\left(x\right)}{dx}\bigg|_{x=b} = 0, \tag{8.11}$$

where $u(x)$ and $v(x)$ are any two solutions of Equation (8.9) with the same or different λ values. Now the differential equation [Eq. (8.9)] plus the boundary conditions [Eqs. (8.10) and (8.11)] is called a **Sturm-Liouville system**. However, we could also work with something less restrictive as

$$v(x)p(x)u'(x)|_{x=a} = v(x)p(x)u'(x)|_{x=b} \, , \tag{8.12}$$

In general this boundary condition corresponds to one of the following:

1. Cases where the solutions $u(x)$ and the $v(x)$ are zero at the end points; $x = a$ and $x = b$. Such conditions are called the (homogeneous) **Dirichlet conditions.** Boundary conditions for the vibrations of a string fixed at both ends are of this type.

2. Cases where the derivatives $\dfrac{du(x)}{dx}$ and $\dfrac{dv(x)}{dx}$ are zero at the end points; $x = a$ and $x = b$. Acoustic wave problems require this type of boundary conditions, and they are called the (homogeneous) **Neumann conditions.**

3. Cases where $u(x) + \alpha\dfrac{du(x)}{dx}\bigg|_{x=a} = 0$ and $v(x) + \beta\dfrac{dv(x)}{dx}\bigg|_{x=b} = 0$, where α and β are constants independent of the eigenvalues. An example for this type of boundary conditions, which are called **general unmixed**, is the vibrations of a string with elastic connections.

4. Cases where one type of boundary conditions is satisfied at $x = a$ and another type at $x = b$.

A common property of all these conditions is that the value of $\dfrac{du(x)}{dx}$ and the value of $u(x)$ at the end point a are independent of their values at the other end point b; hence they are called **unmixed boundary conditions**. Depending on the problem, it is also possible to impose more complicated boundary conditions.

Even though the $\mathcal{L}$ operator is real, solutions of Equation (8.9) could involve complex functions; thus we write Equation (8.12) as

$$v^{*}pu'|_{x=a} = v^{*}pu'|_{x=b} \tag{8.13}$$

and take its complex conjugate:

$$vpu'^{*}|_{x=a} = vpu'^{*}|_{x=b} \, . \tag{8.14}$$

Since all the eigenfunctions satisfy the same boundary conditions, we interchange u and v to write

$$v'^{*}pu\big|_{x=a} = v'^{*}pu\big|_{x=b}. \tag{8.15}$$

8.3 HERMITIAN OPERATORS

We now show that the self-adjoint operator $\mathcal{L}$ and the differential equation

$$\mathcal{L}u(x) + \lambda\omega(x)u(x) = 0, \tag{8.16}$$

along with the boundary conditions [Eqs. (8.13) and (8.15)] have an interesting property. We first multiply

$$\mathcal{L}u(x)$$

from the left with v^{*} and integrate over $[a, b]$:

$$\int_{a}^{b} v^{*}\mathcal{L}u\,dx = \int_{a}^{b} v^{*}\left(pu'\right)'dx + \int_{a}^{b} v^{*}qu\,dx. \tag{8.17}$$

Integrating the first term on the right-hand side by parts gives us

$$\int_{a}^{b} v^{*}\left(pu'\right)'dx = v^{*}pu'\big|_{a}^{b} - \int_{a}^{b} \left(v^{*\prime}p\right)u'\,dx. \tag{8.18}$$

Using the boundary condition (8.13) the integrated term is zero. Integrating the second term in Equation (8.18) by parts again and using the boundary condition (8.15) we see that the integrated term is again zero, thus obtaining

$$\int_{a}^{b} v^{*}\left(pu'\right)'dx = \int_{a}^{b} u\left(pv^{*\prime}\right)'dx. \tag{8.19}$$

Substituting this result in Equation (8.17) we obtain

$$\int_{a}^{b} v^{*}\mathcal{L}u\,dx = \int_{a}^{b} u\mathcal{L}v^{*}dx. \tag{8.20}$$

Operators that satisfy this relation are called **Hermitian** with respect to the functions u and v satisfying the boundary conditions (8.13) and (8.15). In other words, hermiticity of an operator is closely tied to the boundary conditions imposed.

8.4 PROPERTIES OF HERMITIAN OPERATORS

Hermitian operators have the following very useful properties:

1. Eigenvalues are real.

2. Eigenfunctions are orthogonal with respect to a weight function $w(x)$.

3. Eigenfunctions form a complete set.

8.4.1   Real Eigenvalues

Let us write the eigenvalue equations for the eigenvalues λ_i and λ_j as

$$\pounds u_i + \lambda_i \omega(x) u_i = 0, \tag{8.21}$$
$$\pounds u_j + \lambda_j \omega(x) u_j = 0. \tag{8.22}$$

In these equations even though the $\pounds$ operator and the weight function $w(x)$ are real, the eigenfunctions and the eigenvalues could be complex. Taking the complex conjugate of Equation (8.22) we write

$$\pounds u_j^* + \lambda_j^* w(x) u_j^* = 0. \tag{8.23}$$

We multiply Equation (8.21) by u_j^* and Equation (8.23) by u_i and subtract to get

$$u_j^* \pounds u_i - u_i \pounds u_j^* = \left(\lambda_j^* - \lambda_i\right) \omega(x) u_i u_j^*. \tag{8.24}$$

We now integrate both sides:

$$\int_a^b u_j^* \pounds u_i dx - \int_a^b u_i \pounds u_j^* dx = \left(\lambda_j^* - \lambda_i\right) \int_a^b u_i u_j^* \omega(x) dx. \tag{8.25}$$

For Hermitian operators the left-hand side of the above equation is zero, thus we obtain

$$\left(\lambda_j^* - \lambda_i\right) \int_a^b u_i u_j^* \omega(x) dx = 0. \tag{8.26}$$

Since $\omega(x) \neq 0$ except for a finite number of isolated points, for $i = j$ we conclude that

$$\lambda_i^* = \lambda_i, \tag{8.27}$$

that is, the eigenvalues of Hermitian operators are real. In quantum mechanics eigenvalues correspond to precisely measured quantities; thus observables like energy and momentum are represented by Hermitian operators.

8.4.2   Orthogonality of Eigenfunctions

When $i \neq j$ and when the eigenfunctions are distinct, $\lambda_i \neq \lambda_j$, Equation (8.26) gives us

$$\int_a^b u_i(x) u_j^*(x) \omega(x) dx = 0, \quad i \neq j. \tag{8.28}$$

We say that the eigenfunctions u_i are orthogonal with respect to the weight function $\omega(x)$ in the interval $[a, b]$. In the case of degenerate eigenvalues, that is, when two different eigenfunctions have the same eigenvalue ($i \neq j$ but $\lambda_i = \lambda_j$), then the integral

$$\int_a^b u_i u_j^* \omega dx$$

does not have to vanish. However, in such cases we can always use the Gram-Schmidt orthogonalization method to choose the eigenfunctions as orthogonal. In summary, in any case we can normalize the eigenfunctions to define an orthonormal set with respect to the weight function $w(x)$ as

$$\int_a^b u_i(x) u_j^*(x) \omega(x) dx = \delta_{ij}. \tag{8.29}$$

8.4.3   Completeness of the Set of Eigenfunctions $\{u_m(x)\}$

Proof of completeness of the set of eigenfunctions is rather technical and can be found in Courant and Hilbert. What is important in most applications is that any sufficiently well-behaved and at least piecewise continuous function can be expressed as an infinite series in terms of $\{u_m(x)\}$ as

$$F(x) = \sum_{m=0}^{\infty} a_m u_m(x). \tag{8.30}$$

For a Sturm-Liouville system using variational analysis it can be shown that the limit

$$\lim_{N \to \infty} \int_a^b \left[F(x) - \sum_{m=0}^{N} a_m u_m(x) \right]^2 \omega(x) dx \to 0 \tag{8.31}$$

is true (Mathews and Walker p. 338). This means that in the interval $[a, b]$ the series

$$\sum_{m=0}^{\infty} a_m u_m(x) \tag{8.32}$$

converges to $F(x)$ in the mean. However, convergence in the mean does not imply uniform (or pointwise) convergence, which requires

$$\lim_{N \to \infty} \sum_{m=0}^{N} a_m u_m(x) \to F(x). \tag{8.33}$$

For most practical situations convergence in the mean accompanies uniform convergence and is sufficient. Note that uniform convergence also implies pointwise convergence but not vice versa. We conclude this section by stating a theorem from Courant and Hilbert (p. 427, vol. I).

The expansion theorem: Any piecewise continuous function defined in the fundamental interval $[a, b]$ with a square integrable first derivative (i.e., sufficiently smooth) could be expanded in an eigenfunction series:

$$F(x) = \sum_{m=0}^{\infty} a_m u_m(x),$$

which converges absolutely and uniformly in all subintervals free of points of discontinuity. At the points of discontinuity this series represents (as in the Fourier series) the arithmetic mean of the right- and the left-hand limits.

In this theorem the function $F(x)$ does not have to satisfy the boundary conditions. This theorem also implies convergence in the mean and pointwise convergence. That the derivative is square integrable means that the integral of the square of the derivative is finite for all the subintervals of the fundamental domain $[a, b]$ in which the function is continuous.

8.5 GENERALIZED FOURIER SERIES

Series expansion of a sufficiently smooth $F(x)$ in terms of the eigenfunction set $\{u_m(x)\}$ can now be written as

$$F(x) = \sum_{m=0}^{\infty} a_m u_m(x), \tag{8.34}$$

which is called the generalized Fourier series of $F(x)$. Expansion coefficients, a_m, are found using the orthogonality relation of $\{u_m(x)\}$ as

$$
\begin{aligned}
\int_a^b F(x) u_m^*(x) \omega(x)\, dx &= \int_a^b \sum_n a_n u_n(x) u_m^*(x) \omega(x)\, dx \\
&= \sum_n a_n \left[\int_a^b u_n(x) u_m^*(x) \omega(x)\, dx \right] \\
&= \sum_n a_n \delta_{nm} \\
&= a_m.
\end{aligned}
\tag{8.35}
$$

Substituting this in Equation (8.34) we get

$$
\begin{aligned}
F(x) &= \sum_{m=0}^{\infty} \int_a^b F(x') u_m^*(x') \omega(x') u_m(x)\, dx' \\
&= \int_a^b F(x') \left[\sum_{m=0}^{\infty} u_m^*(x') \omega(x') u_m(x) \right] dx'.
\end{aligned}
\tag{8.36}
$$

Using the basic definition of the Dirac-delta function, that is,

$$g(x) = \int g(x')\delta(x - x')dx',$$

we can now give a formal expression of the completeness of the set $\{\phi_m(x)\}$ as

$$\sum_{m=0}^{\infty} u_m^*(x')\omega(x')u_m(x) = \delta(x - x').\qquad(8.37)$$

It is needless to say that this is not a proof of completeness.

8.6   TRIGONOMETRIC FOURIER SERIES

Trigonometric Fourier series are defined with respect to the eigenvalue problem

$$\frac{d^2y}{dx^2} + n^2y(x) = 0,\qquad(8.38)$$

with the operator given as $\pounds = d^2/dx^2$. This could correspond to a vibrating string. Using the periodic boundary conditions

$$\left\{\begin{array}{c} u(a) = u(b) \\ v(a) = v(b), \end{array}\right\}\qquad(8.39)$$

we find the eigenfunctions as

$$\left\{\begin{array}{c} u_n = \cos nx, \ n = 0, 1, 2, ... \\ v_m = \sin mx, \ m = 1, 2, ... \ . \end{array}\right\}\qquad(8.40)$$

Orthogonality of the eigenfunctions is expressed as

$$\int_{x_0}^{x_0+2\pi} \sin mx \sin nx dx = A_n \delta_{nm},$$

$$\int_{x_0}^{x_0+2\pi} \cos mx \cos nx dx = B_n \delta_{nm},\qquad(8.41)$$

$$\int_{x_0}^{x_0+2\pi} \sin mx \cos nx dx = 0,$$

where

$$A_n = \begin{cases} \pi & n \neq 0 \\ 0 & n = 0 \end{cases} , \qquad (8.42)$$

$$B_n = \begin{cases} \pi & n \neq 0 \\ 2\pi & n = 0 \end{cases} . \qquad (8.43)$$

Now the trigonometric Fourier series of any sufficiently well-behaved function becomes

$$f(x) = \frac{a_0}{2} + \sum_{n=1}^{\infty} [a_n \cos nx + b_n \sin nx], \qquad (8.44)$$

where the expansion coefficients are given as

$$a_n = \frac{1}{\pi} \int_{-\pi}^{\pi} f(t) \cos nt\, dt, \quad n = 0, 1, 2, \dots \qquad (8.45)$$

and

$$b_n = \frac{1}{\pi} \int_{-\pi}^{\pi} f(t) \sin nt\, dt, \qquad n = 1, 2, \dots \ . \qquad (8.46)$$

Example 8.1. *Trigonometric Fourier series:* Trigonometric Fourier series of a square wave

$$f(x) = \begin{cases} +\dfrac{d}{2} & 0 < x < \pi \\[2mm] -\dfrac{d}{2} & -\pi < x < 0 \end{cases} , \qquad (8.47)$$

can now be written as

$$f(x) = \frac{2d}{\pi} \sum_{n=0}^{\infty} \frac{\sin (2n + 1) x}{(2n + 1)}, \qquad (8.48)$$

where we have substituted the coefficients

$$a_n = 0$$
$$b_n = \frac{d}{n\pi}(1 - \cos n\pi) = \begin{cases} 0 & n = \text{even} \\[2mm] \dfrac{2h}{n\pi} & n = \text{odd} \end{cases} \qquad (8.49)$$

8.7   HERMITIAN OPERATORS IN QUANTUM MECHANICS

In quantum mechanics the state of a system is completely described by a complex valued function, $\Psi(x)$, in terms of the real variable x. Observable quantities are represented by differential operators (not necessarily second order) acting on the wave functions. These operators are usually obtained from their classical expressions by replacing position, momentum, and energy with their operator counterparts as

$$\overrightarrow{x} \to \overrightarrow{x}$$

$$\overrightarrow{p} \to -i\hbar\overrightarrow{\nabla} \tag{8.50}$$

$$E \to i\hbar\frac{\partial}{\partial t}$$

For example, the angular momentum operator is obtained from its classical expression $\overrightarrow{L} = \overrightarrow{r} \times \overrightarrow{p}$ as

$$\mathbf{\overrightarrow{L}} = -i\hbar\left(\overrightarrow{r} \times \overrightarrow{\nabla}\right) .$$

Similarly, the Hamiltonian operator is obtained from its classical expression $H = p^2/2m + V(x)$ as

$$\mathbf{H} = -\frac{1}{2m}\overrightarrow{\nabla}^2 + V(x).$$

The observable value of a physical property is given by the expectation value of the corresponding operator $\mathcal{L}$ as

$$\langle\mathcal{L}\rangle = \int \Psi^*\mathcal{L}\Psi dx. \tag{8.51}$$

Because $\langle\mathcal{L}\rangle$ corresponds to a measurable quantity it has to be real; hence observable properties in quantum mechanics are represented by Hermitian operators. For the real Sturm-Liouville operators Hermitian property [Eq. (8.20)] was defined with respect to the eigenfunctions u and v, which satisfy the boundary conditions (8.13) and (8.15). To accommodate complex operators in quantum mechanics we modify this definition as

$$\int \Psi_1^*\mathcal{L}\Psi_2 dx = \int (\mathcal{L}\Psi_1)^*\Psi_2 dx, \tag{8.52}$$

where Ψ_1 and Ψ_2 do not have to be the eigenfunctions of the operator $\mathcal{L}$. The fact that Hermitian operators have real expectation values can be seen from

$$\langle\mathcal{L}\rangle = \int \Psi^*\mathcal{L}\Psi dx$$

$$= \int (\mathcal{L}\Psi)^*\Psi dx$$

$$= \langle\mathcal{L}\rangle^* . \tag{8.53}$$

A Hermitian Sturm-Liouville operator must be second order. However, in quantum mechanics order of the Hermitian operators is not restricted. Remember that the momentum operator is first order, but it is Hermitian because of the presence of i in its definition:

$$\langle p \rangle = \int_{-\infty}^{\infty} \Psi^*(-i\hbar \frac{\partial}{\partial x})\Psi dx \qquad (8.54)$$

$$= \int_{-\infty}^{\infty} (-i\hbar \frac{\partial}{\partial x}\Psi)^* \Psi dx \qquad (8.55)$$

$$= i\hbar \left. \Psi^* \Psi \right|_{-\infty}^{\infty} - \int_{-\infty}^{\infty} \Psi^*(i\hbar \frac{\partial}{\partial x})\Psi dx \qquad (8.56)$$

$$= \int_{-\infty}^{\infty} \Psi^*(-i\hbar \frac{\partial}{\partial x})\Psi dx. \qquad (8.57)$$

In proving that the momentum operator is Hermitian we have imposed the boundary condition that Ψ is sufficiently smooth and vanishes at large distances.

A general boundary condition that all wave functions must satisfy is that they have to be square integrable, and thus normalizable. Space of all square integrable functions actually forms an infinite dimensional vector space called L_2 or the **Hilbert space**. Functions in this space can be expanded as generalized Fourier series in terms of the complete and orthonormal set of eigenfunctions, $\{u_m(x)\}$, of a Hermitian operator. Eigenfunctions satisfy the eigenvalue equation

$$\pounds u_m(x) = \lambda_m u_m(x), \qquad (8.58)$$

where λ_m represents the eigenvalues. In other words, $\{u_m(x)\}$ spans the infinite dimensional vector space of square integrable functions. The inner product (analog of dot product) in Hilbert space is defined as

$$(\Psi_1, \Psi_2) = \int \Psi_1^*(x)\Psi_2(x)dx, \qquad (8.59)$$

which has the following properties:

$$(\Psi_1, \alpha\Psi_2) = \alpha(\Psi_1, \Psi_2), \qquad (8.60)$$
$$(\alpha\Psi_1, \Psi_2) = \alpha^*(\Psi_1, \Psi_2),$$
$$(\Psi_1, \Psi_2)^* = (\Psi_2, \Psi_1),$$
$$(\Psi_1 + \Psi_2, \Psi_3) = (\Psi_1, \Psi_3) + (\Psi_2, \Psi_3),$$

where α is a complex number. The inner product also satisfies the **triangle inequality:**

$$|\Psi_1 + \Psi_2| \leq |\Psi_1| + |\Psi_2| \qquad (8.61)$$

and the

Schwartz inequality:

$$|\Psi_1||\Psi_2| \geq |(\Psi_1, \Psi_2)|. \tag{8.62}$$

An important consequence of the Schwartz inequality is that convergence of (Ψ_1, Ψ_2) follows from the convergence of (Ψ_1, Ψ_1) and (Ψ_2, Ψ_2).

Problems

8.1   Show that the Laguerre equation

$$x\frac{d^2y}{dx^2} + (1-x)\frac{dy}{dx} + ny = 0$$

can be brought into the self-adjoint form by multiplying it with e^{-x} .

8.2   Write the Chebyshev equation

$$(1-x^2)T_n''(x) - xT_n'(x) + n^2T_n(x) = 0$$

in the self-adjoint form.

8.3   Find the weight function for the associated Laguerre equation

$$x\frac{d^2y}{dx^2} + (k+1-x)\frac{dy}{dx} + ny = 0.$$

8.4   A function $y(x)$ is to be a finite solution of the differential equation

$$x(1-x)\frac{d^2y}{dx^2} + (\frac{3}{2} - 2x)\frac{dy}{dx} + \left[\lambda - \frac{(2+5x-x^2)}{4x(1-x)}\right]y(x) = 0,$$

in the entire interval $x \in [0,1]$.)a Show that this condition can only be satisfied for certain values of λ and write the solutions explicitly for the lowest three values of λ. b) Find the weight function $w(x)$. c) Show that the solution set $\{y_\lambda(x)\}$ is orthogonal with respect to the $w(x)$ found above.

8.5   Show that the Legendre equation can be written as

$$\frac{d}{dx}[(1-x^2)P_l'] + l(l+1)P_l = 0.$$

8.6   For the Sturm-Liouville equation

$$\frac{d^2y}{dx^2} + \lambda y = 0$$

with the boundary conditions

$$y(0) = 0$$
$$y(\pi) - y'(\pi) = 0,$$

find the eigenvalues and the eigenfunctions.

8.7   Find the eigenvalues and the eigenfunctions of the Sturm-Liouville system

$$\frac{d}{dx}[(x^2+1)\frac{dy}{dx}] + \frac{\lambda}{x^2+1}y = 0,$$

$$y(0) = 0,$$
$$y(1) = 0.$$

Hint: Try the substitution $x = \tan t$.

8.8   Show that the Hermite equation can be written as

$$\frac{d}{dx}[e^{-x^2}H_n'] + 2ne^{-x^2}H_n = 0.$$

8.9   Given the Sturm-Liouville equation

$$\frac{d}{dx}\left[p(x)\frac{d}{dx}y(x)\right] + \lambda_n w(x)y(x) = 0.$$

If $y_n(x)$ and $y_m(x)$ are two orthogonal solutions and satisfy the appropriate boundary conditions, then show that $y_n'(x)$ and $y_m'(x)$ are orthogonal with the weight function $p(x)$.

8.10   Show that the Bessel equation can be written in the self-adjoint form as

$$\frac{d}{dx}[xJ_n'] + (x - \frac{n^2}{x})J_n = 0.$$

8.11   Find the trigonometric Fourier expansion of

$$f(x) = \pi \quad -\pi \le x < 0$$
$$= x \quad 0 < x \le \pi.$$

8.12   Show that the angular momentum operator

$$\overrightarrow{\mathbf{L}} = -i\hbar\left(\overrightarrow{r} \times \overrightarrow{\nabla}\right)$$

and its square are Hermitian.

8.13   Write the operators $\overrightarrow{\mathbf{L}}^2$, and L_z in spherical polar coordinates and show that they have the same eigenfunctions.
a) What are their eigenvalues?
b) Write the L_x and L_y operators in spherical polar coordinates.

8.14   For a Sturm-Liouville operator

$$\mathcal{L} = \frac{d}{dx}\left[p\left(x\right)\frac{d}{dx}\right] + q\left(x\right),$$

let $u(x)$ be a nontrivial solution satisfying $\mathcal{L}u = 0$ with the boundary condition at $x = a$, and let $v(x)$ be another nontrivial solution satisfying $\mathcal{L}v = 0$ with the boundary condition at $x = b$. Show that the Wronskian

$$W[u,v] = \begin{vmatrix} u & v \\ u' & v' \end{vmatrix} = uv' - vu'$$

is equal to $A/p(x)$, where A is a constant.

8.15   For the inner product defined as

$$(\Psi_1, \Psi_2) = \int \Psi_1^*(x)\Psi_2(x)dx,$$

prove the following properties, where α is a complex number:

$$(\Psi_1, \alpha\Psi_2) = \alpha(\Psi_1, \Psi_2),$$
$$(\alpha\Psi_1, \Psi_2) = \alpha^*(\Psi_1, \Psi_2),$$
$$(\Psi_1, \Psi_2)^* = (\Psi_2, \Psi_1),$$
$$(\Psi_1 + \Psi_2, \Psi_3) = (\Psi_1, \Psi_3) + (\Psi_2, \Psi_3)$$

8.16   Prove the **triangle inequality:**

$$|\Psi_1 + \Psi_2| \le |\Psi_1| + |\Psi_2|$$

and the **Schwartz inequality:**

$$|\Psi_1||\Psi_2| \ge |(\Psi_1, \Psi_2)|.$$

8.17   Show that the differential equation

$$y'' + p_1(x)y' + [p_2(x) + \lambda r(x)]y(x) = 0$$

can be put into self-adjoint form as

$$\frac{d}{dx}\left[e^{[\int^x p_1(x)dx]}\frac{dy\left(x\right)}{dx}\right] + p_2(x)e^{[\int^x p_1(x)dx]}y\left(x\right)$$
$$+ \lambda r(x)e^{[\int^x p_1(x)dx]}y(x) = 0.$$

9

STURM-LIOUVILLE SYSTEMS and the FACTORIZATION METHOD

The factorization method allows us to replace a Sturm-Liouville equation, which is a second-order differential equation, with a pair of first-order differential equations. For a large class of problems satisfying certain boundary conditions the method immediately yields the eigenvalues and allows us to write the ladder or the step-up-/-down operators for the problem. These operators are then used to construct the eigenfunctions from a base function. Once the base function is normalized, the manufactured eigenfunctions are also normalized and satisfy the same boundary conditions as the base function. First we introduce the method of factorization and its features in terms of five basic theorems. Next, we show how eigenvalues and eigenfunctions are obtained and introduce six basic types of factorization. In fact, factorization of a given second-order differential equation is reduced to identifying the type it belongs to. To demonstrate the usage of the method we discuss the associated Legendre equation and spherical harmonics in detail. We also discuss the radial part of Schrödinger's equation for the hydrogen-like atoms, Gegenbauer polynomials, the problem of the symmetric top, Bessel functions, and the harmonic oscillator problem via the factorization method. Further details and an extensive table of differential equations that can be solved by this technique can be found in Infeld and Hull (1951), where this method was introduced for the first time.

9.1   ANOTHER FORM FOR THE STURM-LIOUVILLE EQUATION

The Sturm-Liouville equation is usually written in the **first canonical form** as

$$\frac{d}{dx}\left[p(x)\frac{d\Psi(x)}{dx}\right] + q(x)\Psi(x) + \lambda w(x)\Psi(x) = 0, \quad x \in [\alpha,\beta], \tag{9.1}$$

where $p(x)$ is different from zero in the open interval (α,β); however, it could have zeroes at the end points of the interval. We also impose the boundary conditions

$$\Psi^* p\Phi' \big|_{x=\alpha} = \Psi^* p\Phi' \big|_{x=\beta} \tag{9.2}$$

and

$$\Psi'^* p\Phi \big|_{x=\alpha} = \Psi'^* p\Phi \big|_{x=\beta}, \tag{9.3}$$

where Φ and Ψ are any two solutions corresponding to the same or different eigenvalue. Solutions also satisfy the orthogonality relation

$$\int_\alpha^\beta dx\, w(x)\Psi_{\lambda_l}^*(x)\Psi_{\lambda_{l'}}(x) = 0, \quad \lambda_{l'} \neq \lambda_l. \tag{9.4}$$

If $p(x)$ and $w(x)$ are never negative and $w(x)/p(x)$ exists everywhere in (α,β), using the transformations

$$y(z) = \Psi(x)\left[w(x)p(x)\right]^{1/4} \tag{9.5}$$

and

$$dz = dx\left[\frac{w(x)}{p(x)}\right]^{1/2}, \tag{9.6}$$

we can cast the Sturm-Liouville equation into another form, also known as the **second canonical form**

$$\frac{d^2 y_\lambda^m(z)}{dz^2} + \left\{\lambda + r(z,m)\right\} y_\lambda^m(z) = 0, \tag{9.7}$$

where

$$\begin{aligned}
r(z,m) = {} & \frac{q}{w} + \frac{3}{16}\left[\frac{1}{w}\frac{dw}{dz} + \frac{1}{p}\frac{dp}{dz}\right]^2 \\
& - \frac{1}{4}\left[\frac{2}{pw}\frac{dp}{dz}\frac{dw}{dz} + \frac{1}{w}\frac{d^2 w}{dz^2} + \frac{1}{p}\frac{d^2 p}{dz^2}\right].
\end{aligned} \tag{9.8}$$

m and λ are two constant parameters that usually enter into our equations through the process of separation of variables. Their values are restricted by the boundary conditions and in most cases take discrete (real) values like

$$\lambda_0, \lambda_1, \lambda_2, ..., \lambda_l, ... \tag{9.9}$$

and

$$m = m_0, m_0 + 1, m_0 + 2, \ldots . \tag{9.10}$$

However, we could take $m_0 = 0$ without any loss of generality. The orthogonality relation is now given as

$$\int_a^b dz y_{\lambda_{l'}}^{*m}(z) y_{\lambda_l}^m(z) = 0, \quad \lambda_{l'} \neq \lambda_l. \tag{9.11}$$

9.2 METHOD OF FACTORIZATION

We can also write Equation (9.7) in operator form as

$$\mathcal{L}(z, m) y_{\lambda_l}^m(z) = -\lambda_l y_{\lambda_l}^m(z), \tag{9.12}$$

where

$$\mathcal{L}(z, m) = \frac{d^2}{dz^2} + r(z, m). \tag{9.13}$$

We now define two operators $O_\pm(z, m)$ as

$$O_\pm(z, m) = \pm\frac{d}{dz} - k(z, m) \tag{9.14}$$

so that

$$\mathcal{L}(z, m) = O_+(z, m) O_-(z, m). \tag{9.15}$$

We say that Equation (9.7) is factorized if we could replace it by either of the equations

$$O_+(z, m) O_-(z, m) y_{\lambda_l}^m(z) = [\lambda - \mu(m)] y_{\lambda_l}^m(z) \tag{9.16}$$

or

$$O_-(z, m + 1) O_+(z, m + 1) y_{\lambda_l}^m(z) = [\lambda - \mu(m + 1)] y_{\lambda_l}^m(z). \tag{9.17}$$

Substituting the definitions of $O_+(z, m)$ and $O_-(z, m)$ into Equations (9.16) and (9.17), we obtain two equations that $k(z, m)$ and $\mu(m)$ should satisfy simultaneously as

$$-\frac{dk(z, m)}{dz} + k^2(z, m) = -r(z, m) - \mu(m), \tag{9.18}$$

$$\frac{dk(z, m + 1)}{dz} + k^2(z, m + 1) = -r(z, m) - \mu(m + 1). \tag{9.19}$$

9.3   THEORY OF FACTORIZATION AND THE LADDER OPERATORS

We now summarize the fundamental ideas of the factorization method in terms of five basic theorems. The first theorem basically tells us how to generate solutions with different m given $y_{\lambda_l}^m(z)$.

Theorem I: If $y_{\lambda_l}^m(z)$ is a solution of Equation (9.12) corresponding to the eigenvalues λ and m, then

$$O_+(z, m+1)y_{\lambda_l}^m(z) = y_{\lambda_l}^{m+1}(z) \qquad (9.20)$$

and

$$O_-(z, m)y_{\lambda_l}^m(z) = y_{\lambda_l}^{m-1}(z) \qquad (9.21)$$

are also solutions corresponding to the same λ but different m as indicated.

Proof: Multiply Equation (9.17) by $O_+(m+1)$:

$$O_+(z, m+1)\left[O_-(z, m+1)O_+(z, m+1)y_{\lambda_l}^m(z)\right] \qquad (9.22)$$
$$= O_+(z, m+1)\left[\lambda - \mu(m+1)\right]y_{\lambda_l}^m(z).$$

This can be written as

$$O_+(z, m+1)O_-(z, m+1)\left[O_+(z, m+1)y_{\lambda_l}^m(z)\right] \qquad (9.23)$$
$$= \left[\lambda - \mu(m+1)\right]\left[O_+(z, m+1)y_{\lambda_l}^m(z)\right].$$

We now let

$$m \rightarrow m+1 \qquad (9.24)$$

in Equation (9.16) to write

$$O_+(z, m+1)O_-(z, m+1)y_{\lambda_l}^{m+1}(z) = \left[\lambda - \mu(m+1)\right]y_{\lambda_l}^{m+1}(z) \quad (9.25)$$

and compare this with Equation (9.23) to get Equation (9.20). Thus the theorem is proven. Proof of Equation (9.21) is accomplished by multiplying Equation (9.16) with $O_-(z, m)$ and by comparing it with the equation obtained by letting

$$m \rightarrow m-1$$

in Equation (9.17).

This theorem says that if we know the solution $y_{\lambda_l}^m(z)$, we can use $O_+(z, m+1)$ to generate the solutions corresponding to the eigenvalues

$$(m+1), (m+2), (m+3), \dots . \qquad (9.26)$$

Similarly, $O_-(z,m)$ can be used to generate the solutions with the eigenvalues

$$..., (m-3), (m-2), (m-1). \tag{9.27}$$

$O_\pm(z,m)$ are also called the **step-up/-down** or **ladder** operators.

Theorem II: If $y_1(z)$ and $y_2(z)$ are two solutions satisfying the boundary condition

$$y_1^* y_2|_b = y_1^* y_2|_a, \tag{9.28}$$

then

$$\int_a^b dz y_1^*(z)[O_-(z,m)y_2(z)] = \int_a^b dz y_2(z)[O_+(z,m)y_1(z)]^*. \tag{9.29}$$

We say that O_- and O_+ are Hermitian, that is $O_- = O_+^\dagger$ with respect to $y_1(z)$ and $y_2(z)$. Note that the boundary condition [Eq. (9.28)] needed for the factorization method is more restrictive than the boundary conditions [Eqs. (9.2) and (9.3)] used for the Sturm-Liouville problem. Condition (9.28) includes the periodic boundary conditions as well as the cases where the solutions vanish at the end points.

Proof: Proof can easily be accomplished by using the definition of the ladder operators and integration by parts:

$$\int_a^b dz y_1^*(z)[O_-(z,m)y_2(z)] \tag{9.30}$$

$$= \int_a^b dz y_1^*(z) \left[\left(-\frac{d}{dz} - k(z,m) \right) y_2(z) \right]$$

$$= -\int_a^b dz y_1^*(z) \frac{dy_2(z)}{dz} - \int_a^b dz y_1^*(z) k(z,m) y_2(z)$$

$$= -y_1^* y_2|_a^b + \int_a^b dz y_2 \frac{dy_1^*}{dz} - \int_a^b dz y_1^* k(z,m) y_2.$$

Finally, using the boundary condition [Eq. (9.28)], we write this as

$$\int_a^b dz y_1^*(z)[O_-(z,m)y_2(z)] \tag{9.31}$$

$$= \int_a^b dz y_2(z) \left[\left(\frac{d}{dz} - k(z,m) \right) y_1(z) \right]^*$$

$$= \int_a^b dz y_2(z)[O_+(z,m)y_1(z)]^*.$$

Theorem III: If

$$\int_a^b dz \ \left[y_{\lambda_l}^m(z)\right]^2 \tag{9.32}$$

exists and if $\mu(m)$ is an increasing function of m $(m > 0)$, then

$$\int_a^b dz \ \left[O_+(z, m+1)y_{\lambda_l}^m(z)\right]^2 \tag{9.33}$$

also exists. If $\mu(m)$ is a decreasing function of m (and $m > 0$), then

$$\int_a^b dz \ \left[O_-(z, m)y_{\lambda_l}^m(z)\right]^2 \tag{9.34}$$

also exists. $O_+(z, m+1)y_{\lambda_l}^m(z)$ and $O_-(z, m)y_{\lambda_l}^m(z)$ also satisfy the same boundary condition as $y_{\lambda_l}^m(z)$.

Proof: We take

$$y_2 = y_{\lambda_l}^m(z) \tag{9.35}$$

and

$$y_1 = y_{\lambda_l}^{m-1}(z) \tag{9.36}$$

in Theorem II to write

$$\int_a^b dz y_{\lambda_l}^{*m-1}(z)[O_-(z, m)y_{\lambda_l}^m(z)] = \int_a^b dz y_{\lambda_l}^m(z)[O_+(z, m)y_{\lambda_l}^{m-1}(z)]^*. \tag{9.37}$$

Solution $O_-(z, m)y_{\lambda_l}^m(z)$ in Equation (9.21) is equal to $y_{\lambda_l}^{m-1}(z)$ only up to a constant factor. Similarly, $O_+(z, m)y_{\lambda_l}^{m-1}(z)$ is only equal to $y_{\lambda_l}^m(z)$ up to another constant factor. Thus we can write

$$\int_a^b dz y_{\lambda_l}^m(z)y_{\lambda_l}^m(z)^* = C(l, m) \int_a^b dz y_{\lambda_l}^{*m-1}(z)y_{\lambda_l}^{m-1}(z) \tag{9.38}$$

or

$$\int_a^b dz \ \left[y_{\lambda_l}^m(z)\right]^2 = C(l, m) \int_a^b dz \ \left[y_{\lambda_l}^{m-1}(z)\right]^2, \tag{9.39}$$

where $C(l, m)$ is a constant independent of z but dependent on l and m. We are interested in differential equations the coefficients of which may have singularities only at the end points of our interval. Square integrability of a solution actually depends on the behavior of the solution near the end points. Thus it is itself a boundary condition. Hence, for a given square integrable eigenfunction $y_{\lambda_l}^m(z)$, the manufactured eigen-function $y_{\lambda_l}^{m-1}(z)$ is also square integrable as long as $C(l, m)$ is different from zero. Because we have used Theorem II, $y_{\lambda_l}^{m-1}(z)$ also satisfies the same boundary condition as $y_{\lambda_l}^m(z)$. A parallel argument is given for $y_{\lambda_l}^{m+1}(z)$. In conclusion, if $y_{\lambda_l}^m(z)$ is a square integrable function satisfying the boundary condition [Eq. (9.28)], then all other eigenfunctions manufactured from it by the ladder operators $O_\pm(z, m)$ are square integrable and satisfy the same boundary condition. For a complete proof $C(l, m)$ must be studied separately for each factorization type. For our purposes it is sufficient to say that $C(l, m)$ is different from zero for all physically meaningful cases.

Theorem IV: If $\mu(m)$ is an increasing function and $m > 0$, then there exists a maximum value for m, say $m_{\max} = l$, and λ is given as $\lambda = \mu(l+1)$. If $\mu(m)$ is a decreasing function and $m > 0$, then there exists a minimum value for m, say $m_{\min} = l'$, and λ is $\lambda = \mu(l')$.

Proof: Assume that we have some function $y_{\lambda_l}^m(z)$, where $m > 0$, which satisfies the boundary condition [Eq. (9.28)]. We can then write

$$\int_a^b dz \left[y_{\lambda_l}^{m+1}(z) \right]^2 \tag{9.40}$$

$$= \int_a^b dz \left[O_+(z, m+1) y_{\lambda_l}^m(z) \right] \left[O_+(z, m+1) y_{\lambda_l}^m(z) \right]^*$$

$$= \int_a^b dz y_{\lambda_l}^{*m}(z) \left[O_-(z, m+1) O_+(z, m+1) y_{\lambda_l}^m(z) \right]$$

$$= \left[\lambda - \mu(m+1) \right] \int_a^b dz \left[y_{\lambda_l}^m(z) \right]^2,$$

where we have first used Equation (9.29) and then Equation (9.17). Continuing this process k times we get

$$\int_a^b dz \left[y_{\lambda_l}^{m+k}(z) \right]^2 \tag{9.41}$$

$$= \left[\lambda - \mu(m+k) \right] \cdots \left[\lambda - \mu(m+2) \right] \left[\lambda - \mu(m+1) \right] \int_a^b dz \left[y_{\lambda_l}^m(z) \right]^2.$$

If $\mu(m)$ is an increasing function of m, eventually we are going to reach a value of m, say $m_{\max} = l$, that leads us to the contradiction

$$\int_a^b dz \left[y_{\lambda_l}^{l+1}(z) \right]^2 < 0, \tag{9.42}$$

unless

$$y_{\lambda_l}^{l+1}(z) \equiv 0, \tag{9.43}$$

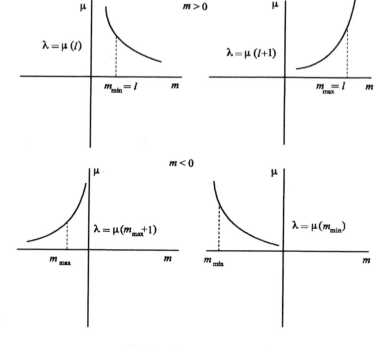

Fig. 9.1   Different cases for $\mu(m)$

that is,

$$O_+(z, l + 1)y^l_{\lambda_l}(z) \equiv 0. \tag{9.44}$$

Since $\int_a^b dz \left[y^l_{\lambda_l}(z) \right]^2 \neq 0$, using the last equation in (9.40) with $m = l$ we determine λ as

$$\lambda = \lambda_l = \mu(l + 1). \tag{9.45}$$

Similarly, it could be shown that if $\mu(m)$ is a decreasing function of m, then there exists a minimum value of m, say $m_{\min} = l$, such that

$$O_-(z, l)y^l_{\lambda_l}(z) \equiv 0. \tag{9.46}$$

λ in this case is determined as

$$\lambda = \lambda_l = \mu(l). \tag{9.47}$$

Cases for $m < 0$ are also shown in Figure 9.1.

We have mentioned that the square integrability of the solutions is itself a boundary condition, which is usually related to the symmetries of the problem. For example, in the case of the associated Legendre equation the end points of our interval correspond to the north and south poles of a sphere. For a spherically symmetric problem, location of the poles is arbitrary. Hence useful solutions should be finite everywhere on a sphere. In the Frobenius method this forces us to restrict λ to certain integer values (Chapter 2). In the factorization method we also have to restrict λ, this time through equation (9.40) to ensure the square integrability of the solutions for a given $\mu(m)$.

Theorem V: When Theorem III holds, we can arrange the ladder operators to preserve not just the square integrability but also the normalization of the eigenfunctions. When $\mu(m)$ is an increasing function of m, we can define new normalized ladder operators

$$\mathcal{L}_\pm(z, l, m) = [\mu(l+1) - \mu(m)]^{-1/2} O_\pm(z, m), \qquad (9.48)$$

which ensures us the normalization of the manufactured solutions.

When $\mu(m)$ is a decreasing function, normalized ladder operators are defined as

$$\mathcal{L}_\pm(z, l, m) = [\mu(l) - \mu(m)]^{-1/2} O_\pm(z, m). \qquad (9.49)$$

Proof: Using the last equation in Equation (9.40) we write

$$\int_a^b \left[y_{\lambda_l}^{m+1}(z)\right]^2 dz = [\lambda - \mu(m+1)] \int_a^b \left[y_{\lambda_l}^m(z)\right]^2 dz,$$

$$\int_a^b \frac{\left[y_{\lambda_l}^{m+1}(z)\right]^2}{[\lambda - \mu(m+1)]} dz = \int_a^b \left[y_{\lambda_l}^m(z)\right]^2 dz. \qquad (9.50)$$

Since

$$y_{\lambda_l}^{m+1}(z) = O_+(z, m+1)y_{\lambda_l}^m(z),$$

we write

$$\int_a^b \left[\frac{O_+(z, m+1)}{[\lambda - \mu(m+1)]^{1/2}} y_{\lambda_l}^m(z)\right]^2 dz \qquad (9.51)$$

$$= \int_a^b \left[y_{\lambda_l}^m(z)\right]^2 dz.$$

Define a new operator $\mathcal{L}_+(z, l, m)$; then Equation (9.51) becomes

$$\int_a^b \left[\mathcal{L}_+(z, l, m+1)y_{\lambda_l}^m(z)\right]^2 dz = \int_a^b \left[y_{\lambda_l}^m(z)\right]^2 dz.$$

Thus, if $y_{\lambda_l}^m(z)$ is normalized, then the eigenfunction manufactured from $y_{\lambda_l}^m(z)$ by the operator $\mathcal{L}_+$ is also normalized. Similarly, one could show that

$$
\int_a^b \left[\frac{O_-(z,l,m)}{[\lambda - \mu(m)]^{1/2}} y_{\lambda_l}^m(z) \right]^2 dz \tag{9.52}
$$
$$
= \int_a^b \left[\mathcal{L}_-(z,l,m) y_{\lambda_l}^m(z) \right]^2 dz
$$
$$
= \int_a^b \left[y_{\lambda_l}^{m-1}(z) \right]^2 dz
$$
$$
= \int_a^b \left[y_{\lambda_l}^m(z) \right]^2 dz.
$$

In conclusion, once $y_{\lambda_l}^m(z)$ is normalized, the manufactured eigenfunctions

$$
y_{\lambda_l}^{m+1}(z) = \mathcal{L}_+(z,l,m+1) y_{\lambda_l}^m(z) \tag{9.53}
$$

and

$$
y_{\lambda_l}^{m-1}(z) = \mathcal{L}_-(z,l,m) y_{\lambda_l}^m(z)
$$

are also normalized. Depending on the functional forms of $\mu(m)$, $\mathcal{L}_\pm(z,l,m)$ are given in Equations (9.48) and (9.49).

9.4   SOLUTIONS VIA THE FACTORIZATION METHOD

We can now manufacture the eigenvalues and the eigenfunctions of an equation once it is factored, that is, once the $k(z,m)$ and the $\mu(m)$ functions corresponding to a given $r(z,m)$ are known. For $m > 0$, depending on whether $\mu(m)$ is an increasing or a decreasing function, there are two cases.

9.4.1   Case I ($m > 0$ and $\mu(m)$ is an increasing function)

In this case, from Theorem IV there is a maximum value for m,

$$
m = 0, 1, 2, ..., l, \tag{9.54}
$$

and the eigenvalues λ_l are given as

$$
\lambda = \lambda_l = \mu(l+1). \tag{9.55}
$$

Since there is no eigenstate with $m > l$, we can write

$$
O_+(z, l+1) y_l^l(z) \equiv 0. \tag{9.56}
$$

Thus we obtain the differential equation

$$\left\{ \frac{d}{dz} - k(z, l+1) \right\} y_l^l(z) = 0. \tag{9.57}$$

Note that we have written $y_{\lambda_l}^l(z) = y_l^l(z)$. Integrating Equation (9.57) we get

$$\frac{dy_l^l}{y_l^l} = k(z, l+1)dz, \tag{9.58}$$

$$\ln y_l^l(z) = \int^z k(z, l+1)dz \tag{9.59}$$

or

$$y_l^l(z) = C \exp\left\{ \int^z k(z, l+1)dz \right\}. \tag{9.60}$$

C is a constant to be determined from the normalization condition

$$\int_a^b dz \left[y_l^l(z) \right]^2 = 1. \tag{9.61}$$

For a given l, once $y_l^{m=l}(z)$ is found, all the other normalized eigenfunctions with $m = l, l-1, l-2, ..., 2, 1, 0$, can be constructed by repeated applications of the step down operator $\mathcal{L}_-(z, l, m)$ as

$$y_l^{m-1}(z) = [\mu(l+1) - \mu(m)]^{-1/2} O_-(z, m) y_l^m(z) \tag{9.62}$$
$$= \mathcal{L}_-(z, l, m) y_l^m(z). \tag{9.63}$$

9.4.2 Case II ($m > 0$ and $\mu(m)$ is a decreasing function)

In this case, from Theorem IV there is a minimum value for m, where

$$m = l, l+1, l+2, ... \tag{9.64}$$

For this case we can write

$$O_-(z, l) y_l^l(z) = 0, \tag{9.65}$$

$$\left\{ -\frac{d}{dz} - k(z, l) \right\} y_l^l(z) = 0. \tag{9.66}$$

Thus

$$y_l^{m=l}(z) = C \exp\left\{ -\int^z k(z, l)dz \right\}. \tag{9.67}$$

C is again determined from the normalization condition

$$\int_a^b dz \left[y_l^l(z) \right]^2 = 1. \tag{9.68}$$

Now all the other normalized eigenfunctions for $m = l, l + 1, l + 2, \ldots$ are obtained from $y_l^l(z)$ by repeated applications of the formula

$$y_l^{m+1}(z) = [\mu(l) - \mu(m+1)]^{-1/2} O_+(z, m+1) y_l^m(z) \tag{9.69}$$
$$= \mathcal{L}_+(z, l, m) y_l^m(z). \tag{9.70}$$

Cases with $m < 0$ are handled similarly. In Section 9.6 we see how such a case is treated with spherical harmonics.

9.5   TECHNIQUE AND THE CATEGORIES OF FACTORIZATION

In Section 9.2 we saw that in order to accomplish factorization we need to determine the two functions $k(z, m)$ and $\mu(m)$, which satisfy the two equations

$$\frac{dk(z, m+1)}{dz} + k^2(z, m+1) = -r(z, m) - \mu(m+1), \tag{9.71}$$

$$-\frac{dk(z, m)}{dz} + k^2(z, m) = -r(z, m) - \mu(m). \tag{9.72}$$

$r(z, m)$ is known from the equation the factorization of which is sought, that is, from

$$\left[\frac{d^2}{dz^2} + r(z, m) \right] y_{\lambda_l}^m(z) = -\lambda_l y_{\lambda_l}^m(z). \tag{9.73}$$

However, following Infeld and Hull (1951) we subtract Equation (9.72) from Equation (9.71) to obtain the difference equation

$$-k^2(z, m) + k^2(z, m+1) + \frac{dk(z, m)}{dz} + \frac{dk(z, m+1)}{dz} \tag{9.74}$$
$$= \mu(m) - \mu(m+1).$$

This is the necessary equation that $k(z, m)$ and $\mu(m)$ should satisfy. This is also a sufficient condition, because $k(z, m)$ and $\mu(m)$ satisfying this equation give a unique $r(z, m)$ from Equation (9.71) or (9.72). We now categorize all possible forms of $k(z, m)$ and $\mu(m)$ that satisfy Equation (9.74).

9.5.1 Possible Forms for $k(z, m)$

9.5.1.1 **Positive powers of m:** We first consider $k(z, m)$ with the m dependence given as

$$k(z, m) = k_0(z) + m k_1(z). \tag{9.75}$$

To find $\mu(m)$ we write Equation (9.74) for successive values of m as (we suppress the z dependence of $k(z, m)$)

$$k^2(m) - k^2(m-1) + k'(m) + k'(m-1) = \mu(m-1) - \mu(m)$$
$$k^2(m-1) - k^2(m-2) + k'(m-1) + k'(m-2) = \mu(m-2) - \mu(m-1)$$
$$k^2(m-2) - k^2(m-3) + k'(m-2) + k'(m-3) = \mu(m-3) - \mu(m-2)$$

$$\vdots$$

$$k^2(1) - k^2(0) + k'(1) + k'(0) = \mu(0) - \mu(1). \tag{9.76}$$

Addition of these equations gives us

$$k^2(m) - k^2(0) + 2m k_0' + k_1' \left[\sum_{m'=1}^{m} m' + \sum_{m'=0}^{m-1} m' \right] = \mu(0) - \mu(m). \tag{9.77}$$

We have used

$$k'(z, m) = k_0'(z) + m k_1'(z). \tag{9.78}$$

Also using

$$\sum_{m'=1}^{m} m' + \sum_{m'=0}^{m-1} m' = \frac{m(m+1)}{2} + \frac{m(m-1)}{2} \tag{9.79}$$

$$= m^2$$

and, since from Equation (9.75) we can write

$$k^2(m) - k^2(0) = [k_0 + m k_1]^2 - k_0^2, \tag{9.80}$$

we finally obtain

$$\mu(m) - \mu(0) = -m^2(k_1^2 + k_1') - 2m(k_0 k_1 + k_0'). \tag{9.81}$$

Since $\mu(m)$ is only a function of m, this could only be satisfied if the coefficients of m are constants.

$$k_1^2 + k_1' = \text{const.} = -a^2 \tag{9.82}$$

and

$$k_0 k_1 + k_0' = \text{const.} = -a^2 c \quad \text{if } a \neq 0 \tag{9.83}$$
$$k_0 k_1 + k_0' = \text{const.} = b \qquad \text{if } a = 0. \tag{9.84}$$

This determines $\mu(m)$ as

$$\mu(m) = \mu(0) + a^2(m^2 + 2mc) \quad \text{for} \quad a \neq 0 \tag{9.85}$$

and

$$\mu(m) = \mu(0) - 2mb \quad \text{for } a = 0. \tag{9.86}$$

In these equations we could take $\mu(0) = 0$ without any loss of generality.
Using these results we now obtain the following categories:

A) For $a \neq 0$ Equation (9.82) gives

$$\frac{dk_1}{k_1^2 + a^2} = -dz, \tag{9.87}$$

$$k_1 = a \cot a(z + p). \tag{9.88}$$

Substituting this into Equation (9.83) and integrating gives us

$$k_0(z) = ca \cot a(z + p) + \frac{d}{\sin a(z + p)}, \tag{9.89}$$

where p and d are integration constants.

With these k_0 and k_1 functions in Equation (9.75) and the $\mu(m)$ given
in Equation (9.85) we obtain $r(z, m)$ from Equation (9.71) or (9.72) as

$$r(z, m) = -\frac{a^2(m + c)(m + c + 1) + d^2 + 2ad(m + c + \frac{1}{2}) \cos a(z + p)}{\sin^2 a(z + p)}. \tag{9.90}$$

We now obtain our first factorization type as

$$k(z, m) = (m + c)a \cot a(z + p) + \frac{d}{\sin a(z + p)}, \tag{9.91}$$

$$\mu(m) = a^2(m + c)^2, \quad \text{we set } \mu(0) = a^2c^2. \tag{9.92}$$

B)

$$k_1 = \text{const.} = ia \tag{9.93}$$

$$k_0 = ica + de^{-iaz}. \tag{9.94}$$

For this type, after writing a instead of ia and adding $-a^2c^2$ to $\mu(m)$
we get

$$r(z, m) = -d^2 e^{2az} + 2ad\left(m + c + \frac{1}{2}\right)e^{az}, \tag{9.95}$$

$$k(z, m) = de^{az} - m - c, \tag{9.96}$$

$$\mu(m) = -a^2(m + c)^2. \tag{9.97}$$

C)

$$k_1 = \frac{1}{z}, \quad a = 0 \tag{9.98}$$

$$k_0 = \frac{b}{2}z + \frac{d}{z}. \tag{9.99}$$

After writing c for d and adding $b/2$ to $\mu(m)$ we obtain

$$r(z, m) = -\frac{(m+c)(m+c+1)}{z^2} - \frac{b^2 z^2}{4} + b(m-c), \tag{9.100}$$

$$k(z, m) = (m+c)/z + bz/2, \tag{9.101}$$

$$\mu(m) = -2bm + b/2. \tag{9.102}$$

D)

$$k_1 = 0, \quad a = 0 \tag{9.103}$$

$$k_0 = bz + d. \tag{9.104}$$

In this case, the operators $O_\pm$ are independent of m. $r(z,m)$, $k(z,m)$, and $\mu(m)$ are now given as

$$r(z, m) = -(bz + d)^2 + b(2m + 1), \tag{9.105}$$

$$k(z, m) = bz + d, \tag{9.106}$$

$$\mu(m) = -2bm. \tag{9.107}$$

We can also try higher positive powers of m in $k(z, m)$ as

$$k(z, m) = k_0(z) + mk_1(z) + m^2 k_2(z) + \cdots . \tag{9.108}$$

However, no new categories result (see Problems 9.5 and 9.6). Also note that the types B, C, and D can be viewed as the limiting forms of type A.

9.5.1.2 **Negative powers of m:**   We now try negative powers of m as

$$k(z, m) = \frac{k_{-1}(z)}{m} + k_0(z) + k_1(z)m. \tag{9.109}$$

We again write Equation (9.74) for successive values of m as (we suppress the z dependence of $k(z, m)$)

$$k^2(m) - k^2(m-1) + k'(m) + k'(m-1) = \mu(m-1) - \mu(m)$$

$$k^2(m-1) - k^2(m-2) + k'(m-1) + k'(m-2) = \mu(m-2) - \mu(m-1)$$

$$k^2(m-2) - k^2(m-3) + k'(m-2) + k'(m-3) = \mu(m-3) - \mu(m-2)$$

$$\vdots$$

$$k^2(2) - k^2(1) + k'(2) + k'(1) = \mu(1) - \mu(2). \tag{9.110}$$

Adding these equations and using

$$k'(z, m) = \frac{k'_{-1}(z)}{m} + k'_0(z) + k'_1(z)m \tag{9.111}$$

give

$$k^2(m) - k^2(1) + k'_{-1} \left[\sum_{m'=2}^{m} \frac{1}{m'} + \sum_{m'=1}^{m-1} \frac{1}{m'} \right]$$
$$+ k'_0 [2m - 2] + k'_1 \left[\sum_{m'=2}^{m} m' + \sum_{m'=1}^{m-1} m' \right]$$
$$= \mu(1) - \mu(m). \tag{9.112}$$

Since the series $\left[\sum_{m'=2}^{m} \frac{1}{m'} + \sum_{m'=1}^{m-1} \frac{1}{m'} \right]$, which is the coefficient of k'_{-1}, contains a logarithmic dependence on m, we set k_{-1} to a constant

$$k_{-1} = q \neq 0. \tag{9.113}$$

Also using

$$\sum_{m'=2}^{m} m' + \sum_{m'=1}^{m-1} m' = m^2 - 1 \tag{9.114}$$

and Equation (9.109) we write

$$k^2(m) - k^2(1) \tag{9.115}$$
$$= \frac{k^2_{-1}}{m^2} + k_1^2 m^2 + \frac{2k_{-1}k_0}{m} + 2k_0 k_1 m - k^2_{-1} - k_1^2 - 2k_{-1}k_0 - 2k_0 k_1.$$

Now Equation (9.112) becomes

$$\frac{k^2_{-1}}{m^2} + k_1^2 m^2 + \frac{2k_{-1}k_0}{m} + 2k_0 k_1 m$$
$$- k^2_{-1} - k_1^2 - 2k_{-1}k_0 - 2k_0 k_1$$
$$+ k'_0 [2m - 2]$$
$$+ k'_1 [m^2 - 1]$$
$$= \mu(1) - \mu(m). \tag{9.116}$$

After some simplification and setting $\mu(1) = 0$, which we can do without any loss of generality, Equation (9.116) gives

$$\frac{k^2_{-1}}{m^2} + \frac{2k_0 k_{-1}}{m} + m(2k_0 k_1 + 2k'_0) + m^2(k_1^2 + k'_1) \tag{9.117}$$
$$+ \left[-(k_1^2 + k'_1) - k^2_{-1} - 2k'_0 - 2(k_1 + k_{-1})k_0 \right] = -\mu(m).$$

We know have two new categories corresponding to the cases $a \neq 0$ and $a = 0$ with

$$k_{-1} = q, \tag{9.118}$$
$$k_0 = 0, \tag{9.119}$$
$$k_1^2 + k_1' = -a^2. \tag{9.120}$$

E)

$$k_1 = a \cot a(z + p), \quad k_0 = 0, \quad k_{-1} = q \text{ for } a \neq 0. \tag{9.121}$$

$r(z, m)$, $k(z, m)$, and $\mu(m)$ are now given as

$$r(z, m) = -\frac{m(m + 1)a^2}{\sin^2 a(z + p)} - 2aq \cot a(z + p),$$
$$k(z, m) = ma \cot a(z + p) + q/m, \tag{9.122}$$
$$\mu(m) = a^2 m^2 - q^2/m^2.$$

F) Our final category is obtained for $a = 0$ as

$$k_1 = 1/z, \quad k_0 = 0, \quad k_{-1} = q, \tag{9.123}$$

where

$$r(z, m) = -2q/z - m(m + 1)/z^2, \tag{9.124}$$
$$k(z, m) = m/z + q/m, \tag{9.125}$$
$$\mu(m) = -q^2/m^2. \tag{9.126}$$

Further generalization of these cases by considering higher negative powers of m leads to no new categories as long as we have a finite number of terms with negative powers in $k(z, m)$. Type F can also be viewed as the limiting form of type E with $a \to 0$. Entries in the table of factorizations given by Infeld and Hull (1951) can be used, with our notation with the replacements $x \to z$ and $L(m) = \mu(m)$.

9.6   ASSOCIATED LEGENDRE EQUATION (TYPE A)

The Legendre equation is given as

$$\frac{d^2\Theta(\theta)}{d\theta^2} + \cot\theta \frac{d\Theta(\theta)}{d\theta} + \left[\lambda_l - \frac{m^2}{\sin^2\theta}\right]\Theta(\theta) = 0, \tag{9.127}$$

where $\theta \in [0, \pi]$ and $m = 0, \pm1, \pm2, \ldots$. We can put this into the first canonical form by the substitution

$$x = \cos\theta, \ \Theta(\theta) = P(x), \ x \in [-1, 1], \tag{9.128}$$

as

$$\left(1 - x^2\right) \frac{d^2 P(x)}{dx^2} - 2x \frac{dP(x)}{dx} + \left[\lambda_l - \frac{m^2}{(1 - x^2)}\right] P(x) = 0, \tag{9.129}$$

$$\frac{d}{dx}\left[\left(1 - x^2\right)\frac{dP(x)}{dx}\right] + \left[\lambda_l - \frac{m^2}{(1 - x^2)}\right] P(x) = 0. \tag{9.130}$$

We now first make the substitutions

$$w(x) = 1, \ \ p(x) = \left(1 - x^2\right), \ dz = \frac{dx}{(1 - x^2)^{1/2}}, \ y(x) = P(x)(1 - x^2)^{1/4}, \tag{9.131}$$

which in terms of θ means that

$$w(x) = 1, \ \ p(x) = \sin^2\theta, \ dz = -d\theta, \ y(\theta) = P(\cos\theta)\sin^{1/2}\theta, \tag{9.132}$$

and thus leads us to the second canonical form

$$\frac{d^2 y(\theta)}{d\theta^2} + \left[(\lambda_l + \frac{1}{4}) - \frac{(m^2 - \frac{1}{4})}{\sin^2\theta}\right] y(\theta) = 0. \tag{9.133}$$

If we call

$$\lambda = (\lambda_l + \frac{1}{4}) \tag{9.134}$$

and compare with

$$\frac{d^2 y_\lambda^m(z)}{dz^2} + \{\lambda + r(z, m)\} y_\lambda^m(z) = 0, \tag{9.135}$$

we obtain

$$r(z, m) = \frac{(m^2 - \frac{1}{4})}{\sin^2 z}. \tag{9.136}$$

This is exactly type A with the coefficients read from Equation (9.90) as

$$a = 1, \ c = -1/2, \ d = 0, \ p = 0, \ z = \theta. \tag{9.137}$$

Thus from Equations (9.91) and (9.92) we obtain the factorization of the associated Legendre equation as

$$k(z, m) = (m - \frac{1}{2})\cot\theta, \tag{9.138}$$

$$\mu(m) = (m - \frac{1}{2})^2. \tag{9.139}$$

For convenience we have taken $\mu(0) = a^2 c^2$ rather than zero in Equation (9.92).

9.6.1   Determining the Eigenvalues λ_l

For $m > 0$

$$\mu(m) = (m - \frac{1}{2})^2. \qquad (9.140)$$

Thus $\mu(m)$ is an increasing function of m and from Theorem IV we know that there exists a maximum value for m, say $m_{max} = l$. This determines λ as

$$\lambda = \mu(l+1) \qquad (9.141)$$

$$= (l + \frac{1}{2})^2.$$

On the other hand, for $m < 0$ we could write

$$\mu(m) = (|m| + \frac{1}{2})^2. \qquad (9.142)$$

Again from the conclusions of Theorem IV there exists a minimum value, m_{min}, thus determining λ as

$$\lambda = m_{min} \qquad (9.143)$$

$$= \left(|m_{min}| + \frac{1}{2}\right)^2.$$

To find m_{min} we equate the two expressions [Eqs. (9.141) and (9.143)] for λ to obtain

$$\left(l + \frac{1}{2}\right)^2 = \left(|m_{min}| + \frac{1}{2}\right)^2, \qquad (9.144)$$

$$|m_{min}| = l, \qquad (9.145)$$

$$m_{min} = -l. \qquad (9.146)$$

Since m changes by integer amounts, we could write

$$m_{min} = m_{max} - \text{integer}$$

$$-l = l - \text{integer}$$

$$2l = \text{integer}. \qquad (9.147)$$

This equation says that l could only take integer values $l = 0, 1, 2, \dots$. We can now write the eigenvalues λ_l as

$$\lambda_l + \frac{1}{4} = \lambda \qquad (9.148)$$

$$\lambda_l + \frac{1}{4} = \left(l + \frac{1}{2}\right)^2 \qquad (9.149)$$

$$\lambda_l + \frac{1}{4} = l^2 + l + \frac{1}{4} \qquad (9.150)$$

$$\lambda_l = l(l+1). \qquad (9.151)$$

Note that Equation (9.147) also has the solution $l =$ integer $/2$. We will elaborate this case in the next chapter in Problem 11.11.

9.6.2 Construction of the Eigenfunctions

Since $m_{\max} = l$, there are no states with $m > l$. Thus

$$O_+(z, l+1)y^l_{\lambda_l}(z) = 0 \tag{9.152}$$

$$\left\{ \frac{d}{dz} - k(z, l+1) \right\} y^l_{\lambda_l}(z) = 0. \tag{9.153}$$

This gives

$$\ln y^l_{\lambda_l}(z) - \ln N = \int^z k(z', l+1)dz'$$

$$= \left(l + \frac{1}{2} \right) \int \cot\theta d\theta$$

$$= \left(l + \frac{1}{2} \right) \ln(\sin\theta)$$

$$= \ln(\sin\theta)^{(l+\frac{1}{2})}. \tag{9.154}$$

Hence the state with $m_{\max} = l$ is determined as

$$y^l_{\lambda_l}(\theta) = N(\sin\theta)^{(l+\frac{1}{2})}. \tag{9.155}$$

N is a normalization constant to be determined from

$$\int_0^\pi \left[y^l_{\lambda_l}(\theta) \right]^2 d\theta = 1, \tag{9.156}$$

$$N^2 \int_0^\pi (\sin\theta)^{2l+1} d\theta = 1 \tag{9.157}$$

which gives

$$N = (-1)^l \left[\frac{(2l+1)!}{2^{2l+1}l!^2} \right]^{1/2}. \tag{9.158}$$

The factor of $(-1)^l$, which is called the **Condon-Shortley phase**, is introduced for convenience. Thus the normalized eigenfunction corresponding to $m_{\max} = l$ is

$$y^l_{\lambda_l}(\theta) = (-1)^l \left[\frac{(2l+1)!}{2^{2l+1}l!^2} \right]^{1/2} (\sin\theta)^{(l+\frac{1}{2})}. \tag{9.159}$$

Using this eigenfunction (eigenstate) we can construct the remaining eigenstates by using the normalized ladder operators [Eqs.(9.48) and (9.49)]. For moving down the ladder we use

$$\mathcal{L}_-(\theta,m) = \frac{O_-(\theta,m)}{\sqrt{\mu(l+1)-\mu(m)}} \tag{9.160}$$

$$= \frac{1}{\sqrt{\left(l+\frac{1}{2}\right)^2 - \left(m-\frac{1}{2}\right)^2}} \left[-\frac{d}{d\theta} - \left(m-\frac{1}{2}\right)\cot\theta\right]$$

$$= \frac{1}{\sqrt{(l+m)(l-m+1)}} \left[-\frac{d}{d\theta} - \left(m-\frac{1}{2}\right)\cot\theta\right]$$

and for moving up the ladder

$$\mathcal{L}_+(\theta,m+1) = \frac{O_+(\theta,m+1)}{\sqrt{\mu(l+1)-\mu(m+1)}} \tag{9.161}$$

$$= \frac{1}{\sqrt{(l+\frac{1}{2})^2 - (m+\frac{1}{2})^2}} \left[\frac{d}{d\theta} - \left(m+\frac{1}{2}\right)\cot\theta\right]$$

$$= \frac{1}{\sqrt{(l-m)(l+m+1)}} \left[\frac{d}{d\theta} - \left(m+\frac{1}{2}\right)\cot\theta\right].$$

Needless to say, the eigenfunctions generated by the operators $\mathcal{L}_\pm$ are also normalized (Theorem V).

Now the **normalized** associated Legendre polynomials are related to $y^l_{\lambda_l}(\theta)$ by

$$P_l^m(\cos\theta) = \frac{y^m_{\lambda_l}(\theta)}{\sqrt{\sin\theta}}. \tag{9.162}$$

9.6.3   Ladder Operators for the Spherical Harmonics

Spherical harmonics are defined as

$$Y_l^m(\theta,\phi) = P_l^m(\cos\theta)\frac{e^{im\phi}}{\sqrt{2\pi}}. \tag{9.163}$$

Using Equation (9.162) we write

$$Y_l^m(\theta,\phi) = \frac{y^m_{\lambda_l}(\theta)}{\sqrt{\sin\theta}}\frac{e^{im\phi}}{\sqrt{2\pi}}. \tag{9.164}$$

Using

$$y^{m-1}_{\lambda_l}(\theta) = \mathcal{L}_-(\theta,m)y^m_{\lambda_l}(\theta) \tag{9.165}$$

and Equation (9.160) we could also write

$$\sqrt{\sin\theta}\,P_l^{m-1}(\theta)\frac{e^{i(m-1)\phi}}{\sqrt{2\pi}} \tag{9.166}$$

$$= e^{-i\phi}\mathcal{L}_-(\theta,m)\left[\sqrt{\sin\theta}\,P_l^m(\theta)\frac{e^{im\phi}}{\sqrt{2\pi}}\right]$$

$$= \frac{e^{-i\phi}\left[-\dfrac{d}{d\theta}-(m-\tfrac{1}{2})\cot\theta\right]\left[\sqrt{\sin\theta}\,P_l^m(\theta)\dfrac{e^{im\phi}}{\sqrt{2\pi}}\right]}{\sqrt{(l+m)(l-m+1)}}$$

$$= \frac{e^{-i\phi}\dfrac{e^{im\phi}}{\sqrt{2\pi}}\left[-\dfrac{d\sqrt{\sin\theta}\,P_l^m(\theta)}{d\theta}-(m-\tfrac{1}{2})\cot\theta\sqrt{\sin\theta}\,P_l^m(\theta)\right]}{\sqrt{(l+m)(l-m+1)}}$$

$$= \frac{\sqrt{\sin\theta}\,e^{-i\phi}\dfrac{e^{im\phi}}{\sqrt{2\pi}}\left[-\dfrac{d}{d\theta}-m\cot\theta\right]P_l^m(\theta)}{\sqrt{(l+m)(l-m+1)}}$$

$$= \frac{\sqrt{\sin\theta}\,e^{-i\phi}}{\sqrt{(l+m)(l-m+1)}}\left[-\dfrac{d}{d\theta}-m\cot\theta\right]\left(P_l^m(\theta)\dfrac{e^{im\phi}}{\sqrt{2\pi}}\right).$$

Cancelling $\sqrt{\sin\theta}$ on both sides and noting that

$$\frac{\partial Y_l^m(\theta,\phi)}{\partial\phi} = imP_l^m(\theta)\frac{e^{im\phi}}{\sqrt{2\pi}}, \tag{9.167}$$

and using Equation (9.164) we finally write

$$Y_l^{m-1}(\theta,\phi) = \frac{e^{-i\phi}}{\sqrt{(l+m)(l-m+1)}}\left[-\frac{\partial}{\partial\theta}+i\cot\theta\frac{\partial}{\partial\phi}\right]Y_l^m(\theta,\phi). \tag{9.168}$$

Similarly

$$Y_l^{m+1}(\theta,\phi) = \frac{e^{i\phi}}{\sqrt{(l-m)(l+m+1)}}\left[\frac{\partial}{\partial\theta}+i\cot\theta\frac{\partial}{\partial\phi}\right]Y_l^m(\theta,\phi). \tag{9.169}$$

We now define the ladder operators L_+ and L_- for the m index of the spherical harmonics as

$$L_- = e^{-i\phi}\left[-\frac{\partial}{\partial\theta}+i\cot\theta\frac{\partial}{\partial\phi}\right], \tag{9.170}$$

$$L_+ = e^{i\phi}\left[+\frac{\partial}{\partial\theta}+i\cot\theta\frac{\partial}{\partial\phi}\right], \tag{9.171}$$

where

$$Y_l^{m-1}(\theta, \phi) = \frac{L_- Y_l^m(\theta, \phi)}{\sqrt{(l+m)(l-m+1)}} \tag{9.172}$$

and

$$Y_l^{m+1}(\theta, \phi) = \frac{L_+ Y_l^m(\theta, \phi)}{\sqrt{(l-m)(l+m+1)}}. \tag{9.173}$$

We can now construct the spherical harmonics from the eigenstate,

$$Y_l^0(\theta, \phi) = \sqrt{\frac{2l+1}{2}} P_l(\cos\theta) \frac{1}{\sqrt{2\pi}}, \tag{9.174}$$

by successive operations of the ladder operators as

$$Y_l^m(\theta, \phi) = \sqrt{\frac{2l+1}{2} \frac{(l-m)!}{(l+1)!}} \frac{1}{2\pi} [L_+]^m P_l(\cos\theta) \tag{9.175}$$

and

$$Y_l^{-m}(\theta, \phi) = \sqrt{\frac{2l+1}{2} \frac{(l-m)!}{(l+1)!}} \frac{1}{2\pi} [L_-]^m P_l(\cos\theta). \tag{9.176}$$

$P_l^{m=0}(\cos\theta) = P_l(\cos\theta)$ is the Legendre polynomial. Note that

$$[L_-]^* = -[L_+] \tag{9.177}$$

and

$$Y_l^{*m}(\theta, \phi) = (-1)^m Y_l^{-m}(\theta, \phi). \tag{9.178}$$

9.6.4   Interpretation of the $L_\pm$ Operators

In quantum mechanics the angular momentum operator (we set $\hbar = 1$) is given as

$$\overrightarrow{L} = -i\overrightarrow{r} \times \overrightarrow{\nabla}. \tag{9.179}$$

We write this in spherical polar coordinates to get

$$\overrightarrow{L} = -i \begin{pmatrix} \widehat{e}_r & \widehat{e}_\theta & \widehat{e}_\phi \\ r & 0 & 0 \\ \dfrac{\partial}{\partial r} & \dfrac{1}{r}\dfrac{\partial}{\partial \theta} & \dfrac{1}{r\sin\theta}\dfrac{\partial}{\partial \phi} \end{pmatrix} \tag{9.180}$$

$$= -i \left[-\widehat{\mathbf{e}}_\theta \left(\frac{1}{\sin\theta} \frac{\partial}{\partial\phi} \right) + \widehat{\mathbf{e}}_\phi \left(\frac{\partial}{\partial\theta} \right) \right]. \tag{9.181}$$

The basis vectors $\widehat{\mathbf{e}}_\theta$ and $\widehat{\mathbf{e}}_\phi$ in spherical polar coordinates are written in terms of the basis vectors $(\widehat{\mathbf{e}}_x, \widehat{\mathbf{e}}_y, \widehat{\mathbf{e}}_z)$ of the Cartesian coordinates as

$$\widehat{\mathbf{e}}_\theta = (\cos\theta \cos\phi)\widehat{\mathbf{e}}_x + (\cos\theta \sin\phi)\widehat{\mathbf{e}}_y - (\sin\theta)\widehat{\mathbf{e}}_z, \tag{9.182}$$
$$\widehat{\mathbf{e}}_\phi = -(\sin\theta)\widehat{\mathbf{e}}_x + (\cos\phi)\widehat{\mathbf{e}}_y. \tag{9.183}$$

Thus the angular momentum operator in Cartesian coordinates becomes

$$\overrightarrow{L} = L_x \widehat{\mathbf{e}}_x + L_y \widehat{\mathbf{e}}_y + L_z \widehat{\mathbf{e}}_z \tag{9.184}$$
$$= \widehat{\mathbf{e}}_x \left(i \cot\theta \cos\phi \frac{\partial}{\partial\phi} + i \sin\phi \frac{\partial}{\partial\theta} \right)$$
$$+ \widehat{\mathbf{e}}_y \left(i \cot\theta \sin\phi \frac{\partial}{\partial\phi} - i \cos\phi \frac{\partial}{\partial\theta} \right) + \widehat{\mathbf{e}}_z \left(-i \frac{\partial}{\partial\phi} \right).$$

It is now clearly seen that

$$L_+ = L_x + iL_y, \tag{9.185}$$
$$L_- = L_x - iL_y,$$

and

$$L_z = -i\frac{\partial}{\partial\phi}.$$

Also note that

$$\overrightarrow{L}^2 = L_x^2 + L_y^2 + L_z^2 \tag{9.186}$$
$$= \frac{1}{2}(L_+ L_- + L_- L_+) + L_z^2$$
$$= -\left[\frac{1}{\sin\theta} \frac{\partial}{\partial\theta} \left[\sin\theta \frac{\partial}{\partial\theta} \right] + \frac{1}{\sin^2\theta} \frac{\partial^2}{\partial\phi^2} \right].$$

From the definition of L_z it is seen that

$$L_z Y_l^m = m Y_l^m, \quad m = -l, ..., 0, ..., l . \tag{9.187}$$

Also, using the $L_\pm$ operators defined in Equations (9.170-171) and Equations (9.172-173) we can write

$$\overrightarrow{L}^2 Y_l^m = \frac{1}{2}(L_+ L_- + L_- L_+) Y_l^m + L_z^2 Y_l^m \tag{9.188}$$
$$= l(l+1)Y_l^m, \quad l = 0, 1, 2, 3... .$$

Thus Y_l^m are the simultaneous eigenfunctions of the $\overrightarrow{L}^2$ and the L_z operators. To understand the physical meaning of the angular momentum operators, consider a scalar function $\Psi(r, \theta, \phi)$, which may represent some physical system (could be a wave function). We now operate on this function with an operator R the effect of which is to rotate a physical system by α counterclockwise about the z-axis. $R\Psi(r, \theta, \phi)$ is now a new function representing the physical system after it has been rotated. This is equivalent to replacing ϕ by $\phi + \alpha$ in $\Psi(r, \theta, \phi)$. After making a Taylor series expansion about $\alpha = 0$ we get

$$R\Psi(r, \theta, \phi) \tag{9.189}$$
$$= \Psi(r, \theta, \phi')$$
$$= \Psi(r, \theta, \phi + \alpha)$$
$$= \Psi(r, \theta, \phi) + \left.\frac{\partial \Psi}{\partial \alpha}\right|_{\alpha=0} \alpha + \frac{1}{2!} \left.\frac{\partial^2 \Psi}{\partial \alpha^2}\right|_{\alpha=0} \alpha^2 + \cdots + \frac{1}{n!} \left.\frac{\partial^n \Psi}{\partial \alpha^n}\right|_{\alpha=0} \alpha^n \cdots .$$

In terms of the coordinate system (r, θ, ϕ), this corresponds to a rotation about the z-axis by $-\alpha$. Thus with the replacement $d\alpha \rightarrow -d\phi$ we get

$$R\Psi(r, \theta, \phi) \tag{9.190}$$
$$= \Psi(r, \theta, \phi) - \left.\frac{\partial \Psi}{\partial \phi}\right|_{\alpha=0} \alpha + \frac{1}{2!} \left.\frac{\partial^2 \Psi}{\partial \phi^2}\right|_{\alpha=0} \alpha^2 + \cdots + \frac{(-1)^n}{n!} \left.\frac{\partial^n \Psi}{\partial \phi^n}\right|_{\alpha=0} \alpha^n \cdots$$
$$= \left[1 - \left.\frac{\partial}{\partial \phi}\right|_{\alpha=0} \alpha + \frac{1}{2!} \left.\frac{\partial^2}{\partial \phi^2}\right|_{\alpha=0} \alpha^2 + \cdots + \frac{(-1)^n}{n!} \left.\frac{\partial^n}{\partial \phi^n}\right|_{\alpha=0} \alpha^n \cdots \right] \Psi(r, \theta, \phi)$$
$$= \left[\exp\left(-\alpha \frac{\partial}{\partial \phi}\right) \right] \Psi(r, \theta, \phi)$$
$$= \left[\exp(-i\alpha L_z) \right] \Psi(r, \theta, \phi).$$

For a rotation about an arbitrary axis along the unit vector $\hat{\mathbf{n}}$ this becomes

$$R\Psi(r, \theta, \phi) = \left[\exp(-i\alpha \overrightarrow{L} \cdot \hat{\mathbf{n}}) \right] \Psi(r, \theta, \phi). \tag{9.191}$$

Thus the angular momentum operator $\overrightarrow{L}$ is related to the rotation operator R by

$$R = \exp(-i\alpha \overrightarrow{L} \cdot \hat{\mathbf{n}}). \tag{9.192}$$

9.6.5 Ladder Operators for the l Eigenvalues

We now write $\lambda_l = l(l+1)$ and $-m^2 = \lambda$ in Equation (9.127) to obtain

$$\frac{d^2 \Theta(\theta)}{d\theta^2} + \cot \theta \frac{d\Theta(\theta)}{d\theta} + \left[l(l+1) + \frac{\lambda}{\sin^2 \theta} \right] \Theta(\theta) = 0. \tag{9.193}$$

We can put this equation into the second canonical form by the transformation

$$z = \ln\left(\tan \frac{\theta}{2} \right), \quad \Theta(\theta) = V(z), \quad z \in [-\infty, \infty] \tag{9.194}$$

as

$$\frac{d^2V(z)}{dz^2} + \left[\lambda + \frac{l(l+1)}{\cosh^2\theta}\right]V(z) = 0. \tag{9.195}$$

Because the roles of l and m are interchanged, we can vary l for fixed m. Comparing Equation (9.195) with

$$\frac{d^2V(z)}{dz^2} + [\lambda + r(z,l)]\,V(z) = 0 \tag{9.196}$$

and Equation (9.90) we see that this is of type A with

$$a = i, \ c = 0, \ p = i\pi/2, \ \text{and } d = 0.$$

Its factorization is therefore obtained as

$$O_+(z,l)O_-(z,l)V_l^{\lambda m}(z) = [\lambda_m - \mu(l)]V_l^{\lambda m}(z) \tag{9.197}$$

with

$$k(z,l) = l\tanh z, \tag{9.198}$$

$$\mu(l) = -l^2. \tag{9.199}$$

Thus the ladder operators are

$$O_\pm(z,l) = \pm\frac{d}{dz} - l\tanh z. \tag{9.200}$$

Because $\mu(l)$ is a decreasing function, from Theorem IV we obtain the top of the ladder for some minimum value of l, say m, thus

$$\lambda = -m^2. \tag{9.201}$$

We can now write

$$O_+(z,l)O_-(z,l)V_l^{\lambda m}(z) = [-m^2 + l^2]V_l^{\lambda m}(z) \tag{9.202}$$

$$O_-(z,l+1)O_+(z,l+1)V_l^{\lambda m}(z) = [-m^2 + (l+1)^2]V_l^{\lambda m}(z). \tag{9.203}$$

Using

$$\int_{-\infty}^{+\infty}\left[V_{l-1}^{\lambda m}(z)\right]^2 dz = \int_{-\infty}^{+\infty}\left[O_-(z,l)V_l^{\lambda m}(z)\right]^2 dz \tag{9.204}$$

$$= \int_{-\infty}^{+\infty}V_l^{\lambda m}(z)\left[O_+(z,l)O_-(z,l)V_l^{\lambda m}(z)\right]dz$$

$$= [-m^2 + l^2]\int_{-\infty}^{+\infty}\left[V_l^{\lambda m}(z)\right]^2 dz.$$

We again see that $l_{\min} = m$, so that

$$O_-(z,l)V_m^{\lambda_m}(z) = 0. \tag{9.205}$$

Because we do not have a state lower than $l = m$, using the definition of $O_-(z,l)$, we can use Equation (9.66) to find $V_l^{\lambda_l}(z)$ as

$$\left[-\frac{d}{dz} - m \tanh z \right] V_m^{\lambda_m}(z) = 0 \tag{9.206}$$

$$\int \frac{dV_m^{\lambda_m}(z)}{V_m^{\lambda_m}(z)} = -m \int \frac{\sinh z}{\cosh z} dz \tag{9.207}$$

$$V_m^{\lambda_m}(z) = N' \frac{1}{\cosh^m z}, \tag{9.208}$$

where N' is a normalization constant in the z-space. Using the transformation given in Equation (9.194) and, since $l = m$, we write $V_m^{\lambda_m}(z)$ as

$$V_m^m(\theta) = V_l^l(\theta) = N \sin^l \theta. \tag{9.209}$$

From Equations (9.162) and (9.155) we note that for $m = l$

$$y_l^l(\theta) = \sqrt{\sin\theta} P_l^l \propto (\sin\theta)^{(l+\frac{1}{2})}, \tag{9.210}$$
$$V_l^l(\theta) \propto y_l^l(\theta)/\sqrt{\sin\theta}.$$

Thus for a general m

$$V_l^m(\theta) = C_{lm} \frac{y_l^m(\theta)}{\sqrt{\sin\theta}}. \tag{9.211}$$

C_{lm} is needed to ensure normalization in the θ-space. Using Equation (9.49) of Theorem V and Equation (9.20) we now find the step-up operator for the l index as

$$V_{l+1}^m(\theta) = \frac{1}{\sqrt{(l+1)^2 - m^2}} \left\{ \frac{d}{dz} - (l+1)\tanh z \right\} V_l^m(\theta), \tag{9.212}$$

$$C_{l+1,m} \frac{y_{l+1}^m(\theta)}{\sqrt{\sin\theta}} = \frac{C_{lm}}{\sqrt{(l+1)^2 - m^2}} \left\{ \frac{d}{dz} - (l+1)\tanh z \right\} \frac{y_l^m(\theta)}{\sqrt{\sin\theta}}. \tag{9.213}$$

Taking tanh of both sides in Equation (9.194), we write

$$\tanh z = -\cos\theta,$$
$$\frac{d}{dz} = \sin\theta \frac{d}{d\theta} \tag{9.214}$$

and obtain

$$y_{l+1}^m(\theta)C_{l+1,m} \tag{9.215}$$

$$= \frac{C_{lm}}{\sqrt{(l+1+m)(l+1-m)}}\left\{\sin\theta\frac{d}{d\theta} + \left(l+\frac{1}{2}\right)\cos\theta\right\}y_l^m(\theta),$$

Similarly for the step-down operator, we find

$$y_{l-1}^m(\theta)C_{l-1,m} \tag{9.216}$$

$$= \frac{C_{lm}}{\sqrt{(l-m)(l+m)}}\left\{-\sin\theta\frac{d}{d\theta} + \left(l+\frac{1}{2}\right)\cos\theta\right\}y_l^m(\theta).$$

Using our previous results [Eqs. (9.160) and (9.161)], ladder operators for the m index in $y_l^m(\theta)$ can be written as

$$y_l^{m+1}(\theta) = \frac{1}{\sqrt{(l-m)(l+m+1)}} \tag{9.217}$$

$$\times \left\{\frac{d}{d\theta} - \left(m+\frac{1}{2}\right)\cot\theta\right\}y_l^m(\theta),$$

$$y_l^{m-1}(\theta) = \frac{1}{\sqrt{(l+m)(l-m+1)}} \tag{9.218}$$

$$\times \left\{-\frac{d}{d\theta} - \left(m-\frac{1}{2}\right)\cot\theta\right\}y_l^m(\theta).$$

To evaluate the normalization constant in θ-space, first we show that the ratio $C_{lm}/C_{l+1,m}$ is independent of m. Starting with the state (l,m) we can reach $(l+1,m+1)$ in two ways.

Path I $(l,m) \rightarrow (l,m+1) \rightarrow (l+1,m+1)$: For this path, using Equations (9.215) and (9.217) we write

$$y_{l+1}^{m+1}(\theta)\frac{C_{l+1,m+1}}{C_{l,m+1}} \tag{9.219}$$

$$= \frac{\left\{\sin\theta\frac{d}{d\theta} + (l+\frac{1}{2})\cos\theta\right\}\left\{\frac{d}{d\theta} - (m+\frac{1}{2})\cot\theta\right\}y_l^m(\theta)}{\sqrt{(l-m)^2(l+m+1)(l+m+2)}}.$$

The numerator on the right-hand side is

$$\left\{\left[\sin\theta\frac{d^2y_l^m}{d\theta^2}\right] + (l-m)\cos\theta\frac{dy_l^m}{d\theta} + \frac{m+\frac{1}{2}}{\sin\theta}\left[1 - (l+\frac{1}{2})\cos^2\theta\right]y_l^m\right\}. \tag{9.220}$$

Using Equation (9.133) with $\lambda_l = l(l+1)$ and simplifying, we obtain

$$y_{l+1}^{m+1}(\theta) \tag{9.221}$$

$$= \frac{C_{l,m+1}}{C_{l+1,m+1}} \frac{1}{(l-m)} \frac{(l-m)}{\sqrt{(l+m+1)(l+m+2)}}$$

$$\cdot \left\{ \cos\theta \frac{d}{d\theta} - \frac{m+\frac{1}{2}}{\sin\theta} - \left(l+\frac{1}{2}\right)\sin\theta \right\} y_l^m(\theta).$$

Path II $(l,m) \to (l+1,m) \to (l+1,m+1)$: Following the same procedure as in the first path we obtain

$$y_{l+1}^{m+1}(\theta)$$

$$= \frac{C_{l,m}}{C_{l+1,m}} \frac{1}{(l+1-m)} \frac{(l+1-m)}{\sqrt{(l+m+1)(l+m+2)}}$$

$$\cdot \left\{ \cos\theta \frac{d}{d\theta} - \frac{m+\frac{1}{2}}{\sin\theta} - \left(l+\frac{1}{2}\right)\sin\theta \right\} y_l^m(\theta). \tag{9.222}$$

Thus we obtain

$$\frac{C_{l,m+1}}{C_{l+1,m+1}} = \frac{C_{l,m}}{C_{l+1,m}}, \tag{9.223}$$

which means that

$$\frac{C_{l,m}}{C_{l+1,m}} \tag{9.224}$$

is independent of m.

Using this result, we can now evaluate $C_{lm}/C_{l+1,m}$. First using Equation (9.159) we write

$$y_{l+1}^l(\theta)$$

$$= \frac{C_{ll}}{C_{l+1,l}} \left\{ \sin\theta \frac{d}{d\theta} + \left(l+\frac{1}{2}\right)\cos\theta \right\} \frac{\sin^{l+1/2}\theta}{\sqrt{2l+1}}(-1)^l \sqrt{\frac{1\cdot3\cdot5\cdots(2l+1)}{2[2\cdot4\cdot6\cdots(2l)]}}$$

$$= \frac{C_{ll}}{C_{l+1,l}} \sin^{l+1/2}\theta \cos\theta \sqrt{2l+1}(-1)^l \sqrt{\frac{1\cdot3\cdot5\cdots(2l+1)}{2[2\cdot4\cdot6\cdots(2l)]}}. \tag{9.225}$$

Using Equations (9.159) and (160) we can also write

$$y_{l+1}^l(\theta) = \mathcal{L}_-(\theta, l+1)y_{l+1}^{l+1}(\theta)$$

$$= \mathcal{L}_-(\theta, l+1)\left[\sin^{l+3/2}\theta(-1)^{l+1}\sqrt{\frac{1\cdot3\cdot5\cdots(2l+3)}{2[2\cdot4\cdot6\cdots2(l+1)]}} \right]$$

$$= \sqrt{(2l+3)}(-1)^l \sqrt{\frac{1\cdot3\cdot5\cdots(2l+1)}{2[2\cdot4\cdot6\cdots(2l)]}} \sin^{l+1/2}\theta \cos\theta. \tag{9.226}$$

Comparing these we get

$$\frac{C_{ll}}{C_{l+1,l}} = \frac{C_{lm}}{C_{l+1,m}} = \sqrt{\frac{2l+3}{2l+1}}. \tag{9.227}$$

Finally, using

$$Y_l^m(\theta,\phi) = \frac{y_l^m(\theta)\, e^{im\phi}}{\sqrt{\sin\theta}\,\sqrt{2\pi}}, \tag{9.228}$$

we now write the complete set of normalized ladder operators of the spherical harmonics for the indices l and m as

$$Y_{l+1}^m(\theta,\phi) = \sqrt{\frac{(2l+3)}{(2l+1)(l+1+m)(l+1-m)}}$$

$$\cdot \left\{ \sin\theta \frac{\partial}{\partial\theta} + (l+1)\cos\theta \right\} Y_l^m(\theta,\phi), \tag{9.229}$$

$$Y_{l-1}^m(\theta,\phi) = \sqrt{\frac{(2l-1)}{(2l+1)(l-m)(l+m)}}$$

$$\cdot \left\{ -\sin\theta \frac{\partial}{\partial\theta} + l\cos\theta \right\} Y_l^m(\theta,\phi). \tag{9.230}$$

and

$$Y_l^{m-1}(\theta,\phi) = \frac{e^{-i\phi} \left[-\dfrac{\partial}{\partial\theta} + i\cot\theta \dfrac{\partial}{\partial\phi} \right]}{\sqrt{(l+m)(l-m+1)}} Y_l^m(\theta,\phi), \tag{9.231}$$

$$Y_l^{m+1}(\theta,\phi) = \frac{e^{+i\phi} \left[\dfrac{\partial}{\partial\theta} + i\cot\theta \dfrac{\partial}{\partial\phi} \right]}{\sqrt{(l-m)(l+m+1)}} Y_l^m(\theta,\phi). \tag{9.232}$$

Adding Equations (9.229) and (9.230) we also obtain a useful relation

$$\cos\theta Y_l^m(\theta,\phi)$$

$$= \sqrt{\frac{(l+1+m)(l+1-m)}{(2l+1)(2l+3)}} Y_{l+1}^m(\theta,\phi)$$

$$+ \sqrt{\frac{(l-m)(l+m)}{(2l+1)(2l-1)}} Y_{l-1}^m(\theta,\phi). \tag{9.233}$$

7 SCHRÖDINGER EQUATION FOR A SINGLE-ELECTRON ATOM AND THE FACTORIZATION METHOD (TYPE F)

he radial part of the Schrödinger equation for a single-electron atom is given

$$\frac{d}{dr}\left(r^2 \frac{dR_l(r)}{dr}\right) + r^2 k^2(r) R_l(r) - l(l+1) R_l(r) = 0, \tag{9.234}$$

here

$$k^2(r) = \frac{2m}{\hbar^2}\left[E + \frac{Ze^2}{r}\right]. \tag{9.235}$$

is the atomic number and e is the electron's charge. Because the electrons
an atom are bounded, their energy values should satisfy $E < 0$. In this
uation if we change the dependent variable as

$$R_l(r) = \frac{u_{E,l}(r)}{r}, \tag{9.236}$$

e differential equation to be solved for $u_{E,l}(r)$ becomes

$$-\frac{\hbar^2}{2m}\frac{d^2 u_{E,l}}{dr^2} + \left[\frac{\hbar^2 l(l+1)}{2mr^2} - \frac{Ze^2}{r}\right] u_{E,l}(r) = E u_{E,l}(r). \tag{9.237}$$

e have seen in Chapter 3 that the conventional method allows us to express
e solutions of this equation in terms of the Laguerre polynomials. To solve
is problem with the factorization method we first write Equation (9.237) as

$$\frac{d^2 u_{E,l}}{d(r/\frac{\hbar^2}{mZe^2})^2} + \left[\frac{2}{(r/\frac{\hbar^2}{mZe^2})} - \frac{l(l+1)}{(r/\frac{\hbar^2}{mZe^2})^2}\right] u_{E,l}(r) + \left(\frac{2\hbar^2 E}{mZ^2 e^4}\right) u_{E,l}(r) = 0. \tag{9.238}$$

aking the unit of length as

$$\frac{\hbar^2}{mZe^2} \tag{9.239}$$

d defining

$$\lambda = \frac{2\hbar^2 E}{mZ^2 e^4}, \tag{9.240}$$

quation (9.238) becomes

$$\frac{d^2 u_\lambda^l}{dr^2} + \left[\lambda + \left(\frac{2}{r} - \frac{l(l+1)}{r^2}\right)\right] u_\lambda^l = 0. \tag{9.241}$$

This is Type F with

$$q = -1 \text{ and } m = l.$$

Thus we determine $k(r, l)$ and $\mu(l)$ as

$$k(r, l) = \frac{l}{r} - \frac{1}{l}, \tag{9.242}$$

$$\mu(l) = -\frac{1}{l^2}. \tag{9.243}$$

Because $\mu(l)$ is an increasing function, we have $l_{\max}$ say n'; thus we obtain λ as

$$\lambda = -\frac{1}{(n'+1)^2}, \quad n' = 0, 1, 2, 3, ..., \tag{9.244}$$

or

$$\lambda = -\frac{1}{n^2}, \quad n = 1, 2, 3, ...$$

Note that $l \leq n = 1, 2, 3, ...$. We also have

$$u_n^{l=n} = (2/n)^{n+1/2} [(2n)!]^{-1/2} r^n \exp(-\frac{r}{n}), \tag{9.245}$$

where

$$u_n^{l-1} = [\mathcal{L}_- (r, l)] \, u_n^n \tag{9.246}$$

and

$$u_n^{l+1} = [\mathcal{L}_+ (r, l+1)] \, u_n^n. \tag{9.247}$$

The normalized ladder operators are defined by Equation (9.48) as

$$\mathcal{L}_\pm(r, l) = \left[-\frac{1}{n^2} + \frac{1}{l^2} \right]^{-1/2} \left\{ \pm \frac{d}{dr} - \frac{l}{r} + \frac{1}{l} \right\}. \tag{9.248}$$

Using (9.240) the energy levels are obtained as

$$E_n = -\frac{mZ^2 e^4}{2\hbar^2 n^2}, \quad n = 1, 2, 3, ..., \tag{9.249}$$

which are the quantized Bohr energy levels.

9.8 GEGENBAUER FUNCTIONS (TYPE A)

The Gegenbauer equation in general is given as

$$(1-x^2)\frac{d^2 C_n^{\lambda'}(x)}{dx^2} - (2\lambda'+1)x\frac{dC_n^{\lambda'}(x)}{dx} + n(n+2\lambda')C_n^{\lambda'}(x) = 0. \quad (9.250)$$

For $\lambda = 1/2$ this equation reduces to the Legendre equation. For integer values of n its solutions reduce to the Gegenbauer or Legendre polynomials:

$$C_n^{\lambda'}(x) = \sum_{r=0}^{[n/2]} (-1)^r \frac{\Gamma(n-r+\lambda')}{\Gamma(\lambda')r!(n-2r)!}(2x)^{n-2r}. \quad (9.251)$$

In the study of surface oscillations of a hypersphere one encounters the equation

$$(1-x^2)\frac{d^2 U_\lambda^m(x)}{dx^2} - (2m+3)x\frac{dU_\lambda^m(x)}{dx} + \lambda U_\lambda^m(x) = 0, \quad (9.252)$$

solutions of which could be expressed in terms of the Gegenbauer polynomials as

$$U_\lambda^m(x) = C_{l-m}^{m+1}(x), \quad (9.253)$$

where

$$\lambda = (l-m)(l+m+2). \quad (9.254)$$

Using

$$x = -\cos\theta \quad (9.255)$$
$$U_\lambda^m(x) = Z_\lambda^m(\theta)(\sin\theta)^{-m-1} \quad (9.256)$$

we can put Equation (9.252) into the second canonical form as

$$\frac{d^2 Z_\lambda^m(\theta)}{d\theta^2} + \left[-\frac{m(m+1)}{\sin^2\theta} + (\lambda+(m+1)^2)\right] Z_\lambda^m(\theta) = 0. \quad (9.257)$$

On the introduction of

$$\lambda'' = \lambda + (m+1)^2, \quad (9.258)$$

and comparing with, Equation (9.90), this is of type A with $c = p = d = 0$, $a = 1$, and $z = \theta$, and its factorization is given by

$$k(\theta, m) = m\cot\theta \quad (9.259)$$
$$\mu(m) = m^2. \quad (9.260)$$

The solutions are found by using

$$Z_\lambda^{m=l}(\theta) = \pi^{-1/4} \left[\frac{\Gamma(l+2)}{\Gamma(l+3/2)} \right]^{1/2} \sin^{l+1}\theta \tag{9.261}$$

and the formulas

$$Z_l^{m-1} = \left[(l+1)^2 - m^2 \right]^{-1/2} \left\{ -\frac{d}{d\theta} - m\cot\theta \right\} Z_l^m \tag{9.262}$$

Note that Z_l^m is the eigenfunction corresponding to the eigenvalue

$$\lambda'' = (l+1)^2, \qquad l-m = 0, 1, 2, ..., \tag{9.263}$$

that is, to

$$\begin{aligned} \lambda &= (l+1)^2 - (m+1)^2 \\ &= (l-m)(l+m+2). \end{aligned} \tag{9.264}$$

9.9   SYMMETRIC TOP (TYPE A)

The wave equation for a symmetric top is encountered in the study of simple molecules. If we separate the wave function as

$$U = \Theta(\theta) \exp(i\kappa\phi) \exp(im\psi), \tag{9.265}$$

where θ, ϕ, and ψ are the Euler angles and κ and m are integers, $\Theta(\theta)$ satisfies the second-order ordinary differential equation

$$\frac{d^2\Theta(\theta)}{d\theta^2} + \cot\theta \frac{d\Theta(\theta)}{d\theta} - \frac{(m-\kappa\cos\theta)^2}{\sin^2\theta}\Theta(\theta) + \sigma\Theta(\theta) = 0, \tag{9.266}$$

where

$$\sigma = \frac{8\pi^2 AW}{h^2} - \frac{A\kappa^2}{C}. \tag{9.267}$$

A, W, C, and h are other constants that come from the physics of the problem. With the substitution

$$Y = \Theta(\theta) \sin^{1/2}\theta, \tag{9.268}$$

Equation (9.266) becomes

$$\frac{d^2Y}{d\theta^2} - \left[\frac{(m-1/2)(m+1/2) + \kappa^2 - 2m\kappa\cos\theta}{\sin^2\theta} \right] Y + (\sigma + \kappa^2 + 1/4)Y = 0. \tag{9.269}$$

This equation is of type A, and we identify the parameters in Equation (9.90) as

$$a = 1, \ c = -1/2, \ d = -\kappa, \ p = 0.$$

The factorization is now given by

$$k(\theta, m) = (m - 1/2)\cot\theta - \kappa/\sin\theta, \tag{9.270}$$

$$\mu(m) = (m - 1/2)^2. \tag{9.271}$$

Eigenfunctions can be obtained from

$$Y_{J\kappa}^J = \left[\frac{\Gamma(2J + 2)}{\Gamma(J - \kappa + 1)\Gamma(J + \kappa + 1)}\right]^{1/2} \tag{9.272}$$
$$\times \sin^{J-\kappa+1/2}\frac{\theta}{2}\cos^{J+\kappa+1/2}\frac{\theta}{2}$$

by using

$$Y_{J\kappa}^{m-1} = \left[(J + \frac{1}{2})^2 - (m - \frac{1}{2})^2\right]^{-1/2} \tag{9.273}$$
$$\times \left\{-\frac{d}{d\theta} - (m - 1/2)\cot\theta + \frac{\kappa}{\sin\theta}\right\}Y_{J\kappa}^m.$$

The corresponding eigenvalues are

$$\sigma + \kappa + 1/4 = (J + 1/2)^2 \tag{9.274}$$

$$J - |m| \ \text{and} \ J - |\kappa| = 0, 1, 2, ... \tag{9.275}$$

so that

$$W = \frac{J(J + 1)h^2}{8\pi^2 A} + \left(\frac{1}{C} - \frac{1}{A}\right)\frac{\kappa^2 h^2}{8\pi^2}. \tag{9.276}$$

9.10   BESSEL FUNCTIONS (TYPE C)

Bessel's equation is given as

$$x^2 J_m''(x) + x J_m'(x) + (\lambda x^2 - m^2)J_m(x) = 0. \tag{9.277}$$

Multiplying this equation by $1/x$, we obtain the first canonical form as

$$\frac{d}{dx}\left[x\frac{dJ_m(x)}{dx}\right] + \left(\lambda x - \frac{m^2}{x}\right)J_m(x) = 0, \tag{9.278}$$

where

$$p(x) = x, \ \text{and} \ w(x) = x. \tag{9.279}$$

A second transformation,

$$\frac{dz}{dx} = \sqrt{\frac{w}{p}} = 1, \tag{9.280}$$

$$J_m = \frac{\Psi}{[wp]^{1/4}} = \frac{\Psi}{\sqrt{x}}, \tag{9.281}$$

gives us the second canonical form

$$\frac{d^2\Psi}{dx^2} + \left[\lambda - \frac{(m^2 - 1/4)}{x^2}\right]\Psi = 0. \tag{9.282}$$

This is type C, and its factorization is given as

$$k(x, m) = \frac{(m - \frac{1}{2})}{x}, \tag{9.283}$$

$$\mu(m) = 0. \tag{9.284}$$

Because $\mu(m)$ is neither a decreasing nor an increasing function of m, we have no limit (upper or lower) to the ladder. We have only the recursion relations

$$\Psi_{m+1} = \frac{1}{\sqrt{x}}\left\{\frac{d}{dx} - \frac{(m + 1/2)}{x}\right\}\Psi_m \tag{9.285}$$

and

$$\Psi_{m-1} = \frac{1}{\sqrt{x}}\left\{-\frac{d}{dx} - \frac{(m - 1/2)}{x}\right\}\Psi_m, \tag{9.286}$$

where

$$\Psi_m = x^{1/2}J_m(\lambda^{1/2}x). \tag{9.287}$$

9.11 HARMONIC OSCILLATOR (TYPE D)

The Schrödinger equation for the harmonic oscillator is given as

$$\frac{d^2\Psi}{d\xi^2} - \xi^2\Psi + \lambda\Psi = 0, \tag{9.288}$$

where $\xi = (\hbar/\mu\omega)^{1/2}x$ and $\lambda = 2E/\hbar\omega$ in terms of the physical variables. This equation can be written in either of the two forms (See Problem 9.14)

$$O_-O_+\Psi_\lambda = (\lambda + 1)\Psi_\lambda \tag{9.289}$$

and

$$O_+O_-\Psi_\lambda = (\lambda - 1)\Psi_\lambda, \tag{9.290}$$

where

$$O_\pm = \pm \frac{d}{d\xi} - \xi \, . \tag{9.291}$$

Operating on Equation (9.289) with $O+$ and on Equation (9.290) with O_- we obtain the analog of Theorem I as

$$\Psi_{\lambda+2} \propto O_+ \Psi_\lambda \tag{9.292}$$

and

$$\Psi_{\lambda-2} \propto O_- \Psi_\lambda. \tag{9.293}$$

Moreover, corresponding to Theorem IV, we find that we can not lower the eigenvalue λ indefinitely. Thus we have a bottom of the ladder

$$\lambda = 2n + 1, \quad n = 0, 1, 2, \dots \, . \tag{9.294}$$

Thus the ground state must satisfy

$$O_- \Psi_0 = 0, \tag{9.295}$$

which determines Ψ_0 as

$$\Psi_0 = \pi^{-1/4} \exp(-\xi^2/2). \tag{9.296}$$

Now the other eigenfunctions can be obtained from

$$\Psi_{n+1} = [2n+2]^{-1/2} O_+ \Psi_n, \tag{9.297}$$

$$\Psi_{n-1} = [2n]^{-1/2} O_- \Psi_n. \tag{9.298}$$

Problems

9.1   Starting from the first canonical form of the Sturm-Liouville equation:

$$\frac{d}{dx}\left[p(x)\frac{d\Psi(x)}{dx}\right] + q(x)\Psi(x) + \lambda w(x)\Psi(x) = 0, \quad x \in [a, b],$$

derive the second canonical form:

$$\frac{d^2 y_\lambda^m(z)}{dz^2} + \{\lambda + r(z, m)\} y_\lambda^m(z) = 0,$$

where

$$r(z, m) = \frac{q}{w} + \frac{3}{16}\left[\frac{1}{w}\frac{dw}{dz} + \frac{1}{p}\frac{dp}{dz}\right]^2$$
$$- \frac{1}{4}\left[\frac{2}{pw}\frac{dp}{dz}\frac{dw}{dz} + \frac{1}{w}\frac{d^2 w}{dz^2} + \frac{1}{p}\frac{d^2 p}{dz^2}\right],$$

by using the transformations

$$y(z) = \Psi(x) \left[w(x) p(x) \right]^{1/4}$$

and

$$dz = dx \left[\frac{w(x)}{p(x)} \right]^{1/2}.$$

9.2   Derive the normalization constants in

$$Y_l^m(\theta, \phi) = \sqrt{\frac{2l+1}{2} \frac{(l-m)!}{(l+1)!} \frac{1}{2\pi}} \left[L_+ \right]^m P_l(\cos \theta)$$

and

$$Y_l^{-m}(\theta, \phi) = \sqrt{\frac{2l+1}{2} \frac{(l-m)!}{(l+1)!} \frac{1}{2\pi}} \left[L_- \right]^m P_l(\cos \theta).$$

9.3   Derive the normalization constant in

$$y_l^{\lambda_l}(\theta) = (-1)^l \left[\frac{(2l+1)!}{2^{2l+1} l!^2} \right]^{1/2} \cdot (\sin \theta)^{(l+\frac{1}{2})}.$$

9.4   Derive Equation (9.195), which is given as

$$\frac{d^2 V(z)}{dz^2} + \left[\lambda + \frac{l(l+1)}{\cosh^2 \theta} \right] V(z) = 0.$$

9.5   The general solution of the differential equation

$$\frac{d^2 y}{dx^2} + \lambda y = 0$$

is given as the linear combination

$$y(x) = C_0 \sin \sqrt{\lambda} x + C_1 \cos \sqrt{\lambda} x.$$

Show that factorization of this equation leads to the trivial result with

$$k(x, m) = 0, \quad \mu(m) = 0,$$

and the corresponding ladder operators just produce other linear combinations of $\sin \sqrt{\lambda} x$ and $\cos \sqrt{\lambda} x$.

9.6   Show that taking

$$k(z, m) = k_0(z) + k_1(z) m + k_2(z) m^2$$

does not lead to any new categories, except the trivial solution given in Problem 9.5. A similar argument works for higher powers of m.

9.7   Show that as long as we admit a finite number of negative powers of m in $k(z, m)$, no new factorization types appear.

9.8   Show that

$$\mu(m) + m^2(k_1^2 + k_1') + 2m(k_0 k_1 + k_0')$$

is a periodic function of m with the period one.
    Use this result to verify

$$\mu(m) - \mu(0) = -m^2(k_1^2 + k_1') - 2m(k_0 k_1 + k_0').$$

9.9   Derive the step-down operator in

$$y_m^{l-1}(\theta) C_{l-1,m} = \frac{C_{lm}}{\sqrt{(l-m)(l+m)}} \left\{ -\sin\theta \frac{d}{d\theta} + \left(l + \frac{1}{2}\right)\cos\theta \right\} y_m^l(\theta).$$

9.10   Follow the same procedure used in Path I in Section 9.6.5 to derive the equation

$$y_{l+1}^{m+1}(\theta) = \frac{C_{l,m}}{C_{l+1,m}} \frac{1}{(l+1-m)} \frac{(l+1-m)}{\sqrt{(l+m+1)(l+m+2)}}$$

$$\times \left\{ \cos\theta \frac{d}{d\theta} - \frac{m+\frac{1}{2}}{\sin\theta} - \left(l + \frac{1}{2}\right)\sin\theta \right\} y_l^m(\theta).$$

9.11   Use the factorization method to show that the spherical Hankel functions of the first kind:

$$h_l^{(1)} = j_l + i n_l$$

can be expressed as

$$h_l^{(1)}(x) = (-1)^l x^l \left[\frac{1}{x}\frac{d}{dx}\right]^l h_0^{(1)}(x)$$

$$= (-1)^l x^l \left[\frac{1}{x}\frac{d}{dx}\right]^l \left(\frac{-ie^{ix}}{x}\right).$$

Hint: Introduce

$$u_l(x) = y_l(x)/x^{l+1}$$

in

$$y_l'' + \left[1 - \frac{l(l+1)}{x^2}\right] y_l = 0.$$

9.12   Using the factorization method, find a recursion relation relating the normalized eigenfunctions $y(n, l, r)$ of the differential equation

$$\frac{d^2 y}{dr^2} + \left[\frac{2}{r} - \frac{l(l+1)}{r^2}\right] y - \frac{1}{n^2} y = 0$$

to the eigenfunctions with $l \pm 1$.

Hint: First show that

$$l = n - 1, n - 2, ..., \quad l = \text{integer}$$

and the normalization is

$$\int_0^\infty y^2(n, l, r) dr = 1.$$

9.13   The harmonic oscillator equation

$$\frac{d^2 \Psi}{dx^2} + (\varepsilon - x^2)\Psi(x) = 0$$

is a rather special case of the factorization method because the operators $O_\pm$ are independent of any parameter.

i) Show that the above equation factorizes as

$$O_+ = \frac{d}{dx} - x$$

and

$$O_- = -\frac{d}{dx} - x.$$

ii) In particular, show that if $\Psi_\varepsilon(x)$ is a solution for the energy eigenvalue ε, then

$$O_+ \Psi_\varepsilon(x)$$

is a solution for $\varepsilon + 2$, while

$$O_- \Psi_\varepsilon(x)$$

is a solution for $\varepsilon - 2$.

iii) Show that ε has a minimum

$$\varepsilon_{\min} = 1,$$

with

$$\varepsilon_n = 2n + 1, \qquad n = 0, 1, 2, ...$$

and show that the $\varepsilon < 0$ eigenvalues are not allowed.

iv) Using the factorization technique, find the eigenfunction corresponding to $\varepsilon_{\min}$ and then use it to express all the remaining eigenfunctions.

Hint: Use the identity

$$\left[\frac{d}{dx} - x\right]\Phi(x) = e^{x^2/2}\frac{d}{dx}\left(e^{-\frac{x^2}{2}}\Phi(x)\right).$$

9.14 Show that the standard method for the harmonic oscillator problem leads to a single ladder with each function on the ladder corresponding to a different eigenvalue λ. This follows from the fact that $r(z, m)$ is independent of m. The factorization we have introduced in Section 9.11 is simpler, and in fact the method of factorization originated from this treatment of the problem.

9.15 The spherical Bessel functions $j_l(x)$ are related to the solutions of

$$\frac{d^2 y_l}{dx^2} + \left[1 - \frac{l(l+1)}{x^2}\right] y_l(x) = 0,$$

(regular at $x = 0$) by

$$j_l(x) = \frac{y_l(x)}{x}.$$

Using the factorization technique, derive recursion formulae
   i) Relating $j_l(x)$ to $j_{l+1}(x)$ and $j_{l-1}(x)$.
   ii) Relating $j_l'(x)$ to $j_{l+1}(x)$ and $j_{l-1}(x)$.

10

COORDINATES and
TENSORS

Starting with a coordinate system is probably the quickest way to introduce mathematics into the study of nature. There are many different ways to choose a coordinate system. Depending on the symmetries of the physical system, a suitable choice not only simplifies the problem but also makes the interpretation of the solution easier. Once a coordinate system is chosen, we can start studying physical processes in terms of mathematical constructs like scalars, vectors, etc. Naturally the true laws of nature do not depend on the coordinate system we use; thus we need a way to express them in coordinate independent formalism. In this regard tensor equations, which preserve their form under general coordinate transformations, have proven to be very useful.

In this chapter we start with the Cartesian coordinates, their transformations, and Cartesian tensors. We then generalize our discussion to generalized coordinates and general tensors. The next stop in our discussion is the coordinate systems in Minkowski spacetime and their transformation properties. We also introduce four-tensors in spacetime and discuss covariance of laws of nature. We finally discuss Maxwell's equations and their transformation properties.

10.1 CARTESIAN COORDINATES

In three-dimensional Euclidean space a Cartesian coordinate system can be constructed by choosing three mutually orthogonal straight lines. A point is defined by giving its coordinates, (x_1, x_2, x_3), or by using the position vector

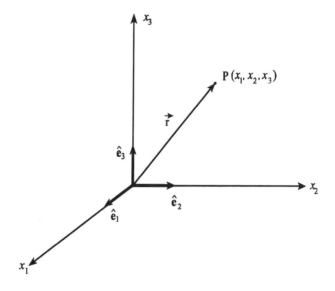

Fig. 10.1 Cartesian coordinate system

$\vec{r}$ as

$$\vec{r} = x_1\widehat{\mathbf{e}}_1 + x_2\widehat{\mathbf{e}}_2 + x_3\widehat{\mathbf{e}}_3 \tag{10.1}$$

$$= (x_1, x_2, x_3), \tag{10.2}$$

where $\widehat{\mathbf{e}}_i$ are unit basis vectors along the coordinate axis (Fig. 10.1). Similarly, an arbitrary vector in Euclidean space can be defined as

$$\vec{a} = a_1\widehat{\mathbf{e}}_1 + a_2\widehat{\mathbf{e}}_2 + a_3\widehat{\mathbf{e}}_3, \tag{10.3}$$

where the magnitude is given as

$$|\vec{a}| = a \tag{10.4}$$

$$= \sqrt{a_1^2 + a_2^2 + a_3^2} \ .$$

10.1.1 Algebra of Vectors

i) Multiplication of a vector with a constant c is done by multiplying each component with that constant:

$$c\,\vec{a} = ca_1\widehat{\mathbf{e}}_1 + ca_2\widehat{\mathbf{e}}_2 + ca_3\widehat{\mathbf{e}}_3, \tag{10.5}$$

$$= (ca_1, ca_2, ca_3).$$

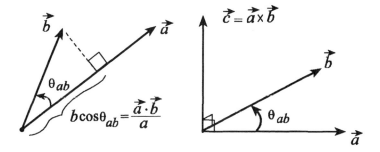

Fig. 10.2 Scalar and vector products

ii) Addition or subtraction is done by adding or subtracting the corresponding components of two vectors:

$$\vec{a} \pm \vec{b} = (a_1 \pm b_1, a_2 \pm b_2, a_3 \pm b_3). \tag{10.6}$$

iii) Multiplication of vectors.
There are two types of vector multiplication:
a) **Dot or scalar product** is defined as

$$(a, b) = \vec{a} \cdot \vec{b} = ab \cos\theta_{ab} \tag{10.7}$$
$$= a_1 b_1 + a_2 b_2 + a_3 b_3,$$

where θ_{ab} is the angle between the two vectors.
b) **Vector product** is defined as

$$\vec{c} = \vec{a} \times \vec{b} \tag{10.8}$$
$$= (a_2 b_3 - a_3 b_2)\hat{e}_1 + (a_3 b_1 - a_1 b_3)\hat{e}_2 + (a_1 b_2 - a_2 b_1)\hat{e}_3 .$$

Using the permutation symbol, we can also write the components of a vector product as

$$c_i = \sum_{j,k=1}^{3} \epsilon_{ijk} a_j b_k, \tag{10.9}$$

where the permutation symbol takes the values

$$\epsilon_{ijk} = \begin{cases} +1 & \text{for cyclic permutations} \\ 0 & \text{when any two indices are equal} \\ -1 & \text{for anticyclic permutations} \end{cases} .$$

The vector product of two vectors is again a vector with the magnitude

$$c = ab \sin\theta_{ab} , \tag{10.10}$$

where the direction is conveniently found by the right-hand rule (Fig. 10.2).

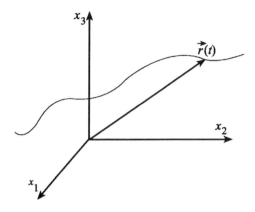

Fig. 10.3 Motion in Cartesian coordinates

10.1.2 Differentiation of Vectors

In a Cartesian coordinate system motion of a particle can be described by giving its position in terms of a parameter, which is usually taken as the time (Fig. 10.3), that is,

$$\vec{r}(t) = (x_1(t), x_2(t), x_3(t)). \tag{10.11}$$

We can now define velocity $\vec{v}$, and acceleration $\vec{a}$ as

$$\vec{v} = \frac{d\vec{r}}{dt} = \lim_{\Delta t \to 0} \frac{\vec{r}(t + \Delta t) - \vec{r}(t)}{\Delta t}, \tag{10.12}$$

$$= \frac{dx_1}{dt}\hat{e}_1 + \frac{dx_2}{dt}\hat{e}_2 + \frac{dx_3}{dt}\hat{e}_3, \tag{10.13}$$

$$\vec{a} = \frac{d\vec{v}}{dt} = \lim_{\Delta t \to 0} \frac{\vec{v}(t + \Delta t) - \vec{v}(t)}{\Delta t}, \tag{10.14}$$

$$= \frac{d^2\vec{r}}{dt^2}. \tag{10.15}$$

The derivative of a general vector is defined similarly. Generalization of these equations to n dimensions is obvious.

10.2 ORTHOGONAL TRANSFORMATIONS

There are many ways to chose the orientation of the Cartesian axes. Symmetries of the physical system often make certain orientations more advantageous than others. In general, we need a dictionary to translate the coordinates

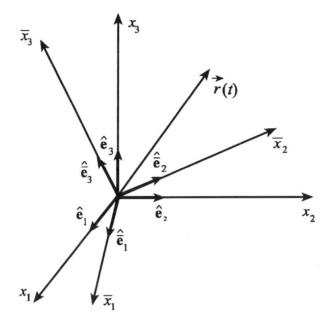

Fig. 10.4 Direction cosines

assigned in one Cartesian system to another. A connection between the coordinates of the position vector assigned by the two sets of Cartesian axes with a common origin can be obtained as (Fig. 10.4)

$$\vec{r} = x_1\hat{e}_1 + x_2\hat{e}_2 + x_3\hat{e}_3 \tag{10.16}$$

$$= \overline{x}_1\hat{\overline{e}}_1 + \overline{x}_2\hat{\overline{e}}_2 + \overline{x}_3\hat{\overline{e}}_3, \tag{10.17}$$

where

$$\overline{x}_1 = \left(\hat{\overline{e}}_1 \cdot \vec{r}\right) = x_1\left(\hat{\overline{e}}_1 \cdot \hat{e}_1\right) + x_2\left(\hat{\overline{e}}_1 \cdot \hat{e}_2\right) + x_3\left(\hat{\overline{e}}_1 \cdot \hat{e}_3\right)$$

$$\overline{x}_2 = \left(\hat{\overline{e}}_2 \cdot \vec{r}\right) = x_1\left(\hat{\overline{e}}_2 \cdot \hat{e}_1\right) + x_2\left(\hat{\overline{e}}_2 \cdot \hat{e}_2\right) + x_3\left(\hat{\overline{e}}_2 \cdot \hat{e}_3\right) \tag{10.18}$$

$$\overline{x}_3 = \left(\hat{\overline{e}}_3 \cdot \vec{r}\right) = x_1\left(\hat{\overline{e}}_3 \cdot \hat{e}_1\right) + x_2\left(\hat{\overline{e}}_3 \cdot \hat{e}_2\right) + x_3\left(\hat{\overline{e}}_3 \cdot \hat{e}_3\right).$$

This can also be written as

$$\overline{x}_1 = (\cos\theta_{11})\,x_1 + (\cos\theta_{12})x_2 + (\cos\theta_{13})x_3$$

$$\overline{x}_2 = (\cos\theta_{21})\,x_1 + (\cos\theta_{22})x_2 + (\cos\theta_{23})x_3 \tag{10.19}$$

$$\overline{x}_3 = (\cos\theta_{31})\,x_1 + (\cos\theta_{32})x_2 + (\cos\theta_{33})x_3,$$

where $\cos\theta_{ij}$ are called the direction cosines defined as

$$\cos\theta_{ij} = \widehat{\overline{\mathbf{e}}}_i \cdot \widehat{\mathbf{e}}_j. \tag{10.20}$$

Note that the first unit basis vector is always taken as the barred system, that is,

$$\cos\theta_{ji} = \widehat{\overline{\mathbf{e}}}_j \cdot \widehat{\mathbf{e}}_i. \tag{10.21}$$

The transformation equations obtained for the position vector are also true for an arbitrary vector $\overrightarrow{a}$ as

$$
\begin{aligned}
\overline{a}_1 &= (\cos\theta_{11})\,a_1 + (\cos\theta_{12})a_2 + (\cos\theta_{13})a_3 \\
\overline{a}_2 &= (\cos\theta_{21})\,a_1 + (\cos\theta_{22})a_2 + (\cos\theta_{23})a_3 \\
\overline{a}_3 &= (\cos\theta_{31})\,a_1 + (\cos\theta_{32})a_2 + (\cos\theta_{33})a_3 \ .
\end{aligned}
\tag{10.22}
$$

The transformation equations given in Equation (10.22) are the special case of general linear transformation, which can be written as

$$
\begin{aligned}
\overline{x}_1 &= a_{11}x_1 + a_{12}x_2 + a_{13}x_3 \\
\overline{x}_2 &= a_{21}x_1 + a_{22}x_2 + a_{23}x_3 \\
\overline{x}_3 &= a_{31}x_1 + a_{32}x_2 + a_{33}x_3 \ .
\end{aligned}
\tag{10.23}
$$

a_{ij} are constants independent of $\overrightarrow{r}$ and $\overrightarrow{\overline{r}}$. A convenient way to write Equation (10.23) is

$$\overline{x}_i = a_{ij}x_j, \quad i,j = 1,2,3\ , \tag{10.24}$$

where we have used the Einstein summation convention, which implies summation over the repeated (dummy) indices, that is, Equation (10.24) means

$$\overline{x}_i = \sum_{j=1}^{3} a_{ij}x_j \tag{10.25}$$

$$= a_{ij}x_j, \quad i,j = 1,2,3. \tag{10.26}$$

Unless otherwise stated, we use the Einstein summation convention. Magnitude of $\overrightarrow{r}$ in this notation is shown as

$$r = \sqrt{x_i x_i}. \tag{10.27}$$

Using matrices, transformation Equations (10.23) can also be written as

$$\overline{\mathbf{r}} = \mathbf{A}\mathbf{r}, \tag{10.28}$$

where $\mathbf{r}$ and $\overline{\mathbf{r}}$ are represented by the column matrices

$$
\mathbf{r} = \begin{bmatrix} x_1 \\ x_2 \\ x_3 \end{bmatrix}
\quad \text{and} \quad
\overline{\mathbf{r}} = \begin{bmatrix} \overline{x}_1 \\ \overline{x}_2 \\ \overline{x}_3 \end{bmatrix},
\tag{10.29}
$$

and the transformation matrix $\mathbf{A}$ is represented by the square matrix

$$\mathbf{A} = \begin{bmatrix} a_{11} & a_{12} & a_{13} \\ a_{21} & a_{22} & a_{23} \\ a_{31} & a_{32} & a_{33} \end{bmatrix} . \tag{10.30}$$

We use both boldface letter $\mathbf{r}$ and $\overline{r}$ to denote a vector. Generalization of these formulas to n dimensions is again obvious. Transpose of a matrix is obtained by interchanging its rows and columns as

$$\widetilde{\mathbf{r}} = \begin{bmatrix} x_1 & x_2 & x_3 \end{bmatrix} \tag{10.31}$$

and

$$\widetilde{\mathbf{A}} = \begin{bmatrix} a_{11} & a_{21} & a_{31} \\ a_{12} & a_{22} & a_{32} \\ a_{13} & a_{23} & a_{33} \end{bmatrix} . \tag{10.32}$$

We can now write the magnitude of a vector as

$$\widetilde{\mathbf{r}}\mathbf{r} = \begin{bmatrix} x_1 & x_2 & x_3 \end{bmatrix} \begin{bmatrix} x_1 \\ x_2 \\ x_3 \end{bmatrix} \tag{10.33}$$

$$= x_1^2 + x_2^2 + x_3^2 .$$

The magnitude of $\overline{\mathbf{r}}$ is now given as

$$\widetilde{\overline{\mathbf{r}}}\,\overline{\mathbf{r}} = \widetilde{\mathbf{r}}\left(\widetilde{\mathbf{A}}\mathbf{A}\right)\mathbf{r}, \tag{10.34}$$

where we have used the matrix property

$$\widetilde{\mathbf{A}\mathbf{B}} = \widetilde{\mathbf{B}}\widetilde{\mathbf{A}} . \tag{10.35}$$

From Equation (10.34) it is seen that linear transformations that preserve the length of a vector must satisfy the condition

$$\widetilde{\mathbf{A}}\mathbf{A} = \mathbf{I}, \tag{10.36}$$

where $\mathbf{I}$ is the identity matrix

$$\mathbf{I} = \begin{bmatrix} 1 & 0 & 0 \\ 0 & 1 & 0 \\ 0 & 0 & 1 \end{bmatrix} . \tag{10.37}$$

Such transformations are called orthogonal transformations. In terms of components the orthogonality condition [Eq. (10.36)] can be written as

$$a_{ij}a_{jk} = \delta_{ik}. \tag{10.38}$$

Taking the determinant of the orthogonality relation, we see that the determinant of transformations that preserve the length of a vector satisfies

$$[Det\mathbf{A}]^2 = 1, \tag{10.39}$$

thus

$$Det\mathbf{A} = \pm 1. \tag{10.40}$$

Orthogonal transformations are basically transformations among Cartesian coordinates without a scale change. Transformations with $Det\mathbf{A} = 1$ are called **proper transformations**. They are composed of rotations and translations. Transformations with $Det\mathbf{A} = -1$ are called **improper transformations,** and they involve reflections.

10.2.1 Rotations About Cartesian Axes

For rotations about the x_3-axis the rotation matrix takes the form

$$\mathbf{R}_3 = \begin{bmatrix} a_{11} & a_{12} & 0 \\ a_{21} & a_{22} & 0 \\ 0 & 0 & 1 \end{bmatrix}. \tag{10.41}$$

Using the direction cosines we can write $R_3(\theta)$ for counterclockwise rotations as (Fig. 10.5)

$$\mathbf{R}_3(\theta) = \begin{bmatrix} \cos\theta & \sin\theta & 0 \\ -\sin\theta & \cos\theta & 0 \\ 0 & 0 & 1 \end{bmatrix}. \tag{10.42}$$

Similarly, the rotation matrices corresponding to counterclockwise rotations about the x_1- and x_2-axis can be written, respectively, as

$$\mathbf{R}_1(\phi) = \begin{bmatrix} 1 & 0 & 0 \\ 0 & \cos\phi & \sin\phi \\ 0 & -\sin\phi & \cos\phi \end{bmatrix} \text{ and } \mathbf{R}_2(\psi) = \begin{bmatrix} \cos\psi & 0 & -\sin\psi \\ 0 & 1 & 0 \\ \sin\psi & 0 & \cos\psi \end{bmatrix}. \tag{10.43}$$

10.3 FORMAL PROPERTIES OF THE ROTATION MATRIX

i) Two sequentially performed rotations, say $\mathbf{A}$ and $\mathbf{B}$, is equivalent to another rotation $\mathbf{C}$ as

$$\mathbf{C} = \mathbf{AB}. \tag{10.44}$$

ii) Because matrix multiplications do not commute, the order of rotations is important, that is, in general

$$\mathbf{AB} \neq \mathbf{BA}. \tag{10.45}$$

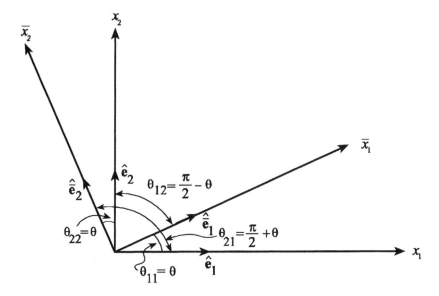

Fig. 10.5 Direction cosines

However, the associative law,

$$A(BC) = (AB)C, \qquad (10.46)$$

holds between any three rotations A, B, and C.

iii) The inverse transformation matrix A^{-1} exists, and from the orthogonality relation [Eq. (10.36)], it is equal to the transpose of A, that is,

$$A^{-1} = \tilde{A} . \qquad (10.47)$$

Thus for orthogonal transformations we can write

$$\tilde{A}A = A\tilde{A} = I. \qquad (10.48)$$

10.4   EULER ANGLES AND ARBITRARY ROTATIONS

The most general rotation matrix has nine components (10.30). However, the orthogonality relation $\mathbf{A}\widetilde{\mathbf{A}} = \mathbf{I}$, written explicitly as

$$
\begin{aligned}
a_{11}^2 + a_{12}^2 + a_{13}^2 &= 1 \\
a_{22}^2 + a_{21}^2 + a_{23}^2 &= 1 \\
a_{33}^2 + a_{31}^2 + a_{32}^2 &= 1 \\
a_{11}a_{21} + a_{12}a_{22} + a_{13}a_{23} &= 0 \\
a_{11}a_{31} + a_{12}a_{32} + a_{13}a_{33} &= 0 \\
a_{21}a_{31} + a_{22}a_{32} + a_{23}a_{33} &= 0,
\end{aligned}
\tag{10.49}
$$

gives six relations among these components. Hence, only three of them can be independent. In the study of rotating systems to describe the orientation of a system it is important to define a set of three independent parameters. There are a number of choices. The most common and useful are the three Euler angles. They correspond to three successive rotations about the Cartesian axes so that the final orientation of the system is obtained. The convention we follow is the most widely used one in applied mechanics, in celestial mechanics, and frequently, in molecular and solid-state physics. For different conventions, we refer the reader to Goldstein et al.

The sequence starts with a counterclockwise rotation by ϕ about the x_3-axis of the initial state of the system as

$$
\mathbf{B}(\phi) : (x_1, x_2, x_3) \rightarrow (x_1', x_2', x_3').
\tag{10.50}
$$

This is followed by a counterclockwise rotation by θ about the x_1' of the intermediate axis as

$$
\mathbf{C}(\theta) : (x_1', x_2', x_3') \rightarrow (x_1'', x_2'', x_3'').
\tag{10.51}
$$

Finally, the desired orientation is achieved by a counterclockwise rotation about the x_3''-axis by ψ as

$$
\mathbf{D}(\psi) : (x_1'', x_2'', x_3'') \rightarrow (\overline{x}_1, \overline{x}_2, \overline{x}_3).
\tag{10.52}
$$

$A(\phi)$, $B(\theta)$, and $C(\psi)$ are the rotation matrices for the corresponding transformations, which are given as

$$B(\phi) = \begin{bmatrix} \cos\phi & \sin\phi & 0 \\ -\sin\phi & \cos\phi & 0 \\ 0 & 0 & 1 \end{bmatrix}, \qquad (10.53)$$

$$C(\phi) = \begin{bmatrix} 1 & 0 & 0 \\ 0 & \cos\theta & \sin\theta \\ 0 & -\sin\theta & \cos\theta \end{bmatrix}, \qquad (10.54)$$

$$D(\psi) = \begin{bmatrix} \cos\psi & \sin\psi & 0 \\ -\sin\psi & \cos\psi & 0 \\ 0 & 0 & 1 \end{bmatrix}. \qquad (10.55)$$

In terms of the individual rotations, elements of the complete transformation matrix can be written as

$$A = DCB, \qquad (10.56)$$

$$A = \qquad (10.57)$$

$$\begin{bmatrix} \cos\psi\cos\phi - \cos\theta\sin\phi\sin\psi & \cos\psi\sin\phi + \cos\theta\cos\phi\sin\psi & \sin\psi\sin\theta \\ -\sin\psi\cos\phi - \cos\theta\sin\phi\cos\psi & -\sin\psi\sin\phi + \cos\theta\cos\phi\cos\psi & \cos\psi\sin\theta \\ \sin\theta\sin\phi & -\sin\theta\cos\phi & \cos\theta \end{bmatrix}.$$

The inverse of A is

$$A^{-1} = \tilde{A}. \qquad (10.58)$$

We can also consider the elements of the rotation matrix as a function of some single parameter t and write

$$\phi = \omega_\phi t, \quad \omega = \omega_\theta t, \quad \psi = \omega_\psi t.$$

If t is taken as time, ω can be interpreted as the constant angular velocity about the corresponding axis. Now, in general the rotation matrix can be written as

$$A(t) = \begin{bmatrix} a_{11}(t) & a_{12}(t) & a_{13}(t) \\ a_{21}(t) & a_{22}(t) & a_{23}(t) \\ a_{31}(t) & a_{32}(t) & a_{33}(t) \end{bmatrix}. \qquad (10.59)$$

Using trigonometric identities it can be shown that

$$A(t_2 + t_1) = A(t_2)A(t_1). \qquad (10.60)$$

Differentiating with respect to t_2 and putting $t_2 = 0$ and $t_1 = t$, we obtain a result that will be useful to us shortly as

$$A'(t) = A'(0)A(t). \qquad (10.61)$$

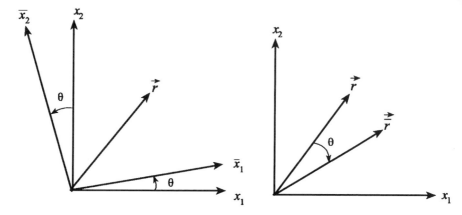

Fig. 10.6   Passive and active views of the rotation matrix

10.5   ACTIVE AND PASSIVE INTERPRETATIONS OF ROTATIONS

It is possible to view the rotation matrix $\mathbf{A}$ in

$$\bar{\mathbf{r}} = \mathbf{A}\mathbf{r} \qquad (10.62)$$

as an operator acting on $\mathbf{r}$ and rotating it in the opposite direction, while keeping the coordinate axes fixed (Fig. 10.6) . This is called the **active view**. The case where the coordinate axes are rotated is called the **passive view**. In principle both the active and passive views lead to the same result. However, as in quantum mechanics, sometimes the active view may offer some advantages in studying the symmetries of a physical system.

In the case of the active view, we also need to know how an operator $\mathbf{A}$ transforms under coordinate transformations. Considering a transformation represented by the matrix $\mathbf{B}$, we multiply both sides of Equation (10.62) by $\mathbf{B}$ to write

$$\mathbf{B}\bar{\mathbf{r}} = \mathbf{B}\mathbf{A}\mathbf{r}. \qquad (10.63)$$

Using

$$\mathbf{B}\mathbf{B}^{-1} = \mathbf{B}^{-1}\mathbf{B} = \mathbf{I}, \qquad (10.64)$$

we now write Equation (10.63) as

$$\mathbf{B}\bar{\mathbf{r}} = \mathbf{B}\mathbf{A}\mathbf{B}^{-1}\mathbf{B}\mathbf{r}, \qquad (10.65)$$

$$\bar{\mathbf{r}}' = \mathbf{A}'\mathbf{r}'. \qquad (10.66)$$

In the new coordinate system $\bar{\mathbf{r}}$ and $\mathbf{r}$ are related by

$$\bar{\mathbf{r}}' = \mathbf{B}\bar{\mathbf{r}} \qquad (10.67)$$

and

$$\mathbf{r}' = \mathbf{B}\mathbf{r}. \tag{10.68}$$

Thus the operator $\mathbf{A}'$ becomes

$$\mathbf{A}' = \mathbf{B}\mathbf{A}\mathbf{B}^{-1}. \tag{10.69}$$

This is called **similarity transformation.** If $\mathbf{B}$ is an orthogonal transformation, we then write

$$\mathbf{A}' = \mathbf{B}\mathbf{A}\widetilde{\mathbf{B}}. \tag{10.70}$$

In terms of components this can also be written as

$$a'_{ij} = b_{ik}b_{lj}a_{kl}. \tag{10.71}$$

10.6   INFINITESIMAL TRANSFORMATIONS

A proper orthogonal transformation depending on a single continuous parameter t can be shown as

$$\mathbf{r}(t) = \mathbf{A}(t)\mathbf{r}(0). \tag{10.72}$$

Differentiating and using Equation (10.61) we obtain

$$\frac{d\mathbf{r}(t)}{dt} = \mathbf{A}'(t)\mathbf{r}(0) \tag{10.73}$$

$$= \mathbf{A}'(0)\mathbf{A}(t)\mathbf{r}(0) \tag{10.74}$$

$$= \mathbf{X}\mathbf{r}(t), \tag{10.75}$$

where

$$\mathbf{X} = \mathbf{A}'(0). \tag{10.76}$$

Differentiating Equation (10.75) we can now obtain the higher-order derivatives as

$$\frac{d^2\mathbf{r}(t)}{dt^2} = \mathbf{X}^2\mathbf{r}(t),$$

$$\frac{d^3\mathbf{r}(t)}{dt^3} = \mathbf{X}^3\mathbf{r}(t), \tag{10.77}$$

$$\vdots$$

Using these in the Taylor series expansion of $\mathbf{r}(t)$ about $t = 0$ we write

$$\mathbf{r}(t) = \mathbf{r}(0) + \frac{d\mathbf{r}(0)}{dt}t + \frac{1}{2!}\frac{d^2\mathbf{r}(0)}{dt^2}t^2 + \cdots, \tag{10.78}$$

thus obtaining

$$\mathbf{r}(t) = (\mathbf{I} + \mathbf{X}t + \frac{1}{2!}\mathbf{X}^2 t^2 + \cdots)\mathbf{r}(0) \ . \tag{10.79}$$

This series converges, yielding

$$\mathbf{r}(t) = \exp(\mathbf{X}t)\mathbf{r}(0) \ . \tag{10.80}$$

This is called the exponential form of the transformation matrix. For infinitesimal transformations t is small; hence we can write

$$\mathbf{r}(t) \simeq (\mathbf{I} + \mathbf{X}t)\mathbf{r}(0), \tag{10.81}$$

$$\mathbf{r}(t) - \mathbf{r}(0) \simeq \mathbf{X}t\mathbf{r}(0), \tag{10.82}$$

$$\delta\mathbf{r} \simeq \mathbf{X}t\mathbf{r}(0), \tag{10.83}$$

where $\mathbf{X}$ is called the generator of the infinitesimal transformation.

Using the definition of $\mathbf{X}$ in Equation (10.76) and the rotation matrices [Eqs. (10.42) and (10.43)] we obtain the generators of infinitesimal rotations about the x_1-, x_2-, and x_3-axes, respectively as

$$\mathbf{X}_1 = \begin{bmatrix} 0 & 0 & 0 \\ 0 & 0 & 1 \\ 0 & -1 & 0 \end{bmatrix}, \ \mathbf{X}_2 = \begin{bmatrix} 0 & 0 & -1 \\ 0 & 0 & 0 \\ 1 & 0 & 0 \end{bmatrix}, \ \mathbf{X}_3 = \begin{bmatrix} 0 & 1 & 0 \\ -1 & 0 & 0 \\ 0 & 0 & 0 \end{bmatrix}. \tag{10.84}$$

An arbitrary infinitesimal rotation by the amounts t_1, t_2, and t_3 about their respective axes can be written as

$$\mathbf{r} = (\mathbf{I} + \mathbf{X}_3 t_3)(\mathbf{I} + \mathbf{X}_2 t_2)(\mathbf{I} + \mathbf{X}_1 t_1)\mathbf{r}(0)$$
$$= (\mathbf{I} + \mathbf{X}_3 t_3 + \mathbf{X}_2 t_2 + \mathbf{X}_1 t_1)\mathbf{r}(0). \tag{10.85}$$

Defining the vector

$$\mathbf{X} = \mathbf{X}_1 \widehat{\mathbf{e}}_1 + \mathbf{X}_2 \widehat{\mathbf{e}}_2 + \mathbf{X}_3 \widehat{\mathbf{e}}_3$$
$$= (\mathbf{X}_1, \mathbf{X}_2, \mathbf{X}_3) \tag{10.86}$$

and the unit vector

$$\widehat{\mathbf{n}} = \frac{1}{\sqrt{t_1^2 + t_2^2 + t_3^2}} \begin{bmatrix} t_1 \\ t_2 \\ t_3 \end{bmatrix}, \tag{10.87}$$

we can write Equation (10.85) as

$$\mathbf{r}(t) = (\mathbf{I} + \mathbf{X}.\widehat{\mathbf{n}}t)\mathbf{r}(0), \tag{10.88}$$

where

$$t = \sqrt{t_1^2 + t_2^2 + t_3^2}. \tag{10.89}$$

This is an infinitesimal rotation about an axis in the direction $\widehat{\mathbf{n}}$ by the amount t. For finite rotations we write

$$\mathbf{r}(t) = e^{\mathbf{X}.\widehat{\mathbf{n}}t}\mathbf{r}(0). \tag{10.90}$$

10.6.1 Infinitesimal Transformations Commute

Two successive infinitesimal transformations by the amounts t_1 and t_2 can be written as

$$\mathbf{r} = (\mathbf{I} + \mathbf{X}_2 t_2)(\mathbf{I} + \mathbf{X}_1 t_1)\mathbf{r}(0) \ , \tag{10.91}$$
$$= [\mathbf{I} + (t_1 \mathbf{X}_1 + t_2 \mathbf{X}_2)] \ \mathbf{r}(0) \ .$$

Because matrices commute with respect to addition and subtraction, infinitesimal transformations also commute, that is

$$\mathbf{r} = [\mathbf{I} + (t_2 \mathbf{X}_2 + t_1 \mathbf{X}_1)] \ \mathbf{r}(0) \tag{10.92}$$
$$= (\mathbf{I} + \mathbf{X}_1 t_1)(\mathbf{I} + \mathbf{X}_2 t_2)\mathbf{r}(0) \ .$$

For finite rotations this is clearly not true. Using Equation (10.43) we can write the rotation matrix for a rotation about the x_2-axis followed by a rotation about the x_1-axis as

$$\mathbf{R}_1 \mathbf{R}_2 = \begin{bmatrix} \cos\psi & 0 & -\sin\psi \\ \sin\phi\sin\psi & \cos\phi & \sin\phi\cos\psi \\ \cos\phi\sin\psi & -\sin\phi & \cos\phi\cos\psi \end{bmatrix} \ . \tag{10.93}$$

Reversing the order we get

$$\mathbf{R}_2 \mathbf{R}_1 = \begin{bmatrix} \cos\psi & \sin\psi\sin\phi & -\sin\psi\cos\phi \\ 0 & \cos\phi & \sin\phi \\ \sin\psi & -\cos\psi\sin\phi & \cos\psi\cos\phi \end{bmatrix} \ . \tag{10.94}$$

It is clear that for finite rotations these two matrices are not equal:

$$\mathbf{R}_1 \mathbf{R}_2 \neq \mathbf{R}_2 \mathbf{R}_1. \tag{10.95}$$

However, for small rotations, say by the amounts $\delta\psi$ and $\delta\phi$, we can use the approximations

$$\sin\delta\psi \simeq \delta\psi, \ \sin\delta\phi \simeq \delta\phi \tag{10.96}$$
$$\cos\delta\psi \simeq 1, \ \cos\delta\phi \simeq 1$$

to find

$$\mathbf{R}_1 \mathbf{R}_2 = \begin{bmatrix} 1 & 0 & -\delta\psi \\ \delta\psi & 1 & \delta\phi \\ \delta\psi & -\delta\phi & 1 \end{bmatrix} = \mathbf{R}_2 \mathbf{R}_1. \tag{10.97}$$

Note that in terms of the generators [Eq. (10.84)] we can also write this as

$$\mathbf{R}_1 \mathbf{R}_2 = \begin{bmatrix} 1 & 0 & 0 \\ 0 & 1 & 0 \\ 0 & 0 & 1 \end{bmatrix} + \delta\psi \begin{bmatrix} 0 & 0 & -1 \\ 0 & 0 & 0 \\ 1 & 0 & 0 \end{bmatrix} + \delta\phi \begin{bmatrix} 0 & 0 & 0 \\ 0 & 0 & 1 \\ 0 & -1 & 0 \end{bmatrix}$$
$$= \mathbf{I} + \delta\psi \mathbf{X}_1 + \delta\phi \mathbf{X}_2$$
$$= \mathbf{I} + \delta\phi \mathbf{X}_2 + \delta\psi \mathbf{X}_1$$
$$= \mathbf{R}_2 \mathbf{R}_1, \tag{10.98}$$

which again proves that infinitesimal rotations commute.

10.7 CARTESIAN TENSORS

Certain physical properties like temperature and mass can be described completely by giving a single number. They are called scalars. Under orthogonal transformations scalars preserve their value. Distance, speed, and charge are other examples of scalars. On the other hand, vectors in three dimensions require three numbers for a complete description, that is, their x_1, x_2, and x_3 components. Under orthogonal transformations we have seen that vectors transform as

$$a'_i = A_{ij} a_j. \tag{10.99}$$

There are also physical properties that in three dimensions require nine components for a complete description. For example, stresses in a solid have nine components that can be conveniently represented as a 3×3 matrix:

$$t_{ij} = \begin{bmatrix} t_{11} & t_{12} & t_{13} \\ t_{21} & t_{22} & t_{23} \\ t_{31} & t_{32} & t_{33} \end{bmatrix}. \tag{10.100}$$

Components t_{ij} correspond to the forces acting on a unit area element, that is, t_{ij} is the ith component of the force acting on a unit area element when its normal is pointing along the jth axis. Under orthogonal transformations stresses transform as

$$t'_{ij} = A_{ik} A_{jl} t_{kl}. \tag{10.101}$$

Stresses, vectors, and scalars are special cases of a more general type of objects called tensors.

Cartesian tensors in general are defined in terms of their transformation properties under orthogonal transformations as

$$T_{ijk...} = A_{ii'} A_{jj'} A_{kk'} ... T_{i'j'k'...}.$$

All indices take the values $1, 2, 3, ..., n$, where n is the dimension of space. An important property of tensors is their **rank**, which is equal to the number of free indices. In this regard, scalars are tensors of zeroth rank, vectors are tensors of first rank, and stress tensor is a second-rank tensor.

10.7.1 Operations with Cartesian Tensors

i) Multiplication with a constant is accomplished by multiplying each component of the tensor with that constant.

ii) Addition/subtraction of tensors of equal rank can be done by adding/subtracting the two tensors term by term.

iii) Rank of a composite tensor is equal to the number of its free indices. For example,

$$A_{ikj}B_{jlm} \qquad (10.102)$$

is a fourth-rank tensor,

$$A_{ijk}B_{ijk} \qquad (10.103)$$

is a scalar, and

$$A_{ijkl}B_{jkl} \qquad (10.104)$$

is a vector.

iv) We can obtain a lower-rank tensor by contracting some of the indices of a tensor or by contracting the indices of a tensor with another tensor. For example,

$$\begin{aligned}
A_{ij} &= A_{ikkj} \\
A_{ijk} &= D_{ijklm}B_{lm} \\
A_{ij} &= C_{ijk}D_k
\end{aligned} \qquad (10.105)$$

For a second-rank tensor, by contracting the two indices we obtain a scalar called the **trace**

$$A = A_{ii} = A_{11} + A_{22} + A_{33} + \cdots + A_{nn}. \qquad (10.106)$$

v) In a tensor equation, rank of both sides must match

$$A_{ij\ldots n} = B_{ij\ldots n} \qquad (10.107)$$

vi) We have seen that tensors are defined with respect to their transformation properties. For example, from two vectors a_i and b_j we can form a second-rank tensor t_{ij} as

$$t_{ij} = a_i b_j. \qquad (10.108)$$

This is also called the **outer product** of two vectors. The fact that t_{ij} is a second-rank tensor can easily be verified by checking its transformation properties under orthogonal transformations.

10.7.2   Tensor Densities or Pseudotensors

Let us now consider the Kronecker delta, which is defined in all coordinates as

$$\delta_{ij} = \begin{cases} 1 & \text{for } i = j \\ 0 & \text{for } i \neq j \end{cases}. \qquad (10.109)$$

To see that it is a second-rank tensor we check how it transforms under orthogonal transformations, that is,

$$\delta'_{ij} = a_{ik}a_{jl}\delta_{kl}$$
$$= a_{ik}a_{kj}. \tag{10.110}$$

From the orthogonality relation [Eq. (10.38)] this gives

$$\delta'_{ij} = \delta_{ij}. \tag{10.111}$$

Hence the Kronecker delta is a second-rank tensor.

Let us now investigate the tensor property of the permutation or the Levi-Civita symbol. It is defined in all coordinates as

$$\epsilon_{ijk} = \begin{cases} +1 & \text{for cyclic permutations} \\ 0 & \text{when any two indices are equal} \\ -1 & \text{for anticyclic permutations} \end{cases} . \tag{10.112}$$

For ϵ_{ijk} to be a third-rank tensor it must transform as

$$\epsilon'_{ijk} = a_{il}a_{jm}a_{kn}\epsilon_{lmn}$$
$$= \epsilon_{ijk}. \tag{10.113}$$

However, using the definition of a determinant, one can show that the right-hand side is $\epsilon_{ijk} \det a$; thus if we admit improper transformations where $\det a = -1$, ϵ_{ijk} is not a tensor. A tensor that transforms according to the law

$$T'_{ijk...} = a_{il}a_{jm}a_{kn}\cdots T_{lmn...}\det a \tag{10.114}$$

is called a pseudotensor or a tensor density.

The cross product of two vectors

$$\overrightarrow{c} = \overrightarrow{a} \times \overrightarrow{b}, \tag{10.115}$$

which in terms of coordinates can be written as

$$c_i = \epsilon_{ijk}a_jb_k, \tag{10.116}$$

is a pseudovector, whereas the triple product

$$\overrightarrow{c} \cdot (\overrightarrow{a} \times \overrightarrow{b}) = \epsilon_{ijk}c_i\,a_jb_k \tag{10.117}$$

is a pseudoscalar.

10.8 GENERALIZED COORDINATES AND GENERAL TENSORS

So far we have confined our discussion to Cartesian tensors, which are defined with respect to their transformation properties under orthogonal transformations. However, the presence of symmetries in the physical system often makes other coordinate systems more practical. For example, in central force problems it is advantageous to work with the spherical polar coordinates, which reflect the spherical symmetry of the system best. For axially symmetric problems use of the cylindrical coordinates simplifies equations significantly. Usually symmetries indicate which coordinate system to use. However, in less obvious cases, discovering the symmetries and their generators can help us to construct the most advantageous coordinate system for the problem at hand. We now extend our discussion of Cartesian coordinates and Cartesian tensors to generalized coordinates and general tensors. These definitions can also be used for defining tensors in spacetime and also for tensors in curved spaces.

A general coordinate transformation can be defined as

$$\overline{x}^i = \overline{x}^i(x^1, x^2, ..., x^n), \quad i = 1, ..., n. \tag{10.118}$$

In short, we write this as

$$\overline{x}^i = \overline{x}^i(x^k), \quad i, k = 1, ..., n. \tag{10.119}$$

The inverse transformation is defined as

$$x^k = x^k(\overline{x}^i), \quad i, k = 1, ..., n. \tag{10.120}$$

For reasons to become clear later we have written all the indices as superscripts. Differentiating Equation (10.119) we can write the transformation law for infinitesimal displacements as

$$d\overline{x}^i = \sum_{k=1}^{n} \frac{\partial \overline{x}^i}{\partial x^k} dx^k \tag{10.121}$$

$$= \sum_{k=1}^{n} \left[\frac{\partial \overline{x}^i}{\partial x^k} \right] dx^k.$$

We now consider a scalar function, $\phi(x^i)$, and differentiate with respect to $\overline{x}^i$ to write

$$\frac{\partial \phi}{\partial \overline{x}^i} = \sum_{k=1}^{n} \frac{\partial \phi}{\partial x^k} \frac{\partial x^k}{\partial \overline{x}^i} \tag{10.122}$$

$$= \sum_{k=1}^{n} \left[\frac{\partial x^k}{\partial \overline{x}^i} \right] \frac{\partial \phi}{\partial x^k}. \tag{10.123}$$

Until we reestablish the Einstein summation convention for general tensors, we write the summation signs explicitly.

10.8.1   Contravariant and Covariant Components

Using the transformation properties of the infinitesimal displacements and the gradient of a scalar function we now define contravariant and covariant components. A contravariant component is defined with respect to the transformation rule

$$\bar{a}^i = \sum_{k=1}^{n} \frac{\partial \bar{x}^i}{\partial x^k} a^k \tag{10.124}$$

$$= \sum_{k=1}^{n} \left[\frac{\partial \bar{x}^i}{\partial x^k} \right] a^k, \tag{10.125}$$

where the inverse transformation is defined as

$$a^k = \sum_{i=1}^{n} \left[\frac{\partial x^k}{\partial \bar{x}^i} \right] \bar{a}^i. \tag{10.126}$$

We also define a covariant component according to the transformation rule

$$\bar{a}_i = \sum_{k=1}^{n} \left[\frac{\partial x^k}{\partial \bar{x}^i} \right] a_k, \tag{10.127}$$

where the components are now shown as subscripts. The inverse transformation is written as

$$a_k = \sum_{i=1}^{n} \left[\frac{\partial \bar{x}^i}{\partial x^k} \right] \bar{a}_i. \tag{10.128}$$

A second-rank tensor can be contravariant, covariant, or with mixed indices with the following transformation properties:

$$\bar{T}^{ij} = \sum_{k=1}^{n} \sum_{l=1}^{n} \left[\frac{\partial \bar{x}^i}{\partial x^k} \frac{\partial \bar{x}^j}{\partial x^l} \right] T^{kl}, \tag{10.129}$$

$$\bar{T}_{ij} = \sum_{k=1}^{n} \sum_{l=1}^{n} \left[\frac{\partial x^k}{\partial \bar{x}^i} \frac{\partial x^l}{\partial \bar{x}^j} \right] T_{kl}, \tag{10.130}$$

$$\bar{T}^i_j = \sum_{k=1}^{n} \sum_{l=1}^{n} \left[\frac{\partial \bar{x}^i}{\partial x^k} \frac{\partial x^l}{\partial \bar{x}^j} \right] T^k_l. \tag{10.131}$$

Similarly, a general tensor can be defined with mixed indices as

$$\bar{T}^{i_1 i_2 \ldots}_{j_1 j_2 \ldots} \tag{10.132}$$

$$= \sum_{k_1=1}^{n} \sum_{k_2=1}^{n} \cdots \sum_{l_1=1}^{n} \sum_{l_2=1}^{n} \cdots \left[\frac{\partial \bar{x}^{i_1}}{\partial x^{k_1}} \frac{\partial \bar{x}^{i_2}}{\partial x^{k_2}} \cdots \frac{\partial x^{l_1}}{\partial \bar{x}^{j_1}} \frac{\partial x^{l_2}}{\partial \bar{x}^{j_2}} \cdots \right] T^{k_1 k_2 \ldots}_{l_1 l_2 \ldots}.$$

Using Equations (10.128) and (10.127) we write

$$a_k = \sum_{i=1}^{n} \left[\frac{\partial \overline{x}^i}{\partial x^k} \right] \overline{a}_i \tag{10.133}$$

$$= \sum_{k'=1}^{n} \left[\sum_{i=1}^{n} \frac{\partial \overline{x}^i}{\partial x^k} \frac{\partial x^{k'}}{\partial \overline{x}^i} \right] u_{k'} \tag{10.134}$$

$$= \sum_{k'=1}^{n} \delta_k^{k'} a_{k'} \tag{10.135}$$

$$= a_k, \tag{10.136}$$

where $\delta_k^{k'}$ is the Kronecker delta, which is a second-rank tensor with the transformation property

$$\overline{\delta}_j^i = \sum_{k=1}^{n} \sum_{l=1}^{n} \left[\frac{\partial \overline{x}^i}{\partial x^k} \frac{\partial x^l}{\partial \overline{x}^j} \right] \delta_l^k \tag{10.137}$$

$$= \delta_j^i. \tag{10.138}$$

It is the only second-rank tensor with this property.

10.8.2   Metric Tensor and the Line Element

Let us now see how the distance between two infinitesimally close points transforms under general coordinate transformations. We take our unbarred coordinate system as the Cartesian coordinates; hence the line element that gives the square of the distance between two infinitesimally close points is

$$ds^2 = \sum_{k=1}^{n} dx^k dx^k. \tag{10.139}$$

Because distance is a scalar, its value does not change under general coordinate transformations; thus we can write

$$d\overline{s}^2 = ds^2 \tag{10.140}$$

$$= \sum_{k=1}^{n} \left[\sum_{i=1}^{n} \frac{\partial x^k}{\partial \overline{x}^i} d\overline{x}^i \right] \left[\sum_{j=1}^{n} \frac{\partial x^k}{\partial \overline{x}^j} d\overline{x}^j \right] \tag{10.141}$$

$$= \sum_{i=1}^{n} \sum_{j=1}^{n} \left[\sum_{k=1}^{n} \frac{\partial x^k}{\partial \overline{x}^i} \frac{\partial x^k}{\partial \overline{x}^j} \right] d\overline{x}^i d\overline{x}^j \tag{10.142}$$

$$= \sum_{i=1}^{n} \sum_{j=1}^{n} g_{ij} d\overline{x}^i d\overline{x}^j, \tag{10.143}$$

where g_{ij} is a second rank-symmetric tensor defined as

$$g_{ij} = \sum_{k=1}^{n} \frac{\partial x^k}{\partial \overline{x}^i} \frac{\partial x^k}{\partial \overline{x}^j}. \tag{10.144}$$

It is called the **metric tensor** or the **fundamental tensor**. The metric tensor is very important in the study of curved spaces and spacetimes.

If we write the line element [Eq. (10.139)] in Cartesian coordinates as

$$ds^2 = \sum_{i=1}^{n} \sum_{j=1}^{n} \delta_{ij} dx^i dx^j, \tag{10.145}$$

we see that the metric tensor is the identity matrix

$$g_{ij} = \mathbf{I} = \delta_{ij}. \tag{10.146}$$

Given an arbitrary contravariant vector u^i, let us see how

$$\left[\sum_{j=1}^{n} g_{ij} u^j \right] \tag{10.147}$$

transforms. We first write $\sum_{j=1}^{n} \left[\overline{g}_{ij} \overline{u}^j \right]$ as

$$
\begin{aligned}
\sum_{j=1}^{n} \left[\overline{g}_{ij} \overline{u}^j \right] &= \sum_{j=1}^{n} \left[\sum_{k=1}^{n} \sum_{l=1}^{n} \frac{\partial x^k}{\partial \overline{x}^i} \frac{\partial x^l}{\partial \overline{x}^j} g_{kl} \right] \left[\sum_{m=1}^{n} \frac{\partial \overline{x}^j}{\partial x^m} u^m \right] \\
&= \sum_{m=1}^{n} \sum_{k=1}^{n} \sum_{l=1}^{n} \left[\frac{\partial x^k}{\partial \overline{x}^i} \left(\sum_{j=1}^{n} \frac{\partial x^l}{\partial \overline{x}^j} \frac{\partial \overline{x}^j}{\partial x^m} \right) \right] g_{kl} u^m \\
&= \sum_{m=1}^{n} \sum_{k=1}^{n} \sum_{l=1}^{n} \left[\frac{\partial x^k}{\partial \overline{x}^i} \delta^l_m \right] g_{kl} u^m \\
&= \sum_{k=1}^{n} \left[\frac{\partial x^k}{\partial \overline{x}^i} \right] \left[\sum_{m=1}^{n} g_{km} u^m \right].
\end{aligned}
\tag{10.148}
$$

Comparing with Equation (10.127), we see that the expression

$$\left[\sum_{m=1}^{n} g_{km} u^m \right] \tag{10.149}$$

transforms like a covariant vector; thus we define the covariant components of u^i as

$$u_i = \sum_{j=1}^{n} g_{ij} u^j. \tag{10.150}$$

Similarly, we can define the metric tensor with contravariant components as

$$g^{kl} = \sum_{i=1}^{n} \left[\frac{\partial x^k}{\partial \overline{x}^i} \frac{\partial x^l}{\partial \overline{x}^i} \right], \tag{10.151}$$

where

$$
\begin{aligned}
\sum_{k=1}^{n} g_{kl} g^{kl'} &= \sum_{k=1}^{n} \left[\sum_{i=1}^{n} \frac{\partial \overline{x}^i}{\partial x^k} \frac{\partial \overline{x}^i}{\partial x^l} \right] \left[\sum_{i'=1}^{n} \frac{\partial x^k}{\partial \overline{x}^{i'}} \frac{\partial x^{l'}}{\partial \overline{x}^{i'}} \right] \\
&= \sum_{i=1}^{n} \sum_{i'=1}^{n} \left[\sum_{k=1}^{n} \frac{\partial \overline{x}^i}{\partial x^k} \frac{\partial x^k}{\partial \overline{x}^{i'}} \right] \left[\frac{\partial \overline{x}^i}{\partial x^l} \frac{\partial x^{l'}}{\partial \overline{x}^{i'}} \right] \\
&= \sum_{i=1}^{n} \sum_{i'=1}^{n} \delta_{i'}^{i} \left[\frac{\partial \overline{x}^i}{\partial x^l} \frac{\partial x^{l'}}{\partial \overline{x}^{i'}} \right] \\
&= \sum_{i=1}^{n} \left[\frac{\partial \overline{x}^i}{\partial x^l} \frac{\partial x^{l'}}{\partial \overline{x}^i} \right] \\
&= \delta_l^{l'}. \tag{10.152}
\end{aligned}
$$

Using the symmetry of the metric tensor we can write

$$\sum_{k=1}^{n} g_{lk} g^{kl'} = g_l^{l'} = \delta_l^{l'}. \tag{10.153}$$

We see that the metric tensor can be used to lower and raise indices of a given tensor. Thus a given vector $\overrightarrow{u}$ can be expressed in terms of either its covariant or its contravariant components. In general the two types of components are different, and they are related by

$$u_i = \sum_{j=1}^{n} g_{ij} u^j \quad \text{or} \quad u^i = \sum_{j=1}^{n} g^{ij} u_j. \tag{10.154}$$

For the Cartesian coordinates the metric tensor is the Kronecker delta; thus we can write

$$g_{ij} = \delta_{ij} = \delta_i^j = \delta_j^i = \delta^{ij} = g^{ij}. \tag{10.155}$$

Hence both the covariant and the contravariant components are equal in Cartesian coordinates, and there is no need for distinction between them.

Contravariant components of the metric tensor are also given as (Gantmacher)

$$g^{ij} = \frac{\Delta^{ji}}{g}, \tag{10.156}$$

where

$$\Delta^{ji} = \text{cofactor of } g_{ji} \tag{10.157}$$

and

$$g = \det g_{ij}. \tag{10.158}$$

10.8.3   Geometric Interpretation of Covariant and Contravariant Components

Covariant and contravariant indices can be geometrically interpreted in terms of oblique axis. A vector $\vec{a}$ in the coordinate system shown in Figure 10.7 can be written as

$$\vec{a} = a^1 \hat{e}_1 + a^2 \hat{e}_2, \tag{10.159}$$

where $\hat{e}_i$ are the unit basis vectors along the coordinate axes. As seen, the contravariant components are found by drawing parallel lines to the coordinate axes. However, we can also define components by dropping perpendiculars to the coordinate axes as

$$a_i = \vec{a} \cdot \hat{e}_i, \quad i = 1, 2. \tag{10.160}$$

The scalar product of two vectors is given as

$$\begin{aligned}
\vec{a} \cdot \vec{b} &= \left(a^1 \hat{e}_1 + a^2 \hat{e}_2\right) \cdot \left(b^1 \hat{e}_1 + b^2 \hat{e}_2\right) \\
&= a^1 b^1 \left(\hat{e}_1 \cdot \hat{e}_1\right) + a^1 b^2 \left(\hat{e}_1 \cdot \hat{e}_2\right) + a^2 b^1 \left(\hat{e}_2 \cdot \hat{e}_1\right) + a^2 b^2 \left(\hat{e}_2 \cdot \hat{e}_2\right).
\end{aligned} \tag{10.161}$$

Defining a symmetric matrix

$$g_{ij} = \hat{e}_i \cdot \hat{e}_j, \quad i, j = 1, 2 \tag{10.162}$$

we can write Equation (10.161) as

$$\vec{a} \cdot \vec{b} = \sum_{i=1}^{2} \sum_{j=1}^{2} g_{ij} a^i b^j. \tag{10.163}$$

We can also write

$$\begin{aligned}
\vec{a} \cdot \vec{b} &= \vec{a} \cdot \left(b^1 \hat{e}_1 + b^2 \hat{e}_2\right) \\
&= b^1 \left(\vec{a} \cdot \hat{e}_1\right) + b^2 \left(\vec{a} \cdot \hat{e}_2\right) \\
&= b^1 a_1 + b^2 a_2 \\
&= \sum_{i=1}^{2} b^i a_i \\
&= \sum_{i=1}^{2} a^i b_i.
\end{aligned} \tag{10.164}$$

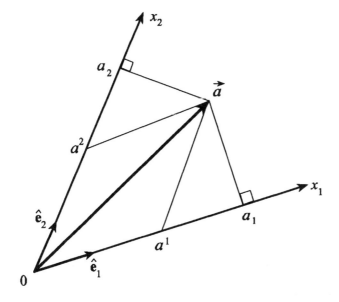

Fig. 10.7 Covariant and contravariant components

All these remind us tensors. To prove that $\overrightarrow{a} \cdot \overrightarrow{b}$ is a tensor equation we have to prove that it has the same form in another coordinate system. It is clear that in another coordinate system with the basis vectors $\widehat{e}'_1$ and $\widehat{e}'_2$, $\overrightarrow{a} \cdot \overrightarrow{b}$ will have the same form as

$$\overrightarrow{a} \cdot \overrightarrow{b} = \sum_{k=1}^{2} \sum_{l=1}^{2} g'_{kl} a'^k b'^l$$

$$= \sum_{i=1}^{2} \sum_{j=1}^{2} g_{ij} a^i b^j, \tag{10.165}$$

where $g'_{kl} = \widehat{e}'_k \cdot \widehat{e}'_l$, thus proving its tensor character. Because $\overrightarrow{a}$ and $\overrightarrow{b}$ are arbitrary vectors, we can take them as the infinitesimal displacement vector as

$$a^i = b^i = dx^i; \tag{10.166}$$

thus

$$ds^2 = \sum_{i=1}^{2} \sum_{j=1}^{2} g_{ij} dx^i dx^j \tag{10.167}$$

gives the line element with the metric

$$g_{ij} = \widehat{e}_i \cdot \widehat{e}_j, \qquad i, j = 1, 2. \tag{10.168}$$

Hence a^i and a_i are indeed the contravariant and the covariant components of an arbitrary vector, and the difference between the covariant and the contravariant components is real.

In curved spaces dx^i corresponds to the coordinate increments on the surface. The metric tensor g_{ij} can now be interpreted as the product $\widehat{e}_i \cdot \widehat{e}_j$ of the unit tangent vectors along the coordinate axis.

10.9   OPERATIONS WITH GENERAL TENSORS

10.9.1   Einstein Summation Convention

Algebraic operations like addition, subtraction, and multiplication are accomplished the same way as in Cartesian tensors. For general tensors the Einstein summation convention, which implies summation over repeated indices, is used by writing one of the indices as covariant and the other as contravariant. For example, the line element can be written in any one of the following forms:

$$ds^2 = g_{ij}dx^i dx^j = dx_j dx^j = dx^i dx_i \qquad (10.169)$$

or

$$ds^2 = g^{ij} dx_i dx_j = dx^j dx_j = dx_i dx^i. \qquad (10.170)$$

From now on, unless otherwise stated, we use this version of the Einstein summation convention.

10.9.2   Contraction of Indices

We can lower the rank of a tensor by contracting some of its indices as

$$E^{ij} = T^{ijk}_k, \qquad (10.171)$$
$$C^i = D^{ijk}_{jk}. \qquad (10.172)$$

We can also lower the rank of a tensor by contracting it with another tensor as

$$F^{ij} = D^{ijk} E_k , \qquad (10.173)$$
$$A = B^i C_i. \qquad (10.174)$$

10.9.3   Multiplication of Tensors

We can obtain tensors of higher-rank by multiplying two lower rank tensors:

$$C_{ijk} = A_{ij}D_k, \tag{10.175}$$

$$T_{ij} - A_iB_j, \tag{10.176}$$

$$F_{ij}^{lm} = B_iC^lD_j^m. \tag{10.177}$$

This is also called the **outer product**.

10.9.4   The Quotient Theorem

A very useful theorem in tensor operations is the quotient theorem. Suppose $T_{j_1\ldots j_m}^{i_1\ldots i_n}$ is a given **matrix** and $A_{i_k\ldots i_n}^{j_l\ldots j_m}$ is an arbitrary tensor. Suppose that it is also known that

$$S_{j_1\ldots j_{l-1}}^{i_1\ldots i_{k-1}} = T_{j_1\ldots j_{l-1}j_l\ldots j_m}^{i_1\ldots i_{k-1}i_k\ldots i_n} A_{i_k\ldots i_n}^{j_l\ldots j_m} \tag{10.178}$$

is a tensor. Then, by the quotient theorem,

$$T_{j_1\ldots j_m}^{i_1\ldots i_n} \tag{10.179}$$

is also a tensor. This could be easily checked by using the transformation properties of tensors.

10.9.5   Equality of Tensors

Two tensors are equal if and only if all their corresponding components are equal, that is,

$$A_k^{ij} = B_k^{ij}, \text{ for all } i, j, \text{ and } k. \tag{10.180}$$

As a consequence of this, a tensor is not zero unless all of its components vanish.

10.9.6   Tensor Densities

A tensor density of weight w transforms according to the law

$$\overline{T}_{j_1 j_2\ldots}^{i_1 i_2\ldots} \tag{10.181}$$

$$= \left[\frac{\partial \overline{x}^{i_1}}{\partial x^{k_1}} \frac{\partial \overline{x}^{i_2}}{\partial x^{k_2}} \cdots \frac{\partial x^{l_1}}{\partial \overline{x}^{j_1}} \frac{\partial x^{l_2}}{\partial \overline{x}^{j_2}} \cdots \right] T_{l_1 l_2\ldots}^{k_1 k_2\ldots} \left| \frac{\partial x}{\partial \overline{x}} \right|^w,$$

where $\left|\dfrac{\partial x}{\partial \overline{x}}\right|$ is the Jacobian of the transformation, that is,

$$\left|\frac{\partial x}{\partial \overline{x}}\right| = \det \begin{bmatrix} \dfrac{\partial x^1}{\partial \overline{x}^1} & \dfrac{\partial x^1}{\partial \overline{x}^2} & \cdots & \dfrac{\partial x^1}{\partial \overline{x}^n} \\ \dfrac{\partial x^2}{\partial \overline{x}^1} & \cdots & \cdots & \cdots \\ \vdots & \vdots & \vdots & \vdots \\ \cdots & \cdots & \cdots & \dfrac{\partial x^n}{\partial \overline{x}^n} \end{bmatrix}. \tag{10.182}$$

The permutation symbol ϵ_{ijk} is a third-rank tensor density of weight -1. The volume element

$$d^n x = dx^1 dx^2 ... dx^n \tag{10.183}$$

transforms as

$$d^n \overline{x} = d^n x \left|\frac{\partial x}{\partial \overline{x}}\right|^{-1}, \tag{10.184}$$

hence it is a scalar density of weight -1.

The metric tensor is a second-rank tensor that transforms as

$$\overline{g}_{ij} = \frac{\partial x^k}{\partial \overline{x}^i} \frac{\partial x^l}{\partial \overline{x}^j} g_{kl}. \tag{10.185}$$

Using matrix multiplication determinant of the metric tensor transforms as

$$\overline{g} = g \left|\frac{\partial x}{\partial \overline{x}}\right|^2 \tag{10.186}$$

or as

$$\sqrt{|\overline{g}|} = \sqrt{|g|} \left|\frac{\partial x}{\partial \overline{x}}\right|. \tag{10.187}$$

In the last equation we have used absolute values in anticipation of applications to relativity, where the metric has signature $(-+++)$ or $(+---)$.

From Equations (10.184) and (10.187) it is seen that

$$\sqrt{|\overline{g}|} d^n \overline{x} = \sqrt{|g|} d^n x. \tag{10.188}$$

Thus, $\sqrt{|g|} d^n x$ is a scalar.

10.9.7 Differentiation of Tensors

We start by taking the derivative of the transform of a covariant vector

$$\overline{u}_i = \frac{\partial x^j}{\partial \overline{x}^i} u_j$$

as

$$\frac{\partial \overline{u}_i}{\partial \overline{x}^k} = \frac{\partial^2 x^j}{\partial \overline{x}^k \partial \overline{x}^i} u_j + \frac{\partial x^j}{\partial \overline{x}^i} \left[\frac{\partial u_j}{\partial x^l} \frac{\partial x^l}{\partial \overline{x}^k} \right]. \tag{10.189}$$

If we write this as

$$\frac{\partial \overline{u}_i}{\partial \overline{x}^k} = \frac{\partial^2 x^j}{\partial \overline{x}^i \partial \overline{x}^k} u_j + \frac{\partial x^j}{\partial \overline{x}^i} \frac{\partial x^l}{\partial \overline{x}^k} \frac{\partial u_j}{\partial x^l} \tag{10.190}$$

and if the first term on the right-hand side was absent, then the derivative of u_j would simply be a second-rank tensor. Rearranging this equation as

$$\frac{\partial \overline{u}_i}{\partial \overline{x}^k} = \frac{\partial}{\partial \overline{x}^i} \left[\frac{\partial x^j}{\partial \overline{x}^k} \right] u_j + \frac{\partial x^j}{\partial \overline{x}^i} \frac{\partial x^l}{\partial \overline{x}^k} \frac{\partial u_j}{\partial x^l} \tag{10.191}$$

$$= \frac{\partial \left[a_k^j \right]}{\partial \overline{x}^i} u_j + \frac{\partial x^j}{\partial \overline{x}^i} \frac{\partial x^l}{\partial \overline{x}^k} \frac{\partial u_j}{\partial x^l}, \tag{10.192}$$

we see that the problem is due to the fact that in general the transformation matrix $\left[a_k^j \right]$ changes with position. For transformations between the Cartesian coordinates the transformation matrix is independent of coordinates; thus this problem does not arise. However, we can still define a covariant derivative that transforms like a tensor.

We first consider the metric tensor, which transforms as

$$\overline{g}_{ij} = \frac{\partial x^k}{\partial \overline{x}^i} \frac{\partial x^l}{\partial \overline{x}^j} g_{kl}, \tag{10.193}$$

and differentiate it with respect to $\overline{x}^m$ as

$$\frac{\partial \overline{g}_{ij}}{\partial \overline{x}^m} = \frac{\partial^2 x^k}{\partial \overline{x}^i \partial \overline{x}^m} \frac{\partial x^l}{\partial \overline{x}^j} g_{kl} + \frac{\partial x^k}{\partial \overline{x}^i} \frac{\partial^2 x^l}{\partial \overline{x}^j \partial \overline{x}^m} g_{kl} + \frac{\partial x^k}{\partial \overline{x}^i} \frac{\partial x^l}{\partial \overline{x}^j} \frac{\partial x^n}{\partial \overline{x}^m} \frac{\partial g_{kl}}{\partial x^n}. \tag{10.194}$$

Permuting the indices, that is, $(ijm) \to (mij) \to (jmi)$, we can obtain two more equations:

$$\frac{\partial \overline{g}_{mi}}{\partial \overline{x}^j} = \frac{\partial^2 x^k}{\partial \overline{x}^m \partial \overline{x}^j} \frac{\partial x^l}{\partial \overline{x}^i} g_{kl} + \frac{\partial x^k}{\partial \overline{x}^m} \frac{\partial^2 x^l}{\partial \overline{x}^i \partial \overline{x}^j} g_{kl} + \frac{\partial x^k}{\partial \overline{x}^m} \frac{\partial x^l}{\partial \overline{x}^i} \frac{\partial x^n}{\partial \overline{x}^j} \frac{\partial g_{kl}}{\partial x^n} \tag{10.195}$$

and

$$\frac{\partial \overline{g}_{jm}}{\partial \overline{x}^i} = \frac{\partial^2 x^k}{\partial \overline{x}^j \partial \overline{x}^i} \frac{\partial x^l}{\partial \overline{x}^m} g_{kl} + \frac{\partial x^k}{\partial \overline{x}^j} \frac{\partial^2 x^l}{\partial \overline{x}^m \partial \overline{x}^i} g_{kl} + \frac{\partial x^k}{\partial \overline{x}^j} \frac{\partial x^l}{\partial \overline{x}^m} \frac{\partial x^n}{\partial \overline{x}^i} \frac{\partial g_{kl}}{\partial x^n}. \tag{10.196}$$

Adding the first two equations and subtracting the last one from the result and after some rearrangement of indices we obtain

$$\frac{1}{2} \left[\frac{\partial \overline{g}_{ik}}{\partial \overline{x}^j} + \frac{\partial \overline{g}_{jk}}{\partial \overline{x}^i} - \frac{\partial \overline{g}_{ij}}{\partial \overline{x}^k} \right] = \frac{\partial x^l}{\partial \overline{x}^i} \frac{\partial x^m}{\partial \overline{x}^j} \frac{\partial x^n}{\partial \overline{x}^k} \frac{1}{2} \left[\frac{\partial g_{ln}}{\partial x^m} + \frac{\partial g_{mn}}{\partial x^l} - \frac{\partial g_{lm}}{\partial x^n} \right]$$

$$+ g_{lm} \frac{\partial x^l}{\partial \overline{x}^k} \frac{\partial^2 x^m}{\partial \overline{x}^i \partial \overline{x}^j}. \tag{10.197}$$

Defining **Christoffel symbols of the first kind** as

$$[ij,k] = \frac{1}{2}\left[\frac{\partial g_{ik}}{\partial x^j} + \frac{\partial g_{jk}}{\partial x^i} - \frac{\partial g_{ij}}{\partial x^k}\right], \tag{10.198}$$

we write Equation (10.197) as

$$\overline{[ij,k]} = \frac{\partial x^l}{\partial \overline{x}^i}\frac{\partial x^m}{\partial \overline{x}^j}\frac{\partial x^n}{\partial \overline{x}^k}\,[lm,n] + g_{lm}\frac{\partial x^l}{\partial \overline{x}^k}\frac{\partial^2 x^m}{\partial \overline{x}^i\partial \overline{x}^j}, \tag{10.199}$$

where

$$\overline{[ij,k]} = \frac{1}{2}\left[\frac{\partial \overline{g}_{ik}}{\partial \overline{x}^j} + \frac{\partial \overline{g}_{jk}}{\partial \overline{x}^i} - \frac{\partial \overline{g}_{ij}}{\partial \overline{x}^k}\right]. \tag{10.200}$$

We can easily solve this equation for the second derivative to obtain

$$\frac{\partial^2 x^h}{\partial \overline{x}^i\partial \overline{x}^j} = \frac{\partial x^h}{\partial \overline{x}^p}\overline{\left\{\begin{matrix}p\\ij\end{matrix}\right\}} - \frac{\partial x^l}{\partial \overline{x}^i}\frac{\partial x^m}{\partial \overline{x}^j}\left\{\begin{matrix}h\\lm\end{matrix}\right\}, \tag{10.201}$$

where we have defined the **Christoffel symbols of the second kind** as

$$\left\{\begin{matrix}i\\jk\end{matrix}\right\} = g^{il}\,[jk,l]. \tag{10.202}$$

Substituting Equation (10.201) in Equation (10.190), we get

$$\frac{\partial \overline{u}_i}{\partial \overline{x}^k} = \frac{\partial x^j}{\partial \overline{x}^l}\overline{\left\{\begin{matrix}l\\ik\end{matrix}\right\}}u_j - \frac{\partial x^l}{\partial \overline{x}^i}\frac{\partial x^m}{\partial \overline{x}^k}\left\{\begin{matrix}j\\lm\end{matrix}\right\}u_j + \frac{\partial x^j}{\partial \overline{x}^i}\frac{\partial x^l}{\partial \overline{x}^k}\frac{\partial u_j}{\partial x^l}. \tag{10.203}$$

Rearranging, and using the symmetry property of the Christoffel symbol of the second kind:

$$\left\{\begin{matrix}i\\jk\end{matrix}\right\} = \left\{\begin{matrix}i\\kj\end{matrix}\right\}, \tag{10.204}$$

this becomes

$$\frac{\partial \overline{u}_i}{\partial \overline{x}^k} - \overline{\left\{\begin{matrix}l\\ik\end{matrix}\right\}}\left[\frac{\partial x^j}{\partial \overline{x}^l}u_j\right] = \frac{\partial x^j}{\partial \overline{x}^i}\frac{\partial x^l}{\partial \overline{x}^k}\left[\frac{\partial u_j}{\partial x^l} - \left\{\begin{matrix}m\\jl\end{matrix}\right\}u_m\right], \tag{10.205}$$

$$\left[\frac{\partial \overline{u}_i}{\partial \overline{x}^k} - \overline{\left\{\begin{matrix}l\\ik\end{matrix}\right\}}\overline{u}_l\right] = \frac{\partial x^j}{\partial \overline{x}^i}\frac{\partial x^l}{\partial \overline{x}^k}\left[\frac{\partial u_j}{\partial x^l} - \left\{\begin{matrix}m\\jl\end{matrix}\right\}u_m\right]. \tag{10.206}$$

The above equation shows that $\left[\dfrac{\partial u_j}{\partial x^l} - \left\{\begin{matrix}m\\jl\end{matrix}\right\}u_m\right]$ transforms like a covariant second-rank tensor. Thus we define the covariant derivative of a covariant vector u_i as

$$u_{i;j} = \left[\frac{\partial u_i}{\partial x^j} - \left\{\begin{matrix}k\\ij\end{matrix}\right\}u_k\right]. \tag{10.207}$$

Similarly, the covariant derivative of a contravariant vector is defined as

$$u^i_{;j} = \left[\frac{\partial u^i}{\partial x^j} + \left\{ \begin{matrix} i \\ jk \end{matrix} \right\} u^k \right]. \tag{10.208}$$

The covariant derivative is also shown as ∂_i, that is, $\partial_j u_i = u_{i;j}$. The covariant derivative of a higher-rank tensor is obtained by treating each index at a time as

$$T^{i_1 i_2 \ldots}_{j_1 j_2 \ldots;\, k} = \frac{\partial T^{i_1 i_2 \ldots}_{j_1 j_2 \ldots}}{\partial x^k} + \left\{ \begin{matrix} i_1 \\ kl \end{matrix} \right\} T^{l i_2 \ldots}_{j_1 j_2 \ldots} + \left\{ \begin{matrix} i_2 \\ kl \end{matrix} \right\} T^{i_1 l \ldots}_{j_1 j_2 \ldots} + \cdots$$
$$- \left\{ \begin{matrix} m \\ kj_1 \end{matrix} \right\} T^{i_1 i_2 \ldots}_{m j_2 \ldots} - \left\{ \begin{matrix} m \\ kj_2 \end{matrix} \right\} T^{i_1 i_2 \ldots}_{j_1 m \ldots} - \cdots . \tag{10.209}$$

Covariant derivatives distribute like ordinary derivatives, that is,

$$(AB)_{;i} = A_{;i}B + AB_{;i} \tag{10.210}$$

and

$$(aA + bB)_{;i} = aA_{;i} + bB_{;i} \tag{10.211}$$

where A and B are tensors of arbitrary rank and a and b are scalars.

10.9.8 Some Covariant Derivatives

In the following we also show equivalent ways of writing these operations commonly encountered in the literature.

1. Using definition Equation (10.123) we can write the covariant derivative of a scalar function Ψ as an ordinary derivative:

$$\vec{\nabla}\Psi = \Psi_{;j} = \partial_j \Psi = \frac{\partial \Psi}{\partial x^j}. \tag{10.212}$$

This is also the covariant component of the gradient

$$\left(\vec{\nabla}\Psi \right)_i . \tag{10.213}$$

2. Using the symmetry of Christoffel symbols, the curl of a vector field $\vec{v}$ can be defined as the second-rank tensor

$$\left(\vec{\nabla} \times \vec{v} \right)_{ij} = \partial_j v_i - \partial_i v_j = v_{i;j} - v_{j;i} \tag{10.214}$$

$$= \frac{\partial v_i}{\partial x^j} - \frac{\partial v_j}{\partial x^i}. \tag{10.215}$$

Note that because we have used the symmetry of the Christoffel symbols, the curl operation can only be performed on the covariant components of a vector.

3. The covariant derivative of the metric tensor is zero:

$$\partial_k g_{ij} = g_{ij;k} = 0, \tag{10.216}$$

with Equation (10.209) and the definition of Christoffel symbols the proof is straightforward.

4. A frequently used property of the Christoffel symbol of the second kind is

$$\left\{ \begin{matrix} i \\ ik \end{matrix} \right\} = \frac{1}{2} g^{il} \frac{\partial g_{il}}{\partial x^k} = \frac{\partial (\ln \sqrt{|g|})}{\partial x^k}. \tag{10.217}$$

In the derivation we use the result

$$\frac{\partial g}{\partial x^k} = g g^{il} \frac{\partial g_{il}}{\partial x^k} \tag{10.218}$$

from the theory of matrices, where $g = \det g_{ij}$.

5. We can now define covariant divergence as

$$\vec{\nabla} \cdot \vec{v} = \partial_i v^i = v^i_{;i} \tag{10.219}$$

$$= \frac{\partial v^i}{\partial x^i} + \left\{ \begin{matrix} i \\ ik \end{matrix} \right\} v^k \tag{10.220}$$

$$= \frac{1}{|g|^{1/2}} \frac{\partial}{\partial x^k} \left[|g|^{1/2} v^k \right]. \tag{10.221}$$

If v^i is a tensor density of weight $+1$, divergence becomes

$$\vec{\nabla} \cdot \vec{v} = v^i_{;i} \ (= \partial_i v^i), \tag{10.222}$$

which is again a scalar density of weight $+1$.

6. Using Equation (10.213) we write the contravariant component of the gradient of a scalar function as

$$\left(\vec{\nabla} \Psi \right)^i = g^{ij} \Psi_{;j} \tag{10.223}$$

$$= g^{ij} \frac{\partial \Psi}{\partial x^j}. \tag{10.224}$$

We can now define the Laplacian as a scalar field:

$$\vec{\nabla}^2 = \vec{\nabla} \cdot (\vec{\nabla} \Psi) = (\vec{\nabla} \Psi)^i_{;i} = \partial_i \partial^i \Psi = \frac{1}{|g|^{1/2}} \frac{\partial}{\partial x^i} \left[|g|^{1/2} g^{ik} \frac{\partial \Psi}{\partial x^k} \right]. \tag{10.225}$$

10.9.9   Riemann Curvature Tensor

Let us take the covariant derivative of v_i twice. The difference

$$v_{i;jk} - v_{i;kj} \tag{10.226}$$

can be written as

$$v_{i;jk} - v_{i;kj} = R^l_{ijk} v_l , \tag{10.227}$$

where R^l_{ijk} is the fourth-rank **Riemann curvature tensor**, which plays a central role in the structure of Riemann spaces:

$$R^l_{ijk} = \begin{Bmatrix} l \\ mj \end{Bmatrix} \begin{Bmatrix} m \\ ik \end{Bmatrix} - \begin{Bmatrix} l \\ mk \end{Bmatrix} \begin{Bmatrix} m \\ ij \end{Bmatrix} + \frac{\partial}{\partial x^j} \begin{Bmatrix} l \\ ik \end{Bmatrix} - \frac{\partial}{\partial x^k} \begin{Bmatrix} l \\ ij \end{Bmatrix}. \tag{10.228}$$

Three of the symmetries of the Riemann curvature tensor can be summarized as

$$R_{ijkl} = -R_{ijlk} \tag{10.229}$$
$$R_{ijkl} = -R_{jikl} \tag{10.230}$$
$$R_{ijkl} = R_{klij}. \tag{10.231}$$

Actually, there is one more symmetry that we will not discuss. The significance of the Riemann curvature tensor is, that all of its components vanish only in flat space, that is we cannot find a coordinate system where

$$R_{ijkl} = 0 \tag{10.232}$$

unless the space is truly flat.

An important scalar in Riemann spaces is the **Riemann curvature scalar,** which is obtained from R_{ijkl} by contracting its indices as

$$R = g^{jl} g^{ik} R_{ijkl} = g^{jl} R^i_{\ jil} = R^i_{\ ji}{}^j . \tag{10.233}$$

Note that $R_{ijkl} = 0$ implies $R = 0$, but not vice versa.

Example 10.1. *Laplacian as a scalar field:* We consider the line element

$$ds^2 = dr^2 + r^2 d\theta^2 + r^2 \sin^2 \theta d\phi^2, \tag{10.234}$$

where

$$x^1 = r, \ x^2 = \theta, \ x^3 = \phi \tag{10.235}$$

and

$$g_{11} = 1, \ g_{22} = r^2, \ g_{33} = r^2 \sin^2 \theta. \tag{10.236}$$

Contravariant components g^{ij} are:

$$g^{11} = 1, \ g^{22} = \frac{1}{r^2}, \ g^{33} = \frac{1}{r^2 \sin^2 \theta}. \tag{10.237}$$

Using Equation (10.225) and $g = r^4 \sin^2 \theta$, we can write the Laplacian as

$$\partial_i \partial^i \Psi = \frac{1}{|g|^{1/2}} \frac{\partial}{\partial x^i} \left[|g|^{1/2} \, g^{ik} \frac{\partial \Psi}{\partial x^k} \right] \tag{10.238}$$

$$= \frac{1}{|g|^{1/2}} \frac{\partial}{\partial x^i} \left[|g|^{1/2} \left(g^{i1} \frac{\partial \Psi}{\partial x^1} + g^{i2} \frac{\partial \Psi}{\partial x^2} + g^{i3} \frac{\partial \Psi}{\partial x^3} \right) \right]$$

$$= \frac{1}{r^2 \sin \theta} \left[\frac{\partial}{\partial r} \left(r^2 \sin \theta \frac{\partial \Psi}{\partial r} \right) + \frac{\partial}{\partial \theta} \left(\frac{r^2 \sin \theta}{r^2} \frac{\partial \Psi}{\partial \theta} \right) + \frac{\partial}{\partial \phi} \left(\frac{r^2 \sin \theta}{r^2 \sin^2 \theta} \frac{\partial \Psi}{\partial \phi} \right) \right].$$

After simplifying, the Laplacian is obtained as

$$\partial_i \partial^i \Psi = \frac{1}{r^2} \frac{\partial}{\partial r} \left(r^2 \frac{\partial \Psi}{\partial r} \right) + \frac{1}{r^2 \sin \theta} \frac{\partial}{\partial \theta} \left(\sin \theta \frac{\partial \Psi}{\partial \theta} \right) + \frac{1}{r^2 \sin^2 \theta} \frac{\partial^2 \Psi}{\partial \phi^2}. \tag{10.239}$$

Here we have obtained a well-known formula in a rather straightforward manner, demonstrating the advantages of the tensor formalism. Note that even though the components of the metric tensor depend on position [Eq. (10.236)], the curvature tensor is zero,

$$R_{ijkl} = 0; \tag{10.240}$$

thus the space of the line element [Eq. (10.234)] is flat. However, for the metric

$$ds^2 = \left[\frac{1}{1 - r^2/R_0^2} \right] dr^2 + r^2 d\theta^2 + r^2 \sin^2 \theta d\phi^2, \tag{10.241}$$

it can be shown that not all the components of R_{ijkl} vanish. In fact, this line element gives the distance between two infinitesimally close points on the surface of a hypersphere (S-3) with constant radius R_0.

10.9.10   Geodesics

Geodesics are defined as the shortest paths between two points in a given geometry. In flat space they are naturally the straight lines. We can generalize the concept of straight lines as curves whose tangents remain constant along the curve. However, the constancy is now with respect to the covariant derivative. If we parametrize an arbitrary curve in terms of arclength s as

$$x^i(s), \tag{10.242}$$

its tangent vector will be given as

$$t^i = \frac{dx^i}{ds}.$$ (10.243)

For geodesics the covariant derivative of t^i must be zero; thus we obtain the equation of geodesics as

$$t^i_{;j} \frac{dx^j}{ds} = \left[\frac{dt^i}{dx^j} + \left\{ \begin{matrix} i \\ jk \end{matrix} \right\} t^k \right] \frac{dx^j}{ds} = 0$$ (10.244)

or as

$$\frac{d^2 x^i}{ds^2} + \left\{ \begin{matrix} i \\ jk \end{matrix} \right\} \frac{dx^j}{ds} \frac{dx^k}{ds} = 0.$$ (10.245)

10.9.11   Invariance and Covariance

We have seen that scalars preserve their value under general coordinate transformations. Certain other properties like the magnitude of a vector and the trace of a second-rank tensor also do not change under general coordinate transformations. Such properties are called **invariants**. They are very important in the study of the coordinate-independent properties of a system.

An important property of tensors is that tensor equations preserve their form under coordinate transformations. For example, the tensor equation

$$A_{ij} = B_{ijkl} C_{kl} + k D_{ij} + E^k_{i;\, jk} + \cdots$$ (10.246)

transforms into

$$A'_{ij} = B'_{ijkl} C'_{kl} + k D'_{ij} + E'^k_{i;\, jk} + \cdots .$$ (10.247)

This is called **covariance**. Under coordinate transformations individual components of tensors change; however, the form of the tensor equation remains the same. One of the early uses for tensors in physics was in searching and expressing the coordinate independent properties of crystals. However, the covariance of tensor equations reaches its full potential only with the introduction of the spacetime concept and the special and the general theories of relativity.

10.10   SPACETIME AND FOUR-TENSORS

10.10.1   Minkowski Spacetime

In Newton's theory, the energy of a freely moving particle is given by the well-known expression for kinetic energy:

$$E = \frac{1}{2} m v^2.$$ (10.248)

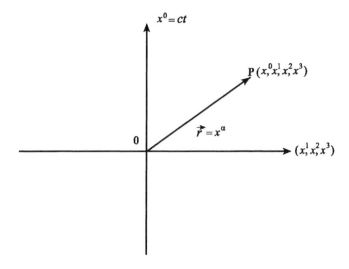

Fig. 10.8 Minkowski spacetime

Because there is no limit to the energy that one could pump into a system, this formula implies that in principle one could accelerate particles to any desired velocity. In classical physics this makes it possible to construct infinitely fast signals to communicate with the other parts of the universe. Another property of Newton's theory is that time is universal (or absolute), that is, identical clocks carried by moving observers, uniform or accelerated, run at the same rate. Thus once two observers synchronize their clocks, they will remain synchronized for ever. In Newton's theory this allows us to study systems with moving parts in terms of a single (universal) time parameter. With the discovery of the special theory of relativity it became clear that clocks carried by moving observers run at different rates; thus using a single time parameter for all observers is not possible.

After Einstein's introduction of the special theory of relativity another remarkable contribution toward the understanding of time came with the introduction of the spacetime concept by Minkowski. Spacetime not only strengthened the mathematical foundations of special relativity but also paved the way to Einstein's theory of gravitation .

Minkowski spacetime is obtained by simply adding a time axis orthogonal to the Cartesian axis, thus treating time as another coordinate (Fig. 10.8). A point in spacetime corresponds to an event. However, space and time are also fundamentally different and cannot be treated symmetrically. For example, it is possible to be present at the same place at two different times; however, if we reverse the roles of space and time, and if space and time were symmetric, then it would also mean that we could be present at two different places at the same time. So far there is no evidence for this, neither in the micro- nor in the macro-realm. Thus, in relativity even though space and time are treated on equal footing as independent coordinates, they are not treated symmetrically. This is evident in the Minkowski line element:

$$ds^2 = c^2 dt^2 - dx^2 - dy^2 - dz^2, \qquad (10.249)$$

where the signs of the spatial and the time coordinates are different. It is for this reason that Minkowski spacetime is called **pseudo-Euclidean.** In this line element c is the speed of light representing the maximum velocity in nature. An interesting property of the Minkowski spacetime is that two events connected by light rays, like the emission of a photon from one galaxy and its subsequent absorption in another, have zero distance between them even though they are widely separated in spacetime.

10.10.2   Lorentz Transformation and the Theory of Special Relativity

In Minkowski spacetime there are many different ways to choose the orientation of the coordinate axis. However, a particular group of coordinate systems, which are related to each other by linear transformations of the form

$$\overline{x}^0 = a_0^0 x^0 + a_1^0 x^1 + a_2^0 x^2 + a_3^0 x^3 \qquad (10.250)$$
$$\overline{x}^1 = a_0^1 x^0 + a_1^1 x^1 + a_2^1 x^2 + a_3^1 x^3$$
$$\overline{x}^2 = a_0^2 x^0 + a_1^2 x^1 + a_2^2 x^2 + a_3^2 x^3$$
$$\overline{x}^3 = a_0^3 x^0 + a_1^3 x^1 + a_2^3 x^2 + a_3^3 x^3$$

and which also preserve the quadratic form

$$\left(x^0\right)^2 - \left(x^1\right)^2 - \left(x^2\right)^2 - \left(x^3\right)^2, \qquad (10.251)$$

have been extremely useful in special relativity. In these equations we have written $x^0 = ct$ to emphasize the fact that time is treated as another coordinate.

In 1905 Einstein published his celebrated paper on the special theory of relativity, which is based on two postulates:

First postulate of relativity: It is impossible to detect or measure uniform translatory motion of a system in free space.

Second postulate of relativity: The speed of light in free space is the maximum velocity in the universe, and it is the same for all uniformly moving observers.

In special relativity two inertial observers K and $\overline{K}$, where $\overline{K}$ is moving uniformly with the velocity v along the common direction of the x^1- and

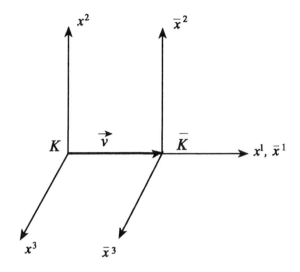

Fig. 10.9 Lorentz transformations

$\overline{x}^1$-axes are related by the Lorentz transformation (Fig. 10.9):

$$\overline{x}^0 = \frac{1}{\sqrt{1 - v^2/c^2}} \left[x^0 - \left(\frac{v}{c} \right) x^1 \right] \tag{10.252}$$

$$\overline{x}^1 = \frac{1}{\sqrt{1 - v^2/c^2}} \left[- \left(\frac{v}{c} \right) x^0 + x^1 \right] \tag{10.253}$$

$$\overline{x}^2 = x^2 \tag{10.254}$$

$$\overline{x}^3 = x^3. \tag{10.255}$$

Inverse transformation is obtained by replacing v with $-v$ as

$$x^0 = \frac{1}{\sqrt{1 - v^2/c^2}} \left[\overline{x}^0 + \left(\frac{v}{c} \right) \overline{x}^1 \right] \tag{10.256}$$

$$x^1 = \frac{1}{\sqrt{1 - v^2/c^2}} \left[\left(\frac{v}{c} \right) \overline{x}^0 + \overline{x}^1 \right] \tag{10.257}$$

$$x^2 = \overline{x}^2 \tag{10.258}$$

$$x^3 = \overline{x}^3. \tag{10.259}$$

If the axis in K and $\overline{K}$ remain parallel but the velocity $\overrightarrow{v}$ of frame $\overline{K}$ in frame K is arbitrary in direction, then the Lorentz transformation is generalized as

$$\overline{x}^0 = \gamma \left[x^0 - \left(\overrightarrow{\beta} \cdot \overrightarrow{x} \right) \right] \tag{10.260}$$

$$\overrightarrow{\overline{x}} = \overrightarrow{x} + \frac{(\gamma - 1)}{\beta^2} \left(\overrightarrow{\beta} \cdot \overrightarrow{x} \right) \overrightarrow{\beta} - \gamma \overrightarrow{\beta} x^0. \tag{10.261}$$

We have written $\gamma = 1/\sqrt{1 - v^2/c^2}$ and $\vec{\beta} = \vec{v}/c$.

10.10.3   Time Dilation and Length Contraction

Two immediate and important consequences of the Lorentz transformation equations [Eqs. (10.252–10.255)] are the time dilation and length contraction formulas, which are given as

$$\Delta \bar{t} = \Delta t (1 - \frac{v^2}{c^2})^{1/2} \tag{10.262}$$

and

$$\Delta \bar{x}^1 = \Delta x^1 (1 - \frac{v^2}{c^2})^{1/2}, \tag{10.263}$$

respectively. These formulas relate the time and the space intervals measured by two inertial observers $\overline{K}$ and K. The second formula is also known as the Lorentz contraction. The time dilation formula indicates that clocks carried by moving observers run slower compared to the clocks of the observer at rest. Similarly, the Lorentz contraction indicates that meter sticks carried by a moving observers appear shorter to the observer at rest.

10.10.4   Addition of Velocities

Another important consequence of the Lorentz transformation is the formula for the addition of velocities, which relates the velocities measured in the K and $\overline{K}$ frames by the formula

$$u^1 = \frac{\bar{u}^1 + v}{1 + \bar{u}^1 v/c^2}, \tag{10.264}$$

where $u^1 = \dfrac{dx^1}{dt}$ and $\bar{u}^1 = \dfrac{d\bar{x}^1}{d\bar{t}}$ are the velocities measured in the K and the $\overline{K}$ frames, respectively. In the limit as $c \to \infty$, this formula reduces to the well-known Galilean result

$$u^1 = \bar{u}^1 + v. \tag{10.265}$$

It is clear from Equation (10.264) that even if we go to a frame moving with the speed of light, it is not possible to send signals faster than c.

If the axes in K and $\overline{K}$ remain parallel, but the velocity $\vec{v}$ of frame $\overline{K}$ in frame K is arbitrary in direction, then the parallel and the perpendicular components of velocity transform as

$$u_{\|} = \frac{\bar{u}_{\|} + v}{1 + \vec{v} \cdot \vec{\bar{u}}/c^2}, \tag{10.266}$$

$$\vec{u}_{\perp} = \frac{\vec{\bar{u}}_{\perp}}{\gamma(1 + \vec{v} \cdot \vec{\bar{u}}/c^2)}. \tag{10.267}$$

In this notation $u_\parallel$ and $\vec{u}_\perp$ refer to the parallel and perpendicular components with respect to $\vec{v}$ and $\gamma = (1 - v^2/c^2)^{-1/2}$.

10.10.5   Four-Tensors in Minkowski Spacetime

From the second postulate of relativity, the invariance of the speed of light means

$$(d\bar{x}^0)^2 - \sum_{i=1}^{3}(d\bar{x}^i)^2 = (dx^0)^2 - \sum_{i=1}^{3}(dx^i)^2 = 0. \qquad (10.268)$$

This can also be written as

$$\bar{g}_{\alpha\beta}d\bar{x}^\alpha d\bar{x}^\beta = g_{\alpha\beta}dx^\alpha dx^\beta = 0, \qquad (10.269)$$

where the metric of the Minkowski spacetime is

$$\bar{g}_{\alpha\beta} = g_{\alpha\beta} = \begin{bmatrix} 1 & 0 & 0 & 0 \\ 0 & -1 & 0 & 0 \\ 0 & 0 & -1 & 0 \\ 0 & 0 & 0 & -1 \end{bmatrix}. \qquad (10.270)$$

We use the notation where the Greek indices take the values $0, 1, 2, 3$ and the Latin indices run through $1, 2, 3$. Note that even though the Minkowski spacetime is flat, because of the reversal of sign for the spatial components it is not Euclidean; thus the covariant and the contravariant indices differ in spacetime. Contravariant metric components can be obtained using (Gantmacher)

$$g^{\alpha\beta} = \frac{|g_{\beta\alpha}|_{\text{cofactor}}}{\det g_{\alpha\beta}} \qquad (10.271)$$

as

$$g^{\alpha\beta} = \begin{bmatrix} 1 & 0 & 0 & 0 \\ 0 & -1 & 0 & 0 \\ 0 & 0 & -1 & 0 \\ 0 & 0 & 0 & -1 \end{bmatrix}. \qquad (10.272)$$

Similar to the position vector in Cartesian coordinates we can define a position vector $\mathbf{r}$ in Minkowski spacetime as

$$\mathbf{r} = x^\alpha = (x^0, x^1, x^2, x^3) \qquad (10.273)$$
$$= (x^0, \vec{r}),$$

where $\mathbf{r}$ defines the time and the position of an event. In terms of linear transformations [Eq. (10.250)] x^α transforms as

$$\bar{x}^\alpha = a^\alpha_\beta x^\beta. \qquad (10.274)$$

For the Lorentz transformations [Eqs. (10.252−10.255) and (10.260−10.261)], a_β^α are given respectively as

$$
a_\beta^\alpha = \begin{bmatrix}
\gamma & -\beta\gamma & 0 & 0 \\
-\beta\gamma & \gamma & 0 & 0 \\
0 & 0 & 1 & 0 \\
0 & 0 & 0 & 1
\end{bmatrix}, \tag{10.275}
$$

and

$$
a_\beta^\alpha = \begin{bmatrix}
\gamma & -\beta_1\gamma & -\beta_2\gamma & -\beta_3\gamma \\
-\beta_1\gamma & 1+\dfrac{(\gamma-1)\beta_1^2}{\beta^2} & \dfrac{(\gamma-1)\beta_1\beta_2}{\beta^2} & \dfrac{(\gamma-1)\beta_1\beta_3}{\beta^2} \\
-\beta_2\gamma & \dfrac{(\gamma-1)\beta_2\beta_1}{\beta^2} & 1+\dfrac{(\gamma-1)\beta_2^2}{\beta^2} & \dfrac{(\gamma-1)\beta_2\beta_3}{\beta^2} \\
-\beta_3\gamma & \dfrac{(\gamma-1)\beta_3\beta_1}{\beta^2} & \dfrac{(\gamma-1)\beta_3\beta_2}{\beta^2} & 1+\dfrac{(\gamma-1)\beta_3^2}{\beta^2}
\end{bmatrix}. \tag{10.276}
$$

For the general linear transformation [Eq. (10.250)] matrix elements a_β^α can be obtained by using

$$
a_\beta^\alpha = \frac{d\overline{x}^\alpha}{dx^\beta}. \tag{10.277}
$$

In Minkowski spacetime the distance between two infinitesimally close points (events) can be written as

$$
\begin{aligned}
ds^2 &= dx^\alpha dx_\alpha \\
&= (dx^0)^2 - (dx^1)^2 - (dx^2)^2 - (dx^3)^2 \\
&= g_{\alpha\beta} dx^\alpha dx^\beta.
\end{aligned} \tag{10.278}
$$

In another inertial frame this becomes

$$
d\overline{s}^2 = \overline{g}_{\alpha\beta} d\overline{x}^\alpha d\overline{x}^\beta. \tag{10.279}
$$

Using Equations (10.270) and (10.274) we can write this as

$$
d\overline{s}^2 = \left[g_{\alpha\beta} a_\gamma^\alpha a_\delta^\beta \right] dx^\gamma dx^\delta . \tag{10.280}
$$

If we restrict ourselves to transformations that preserve the length of a vector we obtain the relation

$$
\left[g_{\alpha\beta} a_\gamma^\alpha a_\delta^\beta \right] = g_{\gamma\delta}. \tag{10.281}
$$

This is the analog of the orthogonality relation [Eq. (10.38)]. The position vector in Minkowski spacetime is called a **four-vector,** and its components transform as $\overline{x}^\alpha = a_\beta^\alpha x^\beta$, where its magnitude is a four-scalar.

An arbitrary four-vector

$$\mathbf{A} = A^\alpha = (A^0, A^1, A^2, A^3),$$
(10.282)

is defined as a vector that transforms like the position vector x^α as

$$\overline{A}^\alpha = a^\alpha_\beta A^\beta.$$
(10.283)

For two four-vectors A^α and B^α their scalar product is a **four-scalar**, which is defined as

$$A^\alpha B_\alpha = A_\alpha B^\alpha = \begin{cases} A^0 B_0 + A^1 B_1 + A^2 B_2 + A^3 B_3 \\[2mm] A_0 B^0 + A_1 B^1 + A_2 B^2 + A_3 B^3 \\[2mm] A_0 B_0 - A_1 B_1 - A_2 B_2 - A_3 B_3 \\[2mm] A^0 B^0 - A^1 B^1 - A^2 B^2 - A^3 B^3 \end{cases}.$$
(10.284)

In general all tensor operations defined for the general tensors are valid in Minkowski spacetime with the Minkowski metric [Eq. (10.270)]. Higher-rank **four-tensors** can also be defined as

$$\overline{T}^{\alpha_1 \alpha_2 \cdots}_{\beta_1 \beta_2 \cdots}$$
(10.285)
$$= \frac{d\overline{x}^{\alpha_1}}{dx^{\gamma_1}} \frac{d\overline{x}^{\alpha_2}}{dx^{\gamma_2}} \cdots \frac{\partial x^{\delta_1}}{\partial \overline{x}^{\beta_1}} \frac{\partial x^{\delta_2}}{\partial \overline{x}^{\beta_2}} \cdots T^{\gamma_1 \gamma_2 \cdots}_{\delta_1 \delta_2 \cdots}.$$

10.10.6   Four-Velocity

Paths of observers in spacetime are called **worldlines** (Fig. 10.10). Because spacetime increments form a four-vector dx^α, which transforms as

$$d\overline{x}^0 = \frac{1}{\sqrt{1 - v^2/c^2}} \left[dx^0 - \left(\frac{v}{c}\right) dx^1 \right]$$
(10.286)

$$d\overline{x}^1 = \frac{1}{\sqrt{1 - v^2/c^2}} \left[-\left(\frac{v}{c}\right) dx^0 + dx^1 \right]$$
(10.287)

$$d\overline{x}^2 = dx^2$$
(10.288)

$$d\overline{x}^3 = dx^3,$$
(10.289)

we divide dx^α with a scalar $d\tau = \dfrac{ds}{c}$, called the **proper time,** to obtain the four-velocity vector as

$$u^\alpha = \frac{dx^\alpha}{d\tau}.$$
(10.290)

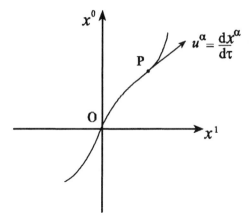

Fig. 10.10   Worldlines and four-velocity

Similarly we can define four-acceleration as

$$a^\alpha = \frac{du^\alpha}{d\tau}$$ (10.291)

$$= \frac{d^2x^\alpha}{d\tau^2}.$$

From the line element [Eq. (10.278)], it is seen that the proper time

$$d\tau = \frac{ds}{c} = \left(1 - \frac{v^2}{c^2}\right)^{\frac{1}{2}} dt,$$ (10.292)

is the time that the clocks carried by moving observers measure.

10.10.7   Four-Momentum and Conservation Laws

Using four-velocity, we can define a four-momentum as

$$p^\alpha = m_0 u^\alpha$$ (10.293)

$$= m_0 \frac{dx^\alpha}{d\tau},$$ (10.294)

where m_0 is the invariant rest mass of the particle. We can now express the energy and momentum conservation laws covariantly as the invariance of the magnitude of the four-momentum as

$$p^\alpha p_\alpha = m_0 u^\alpha u_\alpha = \text{const.}$$ (10.295)

To evaluate the constant value of $p^\alpha p_\alpha$ we use the line element and the definition of the proper time to find $u^\alpha u_\alpha$ as

$$ds^2 = c^2 dt^2 - (dx^1)^2 - (dx^2)^2 - (dx^3)^2, \tag{10.296}$$

$$c^2 = \left(\frac{dx^0}{d\tau}\right)^2 - \left(\frac{dx^1}{d\tau}\right)^2 - \left(\frac{dx^2}{d\tau}\right)^2 - \left(\frac{dx^3}{d\tau}\right)^2, \tag{10.297}$$

$$c^2 = u^\alpha u_\alpha. \tag{10.298}$$

Thus,

$$p^\alpha p_\alpha = m_0^2 c^2. \tag{10.299}$$

Writing the left-hand side of Equation (10.299) explicitly we get

$$p^0 p_0 + p^1 p_1 + p^2 p_2 + p^3 p_3 = m_0^2 c^2 \tag{10.300}$$

$$\left(p^0\right)^2 - \left(p^1\right)^2 - \left(p^2\right)^2 - \left(p^3\right)^2 = m_0^2 c^2. \tag{10.301}$$

Spatial components of the four-momentum are

$$p^i = m_0 \frac{dx^i}{d\tau}, \quad i = 1, 2, 3 \tag{10.302}$$

$$= m_0 \frac{v^i}{\sqrt{1 - v^2/c^2}}. \tag{10.303}$$

Using this in Equation (10.301) we obtain

$$\left(p^0\right)^2 = m_0^2 c^2 + \frac{m_0^2 v^2}{1 - v^2/c^2} \tag{10.304}$$

$$= m_0^2 c^2 \left[1 + \frac{v^2/c^2}{(1 - v^2/c^2)}\right] \tag{10.305}$$

or

$$p^0 = m_0 c \left[1 + \frac{v^2/c^2}{(1 - v^2/c^2)}\right]^{1/2}. \tag{10.306}$$

In order to interpret p^0, we take its classical limit as

$$p^0 = m_0 c \left[1 + \frac{1}{2} \frac{v^2}{c^2} + \cdots\right] \tag{10.307}$$

$$\simeq \frac{1}{c} [m_0 c^2 + \frac{1}{2} m_0 v^2]. \tag{10.308}$$

The second term inside the square brackets is the classical expression for the kinetic energy of a particle; however, the first term is new to Newton's mechanics. It indicates that free particles, even when they are at rest, have energy due to their rest mass. This is the Einstein's famous formula

$$E = m_0 c^2, \tag{10.309}$$

which indicates that mass and energy could be converted into each other. We can now interpret the time component of the four-momentum as E/c, where E is the total energy of the particle; thus the components of p^α become

$$p^\alpha = (\frac{E}{c}, m_0 u^i). \tag{10.310}$$

We now write the conservation of four-momentum equation as

$$p^\alpha p_\alpha = \frac{E^2}{c^2} - \frac{m_0^2 v^2}{(1 - \frac{v^2}{c^2})} = m_0^2 c^2. \tag{10.311}$$

Defining

$$m = \frac{m_0}{(1 - \frac{v^2}{c^2})^{1/2}} \tag{10.312}$$

and calling

$$p^i = m v^i, \tag{10.313}$$

we obtain a relation between the energy and the momentum of a relativistic particle as

$$E^2 = m_0^2 c^4 + p^2 c^2. \tag{10.314}$$

10.10.8 Mass of a Moving Particle

Another important consequence of the special theory of relativity is Equation (10.312), that is,

$$m = \frac{m_0}{(1 - \frac{v^2}{c^2})^{1/2}}. \tag{10.315}$$

This is the mass of a particle moving with velocity v. It says that as the speed of a particle increases its mass (inertia) also increases, thus making it harder to accelerate. As the speed of a particle approaches the speed of light, its inertia approaches infinity, thus making it impossible to accelerate beyond c.

10.10.9 Wave Four-Vector

The phase of a wave

$$\phi = \omega t - k^i x^i, \tag{10.316}$$

where we sum over i, is an invariant. This is so because it is merely a number equal to the number of wave crests getting past a given point; thus we can write

$$\omega t - k^i x^i = \overline{\omega}\overline{t} - \overline{k}^i \overline{x}^i. \tag{10.317}$$

This immediately suggests a wave four-vector as

$$k^\alpha = (k^0, k^i) \tag{10.318}$$

$$= (\frac{\omega}{c}, \frac{2\pi}{\lambda^i}), \tag{10.319}$$

where λ^i is the wavelength along x^i. Because k^α is a four-vector, it transforms as

$$\overline{k}^0 = \gamma(k^0 - \overrightarrow{\beta} \cdot \overrightarrow{k}) \tag{10.320}$$

$$\overline{k}_\parallel = \gamma(k_\parallel - \beta k^0). \tag{10.321}$$

We have written $\gamma = 1/\sqrt{1 - v^2/c^2}$ and $\overrightarrow{\beta} = \overrightarrow{v}/c$.

For light waves

$$\left|\overrightarrow{k}\right| = k^0, \quad \left|\overrightarrow{\overline{k}}\right| = \overline{k}^0; \tag{10.322}$$

thus we obtain the familiar equations for the Doppler shift:

$$\omega = \gamma\overline{\omega}(1 - \beta\cos\theta) \tag{10.323}$$

$$\tan\overline{\theta} = \sin\theta/\gamma(\cos\theta - \beta), \tag{10.324}$$

where θ and $\overline{\theta}$ are the angles of $\overrightarrow{k}$ and $\overrightarrow{\overline{k}}$ with respect to $\overrightarrow{v}$, respectively. Note that because of the presence of γ there is Doppler shift even when $\theta = \pi/2$, that is, when light is emitted perpendicular to the direction of motion.

10.10.10 Derivative Operators in Spacetime

Let us now consider the derivative operator

$$\frac{\partial}{\partial \overline{x}^\alpha}, \tag{10.325}$$

calculated in the $\overline{K}$ frame. In terms of another inertial frame K it will be given as

$$\frac{\partial}{\partial \overline{x}^\alpha} = \frac{\partial x^\beta}{\partial \overline{x}^\alpha} \frac{\partial}{\partial x^\beta}; \tag{10.326}$$

thus $\dfrac{\partial}{\partial x^\beta}$ transforms like a covariant four-vector. In general we write the four-gradient operator as

$$\partial^\alpha \equiv \frac{\partial}{\partial x_\alpha} = \left(\frac{\partial}{\partial x^0}, -\overrightarrow{\nabla} \right) \tag{10.327}$$

or

$$\partial_\alpha = \left(\frac{\partial}{\partial x^0}, \overrightarrow{\nabla} \right). \tag{10.328}$$

Four-divergence of a four-vector is a four-scalar:

$$\partial^\alpha A_\alpha = \partial_\alpha A^\alpha = \frac{\partial^0 A^0}{\partial x^0} + \overrightarrow{\nabla} \cdot \overrightarrow{A}. \tag{10.329}$$

The wave (d'Alembert) operator in space time is written as

$$\Box = \partial^\alpha \partial_\alpha = \frac{\partial^2}{\partial (x^0)^2} - \overrightarrow{\nabla}^2. \tag{10.330}$$

10.10.11   Relative Orientation of Axes in $\overline{K}$ and K Frames

Analogous to the orthogonal coordinates, any four-vector in Minkowski spacetime can be written in terms of basis vectors as

$$\mathbf{A} = (A^0, A^1, A^2, A^3) \tag{10.331}$$
$$= \widehat{\mathbf{e}}_\alpha A^\alpha. \tag{10.332}$$

In terms of another Minkowski frame, the same four-vector can be written as

$$\mathbf{A} = \widehat{\overline{\mathbf{e}}}_\alpha \overline{A}^\alpha, \tag{10.333}$$

where $\widehat{\overline{\mathbf{e}}}_\alpha$ are the new basis vectors of the frame $\overline{K}$, which is moving with respect to K with velocity v along the common direction of the x^1- and $\overline{x}^1$-axes. Both $\widehat{\overline{\mathbf{e}}}_\alpha$ and $\widehat{\mathbf{e}}_\alpha$ are unit basis vectors along their axes in their respective frames. Because $\mathbf{A}$ represents some physical property in Minkowski spacetime, Equations (10.332) and (10.333) are just different representations of $\mathbf{A}$; hence we can write

$$\widehat{\mathbf{e}}_\alpha A^\alpha = \widehat{\overline{\mathbf{e}}}_{\alpha'} \overline{A}^{\alpha'}. \tag{10.334}$$

Using the transformation property of four-vectors we write

$$\overline{A}^{\alpha'} = a^{\alpha'}_\beta A^\beta, \tag{10.335}$$

thus Equation (10.334) becomes

$$\widehat{\mathbf{e}}_\alpha A^\alpha = \widehat{\overline{\mathbf{e}}}_{\alpha'} a^{\alpha'}_\beta A^\beta. \tag{10.336}$$

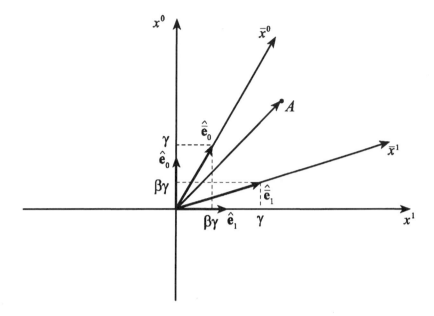

Fig. 10.11 Orientation of the $\overline{K}$ axis with respect to the K frame

We rearrange this as

$$A^\alpha \widehat{\mathbf{e}}_\alpha = A^\beta \left(\widehat{\overline{\mathbf{e}}}_{\alpha'} a_\beta^{\alpha'} \right). \tag{10.337}$$

Since α and β are dummy indices, we can replace β with α to write

$$A^\alpha \widehat{\mathbf{e}}_\alpha = A^\alpha \left(\widehat{\overline{\mathbf{e}}}_{\alpha'} a_\alpha^{\alpha'} \right), \tag{10.338}$$

which gives us the transformation law of the basis vectors as

$$\widehat{\mathbf{e}}_\alpha = \widehat{\overline{\mathbf{e}}}_{\alpha'} a_\alpha^{\alpha'}. \tag{10.339}$$

Note that this is not a component transformation. It gives $\widehat{\mathbf{e}}_\alpha$ as a linear combination of $\widehat{\overline{\mathbf{e}}}_{\alpha'}$. Using

$$a_\alpha^{\alpha'} = \begin{bmatrix} \gamma & -\beta\gamma & 0 & 0 \\ -\beta\gamma & \gamma & 0 & 0 \\ 0 & 0 & 1 & 0 \\ 0 & 0 & 0 & 1 \end{bmatrix}, \tag{10.340}$$

we obtain

$$\widehat{\mathbf{e}}_0 = \gamma\widehat{\overline{\mathbf{e}}}_0 - \beta\gamma\widehat{\overline{\mathbf{e}}}_1 \tag{10.341}$$

$$\widehat{\mathbf{e}}_1 = -\beta\gamma\widehat{\overline{\mathbf{e}}}_0 + \gamma\widehat{\overline{\mathbf{e}}}_1 \tag{10.342}$$

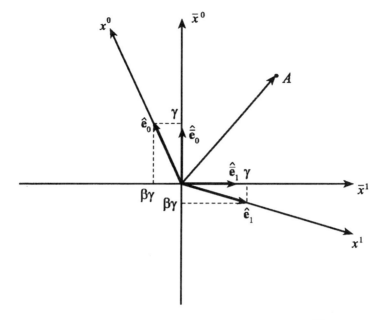

Fig. 10.12 Orientation of the K axis with respect to the $\overline{K}$ frame

and its inverse as

$$\widehat{\overline{e}}_0 = \gamma \widehat{e}_0 + \beta\gamma \widehat{e}_1 \tag{10.343}$$

$$\widehat{\overline{e}}_1 = \beta\gamma \widehat{e}_0 + \gamma \widehat{e}_1. \tag{10.344}$$

The second set gives the orientation of the $\overline{K}$ axis in terms of the K axis. Since $\beta < 1$, relative orientation of the $\overline{K}$ axis with respect to the K axis can be shown as in Figure 10.11.

Similarly, using the first set, we can obtain the relative orientation of the K axis with respect to the $\overline{K}$ axis as shown in Figure 10.12.

10.10.12   Maxwell's Equations in Minkowski Spacetime

Before the spacetime formulation of special relativity, it was known that Maxwell's equations are covariant (form-invariant) under Lorentz transformations. However, their covariance can be most conveniently expressed in terms of four-tensors.

First let us start with the conservation of charge, which can be expressed as

$$\frac{\partial \rho}{\partial t} + \overrightarrow{\nabla} \cdot \overrightarrow{J} = 0, \tag{10.345}$$

where ρ is the charge density and $\overrightarrow{J}$ is the current density in space. Defining a four-current density J^α as

$$J^\alpha = (\rho c, \overrightarrow{J}), \tag{10.346}$$

we can write Equation (10.345) in covariant form as

$$\partial_\alpha J^\alpha = 0, \tag{10.347}$$

where ∂_α stands for covariant derivative [Eq. (10.327)]. Maxwell's field equations, which are given as

$$\overrightarrow{\nabla} \cdot \overrightarrow{E} = 4\pi\rho \tag{10.348}$$

$$\overrightarrow{\nabla} \times \overrightarrow{B} - \frac{1}{c}\frac{\partial \overrightarrow{E}}{\partial t} = \frac{4\pi}{c}\overrightarrow{J} \tag{10.349}$$

$$\overrightarrow{\nabla} \cdot \overrightarrow{B} = 0 \tag{10.350}$$

$$\overrightarrow{\nabla} \times \overrightarrow{E} + \frac{1}{c}\frac{\partial \overrightarrow{B}}{\partial t} = 0, \tag{10.351}$$

determine the electric and magnetic fields for a given charge, and current distribution. We now introduce the **field-strength tensor** $F^{\alpha\beta}$ as

$$F^{\alpha\beta} = \begin{bmatrix} 0 & -E_1 & -E_2 & -E_3 \\ E_1 & 0 & -B_3 & B_2 \\ E_2 & B_3 & 0 & -B_1 \\ E_3 & -B_2 & B_1 & 0 \end{bmatrix}, \tag{10.352}$$

where the covariant components of the field-strength tensor are given as

$$F_{\alpha\beta} = g_{\alpha\gamma}g_{\delta\beta}F^{\gamma\delta} = \begin{bmatrix} 0 & E_1 & E_2 & E_3 \\ -E_1 & 0 & -B_3 & B_2 \\ -E_2 & B_3 & 0 & -B_1 \\ -E_3 & -B_2 & B_1 & 0 \end{bmatrix}. \tag{10.353}$$

Using $F^{\alpha\beta}$, the first two Maxwell's equations can be expressed in covariant form as

$$\partial_\alpha F^{\alpha\beta} = \frac{4\pi}{c}J^\beta. \tag{10.354}$$

For the remaining two Maxwell's equations we introduce the **dual field-strength tensor** $\widehat{F}^{\alpha\beta}$, which is related to the field strength tensor $F^{\alpha\beta}$ through

$$\widehat{F}^{\alpha\beta} = \frac{1}{2}\epsilon^{\alpha\beta\gamma\delta}F_{\gamma\delta} = \begin{bmatrix} 0 & -B_1 & -B_2 & -B_3 \\ B_1 & 0 & E_3 & -E_2 \\ B_2 & -E_3 & 0 & E_1 \\ B_3 & E_2 & -E_1 & 0 \end{bmatrix}, \tag{10.355}$$

where

$$\epsilon^{\alpha\beta\gamma\delta} = \begin{bmatrix} +1 \\ 0 \\ -1 \end{bmatrix} \begin{cases} \text{for even permutations} \\ \text{when any of the two indices are equal} \\ \text{for odd permutations} \end{cases} . \qquad (10.356)$$

Now the remaining two Maxwell's equations can be written as

$$\partial_\alpha \widehat{F}^{\alpha\beta} = 0. \qquad (10.357)$$

The motion of charged particles in an electromagnetic field is determined by the Lorentz force equation:

$$\frac{d\overrightarrow{p}}{dt} = q(\overrightarrow{E} + \frac{\overrightarrow{v}}{c} \times \overrightarrow{B}), \qquad (10.358)$$

where $\overrightarrow{p}$ is the spatial momentum and $\overrightarrow{v}$ is the velocity of the charged particle. We can write this in covariant form by introducing four-momentum

$$p^\alpha = (p^0, \overrightarrow{p}) \qquad (10.359)$$
$$= m_0 u^\alpha, \qquad (10.360)$$

where m_0 is the rest mass, u^α is the four-velocity, and $p^0 = E/c$. Using the derivative in terms of invariant proper time we can write Equation (10.358) as

$$\frac{dp^\alpha}{d\tau} = m_0 \frac{du^\alpha}{d\tau} \qquad (10.361)$$
$$= \frac{q}{c} F^{\alpha\beta} u_\beta . \qquad (10.362)$$

10.10.13   Transformation of Electromagnetic Fields

Because $F^{\alpha\beta}$ is a second-rank four-tensor, it transforms as

$$\overline{F}^{\alpha\beta} = \frac{d\overline{x}^\alpha}{dx^\gamma} \frac{d\overline{x}^\beta}{dx^\delta} F^{\gamma\delta}. \qquad (10.363)$$

Given the values of $F^{\gamma\delta}$ in an inertial frame K, we can find it in another inertial frame $\overline{K}$ as

$$\overline{F}^{\alpha\beta} = a^\alpha_\gamma a^\beta_\delta F^{\gamma\delta}. \qquad (10.364)$$

If $\overline{K}$ corresponds to an inertial frame moving with respect to K with velocity v along the common $\overline{x}_1$- and x_1-axes, the new components of $\overrightarrow{E}$ and $\overrightarrow{B}$ are

$$\overline{E}_1 = E_1 \qquad (10.365)$$
$$\overline{E}_2 = \gamma(E_2 - \beta B_3) \qquad (10.366)$$
$$\overline{E}_3 = \gamma(E_3 + \beta B_2) \qquad (10.367)$$

and

$$\overline{B}_1 = B_1 \tag{10.368}$$

$$\overline{B}_2 = \gamma(B_2 + \beta E_3) \tag{10.369}$$

$$\overline{B}_3 = \gamma(B_3 - \beta E_2). \tag{10.370}$$

If $\overline{K}$ is moving with respect to K with $\overrightarrow{v}$ the transformation equations are given as

$$\overrightarrow{\overline{E}} = \gamma(\overrightarrow{E} + \overrightarrow{\beta} \times \overrightarrow{B}) - \frac{\gamma^2}{1 + \gamma} \overrightarrow{\beta}(\overrightarrow{\beta} \cdot \overrightarrow{E}) \tag{10.371}$$

$$\overrightarrow{\overline{B}} = \gamma(\overrightarrow{B} - \overrightarrow{\beta} \times \overrightarrow{E}) - \frac{\gamma^2}{1 + \gamma} \overrightarrow{\beta}(\overrightarrow{\beta} \cdot \overrightarrow{B}), \tag{10.372}$$

where $\gamma = 1/(1 - \beta^2)^{1/2}$ and $\overrightarrow{\beta} = \overrightarrow{v}/c$. Inverse transformations are easily obtained by interchanging $\overrightarrow{\beta}$ with $-\overrightarrow{\beta}$.

10.10.14   Maxwell's Equations in Terms of Potentials

The Electric and magnetic fields can also be expressed in terms of the potentials $\overrightarrow{A}$ and ϕ as

$$\overrightarrow{E} = -\frac{1}{c}\frac{\partial \overrightarrow{A}}{\partial t} - \overrightarrow{\nabla}\phi \tag{10.373}$$

$$\overrightarrow{B} = \overrightarrow{\nabla} \times \overrightarrow{A}. \tag{10.374}$$

In the Lorentz gauge

$$\frac{1}{c}\frac{\partial \phi}{\partial t} + \overrightarrow{\nabla} \cdot \overrightarrow{A} = 0, \tag{10.375}$$

$\overrightarrow{A}$ and ϕ satisfy

$$\frac{1}{c^2}\frac{\partial^2 \overrightarrow{A}}{\partial t^2} - \overrightarrow{\nabla}^2 \overrightarrow{A} = \frac{4\pi}{c}\overrightarrow{J} \tag{10.376}$$

and

$$\frac{1}{c^2}\frac{\partial^2 \phi}{\partial t^2} - \overrightarrow{\nabla}^2 \phi = 4\pi\rho, \tag{10.377}$$

respectively. Defining a four-potential as

$$A^\alpha = (\phi, \overrightarrow{A}), \tag{10.378}$$

we can write Equations (10.376) and (10.377) in covariant form as

$$\Box A^\alpha = \frac{4\pi}{c}J^\alpha, \tag{10.379}$$

where the d'Alembert operator $\Box$ is defined as $\Box \equiv \dfrac{\partial^2}{d(x^0)^2} - \vec{\nabla}^2$. Now the covariant form of the Lorentz gauge [Eq. (10.375)] becomes

$$\partial_\alpha A^\alpha = 0. \tag{10.380}$$

Field-strength tensor in terms of the four-potential can be written as

$$F^{\alpha\beta} = \partial^\alpha A^\beta - \partial^\beta A^\alpha. \tag{10.381}$$

10.10.15   Covariance of Newton's Dynamical Theory

The concept of relativity was not new to Newton. In fact, it was known that the dynamical equation of Newton:

$$\frac{d\vec{p}}{dt} = \vec{F}, \tag{10.382}$$

is covariant for all uniformly moving (inertial) observers. However, the inertial observers in Newton's theory are related to each other by the Galilean transformation

$$\bar{t} = t \tag{10.383}$$

$$\bar{x}^1 = \left[x^1 - vt\right] \tag{10.384}$$

$$\bar{x}^2 = x^2 \tag{10.385}$$

$$\bar{x}^3 = x^3. \tag{10.386}$$

Note that the Lorentz transformation actually reduces to the Galilean transformation in the limit $c \to \infty$, or $v \ll c$. Before the special theory of relativity it was already known that Maxwell's equations are covariant not under Galilean but under Lorentz transformation. Considering the success of Newton's theory this was a conundrum, which took Einstein's genius to solve by saying that Lorentz transformation is the correct transformation between inertial observers and that all laws of nature should be covariant with respect to it. In this regard we also need to write Newton's dynamical equation as a four-tensor equation in spacetime.

Using the definition of four-momentum

$$p^\alpha = m_0 u^\alpha = (E/c, p^i) \tag{10.387}$$

and differentiating it with respect to invariant proper time, we can write the Newton's dynamical equation in covariant form as

$$\frac{dp^\alpha}{d\tau} = F^\alpha, \tag{10.388}$$

where F^α is now the four-force. Note that the conservation of energy and momentum is now expressed covariantly as the conservation of four-momentum, that is,

$$p^\alpha p_\alpha = \frac{E^2}{c^2} - p^2 = m_0^2 c^2. \qquad (10.389)$$

Problems

10.1 For rotations about the z-axis the transformation matrix is given as

$$\mathbf{R}_z(\theta) = \begin{bmatrix} \cos\theta & \sin\theta & 0 \\ -\sin\theta & \cos\theta & 0 \\ 0 & 0 & 1 \end{bmatrix}.$$

Show that for two successive rotations by the amounts θ_1 and θ_2 about the x_1- and x_2-axes, respectively,

$$\mathbf{R}_z(\theta_1 + \theta_2) = \mathbf{R}_z(\theta_2)\mathbf{R}_z(\theta_1)$$

is true.

10.2 Show that the rotation matrix $\mathbf{R}_z(\theta)$ is a second-rank tensor.

10.3 Using the properties of the permutation symbol and the Kronecker delta, prove the following identities in tensor notation:
   i)

$$\left[\vec{A} \times \vec{B}\right] \cdot \left[\vec{C} \times \vec{D}\right] = (\vec{A} \cdot \vec{C})(\vec{B} \cdot \vec{D}) - (\vec{A} \cdot \vec{D})(\vec{B} \cdot \vec{C}),$$

   ii)

$$\left[\vec{A} \times \left[\vec{B} \times \vec{C}\right]\right] + \left[\vec{B} \times \left[\vec{C} \times \vec{A}\right]\right] + \left[\vec{C} \times \left[\vec{A} \times \vec{B}\right]\right] = 0,$$

   iii)

$$\vec{A} \times \left[\vec{B} \times \vec{C}\right] = \vec{B}(\vec{A} \cdot \vec{C}) - \vec{C}(\vec{A} \cdot \vec{B}).$$

10.4 The trace of a second-rank tensor is defined as

$$tr(A) = A_i^i.$$

Show that trace is invariant under general coordinate transformations.

10.5 Under general coordinate transformations show that the volume element

$$d^n x = dx^1 dx^2 ... dx^n$$

transforms as

$$d^n \overline{x} = d^n x \left| \frac{\partial \overline{x}}{\partial x} \right| = d^n x \left| \frac{\partial x}{\partial \overline{x}} \right|^{-1},$$

where $\left| \frac{\partial x}{\partial \overline{x}} \right|$ is the Jacobian of the transformation, which is defined as

$$\left| \frac{\partial x}{\partial \overline{x}} \right| = \det(\frac{\partial x^i}{\partial \overline{x}^j}).$$

10.6   Show the following relations between the permutation symbol and the Kronecker delta:

$$\sum_{i=1}^{3} \epsilon_{ijk}\epsilon_{ilm} = \delta_{jl}\delta_{km} - \delta_{jm}\delta_{kl}$$

and

$$\sum_{i=1}^{3}\sum_{j=1}^{3} \epsilon_{ijk}\epsilon_{ijm} = 2\delta_{km}.$$

10.7   Evaluate

$$\sum_{ijk}\sum_{lmn} \epsilon_{ijk}\epsilon_{lmn}T_{il}T_{jm}T_{kn},$$

where T is an arbitrary matrix.

10.8   Using the symmetry of the Christoffel symbols show that the curl of a vector $\overrightarrow{v}$ can be written as

$$\left(\overrightarrow{\nabla} \times \overrightarrow{v} \right)_{ij} = v_{i;\,j} - v_{j;\,i} = \frac{\partial v_i}{\partial x^j} - \frac{\partial v_j}{\partial x^i}.$$

10.9   Prove that the covariant derivative of the metric tensor is zero, that is,

$$g_{ij;j} = 0.$$

10.10   Verify that

$$u^i_{;j} = \frac{\partial u^i}{\partial x^j} + \left\{ \begin{matrix} i \\ jk \end{matrix} \right\} u^k$$

transforms like a second-rank tensor.

10.11   Using the following symmetry properties:

$$E_{ijkl} = E_{jikl}, \ E_{ijkl} = E_{ijlk}, \ E_{ijkl} = E_{klij},$$

show that the elasticity tensor E_{ijkl} has 21 independent components.

10.12   Prove the following useful relation of the Christoffel symbol of the second kind:

$$\left\{ \begin{matrix} i \\ ik \end{matrix} \right\} = \frac{\partial(\ln\sqrt{|g|})}{\partial x^k}.$$

10.13   Show the following Christoffel symbols for a diagonal metric, where $g_{ij} = 0$ unless $i = j$:

$$\left\{ \begin{matrix} i \\ jk \end{matrix} \right\} = 0,$$

$$\left\{ \begin{matrix} i \\ jj \end{matrix} \right\} = -\frac{1}{2g_{ii}}\frac{\partial g_{jj}}{\partial x^i},$$

$$\left\{ \begin{matrix} i \\ ji \end{matrix} \right\} = \left\{ \begin{matrix} i \\ ij \end{matrix} \right\} = \frac{\partial}{\partial x^j}(\ln\sqrt{g_{ii}}),$$

$$\left\{ \begin{matrix} i \\ ii \end{matrix} \right\} = \frac{\partial}{\partial x^i}(\ln\sqrt{g_{ii}}).$$

In these equations summation convention is not used and different letters imply different indices.

10.14   Show that the tensor equation

$$A_{ij} = c + B_{ij}^{kl}C_{kl} + E_{i;\ jk}^{k}$$

transforms as

$$A_{ij}' = c + B_{ij}'^{kl}C_{kl}' + E_{i;\ jk}'^{k}.$$

10.15   Find the expressions for the div and the grad operators for the following metrics:

i)

$$ds^2 = dr^2 + \rho^2 d\theta^2 + dz^2$$

ii)

$$ds^2 = \left[\frac{1}{1 - kr^2/R^2}\right]dr^2 + r^2 d\theta^2 + r^2\sin^2\theta d\phi^2,$$

where k takes the values $k = 0, 1, -1$.

10.16   Write the Laplacian operator for the following metrics:

i)

$$ds^2 = R^2(d\chi^2 + \sin^2\chi d\theta^2 + \sin^2\chi \sin^2\theta d\phi^2),$$

where $\chi \in [0, \pi]$, $\theta \in [0, \pi]$, $\phi \in [0, 2\pi]$.

ii)

$$ds^2 = R^2(d\chi^2 + \sinh^2\chi d\theta^2 + \sinh^2\chi \sin^2\theta d\phi^2),$$

where $\chi \in [0, \infty]$, $\theta \in [0, \pi]$, $\phi \in [0, 2\pi]$.

What geometries do these metrics represent?

10.17   Write the line element for the elliptic cylindrical coordinates (u, v, z) :

$$x = a \cosh u \cos v$$
$$y = a \sinh u \sin v$$
$$z = z.$$

10.18   Write the covariant and the contravariant components of the metric tensor for the parabolic cylindrical coordinates (u, v, z) :

$$x = (1/2)(u^2 - v^2)$$
$$y = uv$$
$$z = z.$$

10.19   Write the Laplacian in the parabolic coordinates (u, v, ϕ) :

$$x = uv \cos \phi$$
$$y = uv \sin \phi$$
$$z = (1/2)(u^2 - v^2).$$

10.20   Calculate all the nonzero components of the Christoffel symbols for the metric in the line element

$$ds^2 = dr^2 + r^2 d\theta^2 + r^2 \sin^2\theta d\phi^2,$$

where $r \in [0, \infty]$, $\theta \in [0, \pi]$, $\phi \in [0, 2\pi]$,

10.21   Write the contravariant gradient of a scalar function in spherical polar coordinates.

10.22   Write the divergence operator in spherical polar coordinates.

10.23   Write the Laplace operator in cylindrical coordinates.

10.24   In four dimensions spherical polar coordinates (r, χ, θ, ϕ) are defined as

$$
\begin{aligned}
x &= r \sin \chi \sin \theta \cos \phi \\
y &= r \sin \chi \sin \theta \sin \phi \\
z &= r \sin \chi \cos \theta \\
w &= r \cos \chi.
\end{aligned}
$$

i) Write the line element

$$ds^2 = dx^2 + dy^2 + dz^2 + dw^2$$

in (r, χ, θ, ϕ).

ii) What are the ranges of (r, χ, θ, ϕ)?

iii) Write the metric for the three dimensional surface (S-3) of a hypersphere.

10.25 Which one of the following matrices are Cartesian tensors:

i)

$$\begin{bmatrix} y^2 & -xy \\ -xy & x^2 \end{bmatrix},$$

ii)

$$\begin{bmatrix} xy & y^2 \\ x^2 & -xy \end{bmatrix}.$$

10.26 In cosmology the line element for a closed universe is given as

$$ds^2 = c^2 dt^2 - R_0(t)^2 [d\chi^2 + \sin^2 \chi d\theta^2 + \sin^2 \chi \sin^2 \theta d\phi^2],$$

where t is the universal time and χ, θ, and ϕ are the angular coordinates with the ranges

$$\chi \in [0, \pi], \ \theta \in [0, \pi], \ \phi \in [0, 2\pi].$$

For a static universe with constant radius R_0, the wave equation for the massless conformal scalar field is given as

$$\Box \Phi(t, \chi, \theta, \phi) + \frac{1}{R_0^2} \Phi(t, \chi, \theta, \phi) = 0.$$

$\Box$ is the d'Alembert (wave) operator:

$$\Box = g_{\mu\nu} \partial^\mu \partial_\nu,$$

where ∂_ν stands for the covariant derivative.

Using a separable solution of the form

$$\Phi(t, \chi, \theta, \phi) = T(t)X(\chi)Y(\theta, \phi),$$

show that the wave equation for the massless conformal scalar field can be written as

$$\left[\frac{1}{T(t)} \frac{d^2 T(t)}{c^2 dt^2}\right] - \frac{1}{R_0^2} \left[\frac{1}{X(\chi)} \frac{d^2 X(\chi)}{d\chi^2} + \frac{2 \cos \chi}{\sin \chi} \frac{1}{X(\chi)} \frac{dX(\chi)}{d\chi} - 1\right]$$
$$-\frac{1}{R_0^2 \sin^2 \chi} \left[\frac{1}{Y(\theta, \phi)} \frac{\partial^2 Y(\theta, \phi)}{\partial \theta^2}\right.$$
$$\left. +\frac{\cos \theta}{\sin \theta} \frac{1}{Y(\theta, \phi)} \frac{\partial Y(\theta, \phi)}{\partial \theta} + \frac{1}{\sin^2 \theta} \frac{1}{Y(\theta, \phi)} \frac{\partial^2 Y(\theta, \phi)}{\partial \phi^2}\right] = 0.$$

10.27   Using the four-current

$$J^\alpha = (\rho c, \overrightarrow{J})$$

and the field-strength tensor

$$F^{\alpha\beta} = \begin{bmatrix} 0 & -E_1 & -E_2 & -E_3 \\ E_1 & 0 & -B_3 & B_2 \\ E_2 & B_3 & 0 & -B_1 \\ E_3 & -B_2 & B_1 & 0 \end{bmatrix},$$

show that Maxwell's field equations:

$$\overrightarrow{\nabla} \cdot \overrightarrow{E} = 4\pi\rho$$

$$\overrightarrow{\nabla} \times \overrightarrow{B} - \frac{1}{c}\frac{\partial \overrightarrow{E}}{\partial t} = \frac{4\pi}{c}\overrightarrow{J}$$

$$\overrightarrow{\nabla} \cdot \overrightarrow{B} = 0$$

$$\overrightarrow{\nabla} \times \overrightarrow{E} + \frac{1}{c}\frac{\partial \overrightarrow{B}}{\partial t} = 0$$

can be written in covariant form as

$$\partial_\alpha F^{\alpha\beta} = \frac{4\pi}{c}J^\alpha$$

and

$$\partial_\alpha \widehat{F}^{\alpha\beta} = 0.$$

The dual field-strength tensor $\widehat{F}^{\alpha\beta}$ is defined as

$$\widehat{F}^{\alpha\beta} = \frac{1}{2}\epsilon^{\alpha\beta\gamma\delta}F_{\gamma\delta} = \begin{bmatrix} 0 & -B_1 & -B_2 & -B_3 \\ B_1 & 0 & E_3 & -E_2 \\ B_2 & -E_3 & 0 & E_1 \\ B_3 & E_2 & -E_1 & 0 \end{bmatrix},$$

where $\epsilon^{\alpha\beta\gamma\delta}$ is the permutation symbol.

10.28   If $\overline{K}$ corresponds to an inertial frame moving with respect to K with velocity v along the x_1-axis, show that the new components of $\overrightarrow{E}$ and $\overrightarrow{B}$ become

$$\overline{E}_1 = E_1$$
$$\overline{E}_2 = \gamma(E_2 - \beta B_3)$$
$$\overline{E}_3 = \gamma(E_3 + \beta B_2)$$

and

$$\overline{B}_1 = B_1$$
$$\overline{B}_2 = \gamma(B_2 + \beta E_3)$$
$$\overline{B}_3 = \gamma(B_3 - \beta E_2).$$

10.29 Show that the field-strength tensor can also be written as

$$F^{\alpha\beta} = \partial^\alpha A^\beta - \partial^\beta A^\alpha \ ,$$

where the four-potential is defined as

$$A^\alpha = (\phi, \overrightarrow{A})$$

and

$$\overrightarrow{E} = -\frac{1}{c}\frac{\partial \overrightarrow{A}}{\partial t} - \overrightarrow{\nabla}\phi$$
$$\overrightarrow{B} = \overrightarrow{\nabla} \times \overrightarrow{A}.$$

11

CONTINUOUS GROUPS and REPRESENTATIONS

In everyday language the word "symmetry" is usually associated with "beautiful". It is for this reason that it has been used extensively in architecture and art. Symmetry also allows us to build complex structures or patterns by distributing relatively simple building blocks according to a rule. Most of the symmetries around us are with respect to rotations and reflections. Symmetry in science usually means invariance of a given system under some operation. As in crystals and molecules, symmetries can be discrete, where applications of certain amounts of rotation, translation, or reflections produce the same structure. After the discovery of the Lagrangian formulation of mechanics it became clear that the conservation laws are also due to symmetries in nature. For example, conservation of angular momentum follows from the symmetry of a given system with respect to rotations, whereas the conservation of linear momentum follows from symmetry with respect to translations. In these symmetries a system is carried from one identical state to another continuously; hence they are called continuous (Lie) symmetries. With the discovery of quantum mechanics the relation between conservation laws and symmetries has even become more important. Conservation of isospin and flavor in nuclear and particle physics are important tools in building theories. Gauge symmetry has become an important guide in constructing new models.

Group theory allows us to study symmetries in a systematic way. In this chapter we discuss continuous groups and Lie algebras, which allow us to study symmetries of physical systems and their relation to coordinate transformations. We also discuss representation theory and its applications. In particular, we concentrate on the representations of the rotation group $R(3)$ and the special unitary group $SU(2)$ and their relation. We also discuss the inhomogeneous Lorentz group and introduce its Lie algebra. An advanced treatment of spherical harmonics is given in terms of the rotation group and its representations. We also discuss symmetry of differential equations and the extension (prolongation) of generators.

Even though this chapter could be read independently, occasionally we refer to results from Chapter 10. In particular, we recommend reading the sections on orthogonal transformations and Lorentz transformations before reading this chapter.

11.1   DEFINITION OF A GROUP

The basic properties of rotations in three dimensions are just the properties that make a group. A group is an ensemble of elements

$$G \in \{g_0, g_1, g_2, ...\}, \tag{11.1}$$

with the following properties:

i) For any two elements, $g_a, g_b \in G$, a rule of composition is defined such that

$$g_a g_b \in G. \tag{11.2}$$

ii) With the composition rule defined above, for any three elements, $g_a, g_b, g_c \in G$, the associative rule

$$(g_a g_b) g_c = g_a (g_b g_c) \tag{11.3}$$

is obeyed.

iii) G contains the unit element g_0 such that for any $g \in G$,

$$g g_0 = g_0 g = g. \tag{11.4}$$

iv) For every $g \in G$, there exists an inverse element, $g^{-1} \in G$, such that

$$g g^{-1} = g^{-1} g = g_0. \tag{11.5}$$

11.1.1   Terminology

In n dimensions the set $\{\mathbf{A}\}$ of all linear transformations with $\det \mathbf{A} \neq 0$ forms a group called the **general linear group in n dimensions**:

$$GL(n).$$

We use the letter $\mathbf{A}$ for the transformation and its operator or matrix representation. Matrix elements of $\mathbf{A}$ could be real or complex; thus we also write

$$GL(n, R) \text{ or } GL(n, C).$$

The rotation group in two dimensions is shown as

$$R(2).$$

Elements of $R(2)$ are characterized by a single continuous parameter θ, which is the angle of rotation. A group whose elements are characterized by a number of continuous parameters is called a **continuous or Lie Group**. In a continuous group the group elements can be generated continuously from the identity element. The rotation group in three dimensions is shown as

$$R(3).$$

Elements of $R(3)$ are characterized in terms of three independent parameters, which are usually taken as the three Euler angles. $R(n)$ is a subgroup of $GL(n)$. $R(n)$ is also a subgroup of the group of n-dimensional orthogonal transformations, which is shown as

$$O(n).$$

Elements of $O(n)$ satisfy $|\det \mathbf{A}|^2 = 1$. The set of all linear transformations with $\det \mathbf{A} = 1$ forms a group called the **special linear group**:

$$SL(n, R) \text{ or } SL(n, C).$$

Elements of $O(n)$ satisfying $\det \mathbf{A} = 1$ form the **special orthogonal group** shown as

$$SO(n),$$

which is also a subgroup of $SL(n, R)$. The group of linear transformations acting on vectors in n-dimensional complex space and satisfying $|\det \mathbf{A}|^2 = 1$ is called the **unitary group**:

$$U(n).$$

If the elements of the unitary group also satisfy $\det \mathbf{A} = 1$ we have the **special unitary group**:

$$SU(n).$$

$SU(n)$ is a subgroup of $U(n)$. Groups with infinite number of elements like $R(n)$ are called **infinite groups**. A group with finite number of elements is called a **finite group**, where the number of elements in a finite group is called the **order of the group**. Elements of a group in general do not commute, that is, $g_a g_b$ for any two elements need not be equal to $g_b g_a$. If in a group for every pair $g_a g_b = g_b g_a$ holds, then the group is said to be **commutative or Abelian**.

11.2   INFINITESIMAL RING OR LIE ALGEBRA

For a continuous (Lie) group G if $\mathbf{A}(t) \in G$, we have seen that its generator [Eq. (10.76)] is given as $\mathbf{X} = \mathbf{A}'(0)$. The ensemble $\{\mathbf{A}'(0)\}$ of transformations is called the **infinitesimal ring** or **Lie algebra** of G, and it is denoted by $^r G$.

Differentiating $\mathbf{A}(at) \in G$ we get $\mathbf{A}'(at) = a\mathbf{A}'(at)$, where a is a constant. Substituting $t = 0$ we see that

$$a\mathbf{A}'(0) = a\mathbf{X} \in^r G. \tag{11.6}$$

Also, if $\mathbf{A}(t)$ and $\mathbf{B}(t)$ are any two elements of G, then

$$\mathbf{C}(t) = \mathbf{A}(t)\mathbf{B}(t) \in G. \tag{11.7}$$

Differentiating this and substituting $t = 0$ and using the fact that $\mathbf{A}(0) = \mathbf{B}(0) = \mathbf{I}$, we obtain

$$\begin{aligned} \mathbf{C}'(0) &= \mathbf{A}'(0) + \mathbf{B}'(0), \\ &= \mathbf{X} + \mathbf{Y}. \end{aligned} \tag{11.8}$$

Hence, $\mathbf{X} + \mathbf{Y} \in {}^r G$ if $\mathbf{X}, \mathbf{Y} \in {}^r G$. Lie has proven some very interesting theorems about the relations between continuous groups and their generators. One of these is the relation

$$\begin{aligned} [\mathbf{X}_i, \mathbf{X}_j] &= \mathbf{X}_i\mathbf{X}_j - \mathbf{X}_j\mathbf{X}_i \\ &= \sum_{k=1}^{n} c_{ij}^k \mathbf{X}_k, \end{aligned} \tag{11.9}$$

which says that the commutator of two generators is always a linear combination of generators. The constants c_{ij}^k are called the **structure constants** of the group G, and n is the dimension of G.

We have so far shown the following properties of $^r G$:

If $\mathbf{X}, \mathbf{Y} \in {}^r G$, then

i)

$$a\mathbf{X} \in {}^r G, \text{ where } a \text{ is a real number} \tag{11.10}$$

ii)

$$\mathbf{X} + \mathbf{Y} = \mathbf{Y} + \mathbf{X} \in {}^r G \tag{11.11}$$

iii)

$$\begin{aligned} [\mathbf{X}_i, \mathbf{X}_j] &= \mathbf{X}_i\mathbf{X}_j - \mathbf{X}_j\mathbf{X}_i \\ &= \sum_{k=1}^{n} c_{ij}^k \mathbf{X}_k \in {}^r G \end{aligned} \tag{11.12}$$

Thus, ^rG is a vector space with the multiplication defined in (iii). The dimension of this vector space is equal to the number n of the parameters of the group G; thus it has a basis: $\mathbf{X}_1, \mathbf{X}_2, ..., \mathbf{X}_n$ and every element $\mathbf{X}$ of ^rG can be expressed as a linear combination as

$$\mathbf{X} = x_1\mathbf{X}_1 + x_2\mathbf{X}_2 + \cdots + x_n\mathbf{X}_n. \tag{11.13}$$

From these it is clear that a continuous group completely determines the structure of its Lie algebra. Lie has also proved the converse, that is, the local structure in some neighborhood of the identity of a continuous group is completely determined by the structure constants c_{ij}^k.

11.3 LIE ALGEBRA OF THE ROTATION GROUP $R(3)$

We have seen that the generators of the rotation group are given as [Eq. (10.84)]

$$\mathbf{X}_1 = \begin{bmatrix} 0 & 0 & 0 \\ 0 & 0 & 1 \\ 0 & -1 & 0 \end{bmatrix}, \quad \mathbf{X}_2 = \begin{bmatrix} 0 & 0 & -1 \\ 0 & 0 & 0 \\ 1 & 0 & 0 \end{bmatrix}, \quad \mathbf{X}_3 = \begin{bmatrix} 0 & 1 & 0 \\ -1 & 0 & 0 \\ 0 & 0 & 0 \end{bmatrix}. \tag{11.14}$$

These generators satisfy the commutation relation

$$[\mathbf{X}_i, \mathbf{X}_j] = -\epsilon_{ijk}\mathbf{X}_k, \tag{11.15}$$

where ϵ_{ijk} is the permutation (Levi-Civita) symbol. Using these generators we can write an arbitrary infinitesimal rotation as

$$\mathbf{r} = (\mathbf{I} + \mathbf{X}_1\varepsilon_1 + \mathbf{X}_2\varepsilon_2 + \mathbf{X}_3\varepsilon_3)\mathbf{r}(0), \tag{11.16}$$

$$\delta\mathbf{r} = (\mathbf{X}_1\varepsilon_1 + \mathbf{X}_2\varepsilon_2 + \mathbf{X}_3\varepsilon_3)\mathbf{r}(0). \tag{11.17}$$

Defining two vectors, that is, Θ with the components

$$\Theta = \begin{bmatrix} \varepsilon_1 \\ \varepsilon_2 \\ \varepsilon_3 \end{bmatrix} \tag{11.18}$$

and

$$\mathbf{X} = \mathbf{X}_1\widehat{\mathbf{e}}_1 + \mathbf{X}_2\widehat{\mathbf{e}}_2 + \mathbf{X}_3\widehat{\mathbf{e}}_3, \tag{11.19}$$

we can write

$$\mathbf{r} = (\mathbf{I} + \mathbf{X} \cdot \Theta). \tag{11.20}$$

The operator for finite rotations, where Θ has now the components $(\theta_1, \theta_2, \theta_3)$, can be constructed by successive infinitesimal rotations as

$$\mathbf{r} = \lim_{N \to \infty} (\mathbf{I} + \frac{1}{N}\mathbf{X} \cdot \Theta)(\mathbf{I} + \frac{1}{N}\mathbf{X} \cdot \Theta) \cdots (\mathbf{I} + \frac{1}{N}\mathbf{X} \cdot \Theta)\mathbf{r}(0)$$
$$= e^{\mathbf{X} \cdot \Theta}\mathbf{r}(0). \tag{11.21}$$

Euler's theorem (Goldstein et al.) allows us to look at Equation (11.16) as an infinitesimal rotation about a single axis along the unit normal

$$\widehat{\mathbf{n}} = \frac{1}{\sqrt{\varepsilon_1^2 + \varepsilon_2^2 + \varepsilon_3^2}} \begin{bmatrix} \varepsilon_1 \\ \varepsilon_2 \\ \varepsilon_3 \end{bmatrix} \tag{11.22}$$

by the amount

$$d\theta = \sqrt{\varepsilon_1^2 + \varepsilon_2^2 + \varepsilon_3^2}$$

as

$$\mathbf{r}(\theta) = (\mathbf{I} + \mathbf{X} \cdot \widehat{\mathbf{n}} d\theta)\mathbf{r}(0). \tag{11.23}$$

For finite rotations (Fig. 11.1) we write

$$\mathbf{r}(\theta) = e^{\mathbf{X} \cdot \widehat{\mathbf{n}}\theta}\mathbf{r}(0). \tag{11.24}$$

11.3.1   Another Approach to $^r R(3)$

Let us approach $^r R(3)$ from the operator (active) point of view. We now look for an operator $\mathbf{O}$, which acts on a function $f(\mathbf{r})$ and rotates it clockwise, while keeping the coordinate axis fixed (Fig. 11.2). $f(\mathbf{r})$ could be representing a physical system or a physical property. Instead of the Euler angles we use $\theta_1, \theta_2, \theta_3$, which represent rotations about the x_1- , x_2- , x_3-axes, respectively. For a counterclockwise rotation of the coordinate system about the x_3-axis we write

$$\bar{\mathbf{r}} = \mathbf{R}_3\mathbf{r}. \tag{11.25}$$

This induces the following change in $f(\mathbf{r})$:

$$\overline{f}(\mathbf{r}') = f(\mathbf{R}_3\mathbf{r}). \tag{11.26}$$

If $\mathbf{O}_3$ is an operator acting on $f(\mathbf{r})$, and since both views should agree, we write

$$\mathbf{O}_3 f(\mathbf{r}) = \overline{f}(\mathbf{r}) = f(\mathbf{R}_3^{-1}\mathbf{r}). \tag{11.27}$$

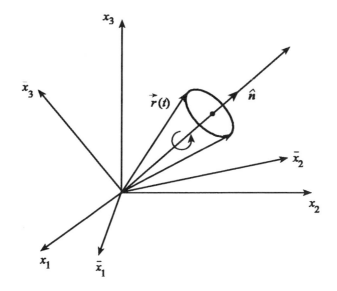

Fig. 11.1   Rotation by θ about an axis along $\hat{\mathbf{n}}$

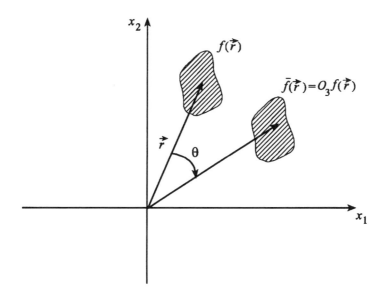

Fig. 11.2   Effect of O_3 on f(r)

We now look for the generator of infinitesimal rotations of $\overline{f}(\mathbf{r})$ about the x_3-axis, which we show by $\overline{\mathbf{X}}_3$. Using Equation (10.42) we write the rotation matrix $\mathbf{R}_3^{-1}$ as

$$\mathbf{R}_3^{-1}(\theta) = \mathbf{R}_3(-\theta). \tag{11.28}$$

For infinitesimal rotations we write $\theta_3 = \varepsilon_3$ and

$$\mathbf{O}_3 = (\mathbf{I} - \overline{\mathbf{X}}_3\varepsilon_3). \tag{11.29}$$

The minus sign in $\mathbf{O}_3$ indicates that the physical system is rotated clockwise (Fig. 11.2), thus

$$(\mathbf{I} - \overline{\mathbf{X}}_3\varepsilon_3)f(\mathbf{r}) = f(x_1\cos\varepsilon_3 - x_2\sin\varepsilon_3, \ x_1\sin\varepsilon_3 + x_2\cos\varepsilon_3, \ x_3), \tag{11.30}$$

$$\overline{\mathbf{X}}_3f(\mathbf{r}) = -\left[\frac{f(x_1\cos\varepsilon_3 - x_2\sin\varepsilon_3, \ x_1\sin\varepsilon_3 + x_2\cos\varepsilon_3, \ x_3) - f(\mathbf{r})}{\varepsilon_3}\right], \tag{11.31}$$

$$\overline{\mathbf{X}}_3f(\mathbf{r}) = -\left[\frac{f(x_1 - x_2\varepsilon_3, \ x_2 + x_1\varepsilon_3, \ x_3) - f(\mathbf{r})}{\varepsilon_3}\right]. \tag{11.32}$$

Using Taylor series expansion about the point (x_1, x_2, x_3) and taking the limit as $\varepsilon_3 \to 0$ we obtain

$$\overline{\mathbf{X}}_3f(\mathbf{r}) = -\left(x_1\frac{\partial}{\partial x_2} - x_2\frac{\partial}{\partial x_1}\right)f(\mathbf{r}). \tag{11.33}$$

Thus

$$\overline{\mathbf{X}}_3 = -\left(x_1\frac{\partial}{\partial x_2} - x_2\frac{\partial}{\partial x_1}\right). \tag{11.34}$$

Similarly, or by cyclic permutations of x_1, x_2, and x_3 we obtain the other operators as

$$\overline{\mathbf{X}}_2 = -\left(x_3\frac{\partial}{\partial x_1} - x_1\frac{\partial}{\partial x_3}\right) \tag{11.35}$$

and

$$\overline{\mathbf{X}}_1 = -\left(x_2\frac{\partial}{\partial x_3} - x_3\frac{\partial}{\partial x_2}\right). \tag{11.36}$$

Note that aside from a minus sign, $\overline{\mathbf{X}}_i$ satisfy the same commutation relations as $\mathbf{X}_i$, that is,

$$\left[\overline{\mathbf{X}}_i, \overline{\mathbf{X}}_j\right] = \epsilon_{ijk}\overline{\mathbf{X}}_k. \tag{11.37}$$

An arbitrary infinitesimal rotation of $f(\mathbf{r})$ can now be written as

$$\overline{f}(\mathbf{r}) = (\mathbf{I} - \overline{\mathbf{X}}_1 \delta\theta_1 - \overline{\mathbf{X}}_2 \delta\theta_2 - \overline{\mathbf{X}}_3 \delta\theta_3) f(\mathbf{r}) \qquad (11.38)$$

$$= (\mathbf{I} - \overline{\mathbf{X}} \cdot \hat{\mathbf{n}} \delta\theta) f(\mathbf{r}), \qquad (11.39)$$

where $\hat{\mathbf{n}}$ and $\delta\theta$ are defined as in Equation (11.23). Now the finite rotation operator becomes

$$\mathbf{O}f(\mathbf{r}) = e^{-\overline{\mathbf{X}} \cdot \hat{\mathbf{n}}\theta} f(\mathbf{r}). \qquad (11.40)$$

For applications in quantum mechanics it is advantageous to adopt the view that the operator $\mathbf{O}$ still rotates the state function $f(\mathbf{r})$ counterclockwise, so that the direction of $\hat{\mathbf{n}}$ is positive when θ is measured with respect to the right-hand rule (Merzbacher, p. 174). In this regard we write

$$\mathbf{O}f(\mathbf{r}) = e^{\overline{\mathbf{X}} \cdot \hat{\mathbf{n}}\theta} f(\mathbf{r}) \qquad (11.41)$$

for the operator that rotates the physical system counterclockwise. We will come back to this point later when we discuss angular momentum and quantum mechanics.

11.4 GROUP INVARIANTS

It is obvious that for $R(3)$ the magnitude of a vector

$$\widetilde{\mathbf{r}}\mathbf{r} = [x_1, x_2, x_3] \begin{bmatrix} x_1 \\ x_2 \\ x_3 \end{bmatrix} \qquad (11.42)$$

$$= x_1^2 + x_2^2 + x_3^2,$$

is not changed through a rotation. A vector function $F(\mathbf{r})$ that remains unchanged by all $g \in G$, that is,

$$F(\mathbf{r}) = F(\overline{\mathbf{r}}), \qquad (11.43)$$

is called a group invariant. We now determine a group whose invariant is

$$x_1^2 + x_2^2 . \qquad (11.44)$$

This group will naturally be a subgroup of $GL(2, R)$. An element of this group can be represented by the transformation

$$\begin{bmatrix} \overline{x}_1 \\ \overline{x}_2 \end{bmatrix} = \begin{bmatrix} a & b \\ c & d \end{bmatrix} \begin{bmatrix} x_1 \\ x_2 \end{bmatrix} . \qquad (11.45)$$

From the group invariant

$$x_1^2 + x_2^2 = \overline{x}_1^2 + \overline{x}_2^2 , \tag{11.46}$$

it follows that the transformation matrix elements must satisfy the relations

$$a^2 + c^2 = 1 , \tag{11.47}$$
$$b^2 + d^2 = 1 , \tag{11.48}$$
$$ab + cd = 0 . \tag{11.49}$$

This means that only one of (a, b, c, d) could be independent. Choosing a new parameter as

$$a = \cos \theta, \tag{11.50}$$

we see that the transformation matrix has the following possible two forms:

$$\begin{bmatrix} \cos \theta & \sin \theta \\ -\sin \theta & \cos \theta \end{bmatrix} \text{ and } \begin{bmatrix} \cos \theta & -\sin \theta \\ -\sin \theta & -\cos \theta \end{bmatrix} . \tag{11.51}$$

The first matrix is familiar; it corresponds to rotations, that is, $R(2)$. However the second matrix has

$$\det \begin{bmatrix} \cos \theta & -\sin \theta \\ -\sin \theta & -\cos \theta \end{bmatrix} = -1 \tag{11.52}$$

and can be written as

$$\begin{bmatrix} \cos \theta & -\sin \theta \\ -\sin \theta & -\cos \theta \end{bmatrix} = \begin{bmatrix} \cos \theta & \sin \theta \\ -\sin \theta & \cos \theta \end{bmatrix} \begin{bmatrix} 1 & 0 \\ 0 & -1 \end{bmatrix} . \tag{11.53}$$

This is reflection, that is,

$$\overline{x}_1 = x_1 \tag{11.54}$$
$$\overline{x}_2 = -x_2 ,$$

followed by a rotation. The group which leaves $x_1^2 + x_2^2$ invariant is called the orthogonal group $O(2)$, where $R(2)$ is its subgroup with the determinant of all of its elements equal to one. $R(2)$ is also a subgroup of $SO(2)$, that includes rotations and translations.

11.4.1   Lorentz Transformation

As another example for a group invariant let us take

$$x^2 - y^2.$$

We can write this as

$$x^2 - y^2 = \begin{bmatrix} x & y \end{bmatrix} \begin{bmatrix} 1 & 0 \\ 0 & -1 \end{bmatrix} \begin{bmatrix} x \\ y \end{bmatrix}. \tag{11.55}$$

For a linear transformation between $(\overline{x}, \overline{y})$ and (x, y) we write

$$\begin{bmatrix} \overline{x} \\ \overline{y} \end{bmatrix} = \begin{bmatrix} a & b \\ c & d \end{bmatrix} \begin{bmatrix} x \\ y \end{bmatrix}. \tag{11.56}$$

Invariance of $(x^2 - y^2)$ can now be expressed as

$$\begin{aligned}
\overline{x}^2 - \overline{y}^2 &= \begin{bmatrix} \overline{x} & \overline{y} \end{bmatrix} \begin{bmatrix} 1 & 0 \\ 0 & -1 \end{bmatrix} \begin{bmatrix} \overline{x} \\ \overline{y} \end{bmatrix} \\
&= \begin{bmatrix} x & y \end{bmatrix} \begin{bmatrix} a & c \\ b & d \end{bmatrix} \begin{bmatrix} 1 & 0 \\ 0 & -1 \end{bmatrix} \begin{bmatrix} a & b \\ c & d \end{bmatrix} \begin{bmatrix} x \\ y \end{bmatrix} \\
&= \begin{bmatrix} x & y \end{bmatrix} \begin{bmatrix} a^2 - c^2 & ab - cd \\ ab - cd & b^2 - d^2 \end{bmatrix} \begin{bmatrix} x \\ y \end{bmatrix} \\
&= \begin{bmatrix} x & y \end{bmatrix} \begin{bmatrix} 1 & 0 \\ 0 & -1 \end{bmatrix} \begin{bmatrix} x \\ y \end{bmatrix} \\
&= x^2 - y^2 .
\end{aligned} \tag{11.57}$$

From above we see that for $(x^2 - y^2)$ to remain invariant under the transformation [Eq. (11.56)], components of the transformation matrix must satisfy

$$\begin{aligned}
a^2 - c^2 &= 1 \\
b^2 - d^2 &= -1 \\
ab - cd &= 0 .
\end{aligned} \tag{11.58}$$

This means that only one of (a, b, c, d) can be independent. Defining a new parameter χ as

$$a = \cosh \chi, \tag{11.59}$$

we see that the transformation matrix in Equation (11.56) can be written as

$$\begin{bmatrix} \cosh \chi & \sinh \chi \\ \sinh \chi & \cosh \chi \end{bmatrix} . \tag{11.60}$$

Introducing

$$\cosh \chi = \gamma \tag{11.61}$$
$$\sinh \chi = -\gamma \beta \tag{11.62}$$
$$\tanh \chi = -\beta, \tag{11.63}$$

where

$$\frac{1}{\sqrt{1-\beta^2}} = \gamma \text{ and } \beta = v/c, \tag{11.64}$$

along with the identification

$$x = ct$$
$$y = x,$$

we obtain

$$\left[\begin{array}{c} c\bar{t} \\ \bar{x} \end{array} \right] = \left[\begin{array}{cc} \gamma & -\beta\gamma \\ -\beta\gamma & \gamma \end{array} \right] \left[\begin{array}{c} ct \\ x \end{array} \right]. \tag{11.65}$$

This is nothing but the Lorentz transformation [Eqs. (10.252–10.255):

$$c\bar{t} = \frac{1}{\sqrt{1-(v/c)^2}}(ct - vx/c) \tag{11.66}$$

$$\bar{x} = \frac{1}{\sqrt{1-(v/c)^2}}(x - vt), \tag{11.67}$$

which leaves distances in spacetime, that is,

$$\left(c^2 t^2 - x^2 \right), \tag{11.68}$$

invariant.

11.5   UNITARY GROUP IN TWO DIMENSIONS: $U(2)$

Quantum mechanics is formulated in complex space. Hence the components of the transformation matrix are in general complex numbers. The scalar or inner product of two vectors in n-dimensional complex space is defined as

$$(\mathbf{x}, \mathbf{y}) = x_1^* y_1 + x_2^* y_2 + \cdots + x_n^* y_n , \tag{11.69}$$

where x^* means the complex conjugate of x. Unitary transformations are linear transformations, which leaves

$$(\mathbf{x}, \mathbf{x}) = |\mathbf{x}|^2 = x_1^* x_1 + x_2^* x_2 + \cdots + x_n^* x_n \tag{11.70}$$

invariant. All such transformations form the unitary group $U(n)$. An element of $U(2)$ can be written as

$$\mathbf{u} = \left[\begin{array}{cc} A & B \\ C & D \end{array} \right], \tag{11.71}$$

where $A, B, C,$ and D are in general complex numbers. Invariance of $(\mathbf{x}, \mathbf{x})$ gives the unitarity condition as

$$\mathbf{u}^\dagger \mathbf{u} = \mathbf{u}\mathbf{u}^\dagger = \mathbf{I} , \tag{11.72}$$

where

$$\mathbf{u}^\dagger = \tilde{\mathbf{u}}^* \tag{11.73}$$

is called the Hermitian conjugate of $\mathbf{u}$. Using the unitarity condition we can write

$$\mathbf{u}^\dagger \mathbf{u} = \begin{bmatrix} A^* & C^* \\ B^* & D^* \end{bmatrix} \cdot \begin{bmatrix} A & B \\ C & D \end{bmatrix} \tag{11.74}$$

$$= \begin{bmatrix} |A|^2 + |C|^2 & A^*B + C^*D \\ AB^* + D^*C & |B|^2 + |D|^2 \end{bmatrix} \tag{11.75}$$

$$= \begin{bmatrix} 1 & 0 \\ 0 & 1 \end{bmatrix} , \tag{11.76}$$

which gives

$$\begin{aligned} |A|^2 + |C|^2 &= 1 \\ |B|^2 + |D|^2 &= 1 \\ A^*B + C^*D &= 0 . \end{aligned} \tag{11.77}$$

From elementary matrix theory (Boas), the inverse of $\mathbf{u}$ can be found as

$$\mathbf{u}^{-1} = \begin{bmatrix} D & -B \\ -C & A \end{bmatrix} . \tag{11.78}$$

Because for $SU(2)$ the inverse of $\mathbf{u}$ is also equal to $\mathbf{u}^\dagger$, we write

$$\mathbf{u}^{-1} = \mathbf{u}^\dagger \tag{11.79}$$

$$\begin{bmatrix} D & -B \\ -C & A \end{bmatrix} = \begin{bmatrix} A^* & C^* \\ B^* & D^* \end{bmatrix} . \tag{11.80}$$

This gives $D = A^*$ and $C = -B^*$; thus $\mathbf{u}$ becomes

$$\mathbf{u} = \begin{bmatrix} A & B \\ -B^* & A^* \end{bmatrix} . \tag{11.81}$$

Taking the determinant of the unitarity condition [Eq. (11.71)] and using the fact that $\det \mathbf{u}^\dagger = \det \mathbf{u}$, we obtain

$$|\det \mathbf{u}|^2 = 1 . \tag{11.82}$$

11.6   SPECIAL UNITARY GROUP $SU(2)$

In quantum mechanics we are particularly interested in $SU(2)$, a subgroup of $U(2)$, where the group elements satisfy the condition

$$\det \mathbf{u} = 1. \tag{11.83}$$

For $SU(2)$, A and B in the transformation matrix

$$\mathbf{u} = \begin{bmatrix} A & B \\ -B^* & A^* \end{bmatrix} \tag{11.84}$$

satisfy

$$\det \mathbf{u} = |A|^2 + |B|^2 = 1. \tag{11.85}$$

Expressing A and B as

$$A = a + id \tag{11.86}$$
$$B = c + ib\,, \tag{11.87}$$

we see that the unitary matrix has the form

$$\mathbf{u} = \begin{bmatrix} a + id & c + ib \\ -c + ib & a - id \end{bmatrix}. \tag{11.88}$$

This can be written as

$$\mathbf{u} = a\mathbf{I} + i(b\sigma_1 + c\sigma_2 + d\sigma_3), \tag{11.89}$$

where σ_i are the Pauli spin matrices:

$$\sigma_1 = \begin{bmatrix} 0 & 1 \\ 1 & 0 \end{bmatrix}, \ \ \sigma_2 = \begin{bmatrix} 0 & -i \\ i & 0 \end{bmatrix}, \ \ \sigma_3 = \begin{bmatrix} 1 & 0 \\ 0 & -1 \end{bmatrix}, \tag{11.90}$$

which satisfy

$$\sigma_i^2 = 1, \tag{11.91}$$
$$\sigma_i\sigma_j = -\sigma_j\sigma_i = i\sigma_k, \tag{11.92}$$

where (i, j, k) are cyclic permutations of $(1, 2, 3)$. Condition (11.83) on the determinant $\mathbf{u}$ gives

$$a^2 + b^2 + c^2 + d^2 = 1. \tag{11.93}$$

This allows us to choose (a, b, c, d) as

$$a = \cos\omega, \ \ b^2 + c^2 + d^2 = \sin^2\omega, \tag{11.94}$$

thus Equation (11.89) becomes

$$\mathbf{u}(\omega) = \mathbf{I}\cos\omega + i\mathbf{S}\sin\omega, \tag{11.95}$$

where we have defined

$$\mathbf{S} - \alpha\sigma_1 + \beta\sigma_2 + \gamma\sigma_3 \tag{11.96}$$

and

$$
\begin{aligned}
\alpha &= \frac{b}{(b^2 + c^2 + d^2)^{1/2}} \,, \\
\beta &= \frac{c}{(b^2 + c^2 + d^2)^{1/2}} \,, \\
\gamma &= \frac{d}{(b^2 + c^2 + d^2)^{1/2}} \,.
\end{aligned}
\tag{11.97}
$$

Note that $\mathbf{u}$ in Equation (11.71) has in general eight parameters. However, among these eight parameters we have five relations, four of which come from the unitarity condition (11.72). We also have the condition fixing the value of the determinant (11.83) for $SU(2)$; thus $SU(2)$ can only have three independent parameters. These parameters can be represented by a point on the three-dimensional surface (S-3) of a unit hypersphere defined by Equation (11.93). In Equation (11.95) we represent the elements of $SU(2)$ in terms of

$$\omega, \text{ and } (\alpha, \beta, \gamma) , \tag{11.98}$$

where (α, β, γ) satisfies

$$\alpha^2 + \beta^2 + \gamma^2 = 1 . \tag{11.99}$$

By changing (α, β, γ) on S-3 we can vary ω in

$$\mathbf{u}(\omega) = \mathbf{I}\cos\omega + \widetilde{\mathbf{X}}\sin\omega, \tag{11.100}$$

where we have defined

$$\widetilde{\mathbf{X}} = i\mathbf{S}, \tag{11.101}$$

hence

$$\widetilde{\mathbf{X}}^2 = -\mathbf{S}^2. \tag{11.102}$$

11.7   LIE ALGEBRA OF $SU(2)$

In the previous section we have seen that the elements of the $SU(2)$ group are given as

$$\mathbf{u}(\omega) = \mathbf{I}\cos\omega + \widetilde{\mathbf{X}}\sin\omega. \tag{11.103}$$

The 2×2 transformation matrix, $\mathbf{u}(\omega)$, transforms complex vectors

$$\mathbf{v} = \begin{bmatrix} v_1 \\ v_2 \end{bmatrix} \tag{11.104}$$

as

$$\overline{\mathbf{v}} = \mathbf{u}(\omega)\mathbf{v}. \tag{11.105}$$

Infinitesimal transformations of $SU(2)$, analogous to $R(3)$, can be written as

$$\mathbf{v}(\omega) = (\mathbf{I} + \widetilde{\mathbf{X}}\delta\omega)\mathbf{v}(0) \tag{11.106}$$

$$\delta\mathbf{v} = \widetilde{\mathbf{X}}\mathbf{v}(0)\delta\omega \tag{11.107}$$

$$\mathbf{v}'(\omega) = \widetilde{\mathbf{X}}\mathbf{v}(0), \tag{11.108}$$

where the generator $\widetilde{\mathbf{X}}$ is obtained in terms of the generators $\widetilde{\mathbf{X}}_1, \widetilde{\mathbf{X}}_2, \widetilde{\mathbf{X}}_3$ as

$$\widetilde{\mathbf{X}} = \mathbf{u}'(0) = i\mathbf{S} \tag{11.109}$$

$$= \alpha \begin{bmatrix} 0 & i \\ i & 0 \end{bmatrix} + \beta \begin{bmatrix} 0 & 1 \\ -1 & 0 \end{bmatrix} + \gamma \begin{bmatrix} i & 0 \\ 0 & -i \end{bmatrix}. \tag{11.110}$$

Writing $\widetilde{\mathbf{X}}$ as

$$\widetilde{\mathbf{X}} = \alpha\,\widetilde{\mathbf{X}}_1 + \beta\widetilde{\mathbf{X}}_2 + \gamma\widetilde{\mathbf{X}}_3,$$

we identify

$$\widetilde{\mathbf{X}}_1 = \begin{bmatrix} 0 & i \\ i & 0 \end{bmatrix}, \; \widetilde{\mathbf{X}}_2 = \begin{bmatrix} 0 & 1 \\ -1 & 0 \end{bmatrix}, \; \widetilde{\mathbf{X}}_3 = \begin{bmatrix} i & 0 \\ 0 & -i \end{bmatrix}. \tag{11.111}$$

$\widetilde{\mathbf{X}}_i$ satisfy the following commutation relation:

$$\left[\widetilde{\mathbf{X}}_i, \widetilde{\mathbf{X}}_j\right] = -2\epsilon_{ijk}\widetilde{\mathbf{X}}_k. \tag{11.112}$$

For $R(3)$ we have seen that the generators satisfy Equation (11.15)

$$[\mathbf{X}_i, \mathbf{X}_j] = -\epsilon_{ijk}\mathbf{X}_k, \tag{11.113}$$

and the exponential form of the transformation matrix for finite rotations was [Eq. (11.24)]

$$\mathbf{r}(t) = e^{\mathbf{X}\cdot\hat{\mathbf{n}}\theta}\mathbf{r}(0). \tag{11.114}$$

If we make the correspondence

$$2\mathbf{X}_i \leftrightarrow \widetilde{\mathbf{X}}_i, \tag{11.115}$$

the two algebras are identical and the groups $SU(2)$ and $R(3)$ are called **isomorphic**. Defining a unit normal vector

$$\hat{\mathbf{n}} = (\alpha, \beta, \gamma), \tag{11.116}$$

we can now use Equation (11.114) to write the exponential form of the transformation matrix for finite rotations in $SU(2)$ as

$$\mathbf{v}(t) = e^{\frac{1}{2}\tilde{\mathbf{X}}\cdot\hat{\mathbf{n}}\theta}\mathbf{v}(0). \tag{11.117}$$

Since

$$\tilde{\mathbf{X}} = i\mathbf{S}, \tag{11.118}$$

This gives us the exponential form of the transformation matrix for $SU(2)$ as

$$\mathbf{v}(t) = e^{\frac{1}{2}i\mathbf{S}\cdot\hat{\mathbf{n}}\theta}\mathbf{v}(0). \tag{11.119}$$

In quantum mechanics the active view, where the vector is rotated counterclockwise, is preferred; thus the operator is taken as

$$e^{-\frac{1}{2}i\mathbf{S}\cdot\hat{\mathbf{n}}\theta}, \tag{11.120}$$

where $\mathbf{S}$ corresponds to the spin angular momentum operator.

$$\mathbf{S} = \alpha\sigma_1 + \beta\sigma_2 + \gamma\sigma_3. \tag{11.121}$$

In Section 11.13 we will see that the presence of the factor $1/2$ in operator (10.120) is very important and it actually indicates that the correspondence between $SU(2)$ and $R(3)$ is two-to-one.

11.7.1   Another Approach to $^r SU(2)$

Using the generators (11.111) and (11.103) we can write

$$\tilde{\mathbf{X}} = \alpha\tilde{\mathbf{X}}_1 + \beta\tilde{\mathbf{X}}_2 + \gamma\tilde{\mathbf{X}} \tag{11.122}$$

and

$$\mathbf{u}(\alpha, \beta, \gamma) = (\mathbf{I}\cos\omega + \tilde{\mathbf{X}}\sin\omega) \tag{11.123}$$

$$= \begin{bmatrix} \cos\omega + i\gamma\sin\omega & (\beta + i\alpha)\sin\omega \\ (-\beta + i\alpha)\sin\omega & \cos\omega - i\gamma\sin\omega \end{bmatrix}. \tag{11.124}$$

The transformation

$$\bar{\mathbf{v}} = \mathbf{u}(\omega)\mathbf{v} \tag{11.125}$$

induces the following change in a function $f(v_1, v_2)$:

$$\overline{f}(\overline{\mathbf{v}}) = f[\mathbf{u}(\alpha, \beta, \gamma)\mathbf{v}]. \tag{11.126}$$

Taking the active view we define an operator $\mathbf{O}$, which acts on $f(\mathbf{v})$. Since both views should agree, we write

$$\mathbf{O}f(\mathbf{r}) = \overline{f}(\mathbf{r}) \tag{11.127}$$

$$= f[\mathbf{u}^{-1}(\alpha, \beta, \gamma)\mathbf{r}] \tag{11.128}$$

$$= f[\mathbf{u}(-\alpha, -\beta, -\gamma)\mathbf{r}]. \tag{11.129}$$

For a given small ω we can write $\mathbf{u}(-\alpha, -\beta, -\gamma)$ in terms of α, β, γ as

$$\mathbf{u}(-\alpha, -\beta, -\gamma) = \begin{bmatrix} 1 - i\gamma\omega & (-\beta - i\alpha)\omega \\ (\beta - i\alpha)\omega & 1 + i\gamma\omega \end{bmatrix}. \tag{11.130}$$

Thus we obtain

$$\overline{v}_1 = (1 - i\gamma\omega)v_1 + (-\beta - i\alpha)\omega v_2 \tag{11.131}$$

$$\overline{v}_2 = (\beta - i\alpha)\omega v_1 + (1 + i\gamma\omega)v_2. \tag{11.132}$$

Writing $\delta v_i = \overline{v}_i - v_i$ this becomes

$$\delta v_1 = -i\gamma\omega v_1 + (-\beta - i\alpha)\omega v_2 \tag{11.133}$$

$$\delta v_2 = (\beta - i\alpha)\omega v_1 + i\gamma\omega v_2. \tag{11.134}$$

We now write the effect of the operator $\mathbf{O}_1$, which induces infinitesimal changes in a function $f(v_1, v_2)$ as

$$(\mathbf{I} - \mathbf{O}_1\omega)f(v_1, v_2) = f(v_1, v_2) + \left[\frac{\partial f(v_1, v_2)}{\partial v_1} \frac{\partial(\delta v_1)}{\partial \alpha} + \frac{\partial f(v_1, v_2)}{\partial v_2} \frac{\partial(\delta v_2)}{\partial \alpha} \right]$$

$$= f(v_1, v_2) + \left[-i\omega v_2 \frac{\partial f(v_1, v_2)}{\partial v_1} - i\omega v_1 \frac{\partial f(v_1, v_2)}{\partial v_2} \right]$$

$$= f(v_1, v_2) - i\omega \left[v_2 \frac{\partial}{\partial v_1} + v_1 \frac{\partial}{\partial v_2} \right] f(v_1, v_2). \tag{11.135}$$

This gives the generator $\mathbf{O}_1$ as

$$\mathbf{O}_1 = i \left[v_2 \frac{\partial}{\partial v_1} + v_1 \frac{\partial}{\partial v_2} \right]. \tag{11.136}$$

Similarly, we write

$$(\mathbf{I} - \mathbf{O}_2\omega)f(v_1, v_2) = f(v_1, v_2) + \left[\frac{\partial f(v_1, v_2)}{\partial v_1} \frac{\partial(\delta v_1)}{\partial \beta} + \frac{\partial f(v_1, v_2)}{\partial v_2} \frac{\partial(\delta v_2)}{\partial \beta} \right]$$

$$= f(v_1, v_2) + \left[-\omega v_2 \frac{\partial f(v_1, v_2)}{\partial v_1} + \omega v_1 \frac{\partial f(v_1, v_2)}{\partial v_2} \right]$$

$$= f(v_1, v_2) + \omega \left[-v_2 \frac{\partial}{\partial v_1} + v_1 \frac{\partial}{\partial v_2} \right] f(v_1, v_2) \tag{11.137}$$

and

$$(\mathbf{I} - \mathbf{O}_3\omega)f(v_1, v_2) = f(v_1, v_2) + \left[\frac{\partial f(v_1, v_2)}{\partial v_1} \frac{\partial(\delta v_1)}{\partial \gamma} + \frac{\partial f(v_1, v_2)}{\partial v_2} \frac{\partial(\delta v_2)}{\partial \gamma} \right]$$

$$= f(v_1, v_2) + \left[-i\omega v_1 \frac{\partial f(v_1, v_2)}{\partial v_1} + i\omega v_2 \frac{\partial f(v_1, v_2)}{\partial v_2} \right]$$

$$= f(v_1, v_2) + \omega i \left[v_2 \frac{\partial}{\partial v_2} - v_1 \frac{\partial}{\partial v_1} \right] f(v_1, v_2). \quad (11.138)$$

These give us the remaining generators as

$$\mathbf{O}_2 = \left[v_2 \frac{\partial}{\partial v_1} - v_1 \frac{\partial}{\partial v_2} \right] \quad (11.139)$$

and

$$\mathbf{O}_3 = i \left[v_1 \frac{\partial}{\partial v_1} - v_2 \frac{\partial}{\partial v_2} \right]. \quad (11.140)$$

$\mathbf{O}_i$ satisfy the commutation relation

$$[\mathbf{O}_i, \mathbf{O}_j] = 2\epsilon_{ijk}\mathbf{O}_k . \quad (11.141)$$

The sign difference with Equation (11.112) is again due to the fact that in the passive view axes are rotated counterclockwise, while in the active view vectors are rotated clockwise.

11.8 LORENTZ GROUP AND ITS LIE ALGEBRA

The ensemble of objects $\left[a_\gamma^\alpha\right]$, which preserve the length of four-vectors in Minkowski spacetime and which satisfy the relation

$$g_{\alpha\beta}a_\gamma^\alpha a_\delta^\beta = g_{\gamma\delta}, \quad (11.142)$$

form the **Lorentz group**. If we exclude reflections and consider only the transformations that can be continuously generated from the identity transformation we have the **homogeneous Lorentz group**. The group that includes reflections as well as the translations is called the **inhomogeneous Lorentz group** or the **Poincare group**. From now on we consider the homogeneous Lorentz group and omit the word homogeneous.

Given the coordinates of the position four-vector x^α in the K frame, elements of the Lorentz group, $\left[a_\beta^\alpha\right]$, give us the components, $\overline{x}^\alpha$, in the $\overline{K}$ frame as

$$\overline{x}^\alpha = a_\beta^\alpha x^\beta. \quad (11.143)$$

In matrix notation we can write this as

$$\overline{\mathbf{x}} = \mathbf{Ax}, \tag{11.144}$$

where

$$\overline{\mathbf{x}} = \begin{bmatrix} \overline{x}^0 \\ \overline{x}^1 \\ \overline{x}^2 \\ \overline{x}^3 \end{bmatrix}, \quad \mathbf{A} = \begin{bmatrix} a_0^0 & a_1^0 & a_2^0 & a_3^0 \\ a_0^1 & a_1^1 & a_2^1 & a_3^1 \\ a_0^2 & a_1^2 & a_2^2 & a_3^2 \\ a_0^3 & a_1^3 & a_2^3 & a_3^3 \end{bmatrix}, \quad \mathbf{x} = \begin{bmatrix} x^0 \\ x^1 \\ x^2 \\ x^3 \end{bmatrix}. \tag{11.145}$$

For transformations preserving the magnitude of four-vectors we write

$$\widetilde{\overline{\mathbf{x}}}\mathbf{g}\overline{\mathbf{x}} = \widetilde{\mathbf{x}}\mathbf{g}\mathbf{x}, \tag{11.146}$$

and after substituting Equation (11.144) we obtain the analogue of the orthogonality condition as

$$\widetilde{\mathbf{A}}\mathbf{g}\mathbf{A} = \mathbf{g}. \tag{11.147}$$

Elements of the Lorentz group are 4×4 matrices, which means they have 16 components. From the orthogonality condition (11.147), which is a symmetric matrix, we have 10 relations among these 16 components; thus only 6 of them are independent. In other words, the Lorentz group is a six-parameter group. These six parameters can be conveniently thought of as the three Euler angles specifying the orientation of the spatial axis and the three components of $\overrightarrow{\beta}$ specifying the relative velocity of the two inertial frames.

Guided by our experience with $R(3)$, to find the generators of the Lorentz group we start with the ansatz that $\mathbf{A}$ can be written in exponential form as

$$\mathbf{A} = e^{\mathbf{L}}, \tag{11.148}$$

where $\mathbf{L}$ is a 4×4 matrix. From the theory of matrices we can write (Gantmacher)

$$\det \mathbf{A} = \det e^{\mathbf{L}} = e^{Tr\mathbf{L}}. \tag{11.149}$$

Using this equation and considering only the proper Lorentz transformations, where $\det \mathbf{A} = 1$, we conclude that $\mathbf{L}$ is traceless. Thus the generator of the proper Lorentz transformations is a real 4×4 traceless matrix.

We now multiply Equation (11.147) from the left by $\mathbf{g}^{-1}$ and from the right by $\mathbf{A}^{-1}$ to write

$$\mathbf{g}^{-1}\widetilde{\mathbf{A}}\mathbf{g}\left[\mathbf{A}\mathbf{A}^{-1}\right] = \mathbf{g}^{-1}\mathbf{g}\mathbf{A}^{-1}, \tag{11.150}$$

which gives

$$\mathbf{g}^{-1}\widetilde{\mathbf{A}}\mathbf{g} = \mathbf{A}^{-1}. \tag{11.151}$$

Since for the Minkowski metric

$$\mathbf{g}^{-1} = \mathbf{g},$$ (11.152)

this becomes

$$\mathbf{g}\widetilde{\mathbf{A}}\mathbf{g} = \mathbf{A}^{-1}.$$ (11.153)

Using Equation (11.153) and the relations $\mathbf{g}^2 = \mathbf{I}$, $\widetilde{\mathbf{A}} = e^{\widetilde{\mathbf{L}}}$, and $\mathbf{A}^{-1} = e^{-\mathbf{L}}$ we can also write

$$\mathbf{g}\widetilde{\mathbf{A}}\mathbf{g} = e^{\mathbf{g}\widetilde{\mathbf{L}}\mathbf{g}} = e^{-\mathbf{L}};$$ (11.154)

thus

$$\mathbf{g}\widetilde{\mathbf{L}}\mathbf{g} = -\mathbf{L}.$$ (11.155)

Since $\widetilde{\mathbf{g}} = \mathbf{g}$ we obtain

$$\widetilde{\mathbf{g}\mathbf{L}} = -\mathbf{g}\mathbf{L}.$$ (11.156)

This equation shows that $\mathbf{g}\mathbf{L}$ is an antisymmetric matrix. Considering that $\mathbf{g}$ is the Minkowski metric and $\mathbf{L}$ is traceless, we can write the general form of $\mathbf{L}$ as

$$\mathbf{L} = \begin{bmatrix} 0 & -L_{01} & -L_{02} & -L_{03} \\ -L_{01} & 0 & L_{12} & L_{13} \\ -L_{02} & -L_{12} & 0 & L_{23} \\ -L_{03} & -L_{13} & -L_{23} & 0 \end{bmatrix}.$$ (11.157)

Introducing six independent parameters $(\beta_1, \beta_2, \beta_3)$ and $(\theta_1, \theta_2, \theta_3)$, this can also be written as

$$\mathbf{L} = \beta_1\mathbf{V}_1 + \beta_2\mathbf{V}_2 + \beta_3\mathbf{V}_3 + \theta_1\mathbf{X}_1 + \theta_2\mathbf{X}_2 + \theta_3\mathbf{X}_3,$$ (11.158)

where

$$\mathbf{V}_1 = \begin{bmatrix} 0 & -1 & 0 & 0 \\ -1 & 0 & 0 & 0 \\ 0 & 0 & 0 & 0 \\ 0 & 0 & 0 & 0 \end{bmatrix}, \quad \mathbf{V}_2 = \begin{bmatrix} 0 & 0 & -1 & 0 \\ 0 & 0 & 0 & 0 \\ -1 & 0 & 0 & 0 \\ 0 & 0 & 0 & 0 \end{bmatrix}, \quad \mathbf{V}_3 = \begin{bmatrix} 0 & 0 & 0 & -1 \\ 0 & 0 & 0 & 0 \\ 0 & 0 & 0 & 0 \\ -1 & 0 & 0 & 0 \end{bmatrix}$$ (11.159)

and

$$\mathbf{X}_1 = \begin{bmatrix} 0 & 0 & 0 & 0 \\ 0 & 0 & 0 & 0 \\ 0 & 0 & 0 & 1 \\ 0 & 0 & -1 & 0 \end{bmatrix}, \quad \mathbf{X}_2 = \begin{bmatrix} 0 & 0 & 0 & 0 \\ 0 & 0 & 0 & -1 \\ 0 & 0 & 0 & 0 \\ 0 & 1 & 0 & 0 \end{bmatrix}, \quad \mathbf{X}_3 = \begin{bmatrix} 0 & 0 & 0 & 0 \\ 0 & 0 & 1 & 0 \\ 0 & -1 & 0 & 0 \\ 0 & 0 & 0 & 0 \end{bmatrix}.$$ (11.160)

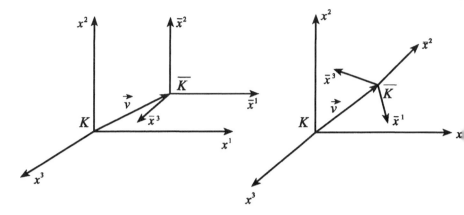

Fig. 11.3 Boost and boost plus rotation

Note that $(\mathbf{X}_1, \mathbf{X}_2, \mathbf{X}_3)$ are the generators of the infinitesimal rotations about the x^1-, x^2-, x^3-axes [Eq. (10.84)], respectively, and $(\mathbf{V}_1, \mathbf{V}_2, \mathbf{V}_3)$ are the generators of the infinitesimal Lorentz transformations or **boosts** from one inertial observer to another moving with respect to each other with velocities $(\beta_1, \beta_2, \beta_3)$ along the x^1-, x^2-, x^3-axes, respectively. These six generators satisfy the commutation relations

$$[\mathbf{X}_i, \mathbf{X}_j] = -\epsilon_{ijk}\mathbf{X}_k, \tag{11.161}$$

$$[\mathbf{X}_i, \mathbf{V}_j] = -\epsilon_{ijk}\mathbf{V}_k, \tag{11.162}$$

$$[\mathbf{V}_i, \mathbf{V}_j] = \epsilon_{ijk}\mathbf{X}_k . \tag{11.163}$$

The first of these three commutators is just the commutation relation for the rotation group $R(3)$; thus the rotation group is also a subgroup of the Lorentz group. The second commutator shows that $\mathbf{V}_i$ transforms under rotation like a vector. The third commutator indicates that boosts in general do not commute, but more important than this, two successive boosts is equal to a boost plus a rotation (Fig. 11.3), that is,

$$\mathbf{V}_i\mathbf{V}_j = \mathbf{V}_j\mathbf{V}_i + \epsilon_{ijk}\mathbf{X}_k. \tag{11.164}$$

Thus boosts alone do not form a group. An important kinematic consequence of this is known as the Thomas precession.

We now define two unit 3-vectors:

$$\hat{\mathbf{n}} = \frac{1}{\sqrt{\theta_1^2 + \theta_2^2 + \theta_3^2}} \begin{bmatrix} \theta_1 \\ \theta_2 \\ \theta_3 \end{bmatrix} \tag{11.165}$$

and

$$\widehat{\beta} = \frac{1}{\sqrt{\beta_1^2 + \beta_2^2 + \beta_3^2}} \begin{bmatrix} \beta_1 \\ \beta_2 \\ \beta_3 \end{bmatrix} \tag{11.166}$$

and introduce the parameters

$$\theta = \sqrt{\theta_1^2 + \theta_2^2 + \theta_3^2} \tag{11.167}$$

and

$$\beta = \sqrt{\beta_1^2 + \beta_2^2 + \beta_3^2} \, , \tag{11.168}$$

so that we can summarize these results as

$$\mathbf{L} = \mathbf{X} \cdot \widehat{\mathbf{n}} \theta + \mathbf{V} \cdot \widehat{\beta} \beta \tag{11.169}$$

and

$$\mathbf{A} = e^{\mathbf{X} \cdot \widehat{\mathbf{n}} \theta + \mathbf{V} \cdot \widehat{\beta} \beta} \, . \tag{11.170}$$

For pure rotations this reduces to the rotation matrix in Equation (11.24)

$$\mathbf{A}_{\text{rot.}} = e^{\mathbf{X} \cdot \widehat{\mathbf{n}} \theta} .$$

For pure boosts it is equal to Equation (10.276)

$$\mathbf{A}_{boost}(\beta) = e^{\mathbf{V} \cdot \widehat{\beta} \beta} \tag{11.171}$$

$$= \begin{bmatrix} \gamma & -\beta_1\gamma & -\beta_2\gamma & -\beta_3\gamma \\ -\beta_1\gamma & 1+\dfrac{(\gamma-1)\beta_1^2}{\beta^2} & \dfrac{(\gamma-1)\beta_1\beta_2}{\beta^2} & \dfrac{(\gamma-1)\beta_1\beta_3}{\beta^2} \\ -\beta_2\gamma & \dfrac{(\gamma-1)\beta_2\beta_1}{\beta^2} & 1+\dfrac{(\gamma-1)\beta_2^2}{\beta^2} & \dfrac{(\gamma-1)\beta_2\beta_3}{\beta^2} \\ -\beta_3\gamma & \dfrac{(\gamma-1)\beta_3\beta_1}{\beta^2} & \dfrac{(\gamma-1)\beta_3\beta_2}{\beta^2} & 1+\dfrac{(\gamma-1)\beta_3^2}{\beta^2} \end{bmatrix} ,$$

where

$$\beta_1 = \frac{v_1}{c}, \ \beta_2 = \frac{v_2}{c}, \ \beta_2 = \frac{v_2}{c}. \tag{11.172}$$

For a boost along the x_1 direction $\beta_2 = \beta_3 = 0$ and $\beta = \beta_1$, thus Equation (11.172) reduces to

$$\mathbf{A}_{boost}(\beta_1) = \begin{bmatrix} \gamma & -\beta\gamma & 0 & 0 \\ -\beta\gamma & \gamma & 0 & 0 \\ 0 & 0 & 1 & 0 \\ 0 & 0 & 0 & 1 \end{bmatrix} . \tag{11.173}$$

Using the parametrization

$$-\beta_1 = \tanh \chi \qquad (11.174)$$

$$\gamma = \cosh \chi \qquad (11.175)$$

$$-\gamma\beta_1 = \sinh \chi \qquad (11.176)$$

Equation (11.173) becomes

$$\mathbf{A}_{boost}(\beta_1) = \begin{bmatrix} \cosh \chi & \sinh \chi & 0 & 0 \\ \sinh \chi & \cosh \chi & 0 & 0 \\ 0 & 0 & 1 & 0 \\ 0 & 0 & 0 & 1 \end{bmatrix}, \qquad (11.177)$$

which is reminiscent of the rotation matrices [Eqs. (10.42–43)] with hyperbolic functions instead of the trigonometric. Notice that in accordance with our previous treatment in Section 11.2, the generator $\mathbf{V}_1$ can also be obtained from

$$\mathbf{V}_1 = \mathbf{A}'_{boost}(\beta_1 = 0). \qquad (11.178)$$

The other generators can also be obtained similarly.

11.9   GROUP REPRESENTATIONS

As defined in Section 11.1, a group with its general element shown with g is an abstract concept. It gains practical meaning only when G is assigned physical operations, $D(g)$, to its elements that act in some space of objects called the **representation space**. These objects could be functions, vectors, and in general tensors. As in the rotation group $R(3)$ group representations can be accomplished by assigning matrices to each element of G, which correspond to rotation matrices acting on vectors. Given a particular representation $D(g)$, another representation can be constructed by a similarity transformation as

$$D'(g) = S^{-1}D(g)S. \qquad (11.179)$$

Representations that are connected by a similarity transformation are called **equivalent representations**. Given two representations $D^{(1)}(g)$ and $D^{(2)}(g)$ we can construct another representation:

$$D(g) = D^{(1)}(g) \oplus D^{(1)}(g) = \begin{bmatrix} D^{(1)}(g) & 0 \\ 0 & D^{(2)}(g) \end{bmatrix}, \qquad (11.180)$$

where $D(g)$ is called the product of $D^{(1)}(g)$ and $D^{(2)}(g)$. If $D^{(1)}(g)$ has dimension n_1, that is, composed of $n_1 \times n_1$ matrices, and $D^{(2)}(g)$ has dimension n_2, the product representation has the dimension $n_1 + n_2$. $D(g)$ is also called a

reducible representation. If $D(g)$ cannot be split into the sums of smaller representations by similarity transformations, it is called an **irreducible representation**. Irreducible representations are very important and they form the building blocks of representation theory. A matrix that commutes with every element of an irreducible representation is a multiple of the unit matrix. We now present without proof an important lemma due to Schur for the criterion of irreducibility of a group representation.

11.9.1 Schur's Lemma

Let $D^{(1)}(g)$ and $D^{(2)}(g)$ be two irreducible representations with dimensions n_1 and n_2, and suppose that a matrix A exists such that

$$AD^{(1)}(g) = D^{(2)}(g)A \tag{11.181}$$

for all g in G. Then either $A = 0$, or $n_1 = n_2$ and $\det A \neq 0$, and the two representations $D^{(1)}(g)$ and $D^{(2)}(g)$ are equivalent.

By a similarity transformation if $D(g)$ can be written as

$$D(g) = \begin{bmatrix} D^{(1)}(g) & 0 & 0 & 0 \\ 0 & D^{(2)}(g) & 0 & 0 \\ 0 & 0 & D^{(2)}(g) & 0 \\ 0 & 0 & 0 & D^{(3)}(g) \end{bmatrix}, \tag{11.182}$$

we write

$$D(g) = D^{(1)}(g) \oplus 2D^{(2)}(g) \oplus D^{(3)}(g). \tag{11.183}$$

If $D^{(1)}(g)$, $D^{(2)}(g)$, and $D^{(3)}(g)$ cannot be reduced further, they are irreducible and $D(g)$ is called a **completely reducible representation**.

Every group has a trivial one-dimensional representation, where each group element is represented by the number one. In an irreducible representation, say $D^{(2)}(g)$ as in the above case, then every element of the representation space is transformed into another element of that space by the action of the group elements $D^{(2)}(g)$. For example, for the rotation group $R(3)$ a three-dimensional representation is given by the rotation matrices and the representation space is the Cartesian vectors. In other words, rotation of Cartesian vectors always results in another Cartesian vector.

11.9.2 Group Character

The characterization of representations by explicitly giving the matrices that represent the group elements is not possible, because by a similarity transformation one could obtain a different set of matrices. Thus we need to identify properties that remain invariant under similarity transformations. One such property is the trace of a matrix. We now define character $\chi^{(i)}(g)$ as the trace of the matrices $D^{(i)}(g)$, that is,

$$\chi^{(i)}(g) = \sum_{j=1}^{n_i} D_{jj}^{(i)}(g). \tag{11.184}$$

11.9.3   Unitary Representation

Representations of a group by unitary (transformation) matrices are called unitary representations. Unitary transformations leave the quadratic form

$$|\mathbf{x}|^2 = \sum_{i=1}^{n} |x_i|^2 \tag{11.185}$$

invariant, which is equivalent to the inner product

$$(\mathbf{x}, \mathbf{y}) = \sum_{i=1}^{n} x_i^* y_i \tag{11.186}$$

in complex space.

11.10   REPRESENTATIONS OF $R(3)$

We now construct the representations of the rotation group. Using Cartesian tensors we can easily construct the irreducible representations as

$$\begin{bmatrix} D^{(1)}(g) & 0 & 0 & 0 \\ 0 & D^{(3)}(g) & 0 & 0 \\ 0 & 0 & D^{(5)}(g) & 0 \\ 0 & 0 & 0 & \ddots \end{bmatrix}, \tag{11.187}$$

where $D^{(1)}(g)$ is the trivial representation, the number one, that acts on scalars. $D^{(3)}(g)$ are given as the 3×3 rotation matrices that act on vectors. The superscript 3 indicates the degrees of freedom, in this case the three independent components of a vector. $D^{(5)}(g)$ is the representation corresponding to the transformation matrices for the symmetric second-rank Cartesian tensors. In this case the dimension of the representation comes from the fact that a second-rank symmetric tensor has six independent components; removing the trace leaves five. In general a symmetric tensor of rank n has $(2n+1)$ independent components; thus the associated representation is $(2n+1)$-dimensional.

11.11   SPHERICAL HARMONICS AND REPRESENTATIONS OF $R(3)$

An elegant and also useful way of obtaining representations of $R(3)$ is to construct them through the transformation properties of the spherical harmonics. The trivial representation $D^{(1)}(g)$ simply consists of the transformation of Y_{00} onto itself. $D^{(3)}(g)$ describes the transformations of $Y_{(l=1)m}(\theta, \phi)$. The three spherical harmonics $(Y_{1-1}, Y_{10}, Y_{11})$ under rotations transform into linear combinations of each other. In general, the transformation properties of the $(2l + 1)$ components of $Y_{lm}(\theta, \phi)$ generate the irreducible representations $D^{(2l+1)}(g)$ of $R(3)$.

11.11.1   Angular Momentum in Quantum Mechanics

In quantum mechanics angular momentum, $\mathbf{L}$, is a differential operator acting on a wave function $\Psi(x, y, z)$. It is obtained from the classical expression for the angular momentum,

$$\overrightarrow{L} = \overrightarrow{r} \times \overrightarrow{p}, \tag{11.188}$$

by replacing position and momentum with their operator counterparts, that is,

$$\overrightarrow{x} \rightarrow \overrightarrow{x},$$
$$\overrightarrow{p} \rightarrow \frac{\hbar}{i} \overrightarrow{\nabla} \tag{11.189}$$

as

$$\mathbf{L} = -i\hbar \overrightarrow{r} \times \overrightarrow{\nabla}. \tag{11.190}$$

Writing $\mathbf{L}$ in Cartesian coordinates we find its components as (we set $\hbar = 1$)

$$L_x = i\overline{\mathbf{X}}_x = -i\left(y\frac{\partial}{\partial z} - z\frac{\partial}{\partial y} \right), \tag{11.191}$$

$$L_y = i\overline{\mathbf{X}}_y = -i\left(z\frac{\partial}{\partial x} - x\frac{\partial}{\partial z} \right), \tag{11.192}$$

$$L_z = i\overline{\mathbf{X}}_z = -i\left(x\frac{\partial}{\partial y} - y\frac{\partial}{\partial x} \right). \tag{11.193}$$

In Section 11.3.1 we have seen that $\overline{\mathbf{X}}_i$ satisfy the commutation relation

$$\left[\overline{\mathbf{X}}_i, \overline{\mathbf{X}}_j\right] = \epsilon_{ijk}\overline{\mathbf{X}}_k, \tag{11.194}$$

thus L_i satisfy

$$[L_i, L_j] = i\epsilon_{ijk}L_k, \tag{11.195}$$

where the indices i, j and k take the values 1,2,3 and they correspond to x, y, z, respectively.

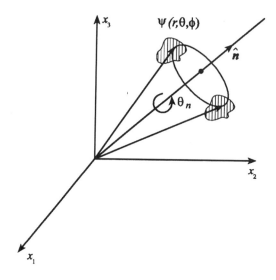

Fig. 11.4   Counterclockwise rotation of the physical system by θ_n about $\hat{\mathbf{n}}$

11.11.2   Rotation of the Physical System

We have seen that the effect of the operator [Eq. (11.40)]

$$e^{-\overline{\mathbf{X}}\cdot\hat{\mathbf{n}}\theta_n} \tag{11.196}$$

is to rotate a function clockwise about an axis pointing in the $\hat{\mathbf{n}}$ direction by θ_n. In quantum mechanics we adhere to the right-handed screw convention, that is, when we curl the fingers of our right hand about the axis of rotation and in the direction of rotation, our thumb points along $\hat{\mathbf{n}}$. Hence we work with the operator

$$e^{\overline{\mathbf{X}}\cdot\hat{\mathbf{n}}\theta_n}, \tag{11.197}$$

which rotates a function counterclockwise by θ_n (Fig. 11.4). Using Equations (11.191−11.193) the quantum mechanical counterpart of the rotation operator now becomes

$$\mathbf{R} = e^{-i\mathbf{L}\cdot\hat{\mathbf{n}}\theta_n}. \tag{11.198}$$

For a rotation about the z-axis this gives

$$\mathbf{R}\Psi(r,\theta,\phi) = \left[e^{-iL_z\phi}\right]\Psi(r,\theta,\phi). \tag{11.199}$$

For a general rotation about an axis in the $\hat{\mathbf{n}}$ direction by θ_n we write

$$\mathbf{R}\Psi(x,y,z) = e^{-i\mathbf{L}\cdot\hat{\mathbf{n}}\theta_n}\Psi(x,y,z). \tag{11.200}$$

11.11.3   Rotation Operator in Terms of the Euler Angles

Using the Euler angles we can write the rotation operator

$$\mathbf{R} = e^{-i\mathbf{L}\cdot\hat{n}\theta_n}, \tag{11.201}$$

as

$$\mathbf{R} = e^{-i\gamma L_{z_2}} e^{-i\beta L_{y_1}} e^{-i\alpha L_z}. \tag{11.202}$$

In this expression we have used another convention commonly used in modern-day quantum mechanical discussions of angular momentum. It is composed of the sequence of rotations, which starts with a counterclockwise rotation by α about the z-axis of the initial state of the system:

$$e^{-i\alpha L_z} : (x, y, z) \rightarrow (x_1, y_1, z_1). \tag{11.203}$$

This is followed by a counterclockwise rotation by β about y_1 of the intermediate axis as

$$e^{-i\beta L_{y_1}} : (x_1, y_1, z_1) \rightarrow (x_2, y_2, z_2). \tag{11.204}$$

Finally the desired orientation is reached by a counterclockwise rotation about the z_2-axis by γ as

$$e^{-i\gamma L_{z_2}} : (x_2, y_2, z_2) \rightarrow (x', y', z'). \tag{11.205}$$

11.11.4   Rotation Operator in Terms of the Original Coordinates

One of the disadvantages of the rotation operator expressed as

$$\mathbf{R} = e^{-i\mathbf{L}\cdot\hat{n}\theta_n} = e^{-i\gamma L_{z_2}} e^{-i\beta L_{y_1}} e^{-i\alpha L_z} \tag{11.206}$$

is that, except for the initial rotation about the z-axis, the remaining two rotations are performed about different sets of axis. Because we are interested in evaluating

$$\mathbf{R}\Psi(x, y, z) = \Psi(x', y', z'), \tag{11.207}$$

where (x, y, z) and (x', y', z') are two points in the same coordinate system, we need to express $\mathbf{R}$ as rotations entirely in terms of the original coordinate axis.

For this we first need to find how the operator $\mathbf{R}$ transforms under coordinate transformations. We now transform to a new coordinate system (x_n, y_n, z_n), where the z_n-axis is aligned with the $\hat{n}$ direction. We show the matrix of this coordinate transformation with the letter R. We are interested in expressing the operator $\mathbf{R}$ in terms of the (x_n, y_n, z_n) coordinates. Action of R on the coordinates induces the following change in $\Psi(x, y, z)$:

$$R\Psi(x, y, z) = \Psi(x_n, y_n, z_n). \tag{11.208}$$

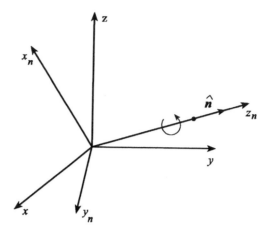

Fig. 11.5   Transformation to the (x_n, y_n, z_n)-axis

Similarly for another point we write

$$R\Psi(x', y', z') = \Psi(x'_n, y'_n, z'_n). \tag{11.209}$$

Inverse transformations are naturally given as (Fig. 11.5)

$$R^{-1}\Psi(x_n, y_n, z_n) = \Psi(x, y, z) \tag{11.210}$$

and

$$R^{-1}\Psi(x'_n, y'_n, z'_n) = \Psi(x', y', z'). \tag{11.211}$$

Operating on Equation (11.210) with **R** we get

$$\mathbf{R}R^{-1}\Psi(x_n, y_n, z_n) = \mathbf{R}\Psi(x, y, z). \tag{11.212}$$

Using Equation (11.207), this becomes

$$\mathbf{R}R^{-1}\Psi(x_n, y_n, z_n) = \Psi(x', y', z'). \tag{11.213}$$

We now operate on this with R to get

$$RRR^{-1}\Psi(x_n, y_n, z_n) = R\Psi(x', y', z') \tag{11.214}$$
$$= \Psi(x'_n, y'_n, z'_n), \tag{11.215}$$

where

$$\mathbf{R} = e^{-i\gamma L_{z2}} e^{-i\beta L_{y1}} e^{-i\alpha L_z}. \tag{11.216}$$

We now observe that (This may take a while to convince oneself. We recommend that the reader first plot all the axes in Equations (11.203) to (11.205) and then, operate on a radial vector drawn from the origin with (11.217). Finally, trace the orbit of the tip separately for each rotation while preserving the order of rotations.)

$$e^{-i\gamma I_{z_2}} = e^{-i\beta L_{y_1}} e^{-i\gamma L_{z_1}} e^{i\beta L_{y_1}} \tag{11.217}$$

to write

$$\mathbf{R} = e^{-i\beta L_{y_1}} e^{-i\gamma L_{z_1}} \left[e^{i\beta L_{y_1}} e^{-i\beta L_{y_1}} \right] e^{-i\alpha L_z}. \tag{11.218}$$

The operator inside the square brackets is the identity operator; thus

$$\mathbf{R} = e^{-i\beta L_{y_1}} e^{-i\gamma L_{z_1}} e^{-i\alpha L_z}. \tag{11.219}$$

We now note the transformation

$$e^{-i\beta L_{y_1}} = e^{-i\alpha L_z} e^{-i\beta L_y} e^{i\alpha L_z} \tag{11.220}$$

to write

$$\mathbf{R} = e^{-i\alpha L_z} e^{-i\beta L_y} e^{i\alpha L_z} e^{-i\gamma L_{z_1}} e^{-i\alpha L_z}. \tag{11.221}$$

Since $z_1 = z$, this becomes

$$\mathbf{R} = e^{-i\alpha L_z} e^{-i\beta L_y} \left[e^{i\alpha L_z} e^{-i\alpha L_z} \right] e^{-i\gamma L_z}. \tag{11.222}$$

Again the operator inside the square brackets is the identity operator, thus giving $\mathbf{R}$ entirely in terms of the original coordinate system (x, y, z) as

$$\mathbf{R} = e^{-i\alpha L_z} e^{-i\beta L_y} e^{-i\gamma L_z}. \tag{11.223}$$

We can now find the effect of $\mathbf{R}(\alpha, \beta, \gamma)$ on $\Psi(x, y, z)$ as

$$\mathbf{R}(\alpha, \beta, \gamma)\Psi(x, y, z) = \Psi(x', y', z') \tag{11.224}$$
$$e^{-i\alpha L_z} e^{-i\beta L_y} e^{-i\gamma L_z} \Psi(x, y, z) = \Psi(x', y', z').$$

In spherical polar coordinates this becomes

$$\mathbf{R}(\alpha, \beta, \gamma)\Psi(r, \theta, \phi) = \Psi(r, \theta,' \phi'). \tag{11.225}$$

Expressing the components of the angular momentum operator in spherical polar coordinates:

$$x + iy = r \sin\theta e^{\pm i\phi} \text{ and} \tag{11.226}$$
$$z = r \cos\theta,$$

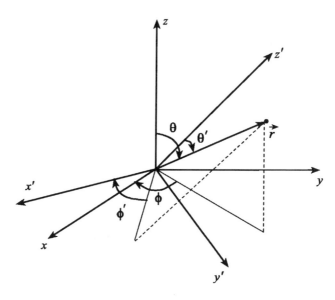

Fig. 11.6 (x, y, z) and the (x', y', z') coordinates

we obtain (Fig. 11.6)

$$L_x = \left(i \cot \theta \cos \phi \frac{\partial}{\partial \phi} + i \sin \phi \frac{\partial}{\partial \theta} \right), \quad \text{(we set } \hbar = 1) \tag{11.227}$$

$$L_y = \left(i \cot \theta \sin \phi \frac{\partial}{\partial \phi} - i \cos \phi \frac{\partial}{\partial \theta} \right), \tag{11.228}$$

$$L_z = -i \frac{\partial}{\partial \phi}, \tag{11.229}$$

$$L_\pm = L_x \pm i L_y = e^{\pm i\phi}(\pm \frac{\partial}{\partial \theta} + i \cot \theta \frac{\partial}{\partial \phi}), \tag{11.230}$$

$$L^2 = L_x^2 + L_y^2 + L_z^2 \tag{11.231}$$

$$= -\left[\frac{1}{\sin \theta} \left(\frac{\partial}{\partial \theta} \sin \theta \frac{\partial}{\partial \theta} \right) + \frac{1}{\sin^2 \theta} \frac{\partial^2}{\partial \phi^2} \right].$$

Using Equations (11.227–11.229) we can now write Equation (11.224) as

$$\left[e^{-\alpha \frac{\partial}{\partial \phi}} e^{-\beta \left(\cos \phi \frac{\partial}{\partial \theta} - \sin \phi \cot \theta \frac{\partial}{\partial \phi} \right)} e^{-\gamma \frac{\partial}{\partial \phi}} \right] \Psi(r, \theta, \phi) = \Psi(r, \theta,' \phi'), \tag{11.232}$$

which is now ready for applications to spherical harmonics $Y_{lm}(\theta, \phi)$.

11.11.5   Eigenvalue Equations for L_z, $L_\pm$, and L^2

In Chapter 9 we have established the following eigenvalue equations [Eqs. (9.172–173) and (9.187–188)]:

$$L_z Y_{lm}(\theta, \phi) = m Y_{lm}(\theta, \phi), \tag{11.233}$$

$$L_- Y_{lm}(\theta, \phi) = \sqrt{(l+m)(l-m+1)} Y_{l,m-1}(\theta, \phi), \tag{11.234}$$

$$L_+ Y_{lm}(\theta, \phi) = \sqrt{(l-m)(l+m+1)} Y_{l,m+1}(\theta, \phi), \tag{11.235}$$

$$L^2 Y_{lm}(\theta, \phi) = l(l+1) Y_{lm}(\theta, \phi). \tag{11.236}$$

Using these and the definition $L_\pm = L_x \pm i L_y$ we can also write

$$L_x Y_{lm}(\theta, \phi) = \frac{1}{2}\sqrt{(l-m)(l+m+1)} Y_{l,m+1}(\theta, \phi) \tag{11.237}$$

$$+ \frac{1}{2}\sqrt{(l+m)(l-m+1)} Y_{l,m-1}(\theta, \phi),$$

$$L_y Y_{lm}(\theta, \phi) = -\frac{i}{2}\sqrt{(l-m)(l+m+1)} Y_{l,m+1}(\theta, \phi) \tag{11.238}$$

$$+ \frac{i}{2}\sqrt{(l+m)(l-m+1)} Y_{l,m-1}(\theta, \phi).$$

11.11.6   Generalized Fourier Expansion in Spherical Harmonics

We can expand a sufficiently smooth function, $F(\theta, \phi)$, in terms of spherical harmonics, which forms a complete and orthonormal set as

$$F(\theta, \phi) = \sum_{l'=0}^{\infty} \sum_{m'=-l'}^{m'=l'} C_{l'm'} Y_{l'm'}(\theta, \phi), \tag{11.239}$$

where the expansion coefficients are given as

$$C_{l'm'} = \int\int d\Omega Y_{l'm'}^*(\theta, \phi) F(\theta, \phi). \tag{11.240}$$

Spherical harmonics satisfy the orthogonality relation

$$\int\int d\Omega Y_{l'm'}^*(\theta, \phi) Y_{lm}(\theta, \phi) = \delta_{ll'}\delta_{mm'} \tag{11.241}$$

and the completeness relation

$$\sum_{l=0}^{\infty} \sum_{m=-l}^{m=l} Y_{lm}^*(\theta', \phi') Y_{lm}(\theta, \phi) = \delta(\cos\theta - \cos\theta')\delta(\phi - \phi'), \tag{11.242}$$

where $d\Omega = \sin\theta d\theta d\phi$. In the study of angular momentum in quantum physics we frequently need expansions of expressions like

$$F_{lm}(\theta, \phi) = f(\theta, \phi) Y_{lm}(\theta, \phi) \tag{11.243}$$

and

$$G_{lm}(\theta,\phi) = O(\frac{\partial}{\partial\theta},\frac{\partial}{\partial\phi},\theta,\phi)Y_{lm}(\theta,\phi), \tag{11.244}$$

where $O(\frac{\partial}{\partial\theta},\frac{\partial}{\partial\phi},\theta,\phi)$ is some differential operator. For $F_{lm}(\theta,\phi)$ we can write

$$F_{lm}(\theta,\phi) = f(\theta,\phi)Y_{lm}(\theta,\phi) = \sum_{l'=0}^{\infty}\sum_{m'=-l'}^{m'=l'} C_{l'm'}Y_{l'm'}(\theta,\phi), \tag{11.245}$$

where the expansion coefficients are given as

$$C_{l'm'} = \int\int d\Omega \, Y_{l'm'}^*(\theta,\phi)f(\theta,\phi)Y_{lm}(\theta,\phi) \tag{11.246}$$

$$= C_{l'm',lm} . \tag{11.247}$$

For $G_{lm}(\theta,\phi)$ we can write

$$G_{lm}(\theta,\phi) = \sum_{l'=0}^{\infty}\sum_{m'=-l'}^{m'=l'} C_{l'm'}Y_{l'm'}(\theta,\phi), \tag{11.248}$$

where

$$C_{l'm'} = \int\int d\Omega \, Y_{l'm'}^*(\theta,\phi)\left[O(\frac{\partial}{\partial\theta},\frac{\partial}{\partial\phi},\theta,\phi)Y_{lm}(\theta,\phi)\right] \tag{11.249}$$

$$= C_{l'm',lm}.$$

Based on these we can also write the expansion

$$f_1(\theta,\phi)\left[f_2(\theta,\phi)Y_{lm}(\theta,\phi)\right]$$

$$= f_1(\theta,\phi)\left[\sum_{l'=0}^{\infty}\sum_{m'=-l'}^{m'=l'} C_{l'm',lm}^{(2)}Y_{l'm'}(\theta,\phi)\right] \tag{11.250}$$

$$= \sum_{l''=0}^{\infty}\sum_{m''=-l''}^{m''=l''}\sum_{l'=0}^{\infty}\sum_{m'=-l'}^{m'=l'} Y_{l''m''}^*(\theta,\phi)C_{l''m'',l'm'}^{(1)}C_{l'm',lm}^{(2)} \tag{11.251}$$

$$= \sum_{l''=0}^{\infty}\sum_{m''=-l''}^{m''=l''} C_{l''m'',lm}^{(1,2)}Y_{l''m''}(\theta,\phi), \tag{11.252}$$

where

$$C_{l''m'',lm}^{(1,2)} = \sum_{l'=0}^{\infty}\sum_{m'=-l'}^{m'=l'} C_{l''m'',l'm'}^{(1)}C_{l'm',lm}^{(2)} . \tag{11.253}$$

11.11.7   Matrix Elements of L_x, L_y, and L_z

Using the result [Eq. (11.249)] of the previous section we can now evaluate $L_y Y_{(l=1)m}$ as

$$L_y Y_{(l=1)m} = \sum_{m'=-1}^{m'=1} C_{l'm',(l=1)m} Y_{l'm'}(\theta, \phi). \qquad (11.254)$$

This gives the following matrix elements for the angular momentum operator $\mathbf{L}_y (l = 1)$ as

$$(Y_{l'=1m'}, L_y Y_{l=1m}) = C_{l'=1m',l=1m} \qquad (11.255)$$

$$= \int\int d\Omega \, Y_{l'=1m'}^*(\theta, \phi) L_y Y_{l=1m}(\theta, \phi).$$

We have dropped brackets in the l indices. We now use Equation (11.238):

$$L_y Y_{lm}(\theta, \phi) = -\frac{i}{2}\sqrt{(l-m)(l+m+1)} Y_{l,m+1}(\theta, \phi) \qquad (11.256)$$

$$+ \frac{i}{2}\sqrt{(l+m)(l-m+1)} Y_{l,m-1}(\theta, \phi)$$

to write

$$[\mathbf{L}_y(l=1)]_{mm'} = C_{l'=1m',l=1m} \qquad (11.257)$$

$$= \begin{bmatrix} 0 & -\frac{i}{\sqrt{2}} & 0 \\ \frac{i}{\sqrt{2}} & 0 & -\frac{i}{\sqrt{2}} \\ 0 & \frac{i}{\sqrt{2}} & 0 \end{bmatrix}.$$

Operating on Equation (11.254) with L_y and using Equation (11.257) we can write

$$L_y^2 Y_{l=1m} = \sum_{m'=-1}^{m'=1} L_y Y_{l'=1m'}(\theta, \phi) [\mathbf{L}_y(l=1)]_{m'm} \qquad (11.258)$$

to obtain the matrix elements of L_y^2 as

$$[\mathbf{L}_y^2(l=1)]_{mm'} \qquad (11.259)$$

$$= \begin{bmatrix} 0 & -\frac{i}{\sqrt{2}} & 0 \\ \frac{i}{\sqrt{2}} & 0 & -\frac{i}{\sqrt{2}} \\ 0 & \frac{i}{\sqrt{2}} & 0 \end{bmatrix} \begin{bmatrix} 0 & -\frac{i}{\sqrt{2}} & 0 \\ \frac{i}{\sqrt{2}} & 0 & -\frac{i}{\sqrt{2}} \\ 0 & \frac{i}{\sqrt{2}} & 0 \end{bmatrix}$$

$$= \begin{bmatrix} 1/2 & 0 & -1/2 \\ 0 & 1 & 0 \\ -1/2 & 0 & 1/2 \end{bmatrix}.$$

Similarly we find the other powers as

$$\left[\mathbf{L}_y(l=1)\right]_{mm'} = \left[\mathbf{L}_y^3(l=1)\right]_{mm'}$$
$$= \left[\mathbf{L}_y^5(l=1)\right]_{mm'} \tag{11.260}$$

$$\vdots$$

and

$$\left[\mathbf{L}_y^2(l=1)\right]_{mm'} = \left[\mathbf{L}_y^4(l=1)\right]_{mm'}$$
$$= \left[\mathbf{L}_y^6(l=1)\right]_{mm'}$$

$$\vdots \tag{11.261}$$

Using Equations (11.233–11.236) and the orthogonality relation (11.241) we can write the following matrix elements:

$$\mathbf{L}_x = (Y_{l'm'}, L_x Y_{lm}) \tag{11.262}$$
$$= \frac{1}{2}\sqrt{(l-m)(l+m+1)}\delta_{ll'}\delta_{m'(m+1)}$$
$$+ \frac{1}{2}\sqrt{(l+m)(l-m+1)}\delta_{ll'}\delta_{m'(m-1)},$$
$$\mathbf{L}_y = (Y_{l'm'}, L_y Y_{lm}) \tag{11.263}$$
$$= -\frac{i}{2}\sqrt{(l-m)(l+m+1)}(Y_{l'm'}, Y_{l,m+1})$$
$$+ \frac{i}{2}\sqrt{(l+m)(l-m+1)}(Y_{l'm'}, Y_{l,m-1})$$
$$= -\frac{i}{2}\sqrt{(l-m)(l+m+1)}\delta_{ll'}\delta_{m'(m+1)}$$
$$+ \frac{i}{2}\sqrt{(l+m)(l-m+1)}\delta_{ll'}\delta_{m'(m-1)},$$
$$\mathbf{L}_z = (Y_{l'm'}, L_z Y_{lm}) = m\delta_{ll'}\delta_{mm'}, \tag{11.264}$$
$$\mathbf{L}^2 = (Y_{l'm'}, L^2 Y_{lm}) = l(l+1)\delta_{ll'}\delta_{mm'}. \tag{11.265}$$

11.11.8   Rotation Matrices for the Spherical Harmonics

Because the effect of the rotation operator $\mathbf{R}(\alpha, \beta, \gamma)$ on the spherical harmonics is to rotate them from (θ, ϕ) to new values $(\theta,' \phi')$, we write

$$\mathbf{R}(\alpha, \beta, \gamma)Y_{lm}(\theta, \phi) = e^{-i\alpha L_z}e^{-i\beta L_y}e^{-i\gamma L_z}Y_{lm}(\theta, \phi) \tag{11.266}$$
$$= Y_{lm}(\theta,' \phi'). \tag{11.267}$$

In spherical polar coordinates this becomes

$$e^{-\alpha\frac{\partial}{\partial\phi}}e^{-\beta(\cos\phi\frac{\partial}{\partial\theta} - \sin\phi\cot\theta\frac{\partial}{\partial\phi})}e^{-\gamma\frac{\partial}{\partial\phi}}Y_{lm}(\theta, \phi) = Y_{lm}(\theta,' \phi'). \tag{11.268}$$

We now express $Y_{lm}(\theta,'\phi')$ in terms of the original $Y_{lm}(\theta,\phi)$ as

$$Y_{lm}(\theta,'\phi') = \sum_{l'm'} Y_{l'm'}(\theta,\phi)C_{l'm',lm}, \tag{11.269}$$

where

$$C_{l'm',lm} = \int\int d\Omega Y_{l'm'}^*(\theta,\phi)\mathbf{R}(\alpha,\beta,\gamma)Y_{lm}(\theta,\phi). \tag{11.270}$$

Since the spherical harmonics are defined as

$$Y_{lm}(\theta,\phi) = \frac{\sin^m\theta}{\sqrt{2\pi}}\frac{d^m}{d(\cos\theta)^m}[P_{lm}(\cos\theta)]\,e^{im\phi}, \tag{11.271}$$

$\mathbf{R}$ does not change their l value. Hence only the coefficients with $l = l'$ are nonzero in Equation (11.270), thus giving

$$C_{lm',lm} = \int\int d\Omega Y_{lm'}^*(\theta,\phi)\mathbf{R}(\alpha,\beta,\gamma)Y_{lm}(\theta,\phi) \tag{11.272}$$

$$= D_{m'm}^l(\alpha,\beta,\gamma), \tag{11.273}$$

where

$$Y_{lm}(\theta,'\phi') = \sum_{m'=-l}^{m'=l} Y_{lm'}(\theta,\phi)D_{m'm}^l(\alpha,\beta,\gamma). \tag{11.274}$$

$D_{m'm}^l(\alpha,\beta,\gamma)$ is called the rotation matrix of the spherical harmonics. Using the definition [Eq. (11.223)] of $\mathbf{R}(\alpha,\beta,\gamma)$ we can construct the rotation matrix as

$$
\begin{aligned}
& D_{m'm}^l(\alpha,\beta,\gamma) \\
&= \int\int d\Omega Y_{lm'}^*(\theta,\phi)e^{-i\alpha L_z}e^{-i\beta L_y}\left[e^{-i\gamma L_z}Y_{lm}(\theta,\phi)\right] \\
&= \int\int d\Omega Y_{lm'}^*(\theta,\phi)e^{-i\alpha L_z}e^{-i\beta L_y}\left[e^{-\gamma\frac{\partial}{\partial\phi}}e^{im\phi}(\text{function of }\theta)\right] \\
&= \int\int d\Omega Y_{lm'}^*(\theta,\phi)e^{-i\alpha L_z}e^{-i\beta L_y}\left[\sum_{n=0}^{\infty}\frac{(-\gamma)^n}{n!}(im)^n Y_{lm}(\theta,\phi)\right] \\
&= \int\int d\Omega Y_{lm'}^*(\theta,\phi)e^{-i\alpha L_z}e^{-i\beta L_y}\left[e^{-i\gamma m}Y_{lm}(\theta,\phi)\right] \\
&= \int\int d\Omega e\left[{}^{-i\alpha L_z}Y_{lm'}(\theta,\phi)\right]^* e^{-i\beta L_y}\left[e^{-i\gamma m}Y_{lm}(\theta,\phi)\right] \\
&= \int\int d\Omega \left[e^{i\alpha L_z}Y_{lm'}^*(\theta,\phi)\right]e^{-i\beta L_y}\left[e^{-i\gamma m}Y_{lm}(\theta,\phi)\right] \\
&= e^{-i\alpha m'}\left[\int\int d\Omega Y_{lm'}^*(\theta,\phi)\,e^{-i\beta L_y}Y_{lm}(\theta,\phi)\right]e^{-i\gamma m}. \tag{11.275}
\end{aligned}
$$

We have used the fact that L_z is a Hermitian operator, that is,

$$\int \Psi_1^* L_z \Psi_2 d\Omega = \int (L_z \Psi_1)^* \Psi_2 d\Omega.$$

Defining the reduced rotation matrix $d_{m'm}^l(\beta)$ as

$$d_{m'm}^l(\beta) = \int \int d\Omega Y_{lm'}^*(\theta, \phi) \, e^{-i\beta L_y} Y_{lm}(\theta, \phi), \qquad (11.276)$$

we finally obtain

$$D_{m'm}^l(\alpha, \beta, \gamma) = e^{-i\alpha m'} d_{m'm}^l(\beta) e^{-i\gamma m}. \qquad (11.277)$$

11.11.9   Evaluation of the $d_{m'm}^l(\beta)$ Matrices

For the low values of l it is relatively easy to evaluate $d_{m'm}^l(\beta)$. For example, for $l = 0$:

$$d_{m'm}^0(\beta) = 1, \qquad (11.278)$$

which is the trivial 1×1 matrix.

For $l = 1$ we can write Equation (11.276) as

$$d_{m'm}^1(\beta) = \int \int d\Omega Y_{1m'}^*(\theta, \phi) \begin{pmatrix} 1+ \\ -i\beta L_y + i\beta^3 L_y^3/3! + \cdots \\ -\beta^2 L_y^2/2! + \beta^4 L_y^4/4! + \cdots \end{pmatrix} Y_{1m}(\theta, \phi). \qquad (11.279)$$

Using the matrix elements of $(L_y)^n$ obtained in Section 11.11.7, we write this as

$$d_{m'm}^1(\beta) = \delta_{mm'} - i \, (L_y)_{mm'} \sin \beta + (L_y^2)_{mm'} (\cos \beta - 1) \qquad (11.280)$$

$$= \begin{bmatrix} 1 & 0 & 0 \\ 0 & 1 & 0 \\ 0 & 0 & 1 \end{bmatrix} - i \sin \beta \begin{bmatrix} 0 & -\frac{i}{\sqrt{2}} & 0 \\ \frac{i}{\sqrt{2}} & 0 & -\frac{i}{\sqrt{2}} \\ 0 & \frac{i}{\sqrt{2}} & 0 \end{bmatrix}$$

$$+ (\cos \beta - 1) \begin{bmatrix} 1/2 & 0 & -1/2 \\ 0 & 1 & 0 \\ -1/2 & 0 & 1/2 \end{bmatrix}.$$

Finally adding these we find

$$d_{m'm}^1(\beta) \qquad (11.281)$$

$$= \begin{array}{c} \\ m' = 1 \\ m' = 0 \\ m' = -1 \end{array} \begin{bmatrix} \frac{1}{2}(1 + \cos \beta) & -\dfrac{\sin \beta}{\sqrt{2}} & \frac{1}{2}(1 - \cos \beta) \\ \dfrac{\sin \beta}{\sqrt{2}} & \cos \beta & -\dfrac{\sin \beta}{\sqrt{2}} \\ \frac{1}{2}(1 - \cos \beta) & \dfrac{\sin \beta}{\sqrt{2}} & \frac{1}{2}(1 + \cos \beta) \end{bmatrix}$$

with column headings $m = 1$, $m = 0$, $m = -1$.

11.11.10   Inverse of the $d^l_{m'm}(\beta)$ Matrices

To find the inverse matrices we invert

$$Y_{lm}(\theta', \phi') = \mathbf{R}(\alpha, \beta, \gamma) Y_{lm}(\theta, \phi) \tag{11.282}$$

to write

$$Y_{lm}(\theta, \phi) = \mathbf{R}^{-1}(\alpha, \beta, \gamma) Y_{lm}(\theta', \phi') \tag{11.283}$$

$$= \mathbf{R}(-\gamma, -\beta, -\alpha) Y_{lm}(\theta', \phi'). \tag{11.284}$$

Note that we have reversed the sequence of rotations because

$$\mathbf{R}^{-1}(\alpha, \beta, \gamma) = [\mathbf{R}(\alpha)\mathbf{R}(\beta)\mathbf{R}(\gamma)]^{-1} = \mathbf{R}(-\gamma)\mathbf{R}(-\beta)\mathbf{R}(-\alpha). \tag{11.285}$$

We can now write $Y_{lm}(\theta, \phi)$ in terms of $Y_{lm}(\theta', \phi')$ as

$$Y_{lm}(\theta, \phi) \tag{11.286}$$

$$= \sum_{m''} Y_{lm''}(\theta', \phi') \left\{ \int \int d\Omega Y^*_{lm''} \left[e^{i\gamma L_z} e^{i\beta L_y} e^{i\alpha L_z} \right] Y_{lm} \right\}$$

$$= \sum_{m''} Y_{lm''}(\theta', \phi') \int \int d\Omega \left[Y^*_{lm''} e^{i\gamma L_z} \right] e^{i\beta L_y} \left[e^{i\alpha L_z} Y_{lm} \right].$$

Using the fact that L_z is Hermitian, this can be written as

$$Y_{lm}(\theta, \phi) = \sum_{m''} Y_{lm''}(\theta', \phi') \int \int d\Omega \left[e^{i\gamma L_z} Y_{lm''} \right]^* e^{i\beta L_y} Y_{lm} e^{i\alpha m}. \tag{11.287}$$

This leads to

$$Y_{lm}(\theta, \phi) = \sum_{m''} Y_{lm''}(\theta', \phi') e^{i\gamma m''} \left[\int \int d\Omega Y^*_{lm''} e^{i\beta L_y} Y_{lm} \right] e^{i\alpha m}$$

$$= \sum_{m''} Y_{lm''}(\theta', \phi') e^{i\gamma m''} \left[\int \int d\Omega Y_{lm} e^{-i\beta L_y} Y^*_{lm''} \right] e^{i\alpha m}$$

$$= \sum_{m''} Y_{lm''}(\theta', \phi') e^{i\gamma m''} \left[d^l_{mm''}(\beta) \right]^* e^{i\alpha m}$$

$$= \sum_{m''} Y_{lm''}(\theta', \phi') \left[e^{i\alpha m} \left[d^l_{mm''}(\beta) \right]^* e^{i\gamma m''} \right]$$

$$= \sum_{m''} Y_{lm''}(\theta', \phi') \left[D^l_{mm''}(\alpha, \beta, \gamma) \right]^*, \tag{11.288}$$

where we have used the fact that L_y is Hermitian and $L^*_y = -L_y$. This result can also be written as

$$Y_{lm}(\theta, \phi) = \sum_{m''} Y_{lm''}(\theta', \phi') \, D^l_{m''m}(-\gamma, -\beta, -\alpha), \tag{11.289}$$

which implies

$$D^l_{m''m}(\mathbf{R}^{-1}) = \left[D^l_{m''m}(\mathbf{R}) \right]^{-1} = \left[D^l_{mm''}(\mathbf{R}) \right]^*.$$

11.11.11   Differential Equation for $d^l_{m'm}(\beta)$

From the definition of the Euler angles (Section 11.11.3) it is clear that the rotations (α, β, γ) are all performed about different sets of axes. Only the first rotation is about the z-axis of our original coordinates, that is,

$$-i\frac{\partial}{\partial \alpha} = L_z, \text{ (we set } \hbar = 1). \tag{11.290}$$

Similarly, we can write the components of the angular momentum vector about the other intermediate axes, that is, y_1 and the z_2-axis, in terms of the components of the angular momentum about the x-, y-, and z-axes as:

$$L_{y_1} = -i\frac{\partial}{\partial \beta} = -\sin\alpha L_x + \cos\alpha L_y \tag{11.291}$$

and

$$L_{z_2} = -i\frac{\partial}{\partial \gamma} = \sin\beta\cos\alpha L_x + \sin\beta\sin\alpha L_y + \cos\beta L_z. \tag{11.292}$$

Inverting these we obtain

$$L_x = -i\left[-\sin\alpha\frac{\partial}{\partial \beta} + \frac{\cos\alpha}{\sin\beta}\frac{\partial}{\partial \gamma} - \cos\alpha\cot\beta\frac{\partial}{\partial \alpha}\right] \tag{11.293}$$

$$L_y = -i\left[\cos\alpha\frac{\partial}{\partial \beta} + \frac{\sin\alpha}{\sin\beta}\frac{\partial}{\partial \gamma} - \sin\alpha\cot\beta\frac{\partial}{\partial \alpha}\right] \tag{11.294}$$

$$L_z = -i\frac{\partial}{\partial \alpha}. \tag{11.295}$$

We now construct L^2 as

$$L^2 = L_x^2 + L_y^2 + L_z^2 \tag{11.296}$$

$$= -\left[\frac{\partial^2}{\partial \beta^2} + \cot\beta\frac{\partial}{\partial \beta} + \frac{1}{\sin^2\beta}\left(\frac{\partial^2}{\partial \alpha^2} + \frac{\partial^2}{\partial \gamma^2} - 2\cos\beta\frac{\partial^2}{\partial \alpha\partial \gamma}\right)\right].$$

We could use the L^2 operator either in terms of (α, β, γ) as $L^2(\frac{\partial}{\partial \alpha}, \frac{\partial}{\partial \beta}, \frac{\partial}{\partial \gamma}, \alpha, \beta, \gamma)$ and act on $D^l_{m'm}(\alpha, \beta, \gamma)$, or in terms of (θ, ϕ) as $L^2(\frac{\partial}{\partial \theta}, \frac{\partial}{\partial \phi}, \theta, \phi)$ and act on $Y_{lm}(\theta, \phi)$. We first write (we suppress derivatives in L^2)

$$L^2(\theta, \phi)Y_{lm}(\theta', \phi') = L^2(\alpha, \beta, \gamma)Y_{lm}(\theta', \phi') \tag{11.297}$$

and use Equation (11.274) and

$$L^2(\theta, \phi)Y_{lm}(\theta, \phi) = l(l+1)Y_{lm}(\theta, \phi) \tag{11.298}$$

to write

$$\sum_{m'=-l}^{m'=l} Y_{lm'}(\theta, \phi) \left[l(l+1) D_{m'm}^l(\alpha, \beta, \gamma) \right] = \tag{11.299}$$

$$\sum_{m'=-l}^{m'=l} Y_{lm'}(\theta, \phi) \left[L^2(\alpha, \beta, \gamma) D_{m'm}^l(\alpha, \beta, \gamma) \right].$$

Since $Y_{lm'}(\theta, \phi)$ are linearly independent, this gives the differential equation that $D_{m'm}^l(\alpha, \beta, \gamma)$ satisfies as:

$$\left[L^2\left(\frac{\partial}{\partial \alpha}, \frac{\partial}{\partial \beta}, \frac{\partial}{\partial \gamma}, \alpha, \beta, \gamma\right) \right] D_{m'm}^l(\alpha, \beta, \gamma) = l(l+1) D_{m'm}^l(\alpha, \beta, \gamma). \tag{11.300}$$

Using Equation (11.296) for $L^2(\frac{\partial}{\partial \alpha}, \frac{\partial}{\partial \beta}, \frac{\partial}{\partial \gamma}, \alpha, \beta, \gamma)$ and the derivatives

$$\frac{\partial^2}{\partial \alpha^2} D_{m'm}^l = -m'^2 D_{m'm}^l$$

$$\frac{\partial^2}{\partial \gamma^2} D_{m'm}^l = -m^2 D_{m'm}^l \tag{11.301}$$

$$\frac{\partial^2}{\partial \alpha \partial \gamma} D_{m'm}^l = -m'm D_{m'm}^l,$$

which follow from Equation (11.277), we obtain the differential equation for $d_{m'm}^l(\beta)$ as

$$\left\{ \frac{d^2}{d\beta^2} + \cot \beta \frac{d}{d\beta} + \left[l(l+1) - \left(\frac{m^2 + m'^2 - 2mm' \cos \beta}{\sin^2 \beta} \right) \right] \right\} d_{m'm}^l(\beta) = 0. \tag{11.302}$$

Note that for

$$m' = 0 \ \text{ or } \ m = 0 \tag{11.303}$$

this reduces to the associated Legendre equation, which has the following solutions:

$$D_{0m}^l \propto Y_{lm}^*(\beta, \gamma), \tag{11.304}$$

$$D_{m'0}^l \propto Y_{lm'}^*(\beta, \alpha). \tag{11.305}$$

Also note that some books call $D_{mm'}^l(\mathbf{R})$ what we call $\left[D_{mm'}^l(\mathbf{R}) \right]^{-1}$. Using the transformation

$$d_{m'm}^l(\beta) = \frac{y(\lambda_l, m', m, \beta)}{\sqrt{\sin \beta}}, \tag{11.306}$$

we can put Equation (11.302) in the second canonical form of Chapter 9 as

$$\frac{d^2 y(\lambda_l, m', m, \beta)}{d\beta^2} \tag{11.307}$$

$$+ \left[\left(l(l+1) + \frac{1}{4} \right) - \left(\frac{m^2 + m'^2 - 2mm' \cos \beta - \frac{1}{4}}{\sin^2 \beta} \right) \right] y(\lambda_l, m', m, \beta) = 0.$$

11.11.12   Addition Theorem for Spherical Harmonics

We have seen that the spherical harmonics transform as

$$Y_{lm}(\theta,' \phi') = \sum_{m'=-l}^{m'=l} Y_{lm'}(\theta, \phi) D^l_{m'm}(\alpha, \beta, \gamma), \tag{11.308}$$

with the inverse transformation given as

$$Y_{lm}(\theta, \phi) = \sum_{m'=-l}^{m'=l} D^{*l}_{mm'}(\alpha, \beta, \gamma) Y_{lm'}(\theta', \phi'), \tag{11.309}$$

where

$$D^l_{m'm}(\alpha, \beta, \gamma) = e^{-i\alpha m'} d^l_{m'm}(\beta) e^{-i\gamma m}. \tag{11.310}$$

We now prove an important theorem about spherical harmonics, which says that the sum

$$\sum_{m=-l}^{m=l} Y^*_{lm}(\theta_1, \phi_1) Y_{lm}(\theta_2, \phi_2) = I_l \tag{11.311}$$

is an invariant. This is the generalization of $\mathbf{r}_1 \cdot \mathbf{r}_2$ and the angles are defined as in Figure 11.7.

Before we prove this theorem let us consider the special case $l = 1$, where

$$Y_{l=1, m=\pm 1} = \mp \sqrt{\frac{3}{8\pi}} \sin \theta e^{\pm i\phi} \tag{11.312}$$

and

$$Y_{l=1, m=0} = \sqrt{\frac{3}{4\pi}} \cos \theta. \tag{11.313}$$

Using Cartesian coordinates we can write also these as

$$Y_{l=1, m=\pm 1} = \mp \sqrt{\frac{3}{8\pi}} \frac{(x \pm iy)}{r} \tag{11.314}$$

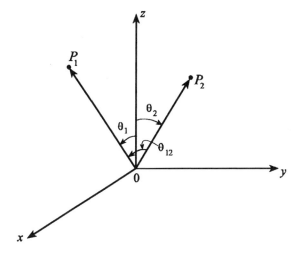

Fig. 11.7 Definition of the angles in the addition theorem of the spherical harmonics

and

$$Y_{l=1,m=0} = \sqrt{\frac{3}{4\pi}} \frac{z}{r}. \tag{11.315}$$

We now evaluate I_1 as

$$I_1 = \frac{3}{4\pi} [\cos\theta_1 \cos\theta_2 + \sin\theta_1 \sin\theta_2 \cos(\phi_1 - \phi_2)] \tag{11.316}$$

$$= \frac{3}{4\pi} \cos\theta_{12} \tag{11.317}$$

$$= \frac{3}{4\pi} \frac{x_1 x_2 + y_1 y_2 + z_1 z_2}{r_1 r_2} = \frac{3}{4\pi} \frac{\mathbf{r}_1 \cdot \mathbf{r}_2}{r_1 r_2}. \tag{11.318}$$

To prove the invariance for a general l we write the expression

$$\sum_{m=-l}^{m=l} Y_{lm}^*(\theta_1', \phi_1') Y_{lm}(\theta_2', \phi_2') \tag{11.319}$$

as

$$\sum_{m=-l}^{m=l} Y_{lm}^*(\theta_1', \phi_1') Y_{lm}(\theta_2', \phi_2') \tag{11.320}$$

$$= \sum_{m} \left[\sum_{m'm''} Y_{lm'}^*(\theta_1, \phi_1) Y_{lm''}(\theta_2, \phi_2) \right] D_{m'm}^{*l}(\alpha, \beta, \gamma) D_{m''m}^{l}(\alpha, \beta, \gamma)$$

$$= \left[\sum_{m'm''} Y_{lm'}^*(\theta_1, \phi_1) Y_{lm''}(\theta_2, \phi_2) \right] \sum_{m} D_{m'm}^{*l}(\alpha, \beta, \gamma) D_{m''m}^{l}(\alpha, \beta, \gamma).$$

Using Equation (11.288) and (11.289) we can write

$$D_{m''m}^{l}(\mathbf{R}^{-1}) = \left[D_{m''m}^{l}(\mathbf{R}) \right]^{-1} = \left[D_{mm''}^{l}(\mathbf{R}) \right]^*,$$

where $\mathbf{R}$ stands for $\mathbf{R}(\alpha, \beta, \gamma)$; hence

$$\sum_{m} D_{m'm}^{*l}(\alpha, \beta, \gamma) D_{m''m}^{l}(\alpha, \beta, \gamma) = \sum_{m} D_{m''m}^{l}(\mathbf{R}) D_{mm'}^{l}(\mathbf{R}^{-1})$$

$$= \sum_{m} D_{m''m}^{l}(\mathbf{R}) \left[D_{mm'}^{l}(\mathbf{R}) \right]^{-1}$$

$$= D_{m''m'}^{l}(\mathbf{R}\mathbf{R}^{-1})$$

$$= \delta_{m''m'}. \tag{11.321}$$

We now use this in Equation (11.320) to write

$$\sum_{m=-l}^{m=l} Y_{lm}^*(\theta_1', \phi_1') Y_{lm}(\theta_2', \phi_2') = \sum_{m'm''} Y_{lm'}^*(\theta_1, \phi_1) Y_{lm''}(\theta_2, \phi_2) \delta_{m''m'}$$

$$= \sum_{m} Y_{lm}^*(\theta_1, \phi_1) Y_{lm}(\theta_2, \phi_2)$$

$$= I_l,$$

thus proving the theorem.

11.11.13   Determination of I_l in the Addition Theorem

Because I_l is an invariant we can choose our axis, and the location of the points P_1 and P_2 conveniently as shown in Figure 11.8. Thus we can write

$$I_l = \sum_{m=-l}^{m=l} Y_{lm}^*(0, -) Y_{lm}(\theta_{12}, 0) \tag{11.322}$$

$$= Y_{l0}^*(0) Y_{l0}(\theta_{12}) \tag{11.323}$$

$$= \left(\sqrt{\frac{2l+1}{4\pi}} \right)^2 P_l(0) P_l(\cos\theta_{12}). \tag{11.324}$$

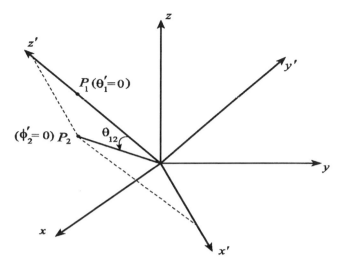

Fig. 11.8 Evaluation of I_l

Using the value $P_l(0) = 1$, we complete the derivation of the addition theorem of the spherical harmonics as

$$\sum_{m=-l}^{m=l} Y_{lm}^*(\theta_1, \phi_1) Y_{lm}(\theta_2, \phi_2) = \frac{2l+1}{4\pi} P_l(\cos\theta_{12}). \qquad (11.325)$$

Example 11.1. *Multipole expansion:* We now consider the electrostatic potential of an arbitrary charge distribution at (r, θ, ϕ) as shown in Figure 11.9. Given the charge density $\rho(r', \theta', \phi')$ we can write the electrostatic potential as

$$\Phi((r, \theta, \phi) = \iiint_{V''} \frac{\rho(r', \theta', \phi') r'^2 \sin\theta' \, dr' \, d\theta' \, d\phi'}{\sqrt{r'^2 + r^2 - 2rr' \cos\theta_{12}}}. \qquad (11.326)$$

The integral is to be taken over the source variables (r', θ', ϕ'), while (r, θ, ϕ) denotes the field point. For a field point outside the source we define a new variable, $t = r'/r$, to write

$$\Phi(r, \theta, \phi) = \iiint_{V'} \frac{\rho(r', \theta', \phi') r'^2 \sin\theta' \, dr' \, d\theta' \, d\phi'}{r\sqrt{1 + t^2 - 2t \cos\theta_{12}}}. \qquad (11.327)$$

Using the generating function definition for the Legendre polynomials, which is given as

$$T(x, t) = \frac{1}{\sqrt{1 + t^2 - 2tx}} = \sum_{l=0}^{\infty} P_l(x) t^l, \quad |t| < 1, \qquad (11.328)$$

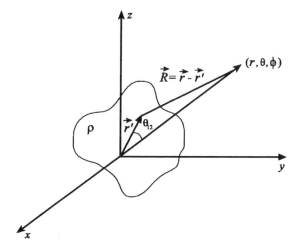

Fig. 11.9 Multipole expansion

equation (11.327) becomes

$$\Phi(r,\theta,\phi) = \sum_{l=0}^{\infty} \frac{1}{r^{l+1}} \iiint_{V'} \rho(r',\theta',\phi')r'^l P_l(\cos\theta_{12})dv'. \quad (11.329)$$

Using the addition theorem this can now be written as

$$\Phi(r,\theta,\phi) = \sum_{l,m} \frac{4\pi}{(2l+1)} \frac{Y_{lm}(\theta,\phi)}{r^{l+1}} \left[\iiint_{V'} \rho(r',\theta',\phi')r'^l Y_{lm}^*(\theta',\phi')dv' \right],$$

$$(11.330)$$

where $\sum_{l,m} = \sum_{l=0}^{\infty} \sum_{m=-l}^{l}$. The expression

$$q_{lm} = \iiint_{V'} \rho(r',\theta',\phi')r'^l Y_{lm}^*(\theta',\phi')dv' \quad (11.331)$$

is called the (lm)th multipole moment.

11.12 IRREDUCIBLE REPRESENTATIONS OF $SU(2)$

From the physical point of view a very important part of the group theory is representing each element of the group with a linear transformation acting in a vector space. We now introduce the irreducible matrix representations of $SU(2)$. There is again the trivial one-dimensional representation $D^{(1)}$, where each group element is represented by the number one. Next we have the two-dimensional representation $D^{(2)}$ provided by the matrices (11.84)

$$\mathbf{u} = \begin{bmatrix} A & B \\ -B^* & A^* \end{bmatrix}, \tag{11.332}$$

where A and B satisfy

$$\det \mathbf{u} = |A|^2 + |B|^2 = 1. \tag{11.333}$$

These act on two-dimensional vectors in the complex plane, which we show as

$$\mathbf{w} = \begin{bmatrix} w_1 \\ w_2 \end{bmatrix}, \tag{11.334}$$

or

$$\mathbf{w} = w_\alpha, \quad \alpha = 1, 2. \tag{11.335}$$

For the higher-order representations we need to define tensors, such that each element of the group corresponds to transformations of the various components of a tensor into each other. Such a representation is called generated by tensors. In this regard, $D^{(1)}$ is generated by scalars, $D^{(2)}$ is generated by vectors, and $D^{(3)}$ is generated by symmetric second-rank tensors $w_{\alpha\beta}$. In general, $D^{(n)}$ is generated by completely symmetric tensors with $(n-1)$ indices,

$$w_{\alpha_1 \alpha_2 .... \alpha_{n-1}}, \tag{11.336}$$

as

$$\begin{bmatrix} D^{(1)} & 0 & 0 & . & 0 \\ 0 & D^{(2)} & 0 & .. & 0 \\ 0 & 0 & D^{(3)} & ... & 0 \\ . & .. & ... & \ddots & 0 \\ 0 & 0 & 0 & 0 & D^{(n)} \end{bmatrix}. \tag{11.337}$$

If $w_{\alpha_1 \alpha_2 ... \alpha_{n-1}}$ is not symmetric with respect to any of the two indices, then we can contract those two indices and obtain a symmetric tensor of rank two less than the original one; thus $w_{\alpha_1 \alpha_2 ... \alpha_{n-1}}$ is not irreducible because it contains the smaller representation generated by the contracted tensor.

11.13   RELATION OF $SU(2)$ AND $R(3)$

We have seen that the general element, $\mathbf{u}$, of $SU(2)$ is given as Equation (11.332). We now define a matrix operator in this space as

$$P = \begin{bmatrix} z & x - iy \\ x + iy & -z \end{bmatrix}, \tag{11.338}$$

where (x, y, z) are real quantities. However, we will interpret them as the coordinates of a point. Under the action of **u**, P transforms as

$$P' = \mathbf{u}P\mathbf{u}^{-1}, \tag{11.339}$$

which is nothing but a similarity transformation. For unitary operators **u** satisfies

$$\mathbf{u}^{-1} = \mathbf{u}^\dagger = \tilde{\mathbf{u}}^*. \tag{11.340}$$

Note that P is Hermitian, that is,

$$P^\dagger = P \tag{11.341}$$

and traceless. Because the Hermitian property, trace, and the determinant of a matrix are invariant under similarity transformations, we can write

$$P' = \begin{bmatrix} z' & x' - iy' \\ x' + iy' & -z' \end{bmatrix}, \tag{11.342}$$

where (x', y', z') are again real and satisfy

$$-\det P' = (x'^2 + y'^2 + z'^2) = (x^2 + y^2 + z^2) = -\det P. \tag{11.343}$$

This is just the orthogonality condition, which requires that the magnitude of a vector,

$$\vec{r} = (x, y, z), \tag{11.344}$$

remain unchanged. In summary, for every element of $SU(2)$ we can associate an orthogonal transformation in three dimensions.

We have seen that the orientation of a system in three dimensions can be completely specified by the three Euler angles (ϕ, θ, ψ). A given orientation can be obtained by three successive rotations (Section 10.4). We now find the corresponding operators in $SU(2)$. For convenience we first define

$$x_+ = x + iy \tag{11.345}$$
$$x_- = x - iy. \tag{11.346}$$

This allows us to write the first transformation, which corresponds to rotation about the z-axis as

$$x'_+ = e^{-i\phi}x_+$$
$$x'_- = e^{i\phi}x_- \tag{11.347}$$
$$z' = z.$$

In $SU(2)$ this corresponds to the transformation

$$P' = \begin{bmatrix} z' & x'_- \\ x'_+ & -z' \end{bmatrix} \tag{11.348}$$

$$= \begin{bmatrix} A & B \\ -B^* & A^* \end{bmatrix} \begin{bmatrix} z & x_- \\ x_+ & -z \end{bmatrix} \begin{bmatrix} A^* & -B \\ B^* & A \end{bmatrix}. \tag{11.349}$$

On performing the multiplications we get

$$\begin{aligned} x'_+ &= -2B^*A^*z - B^{*2}x_- + A^{*2}x_+ \\ x'_- &= -2ABz + A^2x_- - B^2x_+ \\ z' &= (|A|^2 - |B|^2)z + AB^*x_- + BA^*x_+. \end{aligned} \tag{11.350}$$

Comparing this with Equation (11.347) gives

$$B = B^* = 0, \tag{11.351}$$

$$A^2 = e^{i\phi}. \tag{11.352}$$

Thus

$$\mathbf{u}_\phi = \begin{bmatrix} e^{i\phi/2} & 0 \\ 0 & e^{-i\phi/2} \end{bmatrix}. \tag{11.353}$$

Similarly, we obtain the other matrices as

$$\mathbf{u}_\theta = \begin{bmatrix} \cos\theta/2 & i\sin\theta/2 \\ i\sin\theta/2 & \cos\theta/2 \end{bmatrix} \tag{11.354}$$

and

$$\mathbf{u}_\psi = \begin{bmatrix} e^{i\psi/2} & 0 \\ 0 & e^{-i\psi/2} \end{bmatrix}. \tag{11.355}$$

For the complete sequence we can write

$$\mathbf{u} = \mathbf{u}_\psi \mathbf{u}_\theta \mathbf{u}_\phi \tag{11.356}$$

$$= \begin{bmatrix} e^{i\psi/2} & 0 \\ 0 & e^{-i\psi/2} \end{bmatrix} \begin{bmatrix} \cos\theta/2 & i\sin\theta/2 \\ i\sin\theta/2 & \cos\theta/2 \end{bmatrix} \begin{bmatrix} e^{i\phi/2} & 0 \\ 0 & e^{-i\phi/2} \end{bmatrix},$$

which is

$$\mathbf{u} = \begin{bmatrix} e^{i(\psi+\phi)/2}\cos\theta/2 & ie^{i(\psi-\phi)/2}\sin\theta/2 \\ ie^{-i(\psi-\phi)/2}\sin\theta/2 & e^{-i(\psi+\phi)/2}\cos\theta/2 \end{bmatrix}. \tag{11.357}$$

In terms of the three Euler angles the four independent parameters of $\mathbf{u}$ [Eq. (11.88)] are now given as

$$A = a + id = e^{i(\psi+\phi)/2}\cos\theta/2 \tag{11.358}$$

$$B = c + ib = ie^{i(\psi-\phi)/2}\sin\theta/2. \tag{11.359}$$

The presence of half-angles in these matrices is interesting. If we examine $\mathbf{u}_\phi$, for $\phi = 0$ it becomes

$$\mathbf{u}_\phi(0) = \begin{bmatrix} 1 & 0 \\ 0 & 1 \end{bmatrix}, \tag{11.360}$$

which corresponds to the identity matrix in $R(3)$. However, for $\phi = 2\pi$, which also gives the identity matrix in $R(3)$, $\mathbf{u}_\phi$ corresponds to

$$\mathbf{u}_\phi(2\pi) = \begin{bmatrix} -1 & 0 \\ 0 & -1 \end{bmatrix} \tag{11.361}$$

in $SU(2)$. Hence the correspondence (isomorphism) between $SU(2)$ and $R(3)$ is two-to-one. The matrices $(\mathbf{u}, -\mathbf{u})$ in $SU(2)$ correspond to a single matrix in $R(3)$.

The complex two-dimensional vector space is called the **spinor space**. It turns out that in quantum mechanics the wave function, at least parts of it, must be composed of spinors. The double-valued property and the half-angles are associated with the fact that spin is half-integer.

11.14   GROUP SPACES

11.14.1   Real Vector Space

We have seen that the elements of $R(3)$ act in a real vector space and transform vectors into other vectors. A real vector space V is defined as a collection of objects (vectors) with the following properties:

Let $\vec{v}_1$, $\vec{v}_2$, $\vec{v}_3$ be any three elements of V.

1. Addition of vectors results in another vector:

$$\vec{v}_1 + \vec{v}_2 \in V.$$

2. Addition is commutative:

$$\vec{v}_1 + \vec{v}_2 = \vec{v}_2 + \vec{v}_1.$$

3. Addition is associative:

$$(\vec{v}_1 + \vec{v}_2) + \vec{v}_3 = \vec{v}_1 + (\vec{v}_2 + \vec{v}_3).$$

4. There exists a null vector $\vec{0}$ such that for any $\vec{v} \in V$

$$\vec{v} + \vec{0} = \vec{v}.$$

5. For each $\vec{v} \in V$ there exists an inverse $(-\vec{v})$ such that

$$\vec{v} + (-\vec{v}) = 0.$$

6. Multiplication of a vector $\vec{v}$ with the number one leaves it unchanged:

$$1\vec{v} = \vec{v}.$$

7. A vector multiplied with a scalar is another vector:

$$c\vec{v} \in V.$$

A set of vectors $\vec{u}_i \in V$, $i = 1, 2, ..., n$ is said to be **linearly independent** if the equality

$$c_1 \vec{u}_1 + c_2 \vec{u}_2 + \cdots + c_n \vec{u}_n = 0 \tag{11.362}$$

can only be satisfied for the trivial case

$$c_1 = c_2 = \cdots = c_n = 0. \tag{11.363}$$

In an N-dimensional vector space we can find N linearly independent unit basis vectors $\hat{e}_i \in V$, $i = 1, 2, ..., n$, such that any vector in V can be expressed as a linear combination of these vectors as

$$\vec{v} = c_1 \hat{e}_1 + c_2 \hat{e}_2 + \cdots + c_n \hat{e}_n . \tag{11.364}$$

11.14.2   Inner Product Space

Adding to the above properties a scalar or an inner product enriches the vector space concept significantly and makes physical applications easier. In Cartesian coordinates the inner product, also called the dot product, is defined as

$$(\vec{v}_1, \vec{v}_2) = \vec{v}_1 \cdot \vec{v}_2 = v_{1x} v_{2x} + v_{1y} v_{2y} + v_{1z} v_{2z}. \tag{11.365}$$

Generalization to arbitrary dimensions is obvious. The inner product makes it possible to define the **norm** or **magnitude** of a vector as

$$|\vec{v}| = (\vec{v} \cdot \vec{v})^{1/2}, \tag{11.366}$$

where the angle between two vectors is defined as

$$\cos \theta_{12} = \frac{\vec{v}_1 \cdot \vec{v}_2}{|\vec{v}_1| |\vec{v}_2|}. \tag{11.367}$$

Basic properties of the inner product are:

$$\vec{v}_1 \cdot \vec{v}_2 = \vec{v}_2 \cdot \vec{v}_1 \tag{11.368}$$

and

$$\vec{v}_1 \cdot (a\vec{v}_2 + b\vec{v}_3) = a(\vec{v}_1 \cdot \vec{v}_2) + b(\vec{v}_1 \cdot \vec{v}_3), \tag{11.369}$$

where a and b are real numbers. A vector space with the definition of an inner product is also called an inner product space.

11.14.3   Four-Vector Space

In Section 10.10 we have extended the vector concept to Minkowski spacetime as four-vectors, where the elements of the Lorentz group act on four-vectors and transform them into other four-vectors. For four-vector spaces properties (1)–(7) still hold; however, the inner product of two four-vectors A^α and B^α is now defined as

$$A_\alpha B^\alpha = g_{\alpha\beta} A^\alpha B^\beta \tag{11.370}$$
$$= A^0 B^0 - A^1 B^1 - A^2 B^2 - A^3 B^3,$$

where $g_{\alpha\beta}$ is the Minkowski metric.

11.14.4   Complex Vector Space

Allowing complex numbers, we can also define complex vector spaces in the complex plane. For complex vector spaces properties (1)–(7) still hold; however, the inner product is now defined as

$$\overrightarrow{v}_1 \cdot \overrightarrow{v}_2 = \sum_{i=1}^{n} v_{1i}^* v_{2i}, \tag{11.371}$$

where the complex conjugate must be taken to ensure a real value for the norm (magnitude) of a vector, that is,

$$|\overrightarrow{v}| = (\overrightarrow{v} \cdot \overrightarrow{v})^{1/2} = \left(\sum_{i=1}^{n} v_i^* v_i \right)^{1/2}. \tag{11.372}$$

Note that the inner product in the complex plane is no longer symmetric, that is,

$$\overrightarrow{v}_1 \cdot \overrightarrow{v}_2 \neq \overrightarrow{v}_2 \cdot \overrightarrow{v}_1, \tag{11.373}$$

however,

$$(\overrightarrow{v}_1 \cdot \overrightarrow{v}_2) = (\overrightarrow{v}_2 \cdot \overrightarrow{v}_1)^* \tag{11.374}$$

is true.

11.14.5   Function Space and Hilbert Space

We now define a vector space L_2, whose elements are complex valued functions of a real variable x, which are square integrable in the interval $[a, b]$. L_2 is also called the Hilbert space. By square integrable it is meant that the integral

$$\int_a^b |f(x)|^2 \, dx \tag{11.375}$$

exists and is finite. Proof of the fact that the space of square integrable functions satisfies the properties of a vector space is rather technical, and we refer to books like Courant and Hilbert, and Morse and Feshbach. The inner product in L_2 is defined as

$$(f_1, f_2) = \int_a^b f_1^*(x) f_2(x) dx. \tag{11.376}$$

In the presence of a weight function $w(x)$ the inner product is defined as

$$(f_1, f_2) = \int_a^b f_1^*(x) f_2(x) w(x) dx. \tag{11.377}$$

Analogous to choosing a set of basis vectors in ordinary vector space, a major problem in L_2 is to find a suitable complete and orthonormal set of functions, $\{u_m(x)\}$, such that a given $f(x) \in L_2$ can be expanded as

$$f(x) = \sum_{m=0}^{\infty} c_m u_m(x). \tag{11.378}$$

Orthogonality of $\{u_m(x)\}$ is expressed as

$$(u_m, u_n) = \int_a^b u_m^*(x) u_n(x) dx = \delta_{mn}, \tag{11.379}$$

where we have taken $w(x) = 1$ for simplicity. Using the orthogonality relation we can free the expansion coefficients under the summation sign in Equation (11.378) to express them as

$$c_m = (u_m, f) \tag{11.380}$$
$$= \int_a^b u_m^*(x) f(x) dx.$$

In physical applications $\{u_m(x)\}$ is usually taken as the eigenfunction set of a Hermitian operator. Substituting Equation (11.380) back into Equation (11.378) a formal expression for the completeness of the set $\{u_m(x)\}$ is obtained as

$$\sum_{m=0}^{\infty} u_m^*(x') u_m(x) = \delta(x - x'). \tag{11.381}$$

11.14.6   Completeness of the Set of Eigenfunctions $\{u_m(x)\}$

Proof of the completeness of the eigenfunction set is rather technical for our purposes and can be found in Courant and Hilbert (p. 427, vol. 1). What is important for us is that any sufficiently well-behaved and at least piecewise continuous function, $F(x)$, can be expressed as an infinite series in terms of the set $\{u_m(x)\}$ as

$$F(x) = \sum_{m=0}^{\infty} a_m u_m(x). \tag{11.382}$$

Convergence of this series to $F(x)$ could be approached via the variation technique, and it could be shown that for a Sturm-Liouville system the limit (Mathews and Walker, p. 338)

$$\lim_{N \to \infty} \int_a^b \left[F(x) - \sum_{m=0}^{N} a_m u_m(x) \right]^2 \omega(x)\, dx \to 0 \tag{11.383}$$

is true. In this case we say that in the interval $[a, b]$ the series

$$\sum_{m=0}^{\infty} a_m u_m(x) \tag{11.384}$$

converges to $F(x)$ in the mean. Convergence in the mean does not imply point-to-point (uniform) convergence:

$$\lim_{N \to \infty} \sum_{m=0}^{N} a_m u_m(x) \to F(x). \tag{11.385}$$

However, for most practical situations convergence in the mean will accompany point-to-point convergence and will be sufficient. We conclude this section by quoting a theorem from Courant and Hilbert (p. 427).

Expansion Theorem: Any piecewise continuous function defined in the fundamental domain $[a, b]$ with a square integrable first derivative could be expanded in an eigenfunction series $F(x) = \sum_{m=0}^{\infty} a_m u_m(x)$, which converges absolutely and uniformly in all subdomains free of points of discontinuity. At the points of discontinuity it represents the arithmetic mean of the right- and the left-hand limits.

In this theorem the function does not have to satisfy the boundary conditions. This theorem also implies convergence in the mean; however, the converse is not true.

11.15 HILBERT SPACE AND QUANTUM MECHANICS

In quantum mechanics a physical system is completely described by giving its state or wave function, $\Psi(x)$, in Hilbert space. To every physical observable there corresponds a Hermitian differential operator acting on the functions in Hilbert space. Because of their Hermitian nature these operators have real eigenvalues, which are the allowed physical values of the corresponding observable. These operators are usually obtained from their classical definitions by replacing position, momentum, and energy with their operator counterparts. In position space the replacements

$$\overrightarrow{r} \rightarrow \overrightarrow{r},$$
$$\overrightarrow{p} \rightarrow -i\hbar \overrightarrow{\nabla}, \tag{11.386}$$
$$E \rightarrow i\hbar \frac{\partial}{\partial t}$$

have been rather successful. Using these, the angular momentum operator is obtained as

$$\overrightarrow{L} = \overrightarrow{r} \times \overrightarrow{p} \tag{11.387}$$
$$= -i\hbar \overrightarrow{r} \times \overrightarrow{\nabla}.$$

In Cartesian coordinates components of $\overrightarrow{L}$ are given as

$$L_1 = -i\hbar \left(x_2 \frac{\partial}{\partial x_3} - x_3 \frac{\partial}{\partial x_2} \right), \tag{11.388}$$

$$L_2 = -i\hbar \left(x_3 \frac{\partial}{\partial x_1} - x_1 \frac{\partial}{\partial x_3} \right), \tag{11.389}$$

$$L_3 = -i\hbar \left(x_1 \frac{\partial}{\partial x_2} - x_2 \frac{\partial}{\partial x_1} \right), \tag{11.390}$$

where L_i satisfies the commutation relation

$$[L_i, L_j] = i\hbar \epsilon_{ijk} L_k . \tag{11.391}$$

11.16   CONTINUOUS GROUPS AND SYMMETRIES

In everyday language the word symmetry is usually associated with familiar operations like rotations and reflections. In scientific applications we have a broader definition in terms of general operations performed in the parameter space of a given system. Now, symmetry means that a given system is invariant under a certain operation. A system could be represented by a Lagrangian, a state function, or a differential equation. In our previous sections we have discussed examples of continuous groups and their generators. The theory of continuous groups was invented by Lie when he was studying symmetries of differential equations. He also introduced a method for integrating differential equations once the symmetries are known. In what follows we discuss extension (prolongation) of generators of continuous groups so that they could be applied to differential equations.

11.16.1   One-Parameter Point Groups and Their Generators

In two dimensions general point transformations can be defined as

$$\overline{x} = \overline{x}(x, y)$$
$$\overline{y} = \overline{y}(x, y), \tag{11.392}$$

where x and y are two variables that are not necessarily the Cartesian coordinates. All we require is that this transformation form a continuous group so that finite transformations can be generated continuously from the identity element. We assume that these transformations depend on at least on one parameter, ε; hence we write

$$\overline{x} = \overline{x}(x, y; \varepsilon)$$
$$\overline{y} = \overline{y}(x, y; \varepsilon). \tag{11.393}$$

An example is the orthogonal transformation

$$\overline{x} = x \cos \varepsilon + y \sin \varepsilon$$
$$\overline{y} = -x \sin \varepsilon + y \cos \varepsilon, \tag{11.394}$$

which corresponds to counterclockwise rotations about the z-axis by the amount ε. If we expand Equation (11.394) about $\varepsilon = 0$ we get

$$\overline{x}(x, y; \varepsilon) = x + \varepsilon \alpha(x, y) + \cdots$$
$$\overline{y}(x, y; \varepsilon) = y + \varepsilon \beta(x, y) + \cdots, \tag{11.395}$$

where

$$\alpha(x, y) = \left. \frac{\partial \overline{x}}{\partial \varepsilon} \right|_{\varepsilon = 0} \tag{11.396}$$

and

$$\beta(x, y) = \left. \frac{\partial \overline{y}}{\partial \varepsilon} \right|_{\varepsilon = 0}. \tag{11.397}$$

If we define the operator

$$\mathbf{X} = \alpha(x, y) \frac{\partial}{\partial x} + \beta(x, y) \frac{\partial}{\partial y}, \tag{11.398}$$

we can write Equation (11.395) as

$$\overline{x}(x, y; \varepsilon) = x + \varepsilon \mathbf{X} x + \cdots$$
$$\overline{y}(x, y; \varepsilon) = y + \varepsilon \mathbf{X} y + \cdots. \tag{11.399}$$

Operator $\mathbf{X}$ is called the generator of the infinitesimal point transformation. For infinitesimal rotations about the z-axis this agrees with our previous result [Eq. (11.34)] as

$$\mathbf{X}_z = y\frac{\partial}{\partial x} - x\frac{\partial}{\partial y}. \tag{11.400}$$

Similarly, the generator for the point transformation

$$\bar{x} = x + \varepsilon$$
$$\tag{11.401}$$
$$\bar{y} = y,$$

which corresponds to translation along the x-axis, is

$$\mathbf{X} = \frac{\partial}{\partial x}. \tag{11.402}$$

11.16.2   Transformation of Generators and Normal Forms

We have given the generators in terms of the (x, y) variables [Eq. (11.398)]. However, we would also like to know how they look in another set of variables, say

$$u = u(x, y)$$
$$\tag{11.403}$$
$$v = v(x, y).$$

For this we first generalize [Eq. (11.398)] to n variables as

$$\mathbf{X} = a^i(x^j)\frac{\partial}{\partial x^i}, \quad i = 1, 2, ..., n. \tag{11.404}$$

Note that we used the Einstein summation convention for the index i. Defining new variables by

$$\bar{x}^i = \bar{x}^i(x^j), \tag{11.405}$$

we obtain

$$\frac{\partial}{\partial x^i} = \frac{\partial \bar{x}^j}{\partial x^i}\frac{\partial}{\partial \bar{x}^j}. \tag{11.406}$$

When substituted in Equation (11.404) this gives the generator in terms of the new variables as

$$\mathbf{X} = \left[a^i\frac{\partial \bar{x}^j}{\partial x^i}\right]\frac{\partial}{\partial \bar{x}^j} \tag{11.407}$$

$$= \bar{a}^j\frac{\partial}{\partial \bar{x}^j}, \tag{11.408}$$

where

$$\bar{a}^j = \frac{\partial \bar{x}^j}{\partial x^i} a^i. \tag{11.409}$$

Note that if we operate on x^j with $\mathbf{X}$ we get

$$\mathbf{X}x^j = a^i \frac{\partial x^j}{\partial x^i} = a^j. \tag{11.410}$$

Similarly,

$$\mathbf{X}\bar{x}^j = \bar{a}^i \frac{\partial \bar{x}^j}{\partial \bar{x}^i} = \bar{a}^j \tag{11.411}$$

In other words, the coefficients in the definition of the generator can be found by simply operating on the coordinates with the generator; hence we can write

$$\mathbf{X} = (\mathbf{X}x^i)\frac{\partial}{\partial x^i} = (\mathbf{X}\bar{x}^i)\frac{\partial}{\partial \bar{x}^i}. \tag{11.412}$$

We now consider the generator for rotations about the z-axis [Eq. (11.400)] in plane polar coordinates:

$$\rho = (x^2 + y^2)^{1/2}, \tag{11.413}$$

$$\phi = \arctan(y/x). \tag{11.414}$$

Applying Equation (11.412) we obtain the generator as

$$\mathbf{X} = (\mathbf{X}\rho)\frac{\partial}{\partial r} + (\mathbf{X}\phi)\frac{\partial}{\partial \phi} \tag{11.415}$$

$$= \left[\left(y\frac{\partial}{\partial x} - x\frac{\partial}{\partial y}\right)(x^2 + y^2)^{\frac{1}{2}}\right]\frac{\partial}{\partial r} + \left[\left(y\frac{\partial}{\partial x} - x\frac{\partial}{\partial y}\right)\arctan(\frac{y}{x})\right]\frac{\partial}{\partial \phi}$$

$$= [0]\frac{\partial}{\partial r} + [-1]\frac{\partial}{\partial \phi}$$

$$= -\frac{\partial}{\partial \phi}.$$

Naturally, the plane polar coordinates in two dimensions or in general the spherical polar coordinates are the natural coordinates to use in rotation problems. This brings out the obvious question: Is it always possible to find a new definition of variables so that the generator of the one-parameter group of transformations looks like

$$\mathbf{X} = \frac{\partial}{\partial s}? \tag{11.416}$$

We will not go into the proof, but the answer to this question is yes, where the above form of the generator is called the **normal form**.

11.16.3   The Case of Multiple Parameters

Transformations can also depend on multiple parameters. For a group of transformations with m parameters we write

$$\bar{x}^i = \bar{x}^i(x^j; \varepsilon_\mu), \quad i, j = 1, 2, ..., n \text{ and } \mu = 1, 2, ..., m. \quad (11.417)$$

We now associate a generator for each parameter as

$$\mathbf{X}_\mu = a^i_\mu(x^j)\frac{\partial}{\partial x^i}, \quad i = 1, 2, ..., n, \quad (11.418)$$

where

$$a^i_\mu(x^j) = \left.\frac{\partial x^i}{\partial \varepsilon_\mu}\right|_{\varepsilon_\mu = 0}.$$

The generator of a general transformation can now be given as a linear combination of the individual generators as

$$\mathbf{X} = c^\mu \mathbf{X}_\mu, \quad \mu = 1, 2, ..., m. \quad (11.419)$$

We have seen examples of this in $R(3)$ and $SU(2)$. In fact $\mathbf{X}_\mu$ forms the Lie algebra of the m-dimensional group of transformations.

11.16.4   Action of Generators on Functions

We have already seen that the action of the generators of the rotation group $R(3)$ on a function $f(\mathbf{r})$ are given as

$$f'(\mathbf{r}) = (\mathbf{I} - \overline{\mathbf{X}}_1 \delta\theta_1 - \overline{\mathbf{X}}_2 \delta\theta_2 - \overline{\mathbf{X}}_3 \delta\theta_3)f(\mathbf{r}) \quad (11.420)$$

$$= (\mathbf{I} - \overline{\mathbf{X}} \cdot \hat{\mathbf{n}}\delta\theta)\, f(\mathbf{r}), \quad (11.421)$$

where the generators are given as

$$\overline{\mathbf{X}}_1 = -\left(x_2\frac{\partial}{\partial x_3} - x_3\frac{\partial}{\partial x_2}\right)$$

$$\overline{\mathbf{X}}_2 = -\left(x_3\frac{\partial}{\partial x_1} - x_1\frac{\partial}{\partial x_3}\right) \quad (11.422)$$

$$\overline{\mathbf{X}}_3 = -\left(x_1\frac{\partial}{\partial x_2} - x_2\frac{\partial}{\partial x_1}\right).$$

The minus sign in Equation (11.421) means that the physical system is rotated clockwise by θ about an axis pointing in the $\hat{\mathbf{n}}$ direction. Now the change in $f(\mathbf{r})$ is given as

$$\delta f(\mathbf{r}) = -\left(\overline{\mathbf{X}} \cdot \hat{\mathbf{n}}\right) f(\mathbf{r})\delta\theta. \quad (11.423)$$

If a system represented by $f(\mathbf{r})$ is symmetric under the rotation generated by $(\mathbf{X} \cdot \hat{\mathbf{n}})$, that is, it does not change, then we have

$$(\mathbf{X} \cdot \hat{\mathbf{n}}) \, f(\mathbf{r}) = 0. \tag{11.424}$$

For rotations about the z-axis, in spherical polar coordinates this means

$$\frac{\partial}{\partial \phi} f(\mathbf{r}) = 0, \tag{11.425}$$

that is, $f(\mathbf{r})$ does not depend on ϕ explicitly.

For a general transformation we can define two vectors

$$\mathbf{r} = \begin{bmatrix} x^1 \\ x^2 \\ \vdots \\ x^n \end{bmatrix}, \quad \mathbf{e} = \begin{bmatrix} \varepsilon^1 \\ \varepsilon^2 \\ \vdots \\ \varepsilon^m \end{bmatrix}, \tag{11.426}$$

where ε^μ are small, so that

$$
\begin{aligned}
f'(\mathbf{r}) &= (\mathbf{I} - \mathbf{X}_1 \varepsilon^1 - \mathbf{X}_2 \varepsilon^2 - \cdots - \mathbf{X}_m \varepsilon^m) f(\mathbf{r}) \\
&= (\mathbf{I} - \mathbf{X}_\mu \cdot \hat{\mathbf{e}}^\mu \varepsilon) \, f(\mathbf{r}),
\end{aligned} \tag{11.427}
$$

where

$$\hat{\mathbf{e}}^\mu = \mathbf{e}/e, \quad e = |\mathbf{e}| = \left[\left(\varepsilon^1 \right)^2 + \left(\varepsilon^2 \right)^2 + \cdots + \left(\varepsilon^m \right)^2 \right]^{1/2} \tag{11.428}$$

is a unit vector in the direction of $\mathbf{e}$ and the generators are defined as in Equation (11.418).

11.16.5   Infinitesimal Transformation of Derivatives: Extension of Generators

To find the effect of infinitesimal point transformations on a differential equation

$$D(x, y', y'', ..., y^{(n)}) = 0, \tag{11.429}$$

we first need to find how the derivatives $y^{(n)}$ transform. For the point transformation

$$
\begin{aligned}
\overline{x} &= \overline{x}(x, y; \varepsilon) \\
\overline{y} &= \overline{y}(x, y; \varepsilon)
\end{aligned} \tag{11.430}
$$

we can write

$$\overline{y}' = \frac{d\overline{y}}{d\overline{x}} \tag{11.431}$$

$$= \frac{d\overline{y}(x,y;\varepsilon)}{d\overline{x}(x,y;\varepsilon)}$$

$$= \frac{(\partial\overline{y}/\partial x) + (\partial\overline{y}/\partial y)y'}{(\partial\overline{x}/\partial x) + (\partial\overline{x}/\partial y)y'}$$

$$= \overline{y}'(x,y,y';\varepsilon).$$

Other derivatives can also be written as

$$\overline{y}'' = \frac{d\overline{y}'}{d\overline{x}} = \overline{y}''(x,y,y',y'';\varepsilon)$$

$$\vdots \tag{11.432}$$

$$\overline{y}^{(n)} = \frac{d\overline{y}^{(n-1)}}{d\overline{x}} = \overline{y}''(x,y,y',..,y^{(n)};\varepsilon).$$

What we really need is the generators of the following infinitesimal transformations:

$$\overline{x} = x + \varepsilon\alpha(x,y) + \cdots = x + \varepsilon\mathbf{X}x + \cdots, \tag{11.433}$$

$$\overline{y} = y + \varepsilon\beta(x,y) + \cdots = y + \varepsilon\mathbf{X}y + \cdots,$$

$$\overline{y}' = y' + \varepsilon\beta^{[1]}(x,y,y') + \cdots = y' + \varepsilon\mathbf{X}y' + \cdots,$$

$$\vdots$$

$$\overline{y}^{(n)} = y^{(n)} + \varepsilon\beta^{[n]}(x,y,y',..,y^{(n)}) + \cdots = y^{(n)} + \varepsilon\mathbf{X}y^{(n)} + \cdots,$$

where

$$\alpha(x,y) = \left.\frac{\partial\overline{x}}{\partial\varepsilon}\right|_{\varepsilon=0}, \quad \beta(x,y) = \left.\frac{\partial\overline{y}}{\partial\varepsilon}\right|_{\varepsilon=0} \tag{11.434}$$

and

$$\beta^{[1]} = \left.\frac{\partial\overline{y}'}{\partial\varepsilon}\right|_{\varepsilon=0}, ..., \beta^{[n]} = \left.\frac{\partial\overline{y}^{(n)}}{\partial\varepsilon}\right|_{\varepsilon=0}. \tag{11.435}$$

For reasons to become clear shortly we have used $\mathbf{X}$ for all the generators in Equation (11.433). Also note that $\beta^{[n]}$ is not the nth derivative of β.

We now define the **extension** (prolongation) of the generator

$$\mathbf{X} = \alpha(x,y)\frac{\partial}{\partial x} + \beta(x,y)\frac{\partial}{\partial y} \tag{11.436}$$

as

$$\mathbf{X} = \alpha\frac{\partial}{\partial x} + \beta\frac{\partial}{\partial y} + \beta^{[1]}\frac{\partial}{\partial y'} + \cdots + \beta^{[n]}\frac{\partial}{\partial y^{(n)}}. \tag{11.437}$$

To find the coefficients $\beta^{[n]}$ we can use Equation (11.433) in Equation (11.431) and then Equation (11.432) to obtain

$$\bar{y}' = y' + \varepsilon\beta^{[1]} + \cdots \tag{11.438}$$

$$= \frac{d\bar{y}}{d\bar{x}}$$

$$= \frac{y' + \varepsilon(d\beta/dx) + \cdots}{1 + \varepsilon(d\alpha/dx) + \cdots}$$

$$= y' + \varepsilon(\frac{d\beta}{dx} - y'\frac{d\alpha}{dx}) + \cdots$$

We can now write $\beta^{[1]}$ as

$$\beta^{[1]} = (\frac{d\beta}{dx} - y'\frac{d\alpha}{dx}). \tag{11.439}$$

Similarly, we write

$$\bar{y}^{(n)} = y^{(n)} + \varepsilon\beta^{[n]} + \cdots \tag{11.440}$$

$$= \frac{d\bar{y}^{(n-1)}}{d\bar{x}^{(n-1)}}$$

$$= y^{(n)} + \varepsilon\left(\frac{d\beta^{[n-1]}}{dx} - y^{(n)}\frac{d\alpha}{dx}\right) + \cdots$$

and obtain

$$\beta^{[n]} = \frac{d\beta^{[n-1]}}{dx} - y^{(n)}\frac{d\alpha}{dx}. \tag{11.441}$$

This can also be written as

$$\beta^{[n]} = \frac{d^n(\beta - y'\alpha)}{dx^n} + y^{(n+1)}\alpha, \tag{11.442}$$

which for the first two terms gives us

$$\beta^{[1]} = \frac{d\beta(x,y)}{dx} - y'\frac{d\alpha(x,y)}{dx}$$

$$= \frac{\partial\beta}{\partial x} + y'\left(\frac{\partial\beta}{\partial y} - \frac{\partial\alpha}{\partial x}\right) - y'^2\frac{\partial\alpha}{\partial y} \tag{11.443}$$

and

$$\beta^{[2]} = \frac{d^2(\beta - y'\alpha)}{dx^2} + y^{(3)}\alpha$$

$$= \frac{\partial^2\beta}{\partial x^2} + \left(2\frac{\partial^2\beta}{\partial x\partial y} - \frac{\partial^2\alpha}{\partial x^2}\right)y' + \left(\frac{\partial^2\beta}{\partial y^2} - 2\frac{\partial^2\alpha}{\partial x\partial y}\right)y'^2$$

$$- \frac{\partial^2\alpha}{\partial y^2}y'^3 + \left(\frac{\partial\beta}{\partial y} - 2\frac{\partial\alpha}{\partial x} - 3\frac{\partial\alpha}{\partial y}y'\right)y''. \tag{11.444}$$

For the infinitesimal rotations about the z-axis the extended generator can now be written as

$$\mathbf{X} = y\frac{\partial}{\partial x} - x\frac{\partial}{\partial y} - (1+y'^2)\frac{\partial}{\partial y'} - 3y'y''\frac{\partial}{\partial y''}$$
$$- (3y''^2 + 4y'y''')\frac{\partial}{\partial y'''} + \cdots \tag{11.445}$$

For the extension of the generator for translations along the x-axis we obtain

$$\mathbf{X} = \frac{\partial}{\partial x}. \tag{11.446}$$

11.16.6   Symmetries of Differential Equations

We are now ready to discuss symmetry of differential equations under point transformations, which depend on at least one parameter. To avoid some singular cases (Stephani) we confine our discussion to differential equations,

$$D(x, y', y'', ..., y^{(n)}) = 0, \tag{11.447}$$

which can be solved for the highest derivative as

$$D = y^{(n)} - \tilde{D}(x, y', y'', ..., y^{(n-1)}) = 0. \tag{11.448}$$

For example, the differential equation

$$D = 2y'' + y'^2 + y = 0 \tag{11.449}$$

satisfies this property, whereas

$$D = (y'' - y' + x)^2 = 0 \tag{11.450}$$

does not. For the point transformation

$$\overline{x} = \overline{x}(x, y; \varepsilon)$$
$$\overline{y} = \overline{y}(x, y; \varepsilon), \tag{11.451}$$

we say the differential equation is symmetric if the solutions, $y(x)$, of Equation (11.448) are mapped into other solutions, $\overline{y} = \overline{y}(\overline{x})$, of

$$D = \overline{y}^{(n)} - \tilde{D}(\overline{x}, \overline{y}', \overline{y}'', ..., \overline{y}^{(n-1)}) = 0. \tag{11.452}$$

Expanding D with respect to ε about $\varepsilon = 0$ we write

$$D(\overline{x}, \overline{y}', \overline{y}'', ..., \overline{y}^{(n)}; \varepsilon) =$$
$$D(\overline{x}, \overline{y}', \overline{y}'', ..., \overline{y}^{(n)}; \varepsilon)\Big|_{\varepsilon=0} + \frac{\partial D(\overline{x}, \overline{y}', \overline{y}'', ..., \overline{y}^{(n)}; \varepsilon)}{\partial \varepsilon}\Big|_{\varepsilon=0} \varepsilon + \cdots \tag{11.453}$$

For infinitesimal transformations we keep only the linear terms in ε:

$$D(\overline{x}, \overline{y}', \overline{y}'', ..., \overline{y}^{(n)}; \varepsilon) - D(\overline{x}, \overline{y}', \overline{y}'', ..., \overline{y}^{(n)}; \varepsilon)\Big|_{\varepsilon=0} =$$

$$\frac{\partial D}{\partial \overline{x}} \frac{\partial \overline{x}}{\partial \varepsilon} + \frac{\partial D}{\partial \overline{y}} \frac{\partial \overline{y}}{\partial \varepsilon} + \cdots + \frac{\partial D}{\partial \overline{y}^{(n)}} \frac{\partial \overline{y}^{(n)}}{\partial \varepsilon}\Bigg|_{\varepsilon=0} \varepsilon. \qquad (11.454)$$

In the presence of symmetry Equation (11.452) must be true for all ε; thus the left-hand side of Equation (11.454) is zero, and we obtain a formal expression for symmetry as

$$\left[\alpha \frac{\partial D}{\partial x} + \beta \frac{\partial D}{\partial y} + \cdots + \beta^{[n]} \frac{\partial D}{\partial y^{(n)}}\right] = 0,$$

$$\left[\alpha \frac{\partial}{\partial x} + \beta \frac{\partial}{\partial y} + \cdots + \beta^{[n]} \frac{\partial}{\partial y^{(n)}}\right] D = 0,$$

$$\mathbf{X} D = 0. \qquad (11.455)$$

Note that the symmetry of a differential equation is independent of the choice of variables used. Using an arbitrary point transformation only changes the form of the generator. We now summarize these results in terms of a theorem (for special cases and alternate definitions of symmetry we refer the reader to Stephani)

Theorem: An ordinary differential equation, which could be written as

$$D = \overline{y}^{(n)} - \tilde{D}(\overline{x}, \overline{y}', \overline{y}'', ..., \overline{y}^{(n-1)}) = 0, \qquad (11.456)$$

admits a group of symmetries with the generator $\mathbf{X}$ if and only if

$$\mathbf{X} D \equiv 0 \qquad (11.457)$$

holds.

Note that we have written $\mathbf{X} D \equiv 0$ instead of $\mathbf{X} D = 0$ to emphasize the fact that Equation (11.457) must hold for every solution $y(x)$ of $D = 0$. For example, the differential equation

$$D = y'' + a_0 y' + b_0 y = 0 \qquad (11.458)$$

admits the symmetry transformation

$$\overline{x} = 0$$

$$\overline{y} = (1 + \varepsilon)y, \qquad (11.459)$$

since D does not change when we multiply y (also y' and y'') with a constant factor. Using Equation (11.437) the generator of this transformation can be written as

$$\mathbf{X} = y \frac{\partial}{\partial y} + y' \frac{\partial}{\partial y'} + y'' \frac{\partial}{\partial y''}, \qquad (11.460)$$

which gives

$$\mathbf{X}D = \left[y\frac{\partial}{\partial y} + y'\frac{\partial}{\partial y'} + y''\frac{\partial}{\partial y''} \right] (y'' + a_0 y' + b_0 y) \qquad (11.461)$$

$$= (y'' + a_0 y' + b_0 y). \qquad (11.462)$$

Considered with

$$D = 0 \qquad (11.463)$$

this gives

$$\mathbf{X}D = 0. \qquad (11.464)$$

We stated that one can always find a new variable, say $\tilde{x}$, where a generator appears in its normal form as

$$\mathbf{X} = \frac{\partial}{\partial \tilde{x}}. \qquad (11.465)$$

Hence if $\mathbf{X}$ generates a symmetry of a given differential equation, which can be solved for its highest derivative as

$$D = \overline{y}^{(n)} - \tilde{D}(\overline{x}, \overline{y}', \overline{y}'', ..., \overline{y}^{(n-1)}) = 0, \qquad (11.466)$$

then we can write

$$\mathbf{X}D = \frac{\partial D}{\partial \tilde{x}} = 0, \qquad (11.467)$$

which means that in normal coordinates D does not depend explicitly on the independent variable $\tilde{x}$.

Note that restricting our discussion to differential equations that could be solved for the highest derivative guards us from singular cases where all the first derivatives of D are zero. For example, for the differential equation

$$D = (y'' - y' + x)^2 = 0,$$

all the first-order derivatives are zero for $D = 0$:

$$\frac{\partial D}{\partial y''} = 2(y'' - y' + x) = 0,$$

$$\frac{\partial D}{\partial y'} = -2(y'' - y' + x) = 0,$$

$$\frac{\partial D}{\partial y} = 0,$$

$$\frac{\partial D}{\partial x} = 2(y'' - y' + x) = 0.$$

Thus $\mathbf{X}D = 0$ holds for any linear operator, and in normal coordinates even though $\dfrac{\partial D}{\partial \tilde{x}} = 0$, we can no longer say that D does not depend on $\tilde{x}$ explicitly.

Problems

11.1 Consider the linear group in two dimensions

$$x' = ax + by$$
$$y' = cx + dy.$$

Show that the four infinitesimal generators are given as

$$X_1 = x\frac{\partial}{\partial x}, \ X_2 = y\frac{\partial}{\partial x}, \ X_3 = x\frac{\partial}{\partial y}, \ X_4 = y\frac{\partial}{\partial y}$$

and find their commutators.

11.2 Show that

$$\det \mathbf{A} = \det e^{\mathbf{L}} = e^{Tr\mathbf{L}} ,$$

where $\mathbf{L}$ is an $n \times n$ matrix. Use the fact that the determinant and the trace of a matrix are invariant under similarity transformations. Then make a similarity transformation that puts $\mathbf{L}$ into diagonal form.

11.3 Verify the transformation matrix

$$\mathbf{A}_{boost}(\beta)$$
$$= e^{\mathbf{V}\cdot\hat{\beta}\beta}$$
$$= \begin{bmatrix} \gamma & -\beta_1\gamma & -\beta_2\gamma & -\beta_3\gamma \\ -\beta_1\gamma & 1 + \dfrac{(\gamma-1)\beta_1^2}{\beta^2} & \dfrac{(\gamma-1)\beta_1\beta_2}{\beta^2} & \dfrac{(\gamma-1)\beta_1\beta_3}{\beta^2} \\ -\beta_2\gamma & \dfrac{(\gamma-1)\beta_2\beta_1}{\beta^2} & 1 + \dfrac{(\gamma-1)\beta_2^2}{\beta^2} & \dfrac{(\gamma-1)\beta_2\beta_3}{\beta^2} \\ -\beta_3\gamma & \dfrac{(\gamma-1)\beta_3\beta_1}{\beta^2} & \dfrac{(\gamma-1)\beta_3\beta_2}{\beta^2} & 1 + \dfrac{(\gamma-1)\beta_3^2}{\beta^2} \end{bmatrix},$$

where

$$\beta_1 = \frac{v_1}{c}, \ \beta_2 = \frac{v_2}{c}, \ \beta_2 = \frac{v_2}{c}.$$

11.4 Show that the generators $\mathbf{V}_i$ [Eq. (11.59)] can also be obtained from

$$\mathbf{V}_i = \mathbf{A}'_{boost}(\beta_i = 0).$$

11.5 Given the charge distribution

$$\rho(\vec{r}) = r^2 e^{-r} \sin^2\theta,$$

make a multipole expansion of the potential and evaluate all the nonvanishing multipole moments. What is the potential for large distances?

11.6   Show that $d^l_{m'm}(\beta)$ satisfies the differential equation

$$\left\{ \frac{\partial^2}{\partial\beta^2} + \cot\beta\frac{\partial}{\partial\beta} + \left[l(l+1) - \left(\frac{m^2 + m'^2 - 2mm'\cos\beta}{\sin^2\beta} \right) \right] \right\} d^l_{m'm}(\beta) = 0.$$

11.7   Using the substitution

$$d^l_{m'm}(\beta) = \frac{y(\lambda_l, m', m, \beta)}{\sqrt{\sin\beta}},$$

in Problem 11.6 show that the second canonical form of the differential equation for $d^l_{m'm}(\beta)$ (Chapter 9) is given as

$$\frac{\partial^2 y(\lambda_l, m', m, \beta)}{\partial\beta^2} +$$

$$\left[l(l+1) + \frac{1}{4} - \left(\frac{m^2 + m'^2 - 2mm'\cos\beta - \frac{1}{4}}{\sin^2\beta} \right) \right] y(\lambda_l, m', m, \beta) = 0.$$

11.8   Using the result of Problem 11.7, solve the differential equation for $d^l_{mm'}(\beta)$ by the factorization method.

a) Considering m as a parameter, find the normalized step-up and step-down operators $O_+(m+1)$ and $O_-(m)$, which change the index m while keeping the index m' fixed.

b) Considering m' as a parameter, find the normalized step-up and step-down operators $O'_+(m'+1)$ and $O'_-(m')$, which change the index m' while keeping the index m fixed. Show that $|m| \le l$ and $|m'| \le l$.

c) Find the normalized functions with $m = m' = l$.

d) For $l = 2$, construct the full matrix $d^2_{m'm}(\beta)$.

e) By transforming the differential equation for $d^l_{mm'}(\beta)$ into an appropriate form, find the step-up and step-down operators that shift the index l for fixed m and m', giving the **normalized** functions $d^l_{mm'}(\beta)$.

f)Using the result of Problem 11.8.5, derive a recursion relation for $(\cos\beta)\, d^l_{mm'}(\beta)$. That is, express this as a combination of $d^{l'}_{mm'}(\beta)$ with $l' = l \pm 1, \ldots$.

(Note. This is a difficult problem and requires knowledge of the material discussed in Chapter 9.)

11.9   Show that
a)

$$D^l_{m0}(\alpha, \beta, -) = \sqrt{\frac{4\pi}{(2l+1)}}\, Y^*_{lm}(\beta, \alpha)$$

and

b)

$$D^l_{0m}(-,\beta,\gamma) = (-1)^m \sqrt{\frac{4\pi}{(2l+1)}} Y^*_{lm}(\beta,\gamma).$$

Hint. Use the invariant

$$\sum_{m=-l}^{m=l} Y^*_{lm}(\theta_1,\phi_1) Y_{lm}(\theta_2,\phi_2)$$

with $(\theta_1,\phi_1) = (\beta,\alpha)$ and $(\theta_2,\phi_2) = (\theta,\phi)$, $\theta_{12} = \theta'$, and

$$\left[D^l_{mm'}(\alpha,\beta,\gamma)\right]^{-1} = \left[D^l_{m'm}(\alpha,\beta,\gamma)\right]^* = D^l_{mm'}(-\gamma,-\beta,-\alpha).$$

11.10   For $l = 2$ construct the matrices

$$\mathbf{L}^k_y = \left(L^k_y\right)_{mm'}$$

for $k = 0, 1, 2, 3, 4, \dots$ and show that the matrices with $k \geq 5$ can be expressed as linear combinations of these. Use this result to check the result of Problem 11.8.4.

11.11   We have studied spherical harmonics $Y_{lm}(\theta,\phi)$, which are single-valued functions of (θ,ϕ) for $l = 0, 1, 2, \dots$. However, the factorization method also gave us a second family of solutions corresponding to the eigenvalues

$$\lambda = J(J+1)$$

with

$$M = J, (J-1), \dots, 0, \dots, -(J-1), -J,$$

where $J = 0, 1/2, 3/2, \dots$.
  For $J = 1/2$, find the 2×2 matrix of the y component of the angular momentum operator, that is, the generalization of our $[\mathbf{L}_y]_{mm'}$. Show that the matrices for $\mathbf{L}^2_y, \mathbf{L}^3_y, \mathbf{L}^4_y, \dots$ are simply related to the 2×2 unit matrix and the matrix $[\mathbf{L}_y]_{MM'}$. Calculate the d-function for $J = 1/2$:

$$d^{J=1/2}_{MM'}(\beta)$$

with M and M' taking values $+1/2$ or $-1/2$.

11.12   Using the following definition of Hermitian operators:

$$\int \Psi^*_1 \mathcal{L} \Psi_2 dx = \int (\mathcal{L}\Psi_1)^* \Psi_2 dx,$$

show that

$$\int \int d\Omega Y^*_{lm''} e^{i\gamma L_z} e^{i\beta L_y} e^{-i\alpha L_z} Y_{lm}$$
$$= e^{i\gamma m''} \left[\int \int d\Omega Y_{lm} e^{-i\beta L_y} Y^*_{lm''} \right] e^{i\alpha m}.$$

11.13   Convince yourself that the relations

$$e^{-i\beta L_{y1}} = e^{-i\alpha L_z} e^{-i\beta L_y} e^{i\alpha L_z}$$

and

$$e^{-i\gamma L_{z2}} = e^{-i\beta L_{y1}} e^{-i\gamma L_{z1}} e^{i\beta L_{y1}},$$

used in the derivation of the rotation matrix in terms of the original set of axes are true.

11.14   Show that the $D^l_{mm''}(R)$ matrices satisfy the relation

$$\sum_{m''} \left[D^l_{m'm''}(R) \right] \left[D^l_{m''m}(R^{-1}) \right] = \delta_{m'm}.$$

11.15   Show that the extended generator of

$$\mathbf{X} = x \frac{\partial}{\partial x} + y \frac{\partial}{\partial y}$$

is given as

$$\mathbf{X} = x \frac{\partial}{\partial x} + y \frac{\partial}{\partial y} - y'' \frac{\partial}{\partial y''} - 2y''' \frac{\partial}{\partial y'''} + \cdots .$$

11.16   Find the extension of

$$\mathbf{X} = xy \frac{\partial}{\partial x} + y^2 \frac{\partial}{\partial y}$$

up to third order.

11.17   Express the generator

$$\mathbf{X} = x \frac{\partial}{\partial x} + y \frac{\partial}{\partial y}$$

in terms of

$$u = y/x$$
$$v = xy.$$

11.18   Using induction, show that

$$\beta^{[n]} = \frac{d\beta^{[n-1]}}{dx} - y^{(n)}\frac{d\alpha}{dx}$$

can be written as

$$\beta^{[n]} = \frac{d^n(\beta - y'\alpha)}{dx^n} + y^{(n+1)}\alpha.$$

11.19   Does the following transformation form a group?

$$\left\{ \begin{array}{c} \overline{x} = x \\ \overline{y} = ay + a^2 y^2 \end{array} \right\},$$

where a is a constant.

12

COMPLEX VARIABLES and FUNCTIONS

Even though the complex numbers do not exist directly in nature, they are very useful in physics and engineering applications:

1. In the theory of complex functions there are pairs of functions called conjugate harmonic functions, which are very useful in finding solutions of Laplace equation in two dimensions.

2. The method of analytic continuation is very useful in finding solutions of differential equations and evaluating some definite integrals.

3. Infinite series, infinite products, asymptotic solutions, and stability calculations are other areas, in which complex techniques are very useful.

4. Even though complex techniques are very helpful in certain problems of physics and engineering, which are essentially problems defined in the real domain, complex numbers in quantum mechanics appear as an essential part of the physical theory.

12.1 COMPLEX ALGEBRA

A complex number is defined by giving a pair of real numbers

$$(a, b),$$
(12.1)

which could also be written as

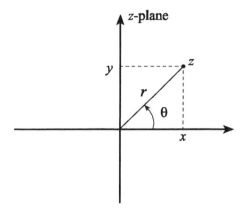

Fig. 12.1   A point in the complex z-plane

$$a + ib \ , \quad i = \sqrt{-1}. \tag{12.2}$$

A convenient way to represent a complex number is to use the concept of the complex z-plane (Fig. 12.1), where a point is shown as

$$z = (x, y) = x + iy.$$

Using plane polar coordinates we can also write a complex number as (Fig. 12.1)

$$x = r \cos \theta, \quad y = r \sin \theta$$

and

$$z = r \left(\cos \theta + i \sin \theta \right) \quad \text{or} \quad z = re^{i\theta}. \tag{12.3}$$

The modulus of z is defined as

$$r = |z| = \sqrt{x^2 + y^2}, \tag{12.4}$$

and θ is the argument of a complex number. Algebraic manipulations with complex numbers can be done according to the following rules:

   i)

$$z_1 + z_2 = (x_1 + iy_1) + (x_2 + iy_2), \tag{12.5}$$
$$= (x_1 + x_2) + i (y_1 + y_2). \tag{12.6}$$

ii)

$$cz = c\left(x + iy\right) \tag{12.7}$$
$$= cx + icy, \tag{12.8}$$

where c is a complex number.

iii)

$$z_1 \cdot z_2 = \left(x_1 + iy_1\right)\left(x_2 + iy_2\right), \tag{12.9}$$
$$= \left(x_1 x_2 - y_1 y_2\right) + i\left(x_1 y_2 + y_1 x_2\right).$$

iv)

$$\frac{z_1}{z_2} = \frac{\left(x_1 + iy_1\right)}{\left(x_2 + iy_2\right)}, \tag{12.10}$$
$$= \frac{\left(x_1 + iy_1\right)\left(x_2 - iy_2\right)}{\left(x_2 + iy_2\right)\left(x_2 - iy_2\right)}, \tag{12.11}$$
$$= \frac{\left[\left(x_1 x_2 + y_1 y_2\right) + i\left(y_1 x_2 - x_1 y_2\right)\right]}{\left(x_2^2 + y_2^2\right)}. \tag{12.12}$$

The **conjugate** of a complex number is defined as

$$z^* = x - iy. \tag{12.13}$$

Thus the modulus of a complex number is given as

$$|z| = zz^* = x^2 + y^2. \tag{12.14}$$

De Moivre's formula

$$e^{in\theta} = \left(\cos\theta + i\sin\theta\right)^n = \cos n\theta + i\sin n\theta \tag{12.15}$$

and the relations

$$|z_1| - |z_2| \le |z_1 + z_2| \le |z_1| + |z_2|,$$
$$|z_1 z_2| = |z_1|\,|z_2|, \tag{12.16}$$
$$\arg\left(z_1 z_2\right) = \arg z_1 + \arg z_2$$

are also very useful in calculation with complex numbers.

12.2 COMPLEX FUNCTIONS

We can define a complex function (Fig. 12.2) as

$$w = f\left(z\right) = u\left(x, y\right) + iv\left(x, y\right). \tag{12.17}$$

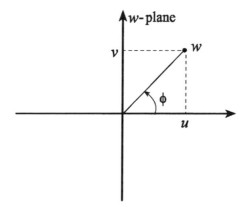

Fig. 12.2 w-plane

As an example for complex functions we can give polynomials like

$$f\left(z\right) = z^2 = \left(x + iy\right)^2 = \left(x^2 - y^2\right) + i\left(2xy\right), \tag{12.18}$$

$$f(z) = 3z^4 + 2z^3 + 2iz. \tag{12.19}$$

Trigonometric functions and some other well-known functions can also be defined in the complex plane as

$$\sin z, \quad \cos z, \quad \ln z, \quad \sinh z. \tag{12.20}$$

However, one must be very careful with multivaluedness.

12.3 COMPLEX DERIVATIVES AND ANALYTIC FUNCTIONS

As in real analysis we can define the derivative of a complex function as

$$\frac{df\left(z\right)}{dz} = \lim_{\triangle z \to 0} \frac{f\left(z + \triangle z\right) - f\left(z\right)}{\triangle z}, \tag{12.21}$$

$$= \lim_{\triangle z \to 0} \left[\frac{\triangle u}{\triangle z} + i\frac{\triangle v}{\triangle z}\right].$$

However, for this derivative to be meaningful it must be independent of the direction in which the limit $\triangle z \to 0$ is taken. If we first approach z parallel to the real axis, that is, when

$$\triangle z = \triangle x, \tag{12.22}$$

we find the derivative as

$$\frac{df}{dz} = \frac{\partial u}{\partial x} + i\frac{\partial v}{\partial x}. \tag{12.23}$$

On the other hand, if we approach z parallel to the imaginary axis, that is, when

$$\Delta z = i\Delta y, \tag{12.24}$$

the derivative becomes

$$\frac{df}{dz} = -i\frac{\partial u}{\partial y} + \frac{\partial v}{\partial y}. \tag{12.25}$$

For the derivative to exist these two expressions must agree; thus we obtain the conditions for the derivative to exist at some point z as

$$\frac{\partial u\,(x,y)}{\partial x} = \frac{\partial v\,(x,y)}{\partial y} \tag{12.26}$$

and

$$\frac{\partial v\,(x,y)}{\partial x} = -\frac{\partial u\,(x,y)}{\partial y}. \tag{12.27}$$

These conditions are called the Cauchy-Riemann conditions, and they are necessary and sufficient for the derivative of $f(z)$ to exist.

12.3.1 Analytic Functions

If the derivative of a function, $f(z)$, exists not only at z_0 but also at every other point in some neighborhood of z_0, then we say that $f(z)$ is analytic at z_0.

Example 12.1. *Analytic functions:* The function

$$f(z) = z^2 + 5z^3, \tag{12.28}$$

like all other polynomials, is analytic in the entire z-plane. On the other hand, even though the function

$$f(z) = \left|z^2\right| \tag{12.29}$$

satisfies the Cauchy-Riemann conditions at $z = 0$, it is not analytic at any other point in the z-plane.

If a function is analytic in the entire z-plane it is called an **entire function**. All polynomials are entire functions. If a function is analytic at every point in the neighborhood of z_0 except at z_0, we call z_0 an **isolated singular point**.

Example 12.2. *Analytic functions:* If we take the derivative of

$$f(z) = \frac{1}{z} \tag{12.30}$$

we find

$$f'(z) = -\frac{1}{z^2},\tag{12.31}$$

which means that $z = 0$ is an isolated singular point of this function. At all other points, this function is analytic.

Theorem: If $f(z)$ is analytic in some domain of the z-plane, then the partial derivatives of all orders of $u(x,y)$ and $v(x,y)$ exist. The $u(x,y)$ and $v(x,y)$ functions of such a function satisfy the Laplace equations

$$\overrightarrow{\nabla}^2_{xy} u(x,y) = \frac{\partial^2 u(x,y)}{\partial x^2} + \frac{\partial^2 u(x,y)}{\partial y^2} = 0,\tag{12.32}$$

and

$$\overrightarrow{\nabla}^2_{xy} v(x,y) = 0.\tag{12.33}$$

Proof. We use the first Cauchy-Riemann condition [Eq. (12.26)] and differentiate with respect to x to get

$$\frac{\partial u}{\partial x} = \frac{\partial v}{\partial y}\tag{12.34}$$

$$\frac{\partial^2 u}{\partial x^2} = \frac{\partial^2 v}{\partial x \partial y}.\tag{12.35}$$

Similarly, we write the second condition [Eq. (12.27)] and differentiate with respect to y to get

$$\frac{\partial v}{\partial x} = -\frac{\partial u}{\partial y}\tag{12.36}$$

$$\frac{\partial^2 u}{\partial y^2} = -\frac{\partial^2 v}{\partial y \partial x}.\tag{12.37}$$

Adding Equations (12.35) and (12.37) gives us

$$\frac{\partial^2 u}{\partial x^2} + \frac{\partial^2 u}{\partial y^2} = \frac{\partial^2 v}{\partial x \partial y} - \frac{\partial^2 v}{\partial x \partial y} = 0.\tag{12.38}$$

One can show Equation (12.33) in exactly the same way. The functions $u(x,y)$ and the $v(x,y)$ are called **harmonic functions**, whereas the pair of functions $(u(x,y), v(x,y))$ are called **conjugate harmonic functions**.

12.3.2 Harmonic Functions

Harmonic functions have very useful properties in applications:

1. The two families of curves defined as $u = c_i$ and $v = d_i$ (c_i and d_i are real numbers) are orthogonal to each other.

   Proof.

$$\vec{\nabla} u \cdot \vec{\nabla} v = \frac{\partial u}{\partial x}\frac{\partial v}{\partial x} + \frac{\partial u}{\partial y}\frac{\partial v}{\partial y}, \tag{12.39}$$

$$\vec{\nabla} u \cdot \vec{\nabla} v = \frac{\partial u}{\partial x}\left(-\frac{\partial u}{\partial y}\right) + \left(\frac{\partial u}{\partial y}\right)\left(\frac{\partial u}{\partial x}\right), \tag{12.40}$$

$$\vec{\nabla} u \cdot \vec{\nabla} v = 0, \tag{12.41}$$

   where we have used the Cauchy-Riemann conditions.

2. If we differentiate an analytic function $w = w(z)$ we get

$$\frac{dw}{dz} = \left(\frac{\partial u}{\partial x} + i\frac{\partial v}{\partial x}\right)\frac{dx}{dz} + \left(\frac{\partial u}{\partial y} + i\frac{\partial v}{\partial y}\right)\left(\frac{dy}{dz}\right)(-i^2), \tag{12.42}$$

$$\frac{dw}{dz} = \left(\frac{\partial u}{\partial x} - i\frac{\partial u}{\partial y}\right)\left(\frac{dx + idy}{dz}\right), \tag{12.43}$$

$$\frac{dw}{dz} = \frac{\partial u}{\partial x} - i\frac{\partial u}{\partial y}. \tag{12.44}$$

The modulus of this gives us

$$\left|\frac{dw}{dz}\right| = \sqrt{\left(\frac{\partial u}{\partial x}\right)^2 + \left(\frac{\partial u}{\partial y}\right)^2}. \tag{12.45}$$

Harmonic functions are very useful in electrostatics. If we take $u(x, y)$ as the potential energy, the electric field will be given by

$$\vec{E} = -\vec{\nabla} u. \tag{12.46}$$

Thus the modulus we have found in Equation (12.45) gives the magnitude of the electric field as

$$\left|\vec{E}\right| = \left|\frac{dw}{dz}\right|. \tag{12.47}$$

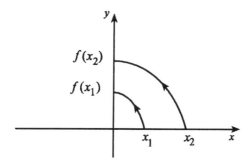

Fig. 12.3 It is not interesting to look at real functions as mappings

3. If $\Psi(u,v)$ satisfies the Laplace equation in the w-plane,

$$\frac{\partial^2 \Psi(u,v)}{\partial u^2} + \frac{\partial^2 \Psi(u,v)}{\partial v^2} = 0, \qquad (12.48)$$

where u and v are conjugate harmonic functions, then $\Psi(x,y)$ will satisfy the Laplace equation in the z-plane as

$$\frac{\partial^2 \Psi(x,y)}{\partial x^2} + \frac{\partial^2 \Psi(x,y)}{\partial y^2} = 0. \qquad (12.49)$$

12.4   MAPPINGS

A real function

$$y = f(x), \qquad (12.50)$$

which defines a curve in the xy-plane, can be interpreted as an operator that maps a point on the x-axis to a point on the y-axis (Fig. 12.3), which is not very interesting. However, in the complex plane a function,

$$w = f(z),$$
$$w = u(x,y) + iv(x,y), \qquad (12.51)$$

maps a point (x,y) in the z-plane to another point (u,v) in the w-plane, which implies that curves and domains in the z-plane are mapped to other curves and domains in the w-plane. This has rather interesting consequences in applications.

Example 12.3. *Translation:* Let us consider the function

$$w = z + z_0. \qquad (12.52)$$

Since this means

$$u = x + x_0 \tag{12.53}$$

and

$$v = y + y_0, \tag{12.54}$$

a point (x, y) in the z-plane is mapped into the translated point $(x + x_0, \ y + y_0)$ in the w-plane.

Example 12.4. *Rotation:* Let us consider the function

$$w = z z_0. \tag{12.55}$$

Using

$$\begin{aligned} w &= \rho e^{i\phi} \\ z &= r e^{i\theta} \\ z_0 &= r_0 e^{i\theta_0}, \end{aligned} \tag{12.56}$$

we write w in plane polar coordinates as

$$\rho e^{i\phi} = r r_0 e^{i(\theta + \theta_0)}. \tag{12.57}$$

In the w-plane this means

$$\rho = r r_0 \tag{12.58}$$
$$\phi = \theta + \theta_0. \tag{12.59}$$

Two things have changed:

i. Modulus r has increased or decreased by a factor r_0.

ii. θ has changed by θ_0.

If we take $z_0 = i$, this mapping (function) corresponds to a pure rotation by $\frac{\pi}{2}$.

Example 12.5. *Inversion:* The function

$$w(z) = \frac{1}{z} \tag{12.60}$$

can be written as

$$\rho e^{i\phi} = \frac{1}{r e^{i\theta}}, \tag{12.61}$$

$$= \frac{1}{r} e^{-i\theta}. \tag{12.62}$$

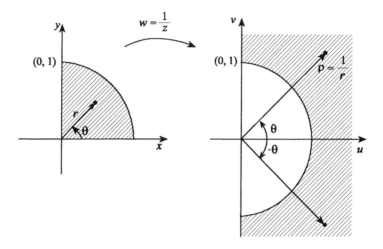

Fig. 12.4   Inversion maps circles to circles

This gives

$$\rho = \frac{1}{r} \tag{12.63}$$

$$\phi = -\theta, \tag{12.64}$$

which means that a point inside the unit circle in the z-plane is mapped to a point outside the unit circle, plus a reflection about the u-axis in the w-plane (Fig. 12.4).

Example 12.6. *Inversion function:* Let us now see how inversion, that is,

$$w(z) = \frac{1}{z} \tag{12.65}$$

maps curves in the z-plane to the w-plane. We first write

$$w = u + iv \tag{12.66}$$

$$= \frac{1}{x + iy} \tag{12.67}$$

$$= \frac{1}{x + iy} \cdot \frac{x - iy}{x - iy} \tag{12.68}$$

$$= \frac{x}{(x^2 + y^2)} - i\frac{y}{(x^2 + y^2)}. \tag{12.69}$$

This gives us the transformation $(x, y) \rightarrow (u, v)$:

$$u = \frac{x}{x^2 + y^2} \qquad (12.70)$$

$$v = \frac{-y}{x^2 + y^2} \qquad (12.71)$$

and its inverse as

$$x = \frac{u}{u^2 + v^2} \qquad (12.72)$$

$$y = \frac{-v}{u^2 + v^2}. \qquad (12.73)$$

We are now ready to see how a circle in the z-plane,

$$x^2 + y^2 = r^2, \qquad (12.74)$$

is mapped to the w-plane by inversion. Using Equations (12.72) and (12.73) we see that this circle is mapped to

$$\frac{u^2}{(u^2 + v^2)^2} + \frac{v^2}{(u^2 + v^2)^2} = r^2, \qquad (12.75)$$

$$u^2 + v^2 = \frac{1}{r^2} \qquad (12.76)$$

$$= \rho^2, \qquad (12.77)$$

which is another circle with the radius $1/r$.

Next, let us consider a straight line in the z-plane as

$$y = c_1. \qquad (12.78)$$

Using Equation (12.73) this becomes

$$-\frac{v}{u^2 + v^2} = c_1 \qquad (12.79)$$

or

$$u^2 + v^2 + \frac{v}{c_1} + \frac{1}{(2c_1)^2} = \frac{1}{(2c_1)^2},$$

$$u^2 + (v + \frac{1}{2c_1})^2 = \frac{1}{(2c_1)^2}. \qquad (12.80)$$

This is nothing but a circle with the radius $\frac{1}{2c_1}$ and with its center located at $\left(0, -\frac{1}{2c_1}\right)$; thus the inversion maps straight lines in the z-plane to circles in the w-plane (Fig. 12.5).

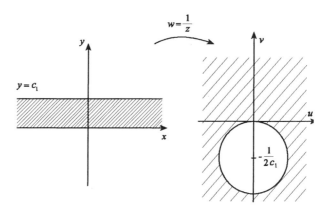

Fig. 12.5   Inversion maps straight lines to circles

All the mappings we have discussed so for are **one-to-one mappings**, that is, a single point in the z-plane is mapped to a single point in the w-plane.

Example 12.7. *Two-to-one mapping:* We now consider the function

$$w = z^2 \tag{12.81}$$

and write it in plane polar coordinates as

$$w = \rho e^{i\theta}. \tag{12.82}$$

Using

$$z = r e^{i\theta}, \tag{12.83}$$

ρ and ϕ become

$$\rho = r^2, \tag{12.84}$$

$$\phi = 2\theta. \tag{12.85}$$

The factor of two in front of the θ is crucial. This means that the first quarter in the z-plane, $0 \leq \theta \leq \frac{\pi}{2}$, is mapped to the upper half of the w-plane, $0 \leq \phi < \pi$. On the other hand, the upper half of the z-plane, $0 \leq \theta < \pi$, is mapped to the entire w-plane, $0 \leq \phi < 2\pi$. In other words,

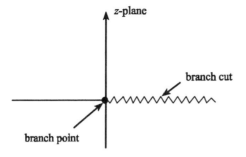

Fig. 12.6   Cut line ends at a branch point

the lower half of the z-plane is mapped to the already covered (used) entire w-plane. Hence, in order to cover the z-plane once we have to cover the w-plane twice. This is called a **two-to-one mapping**. Two different points in the z-plane,

$$z_0 \tag{12.86}$$

and

$$z_0 e^{-i\pi} = -z_0 \tag{12.87}$$

are mapped to the same point in the w-plane as

$$w = z_0^2. \tag{12.88}$$

We now consider

$$w = e^z. \tag{12.89}$$

Writing

$$\rho e^{i\phi} = e^{x+iy}, \tag{12.90}$$

where

$$\rho = e^x \tag{12.91}$$

and

$$\phi = y, \tag{12.92}$$

we see that in the z-plane the $0 \leq y < 2\pi$ band is mapped to the entire w-plane; thus in the z-plane all the other parallel bands given as

$$x + i\left(y + 2n\pi\right), \quad n \text{ integer,} \tag{12.93}$$

are mapped to the already covered w-plane. In this case we say that we have a **many-to-one mapping.**

Let us now consider the function

$$w = \sqrt{z}. \tag{12.94}$$

In plane polar coordinates we can write

$$\rho e^{i\phi} = \sqrt{r}e^{i\theta/2} \tag{12.95}$$

and

$$2\phi = \theta. \tag{12.96}$$

In this case the point

$$r = r_0, \ \theta = 0, \tag{12.97}$$

is mapped to

$$w = \sqrt{r_0},$$

while the point

$$r = r_0, \ \theta = 2\pi, \tag{12.98}$$

is mapped to

$$w = \sqrt{r_0}e^{i\pi} = -\sqrt{r_0} \tag{12.99}$$

in the w-plane. However the coordinates (12.97) and (12.98) represent the same point in the z-plane. In other words, a single point in the z-plane is mapped to two different points, except at the origin, in the w-plane. This is called a **one-to-two mapping**.

To define a square root as a single-valued function so that for a given value of z a single value of w results, all we have to do is to cut out the $\theta = 2\pi$ line from the z-plane. This line is called the **cut line** or the **branch cut,** and the point $z = 0$, where this line ends, is called the **branch point** (Fig. 12.6). What is important here is to find a region in the z-plane where our function is single valued and then extend this region over the entire z-plane without our function becoming multivalued. As seen from Figure 12.7a and Figure 12.7b the problem is at the origin:

$$z = 0. \tag{12.100}$$

For any region that does not include the origin our function will be single valued. However, for any region that includes the origin, where θ changes between $[0, 2\pi]$ we will run into the multivaluedness problem. In order to

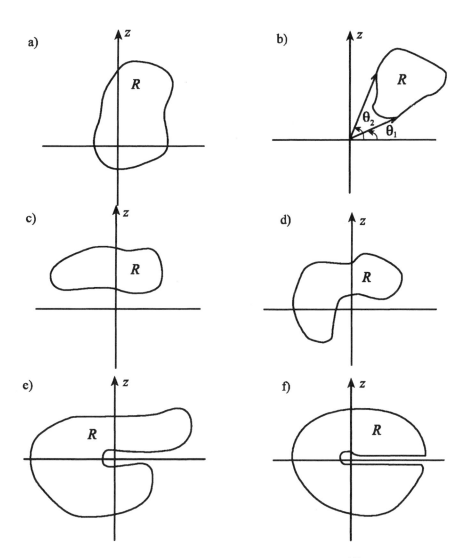

Fig. 12.7 For every region R that does not include the origin $w = z^{1/2}$ is single valued

extend the region in which our function is single valued we start with a region
R, where our function is single valued, and then extend it without including
the origin so that we cover a maximum of the z-plane (Fig. 12.7b, c, d, e,
and f). The only way to do this is to exclude the points on a curve (usually
taken as a straight line), that starts from the origin and then extends all the
way to infinity.

As seen from Figure 12.8, for the square root, $f(z) = \sqrt{z}$, for any path that does not cross the cut line our function is single valued and the value it takes is called the branch I value:

I. branch $\qquad w_1(z) = \sqrt{r}e^{\theta/2}, \qquad\qquad 0 \le \theta < 2\pi.$

For the range $2\pi \le \theta < 4\pi$, since the cut line is crossed once, our function will take the branch II value given as

II. branch $\qquad w_2(z) = -\sqrt{r}e^{\theta/2}, \qquad\qquad 2\pi \le \theta < 4\pi.$

Square root function has two branch values. In cases where θ increases continuously, as in rotation problems, we switch from one branch value to another each time we cross over the cut line. This situation can be conveniently shown by the Riemann sheets (Fig. 12.9).

Riemann sheets for this function are two parallel sheets sewn together along the cut line. As long as we remain in one of the sheets, our function is single valued and takes only one of the branch values. Whenever we cross the cut line we find ourselves on the other sheet and the function switches to the other branch value.

Example 12.8. $w(z)$=lnz function: In the complex plane the ln function is defined as

$$w(z) = \ln z = \ln r + i\theta. \tag{12.101}$$

It has infinitely many branches; thus infinitely many Riemann sheets as

$$
\begin{aligned}
\text{branch } 0 \quad & w_0(z) = \ln r + i\theta \\
\text{branch } 1 \quad & w_1(z) = \ln r + i(\theta + 1.2\pi) \\
\text{branch } 2 \quad & w_2(z) = \ln r + i(\theta + 2.2\pi) \\
\vdots \quad & \qquad\vdots \\
\text{branch } n \quad & w_n(z) = \ln r + i(\theta + n.2\pi)
\end{aligned}
\tag{12.102}
$$

where $0 \le \theta < 2\pi$.

Example 12.9. $w(z)$=$\sqrt{z^2-1}$ function: To investigate the branches of the function

$$w(z) = \sqrt{z^2 - 1}, \tag{12.103}$$

we define

$$(z - 1) = r_1 e^{i\theta_1}, \quad (z + 1) = r_2 e^{i\theta_2} \tag{12.104}$$

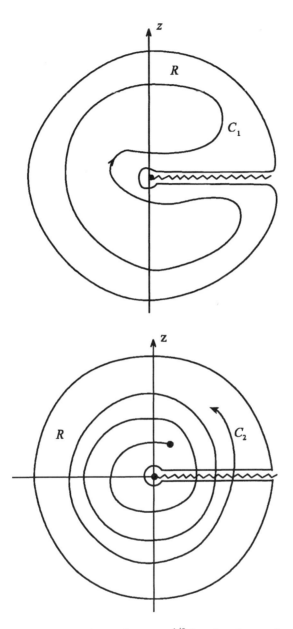

Fig. 12.8   Each time we cross the cut line $w = z^{1/2}$ function changes from one branch value to another

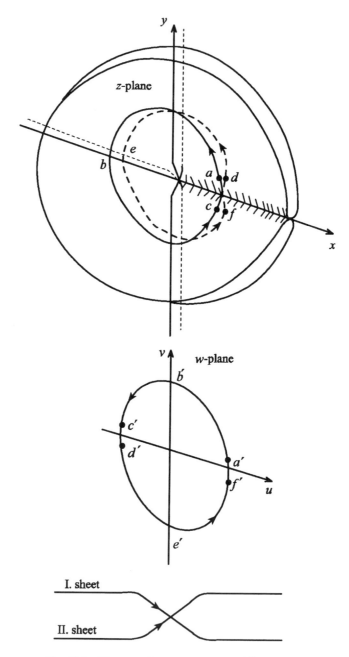

Fig. 12.9   Riemann sheets for the $w = z^{1/2}$ function

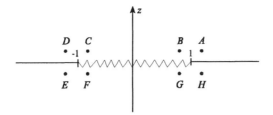

Fig. 12.10 Cut lines for $\sqrt{z^2 - 1}$

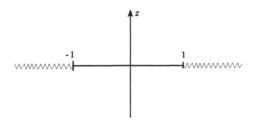

Fig. 12.11 A different Choice for the cut lines of $\sqrt{z^2 - 1}$

and write

$$w(z) = \rho e^{i\phi} \tag{12.105}$$

$$= \sqrt{(z-1)(z+1)} \tag{12.106}$$

$$= \sqrt{r_1 r_2}\, e^{i(\theta_1 + \theta_2)/2}. \tag{12.107}$$

This function has two branch points located at $x = +1$ and $x = -1$. We place the cut lines along the real axis and to the right of the branch points. This choice gives the ranges of θ_1 and θ_2 as

$$0 \le \theta_1 < 2\pi, \tag{12.108}$$

$$0 \le \theta_2 < 2\pi. \tag{12.109}$$

We now investigate the limits of the points A, B, C, D, F, G, and H in the z-plane as they approach the real axis and the corresponding points in the w-plane (Fig. 12.10):

Point	θ_1	θ_2	ϕ	$\sqrt{z^2-1}$
A	0	0	0	single valued
H	2π	2π	2π	single valued
B	π	0	$\pi/2$	double valued
G	π	2π	$3\pi/2$	double valued
C	π	0	$\pi/2$	double valued
F	π	2π	$3\pi/2$	double valued
D	π	π	π	single valued
E	π	π	π	single valued

$$(12.110)$$

Points A and H, which approach the same point in the z-plane, also go to the same point in the w-plane. In other words, where the two cut lines overlap our function is single valued. For pairs (B, G) and (C, F) even though the corresponding points approach the same point in the z-plane, they are mapped to different points in the w-plane. For points D and E the function is again single valued. For this case the cut lines are now shown as in Figure 12.10. The first and second branch values for this function are given as

$$w_1\left(z\right) = \sqrt{r_1 r_2}\, e^{i(\theta_1+\theta_2)/2}, \qquad (12.111)$$

$$w_2\left(z\right) = \sqrt{r_1 r_2}\, e^{i(\theta_1+\theta_2+2\pi)/2}. \qquad (12.112)$$

Riemann sheets for this function will be two parallel sheets sewn together in the middle between points -1 and $+1$.

For this function another choice for the cut lines is given as in Figure 12.11, where

$$0 \leq \theta_1 < 2\pi, \qquad (12.113)$$

$$-\pi \leq \theta_2 < \pi. \qquad (12.114)$$

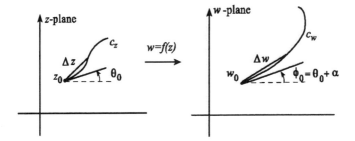

Fig. 12.12 Conformal mapping

12.4.1 Conformal Mappings

To see an interesting property of analytic functions we differentiate

$$w = f(z) \tag{12.115}$$

at z_0, where the modulus and the arguments of the derivative are given as $\left|\frac{df}{dz}\right|_{z_0}$ and α, respectively. We now use polar coordinates to write the modulus

$$\lim_{\Delta z \to 0}\left|\frac{\Delta w}{\Delta z}\right| = \left.\frac{dw}{dz}\right|_{z_0} \tag{12.116}$$

$$= \left.\left|\frac{df}{dz}\right|\right|_{z_0}, \tag{12.117}$$

and the argument (Fig. 12.12) as

$$\arg\left.\frac{df}{dz}\right|_{z_0} = \alpha = \arg\lim_{\Delta z \to 0}\left(\frac{\Delta w}{\Delta z}\right), \tag{12.118}$$

$$\alpha = \lim_{\Delta z \to 0}\arg[\Delta w] - \lim_{\Delta z \to 0}\arg[\Delta z]. \tag{12.119}$$

Since this function, $f(z)$, maps a curve c_z in the z-plane into another curve c_w in the w-plane, from the arguments [Eq. (12.119)] it is seen that if the slope of c_z at z_0 is θ_0, then the slope of c_w at w_0 is $\alpha + \theta_0$. For a pair of curves intersecting at z_0 the angle between their tangents in the w- and z-planes will be equal, that is,

$$\phi_2 - \phi_1 = (\theta_2 + \alpha) - (\theta_1 + \alpha), \tag{12.120}$$

$$= \theta_2 - \theta_1. \tag{12.121}$$

Hence analytic functions preserve angles between the curves they map (Fig. 12.12). For this reason they are also called conformal mappings or transformations.

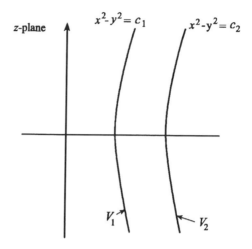

Fig. 12.13   Two plates with hyperbolic cross sections

12.4.2   Electrostatics and Conformal Mappings

Conformal mappings are very useful in electrostatic and laminar (irrotational) flow problems, where the Laplace equation must be solved. Even though the method is restricted to cases with one translational symmetry, it allows one to solve analytically some complex boundary value problems.

Example 12.10.   *Conformal mappings and electrostatics:* Let us consider two conductors held at potentials V_1 and V_2 with hyperbolic cross sections

$$x^2 - y^2 = c_1 \text{ and } x^2 - y^2 = c_2. \tag{12.122}$$

We want to find the equipotentials and the electric field lines. In the complex z-plane the problem can be shown as in Figure 12.13.   We use the conformal mapping

$$w = z^2 \tag{12.123}$$

$$= x^2 - y^2 + i\,(2xy) \tag{12.124}$$

to map these hyperbolae to the straight lines

$$u = c_1 \text{ and } u = c_2 \tag{12.125}$$

in the w-plane (Fig.   12.14).   The problem is now reduced to finding the equipotentials and the electric field lines between two infinitely long

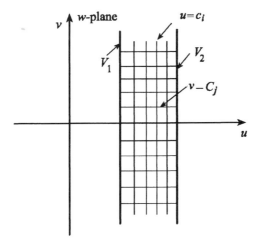

Fig. 12.14 Equipotentials and electric field lines in the w-plane

parallel plates held at potentials V_1 and V_2, where the electric field lines are given by the family of lines

$$v = C_j \tag{12.126}$$

and the equipotentials are given by the lines perpendicular to these as

$$u = c_i. \tag{12.127}$$

Because the problem is in the z-plane, we make the inverse transformation to obtain the electric field lines as

$$(v =) \ 2xy = C_j \tag{12.128}$$

and the equipotentials as

$$(u =) \ x^2 - y^2 = c_i. \tag{12.129}$$

In three dimensions, to find the equipotential surfaces these curves must be extended along the direction of the normal to the plane of the paper.

Example 12.11. *Electrostatics and conformal mappings:* We now find the equipotentials and the electric field lines inside two conductors with semicircular cross sections separated by an insulator and held at potentials $+V_0$ and $-V_0$, respectively (Fig. 12.15). The equation of a circle

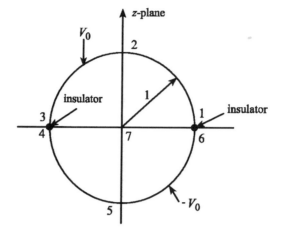

Fig. 12.15   Two conductors with semicircular cross sections

in the z-plane is given as

$$x^2 + y^2 = 1. \tag{12.130}$$

We use the conformal mapping

$$w\left(z\right) = \ln\left(\frac{1+z}{1-z}\right), \tag{12.131}$$

to map these semicircles into straight lines in the w-plane (Fig. 12.16). Using Equation (12.131) we write

$$u + iv = \ln\frac{1+x+iy}{1-x-iy} \tag{12.132}$$

$$= \ln\left[\frac{1 - x^2 - y^2 + 2iy}{1 - 2x + x^2 + y^2}\right] \tag{12.133}$$

and express the argument of the ln function as $Re^{i\alpha}$:

$$u + iv = \ln R + i\alpha. \tag{12.134}$$

Now the v function is found as

$$v = \alpha \tag{12.135}$$

$$= \tan^{-1}\frac{2y}{1 - \left(x^2 + y^2\right)}.$$

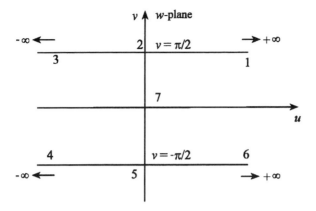

Fig. 12.16 Two semicircular conductors in the w-plane

From the limits

$$\lim_{\substack{x^2+y^2\to 1 \\ y>0}} \left[\tan^{-1}\frac{2y}{1-(x^2+y^2)}\right] = \frac{\pi}{2} \qquad (12.136)$$

and

$$\lim_{\substack{x^2+y^2\to 1 \\ y<0}} \left[\tan^{-1}\frac{2y}{1-(x^2+y^2)}\right] = -\frac{\pi}{2}, \qquad (12.137)$$

we see that the two semicircles in the z-plane are mapped to two straight lines given as

$$v = \frac{\pi}{2} \text{ and } v = -\frac{\pi}{2}. \qquad (12.138)$$

Equipotential surfaces in the w-plane can now be written easily as

$$V(v) = \frac{2V_0}{\pi}v. \qquad (12.139)$$

Using Equation (12.135) we transform this into the z-plane to find the equipotentials as

$$V = \frac{2V_0}{\pi}\tan^{-1}\left[\frac{2y}{1-(x^2+y^2)}\right], \qquad (12.140)$$

$$= \frac{2V_0}{\pi}\tan^{-1}\left[\frac{2r\sin\theta}{1-r^2}\right]. \qquad (12.141)$$

Because this problem has translational symmetry perpendicular to the plane of the paper, equipotential surfaces in three dimensions can be found by extending these curves in that direction. The solution to this problem has been found rather easily and in closed form. Compare this with the separation of variables method, where the solution is given in terms of the Legendre polynomials as an infinite series. However, applications of conformal mapping are limited to problems with one translational symmetry, where the problem can be reduced to two dimensions. Even though there are tables of conformal mappings, it is not always easy as in this case to find an analytic expression for the needed mapping.

12.4.3   Fluid Mechanics and Conformal Mappings

For laminar (irrotational) and frictionless flow, conservation of mass is given as

$$\frac{\partial \rho}{\partial t} + \vec{\nabla} \cdot (\rho \vec{v}) = 0, \tag{12.142}$$

where $\rho(\vec{r}, t)$ and $\vec{v}(\vec{r}, t)$ represent the density and the velocity of a fluid element. For stationary flow we write

$$\frac{\partial \rho}{\partial t} = 0, \tag{12.143}$$

thus Equation (12.142) becomes

$$\vec{\nabla} \cdot (\rho \vec{v}) = 0. \tag{12.144}$$

Also, a lot of realistic situations can be approximated by the incompressible fluid equation of state, that is,

$$\rho = \text{const.} \tag{12.145}$$

This further reduces Equation (12.144) to

$$\vec{\nabla} \cdot \vec{v} = 0. \tag{12.146}$$

This equation alone is not sufficient to determine the velocity field $\vec{v}(\vec{r}, t)$. If the flow is irrotational, it will also satisfy

$$\vec{\nabla} \times \vec{v} = 0, \tag{12.147}$$

thus the two equations

$$\vec{\nabla} \cdot \vec{v} = 0 \tag{12.148}$$

and

$$\vec{\nabla} \times \vec{v} = 0 \tag{12.149}$$

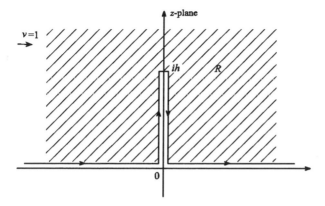

Fig. 12.17 Flow around a wall of height h

completely specify the kinematics of laminar, frictionless flow of incompressible fluids. These equations are also the expressions of linear and angular momentum conservations for the fluid elements. Fluid elements in laminar flow follow streamlines, where the velocity $\vec{v}(\vec{r}, t)$ at a given point is tangent to the streamline at that point.

Equations (12.148) and (12.149) are the same as Maxwell's equations in electrostatics. Following the definition of electrostatic potential, we use Equation (12.149) to define a velocity potential as

$$\vec{v}(\vec{r}, t) = \vec{\nabla}\Phi(\vec{r}, t). \tag{12.150}$$

Substituting this into Equation (12.148) we obtain the Laplace equation

$$\vec{\nabla}^2\Phi(\vec{r}, t) = 0. \tag{12.151}$$

We should note that even though $\Phi(\vec{r}, t)$ is known as the velocity potential it is very different from the electrostatic potential.

Example 12.12. *Flow around an obstacle of height h:* Let us consider laminar flow around an infinitely long and thin obstacle of height h. Since the problem has translational symmetry, we can show it in two dimensions as in Figure 12.17, where we search for a solution of the Laplace equation in the region R.

Even though the velocity potential satisfies the Laplace equation like the electrostatic potential, we have to be careful with the boundary conditions. In electrostatics, electric field lines are perpendicular to the equipotentials; hence the test particles can only move perpendicular to the conducting surfaces. In the laminar flow case, motion perpendicular to the surfaces is not allowed because fluid elements follow the contours of the bounding surfaces. For points far away from the obstacle, we take the flow lines as parallel to the x-axis. As we approach the obstacle, the flow lines follow the contours of the surface. For points away from the obstacle, we set

$$v_\infty = 1. \tag{12.152}$$

We now look for a transformation that maps the region R in the z-plane to the upper half of the w-plane. Naturally, the lower boundary of the region R in Figure 12.17 will be mapped to the real axis of the w-plane. We now construct this transformation in three steps: We first use

$$w_1 = z^2 \tag{12.153}$$

to map the region R to the entire w_1-plane. Here the obstacle is between 0 and $-h^2$. As our second step, we translate the obstacle to the interval between 0 and h^2 by

$$w_2 = z^2 + h^2. \tag{12.154}$$

Finally we map the w_2-plane to the upper half of the w-plane by

$$w = \sqrt{w_2}. \tag{12.155}$$

The complete transformation from the z-plane to the w-plane can be written as (Fig. 12.18)

$$w = \sqrt{z^2 + h^2}.$$

The Laplace equation can now be easily solved in the upper half of the w-plane, and the streamlines are obtained as

$$v = c_j.$$

Curves perpendicular to these will give the velocity equipotentials as

$$u = b_j.$$

Finally transforming back to the z-plane we find the streamlines as the curves

$$c_j = \text{Im}[\sqrt{z^2 + h^2}],$$

and the velocity of the fluid elements that are tangents to the streamlines (Fig. 12.19) are given as

$$|\vec{v}| = \left| \frac{dw}{dz} \right|.$$

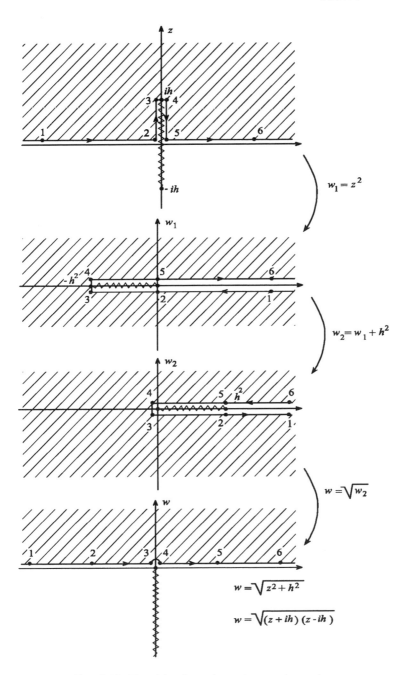

Fig. 12.18 Transition from the *z*-plane to the *w*-plane

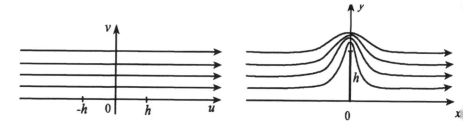

Fig. 12.19   Streamlines in the w and z-planes

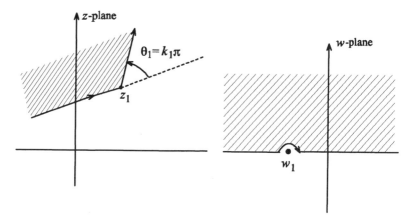

Fig. 12.20   Schwarz-Christoffel transformation maps the inside of a polygon to the upper half of the w-plane

12.4.4   Schwarz-Christoffel Transformations

We have seen that analytic transformations are also conformal mappings, which preserve angles. We now introduce the Schwarz-Christoffel transformations, where the transformation is not analytic at an isolated number of points. Schwarz-Christoffel transformations map the inside of a polygon in the z-plane to the upper half of the w-plane (Fig. 12.20). To construct the Schwarz-Christoffel transformations let us consider the function

$$\frac{dz}{dw} = A\left(w - w_1\right)^{-k_1}, \tag{12.156}$$

where A is complex, k_1 is real, and w_1 is a point on the u-axis. Comparing the arguments of both sides in Equation (12.156) we get

$$\arg\left(\frac{dz}{dw}\right) = \lim_{\Delta w \to 0} [\arg \Delta z - \arg \Delta w] \tag{12.157}$$

$$\lim_{\Delta w \to 0} [\arg \Delta z - \arg \Delta w] = \begin{cases} \arg A - k_1 \pi & w < w_1 \\ \arg A & w > w_1 \end{cases}$$

As we move along the positive u-axis

$$\lim_{\Delta w \to 0} \arg \Delta w = \arg [dw] = 0, \tag{12.158}$$

hence we can write

$$\lim_{\Delta w \to 0} [\arg \Delta z] = \arg[dz] = \begin{cases} \arg A - k_1 \pi & w < w_1 \\ \arg A & w > w_1 \end{cases}. \tag{12.159}$$

For a constant A this means that the transformation [Eq. (12.156)] maps the parts of the u-axis; $w < w_1$ and $w > w_1$, to two line segments meeting at z_0 in the z-plane. Thus

$$A (w - w_1)^{-k_1} \tag{12.160}$$

corresponds to one of the vertices of a polygon with the exterior angle $k_1 \pi$ and located at z_1. For a polygon with n-vertices we can write the Schwarz-Christoffel transformation as

$$\frac{dz}{dw} = A (w - w_1)^{-k_1} (w - w_2)^{-k_2} \cdots (w - w_n)^{-k_n}. \tag{12.161}$$

Because the exterior angles of a polygon add up to 2π, powers k_i should satisfy the condition

$$\sum_{i=1} k_i = 2. \tag{12.162}$$

Integrating Equation (12.161) we get

$$z = A \int^w (w - w_1)^{-k_1} (w - w_2)^{-k_2} \cdots (w - w_n)^{-k_n} \, dw + B,$$

where B is a complex integration constant. In this equation A determines the direction and B determines the location of the polygon in the z-plane. In a Schwarz-Christoffel transformation there are all together $2n + 4$ parameters, that is, n w_is, n k_is, and 4 parameters from the complex constants A and B. A polygon can be specified by giving the coordinates of its n vertices in the z-plane. Along with the constraint [Eq. (12.162)] this determines the $2n+1$ of the parameters in the transformation. This means that we have the freedom to choose the locations of the three w_i on the real axis of the w-plane.

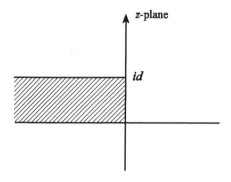

Fig. 12.21   Region we map in Example (12.13)

Example 12.13. *Schwarz-Christoffel transformation:* We now construct a Schwarz-Christoffel transformation that maps the region shown in Figure 12.21 to the upper half of the w-plane. Such transformations are frequently needed in applications. To construct the Schwarz-Christoffel transformation we define a polygon whose inside, in the limit as $z_3 \to -\infty$, goes to the desired region (Fig. 12.22). Using the freedom in defining the Schwarz-Christoffel transformation we map the points z_1, z_2, and z_3 to the points

$$w_3 \to -\infty, \ w_1 = -1, \ w_2 = +1 \qquad (12.163)$$

in the w-plane. We now write the Schwarz-Christoffel transformation as

$$\frac{dz}{dw} = c\,(w+1)^{-k_1}\,(w-1)^{-k_2}\,(w-w_3)^{-k_3}. \qquad (12.164)$$

Powers k_1, k_2, and k_3 are determined from the figure as $\frac{1}{2}, \frac{1}{2}$, and 1, respectively. Note how the signs of k_i are chosen as plus because of the counterclockwise directions shown in Figure 12.22. Because the constant c is still arbitrary, we define a new finite complex number A as

$$\lim_{w_3 \to -\infty} \frac{c}{(-w_3)^{k_3}} \to A, \qquad (12.165)$$

so that the Schwarz-Christoffel transformation becomes

$$\frac{dz}{dw} = A\,(w+1)^{-\frac{1}{2}}\,(w-1)^{-\frac{1}{2}}, \qquad (12.166)$$

$$\frac{dz}{dw} = \frac{A}{\sqrt{w^2 - 1}}. \qquad (12.167)$$

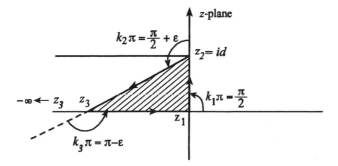

Fig. 12.22 The polygon whose interior goes to the desired region in Example 12.13 in the limit $z_3 \to \infty$

This can be integrated as

$$z = A \cosh^{-1} w + B, \tag{12.168}$$

where the constants A and B are found from the locations of the vertices, that is,

$$z = 0 \to w = -1 \tag{12.169}$$

$$z = id \to w = +1 \tag{12.170}$$

as

$$A = \frac{d}{\pi} \text{ and } B = id. \tag{12.171}$$

Example 12.14. *Semi-infinite parallel plate capacitor:* We now calculate the fringe effects in a semi-infinite parallel plate capacitor. Making use of the symmetry of the problem we can concentrate on the region shown in Figure 12.23. To find a Schwarz-Christoffel transformation that maps this region into the upper half of the w-plane we choose the points on the real w-axis as

$$\begin{bmatrix} z_1 & \to & w_1 \to & -\infty \\ z_4 & \to & w_4 \to & +\infty \\ z_2 & \to & w_2 = & -1 \\ z_3 & \to & w_3 = & 0 \end{bmatrix}. \tag{12.172}$$

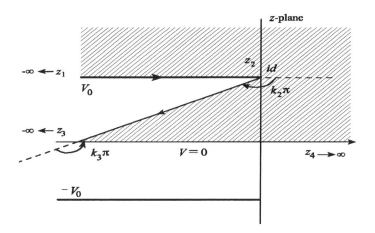

Fig. 12.23   Semi-infinite parallel plate capacitor

Since $k_2 = -1$ and $k_3 = 1$, we can write

$$
\begin{aligned}
\frac{dz}{dw} &= c \left(w + 1\right)^{-k_2} \left(w - 0\right)^{-k_3} \\
&= c \frac{(w + 1)}{w} \\
&= c(1 + \frac{1}{w}).
\end{aligned}
$$
(12.173)

Integrating this we get

$$
z = c \left(w + \ln w\right) + D.
$$
(12.174)

If we substitute

$$
w = |w| \, e^{i\phi},
$$
(12.175)

Equation (12.174) becomes

$$
z = c \left[|w| \, e^{i\phi} + \ln |w| + i\phi\right] + D.
$$
(12.176)

Considering the limit in Figure 12.24 we can write

$$
z_3^{\text{upper}} - z_3^{\text{lower}} = id.
$$

Using Equation (12.176) this becomes

$$
\begin{aligned}
z_3^{\text{upper}} - z_3^{\text{lower}} &= c \left[0 + i \left(\phi_3^{\text{upper}} - \phi_3^{\text{lower}}\right)\right] \\
&= ci \left(\pi - 0\right),
\end{aligned}
$$
(12.177)

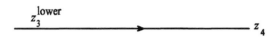

Fig. 12.24   Limit of the point z_3

thus determining the constant c as

$$c = \frac{d}{\pi}.$$ (12.178)

On the other hand, considering that the vertex

$$z_2 = id$$ (12.179)

is mapped to the point -1 in the w-plane, we write

$$id = \frac{d}{\pi}(-1 + i\pi) + D$$ (12.180)

and determine D as

$$D = \frac{d}{\pi}.$$ (12.181)

This determines the Schwarz-Christoffel transformation

$$z = \frac{d}{\pi}[w + \ln w + 1],$$ (12.182)

which maps the region shown in Figure 12.23 to the upper half w-plane shown in Figure 12.25. We now consider the transformation

$$\bar{z} = \frac{d}{\pi}\ln w \quad \text{or} \quad w = e^{\bar{z}\pi/d},$$ (12.183)

which maps the region in Figure 12.25 to the region shown in Figure 12.26 in the $\bar{z}$-plane. In the $\bar{z}$-plane equipotentials are easily written as

$$\bar{y} = \text{const.} = \frac{V}{V_0}d \quad \text{or} \quad V(\bar{y}) = \frac{V_0}{d}\bar{y}.$$ (12.184)

Using the inverse transformation in Equation (12.182), we write

$$z = x + iy$$ (12.185)

$$= \frac{d}{\pi}\left\{e^{\bar{x}\pi/d}\left[\cos\left(\frac{V}{V_0}\pi\right) + i\sin\left(\frac{V}{V_0}\pi\right)\right] + 1\right\} + \bar{x} + i\frac{V}{V_0}d.$$

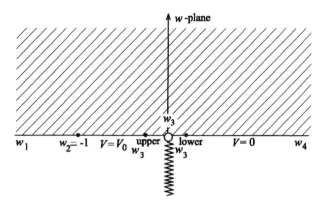

Fig. 12.25 w-Plane for the semi-infinite parallel plate capacitor

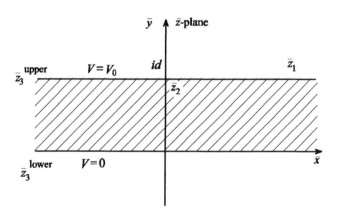

Fig. 12.26 $\bar{z}$-Plane for the semi-infinite parallel plate capacitor

This gives us the parametric expression of the equipotentials in the z-plane (Fig. 12.27) as

$$x = \frac{d}{\pi}\left[e^{\overline{x}\pi/d}\cos\left(\frac{V}{V_0}\pi\right) + 1\right] + \overline{x}, \qquad (12.186)$$

$$y = \frac{d}{\pi}e^{\overline{x}\pi/d}\sin\left(\frac{V}{V_0}\pi\right) + \frac{V}{V_0}d. \qquad (12.187)$$

Similarly, the electric field lines in the $\overline{z}$-plane are written as

$$\overline{x} = \text{const.} \qquad (12.188)$$

Transforming back to the z-plane, with the definitions

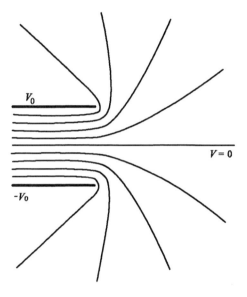

Fig. 12.27   Equipotentials for the semi-infinite parallel plate capacitor

$$\frac{\overline{x}\pi}{d} = \kappa \quad \text{and} \quad \theta = \frac{\overline{y}\pi}{d}, \qquad (12.189)$$

we get

$$x = \frac{d}{\pi}[e^{\kappa}\cos\theta + 1] + \kappa\frac{d}{\pi}, \qquad (12.190)$$

$$y = \frac{d}{\pi}[e^{\kappa}\sin\theta + \theta]. \qquad (12.191)$$

Problems

12.1   For conjugate harmonic pairs show that if $\Psi(u, v)$ satisfies the Laplace equation

$$\frac{\partial^2 \Psi(u, v)}{\partial u^2} + \frac{\partial^2 \Psi(u, v)}{\partial v^2} = 0$$

in the w-plane, then $\Psi(x, y)$ will satisfy

$$\frac{\partial^2 \Psi(x, y)}{\partial x^2} + \frac{\partial^2 \Psi(x, y)}{\partial y^2} = 0$$

in the z-plane.

12.2   Show that

$$u(x, y) = \sin x \cosh y + x^2 - y^2 + 4xy$$

is a harmonic function and find its conjugate.

12.3   Show that

$$u(x, y) = \sin 2x / (\cosh 2y + \cos 2x)$$

can be the real part of an analytic function $f(z)$. Find its imaginary part and express $f(z)$ explicitly as a function of z.

12.4   Using cylindrical coordinates and the method of separation of variables find the equipotentials and the electric field lines inside two conductors with semi-circular cross sections separated by an insulator and held at potentials $+V_0$ and $-V_0$, respectively (Fig. 12.15). Compare your result with Example 12.11 and show that the two methods agree.

12.5   With aid of a computer program plot the equipotentials and the electric field lines found in Example 12.14 for the semi-infinite parallel plate capacitor.

12.6   In a two-dimensional potential problem the surface ABCD is at potential V_0 and the surface EFG is at potential zero. Find the transformation (in differential form) that maps the region R into the upper half of the w-plane (Fig. 12.28). Do not integrate but determine all the constants.

12.7   Given the following two-dimensional potential problem in Figure 12.29, The surface ABC is held at potential V_0 and the surface DEF is at potential zero. Find the transformation that maps the region R into upper half of the w-plane. Do not integrate but determine all the constants in the differential form of the transformation

12.8   Find the Riemann surface on which

$$\sqrt{(z - 1)(z - 2)(z - 3)}$$

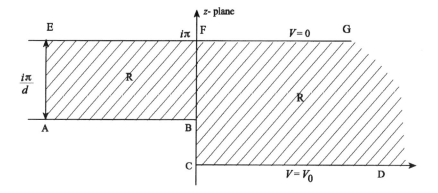

Fig. 12.28   Two-dimensional equipotential problem

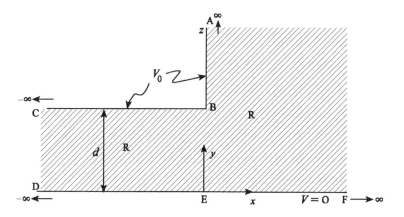

Fig. 12.29   Schwartz-Christoffel transformation

8. Find the Riemann surface on which

$$\sqrt{(z-1)(z-2)(z-3)}$$

is single valued and analytic except at $z = 1, 2, 3$.

9. Find the singularities of

$$f(z) = \tanh z.$$

10. Show that the transformation

$$\frac{w}{2} = \tan^{-1}\left(\frac{iz}{a}\right)$$

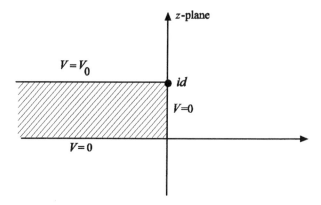

Fig. 12.30   Rectangular region surrounded by metallic plates

or

$$w = -i \ln \left[\frac{1 + \dfrac{z}{a}}{1 - \dfrac{z}{a}} \right]$$

maps the $v =$const. lines into circles in the z-plane.

12.11   Use the transformation given in Problem 12.10 to find the equipotentials and the electric field lines for the electrostatics problem of two infinite parallel cylindrical conductors, each of radius R and separated by a distance of d, and held at potentials $+V_0$ and $-V_0$, respectively.

12.12   Consider the electrostatics problem for the rectangular region surrounded by metallic plates as shown in Figure 12.30. The top plate is held at potential V_0, while the bottom and the right sides are grounded $(V = 0)$. The two plates are separated by an insulator. Find the equipotentials and the electric field lines and plot.

12.13   Map the real w-axis into the triangular region shown in Figure 12.31, in the limit as

$$x_5 \quad \rightarrow \quad \infty$$
$$\text{and}$$
$$x_3 \quad \rightarrow \quad -\infty$$

12.14   Find the equipotentials and the electric field lines for a conducting circular cylinder held at potential V_0 and parallel to a grounded infinite conducting plane (Fig. 12.32). Hint: Use the transformation $z = a \tanh iw/2$.

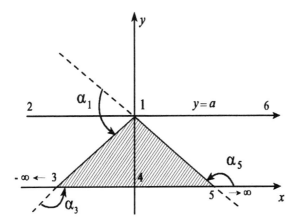

Fig. 12.31   Triangular region

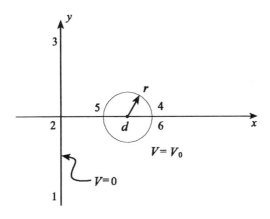

Fig. 12.32   Conducting circular cylinder parallel to infinite metallic plate

13

COMPLEX
INTEGRALS and
SERIES

In this chapter we first introduce the complex integral theorems. Using analytic continuation we discuss how these theorems can be used to evaluate some frequently encountered definite integrals. In conjunction with our discussion of definite integrals, we also introduce the gamma and beta functions. We also introduce complex series and discuss classification of singular points.

13.1   COMPLEX INTEGRAL THEOREMS

I. Cauchy-Goursat Theorem

Let C be a closed contour in a simply connected domain (Fig. 13.1). If a given function, $f(z)$, is analytic in and on this contour, then the integral

$$\oint_C f(z)\,dz = 0 \tag{13.1}$$

is true.

Proof. We write the function $f(z)$ as

$$f(z) = u + iv. \tag{13.2}$$

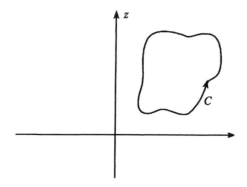

Fig. 13.1   Contour for the Cauchy-Goursat theorem

Integral (13.1) becomes

$$\oint_C (u + iv)(dx + idy)$$
$$= \oint (udx - vdy) + i \oint (vdx + udy).$$ (13.3)

Using the Stokes theorem

$$\oint \vec{A} \cdot d\vec{l} = \iint \left(\vec{\nabla} \times \vec{A} \right) \cdot \hat{n} ds,$$ (13.4)

we can write integral (13.3) as

$$\oint_C (u + iv)(dx + idy)$$
$$= \iint_S \left(-\frac{dv}{dx} - \frac{du}{dy} \right) ds + \iint_S \left(\frac{du}{dx} - \frac{dv}{dy} \right) ds,$$ (13.5)

where S is an oriented surface bounded by the closed path C. Because the Cauchy-Riemann conditions are satisfied in and on the closed path C, the right-hand side of Equation (13.5) is zero, thus proving the theorem.

II. Cauchy Integral Theorem

If $f(z)$ is analytic in and on a closed path C in a simply connected domain (Fig. 13.2) and if z_0 is a point inside the path C, then we can write the integral

$$\frac{1}{2\pi i} \oint_C \frac{f(z) \, dz}{(z - z_0)} = f(z_0).$$ (13.6)

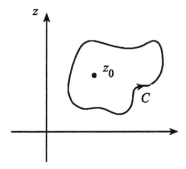

Fig. 13.2 Contour for the Cauchy integral theorem

Proof. To prove this theorem we modify the path in Figure 13.2 and use that in Figure 3.3, where we can use the Cauchy-Goursat theorem to write

$$\oint_{C+L_1+L_2+C_0} \frac{f(z)\,dz}{(z-z_0)} = 0. \qquad (13.7)$$

This integral must be evaluated in the limit as the radius of the path C_0 goes to zero. Integrals over L_1 and L_2 cancel each other. Also noting that the integral over C_0 is taken clockwise, we write

$$\frac{1}{2\pi i}\oint_C \frac{f(z)\,dz}{(z-z_0)} = \frac{1}{2\pi i}\oint_{C_0} \frac{f(z)\,dz}{(z-z_0)}, \qquad (13.8)$$

where both integrals are now taken counterclockwise. The integral on the left-hand side is what we want. For the integral on the right-hand side we can write

$$\frac{1}{2\pi i}\oint_{C_0} \frac{f(z)\,dz}{(z-z_0)}$$
$$= \frac{1}{2\pi i} f(z_0) \oint_{C_0} \frac{dz}{(z-z_0)} + \frac{1}{2\pi i}\oint_{C_0} \frac{f(z)-f(z_0)}{(z-z_0)}dz. \qquad (13.9)$$

Using the substitution $z - z_0 = R_0 e^{i\theta}$, the first integral on the right-hand side can be evaluated easily, giving us

$$\frac{1}{2\pi i} f(z_0) \oint_{C_0} \frac{dz}{(z-z_0)} = \frac{1}{2\pi i} f(z_0)\, i \int_0^{2\pi} \frac{R_0 e^{i\theta}\,d\theta}{R_0 e^{i\theta}} = f(z_0). \qquad (13.10)$$

The second integral in Equation (13.9), which we call I_2, can be bounded

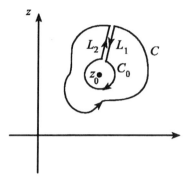

Fig. 13.3   A different path for the Cauchy integral theorem

from above as

$$|I_2| \leq \int_{C_0} \left| \frac{f(z) - f(z_0)}{z - z_0} \right| |dz| \leq M \cdot L, \qquad (13.11)$$

where M is the maximum value of $\left| \dfrac{f(z) - f(z_0)}{z - z_0} \right|$ on C_0 and L is the circumference of C_0, which is

$$L = 2\pi R_0. \qquad (13.12)$$

Now let ϵ be a given small number such that on C_0

$$|f(z) - f(z_0)| < \epsilon \qquad (13.13)$$

is satisfied. Because $f(z)$ is analytic in C, no matter how small an ϵ is chosen, we can always find a sufficiently small radius R_0,

$$|z - z_0| \leq R_0 = \delta, \qquad (13.14)$$

such that condition (13.13) is satisfied; thus we can write

$$I_2 \leq M \cdot L = 2\pi\epsilon. \qquad (13.15)$$

From the limit

$$\lim_{\delta \to 0} \epsilon \to 0, \qquad (13.16)$$

it follows that $I_2 \to 0$; thus the desired result is obtained as

$$\frac{1}{2\pi i} \oint_C \frac{f(z)\, dz}{(z - z_0)} = f(z_0). \qquad (13.17)$$

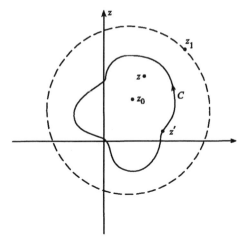

Fig. 13.4 Path for Taylor series

Note that the limit $\lim\limits_{z \to z_0} \dfrac{f(z) - f(z_0)}{z - z_0}$ is actually the definition of the derivative $f'(z)$ evaluated at z_0. Because $f(z)$ is analytic in and on the contour C, it exists and hence M in Equation (13.11) is finite. Thus, $|I_2| \to 0$ as $R_0 \to 0$.

III. Cauchy Theorem

Because the position of the point z_0 is arbitrary in the Cauchy integral theorem, we can treat it as a parameter and differentiate Equation (13.6) with respect to z_0 as

$$f'(z_0) = \frac{df}{dz}\bigg|_{z=z_0} = \frac{1}{2\pi i} \oint_C \frac{f(z)\,dz}{(z - z_0)^2}. \tag{13.18}$$

After n-fold differentiation, we obtain a very useful formula:

$$f^{(n)}(z_0) = \frac{d^n f}{dz^n}\bigg|_{z=z_0} = \frac{n!}{2\pi i} \oint_C \frac{f(z)\,dz}{(z - z_0)^{n+1}}. \tag{13.19}$$

13.2   TAYLOR SERIES

Let us expand a function $f(z)$ about z_0, where it is analytic. Also, let z_1 be the nearest singular point of $f(z)$ to z_0. If $f(z)$ is analytic on and inside a closed contour C, we can use the Cauchy theorem to write

$$f(z) = \frac{1}{2\pi i} \oint_C \frac{f(z')\,dz'}{(z'-z)}, \tag{13.20}$$

where z' is a point on the contour and z is a point inside the contour C (Fig. 13.4). We can now write Equation (13.20) as

$$
\begin{aligned}
f(z) &= \frac{1}{2\pi i} \oint_C \frac{f(z')\,dz'}{[(z'-z_0)-(z-z_0)]} \\
&= \frac{1}{2\pi i} \oint_C \frac{f(z')\,dz'}{(z'-z_0)\left[1-\frac{(z-z_0)}{z'-z_0}\right]}.
\end{aligned} \tag{13.21}
$$

Since the inequality $|z-z_0| < |z'-z_0|$ is satisfied in and on C, we can use the binomial formula

$$\frac{1}{1-t} = 1 + t + t^2 + \cdots = \sum_{n=0}^{\infty} t^n, \qquad |t| < 1 \tag{13.22}$$

to write

$$f(z) = \frac{1}{2\pi i} \oint_C \sum_{n=0}^{\infty} \frac{(z-z_0)^n}{(z'-z_0)^{n+1}} f(z')\,dz'. \tag{13.23}$$

Interchanging the integral and the summation signs we find

$$f(z) = \frac{1}{2\pi i} \sum_{n=0}^{\infty} (z-z_0)^n \oint_C \frac{f(z')\,dz'}{(z'-z_0)^{n+1}}, \tag{13.24}$$

which gives us the Taylor series expansion of $f(z)$ as

$$f(z) = \sum_{n=0}^{\infty} A_n (z-z_0)^n. \tag{13.25}$$

Using Equation (13.19) we can write the expansion coefficients as

$$A_n = \frac{1}{n!} f^{(n)}(z_0). \tag{13.26}$$

This expansion is unique and valid in the region $|z-z_0| < |z_1-z_0|$, where $f(z)$ is analytic.

13.3 LAURENT SERIES

Sometimes $f(z)$ is analytic inside an annular region as shown in Figure 13.5. For a closed contour in the region where our function is analytic (Fig. 13.5),

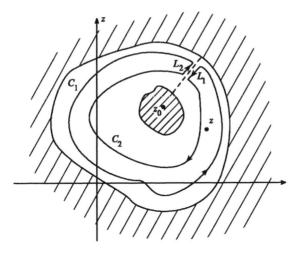

Fig. 13.5 Laurent series are defined in an annular region

integrals over L_1 and L_2 cancel each other, thus we can write

$$f(z) = \frac{1}{2\pi i} \oint_{C_1} \frac{f(z')\,dz'}{(z'-z)} - \frac{1}{2\pi i} \oint_{C_2} \frac{f(z')\,dz'}{(z'-z)}. \tag{13.27}$$

Since the inequality $|z'-z_0| > |z-z_0|$ is satisfied on C_1 and $|z'-z_0| < |z-z_0|$ is satisfied on C_2, we can write the above equation as

$$
\begin{aligned}
f(z) &= \frac{1}{2\pi i} \oint_{C_1} \frac{f(z')\,dz'}{[(z'-z_0)-(z-z_0)]} \\
&\quad - \frac{1}{2\pi i} \oint_{C_2} \frac{f(z')\,dz'}{-[(z-z_0)-(z'-z_0)]} \\
&= \frac{1}{2\pi i} \oint_{C_1} \frac{f(z')\,dz'}{(z'-z_0)\left[1-\dfrac{z-z_0}{z'-z_0}\right]} \\
&\quad + \frac{1}{2\pi i} \oint_{C_2} \frac{f(z')\,dz'}{(z-z_0)\left[1-\dfrac{z'-z_0}{z-z_0}\right]}.
\end{aligned}
$$

$$\tag{13.28}$$

$$\tag{13.29}$$

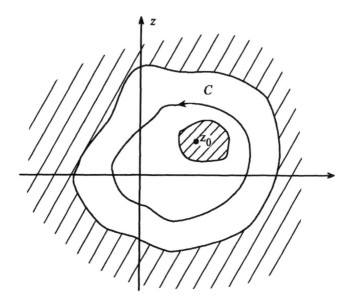

Fig. 13.6 Another contour for the Laurent series

We now use the binomial formula and interchange the integral and the summation signs to obtain the Laurent expansion as

$$f(z) = \sum_{n=0}^{\infty} (z - z_0)^n \left[\frac{1}{2\pi i} \oint_{C_1} \frac{f(z') dz'}{(z' - z_0)^{n+1}} \right]$$
$$+ \sum_{n=0}^{\infty} \frac{1}{(z - z_0)^{n+1}} \left[\frac{1}{2\pi i} \oint_{C_2} \frac{f(z') dz'}{(z' - z_0)^{-n}} \right]. \qquad (13.30)$$

Using the contour in Figure 13.6 we can also write the Laurent series as

$$f(z) = \sum_{n=-\infty}^{\infty} a_n (z - z_0)^n, \qquad (13.31)$$

where

$$a_n = \frac{1}{2\pi i} \oint_C \frac{f(z') dz'}{(z' - z_0)^{n+1}}. \qquad (13.32)$$

Example 13.1. *Taylor series:* We find the series expansion of

$$f(z) = \frac{1}{\sqrt{z^2 - 1}} \qquad (13.33)$$

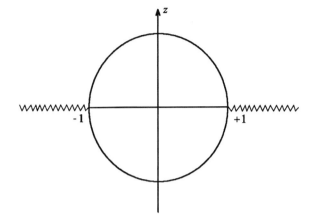

Fig. 13.7   For the $\frac{1}{\sqrt{z^2-1}}$ function we write the Taylor series in the region $|z| < 1$

in the interval $|z| < 1$ about the origin. Since this function is analytic inside the unit circle (Fig. 13.7), we need the Taylor series

$$f(z) = \sum_{n=0}^{\infty} \frac{a_n}{n!} z^n. \tag{13.34}$$

Using

$$a_n = \left[\frac{d^n f}{dz^n} \right]_{z=0}, \tag{13.35}$$

we write the Taylor series as

$$f(z) = \frac{1}{i} \left[1 + \frac{1}{2}z^2 + \frac{3}{8}z^4 + \frac{5}{16}z^6 + \cdots \right]. \tag{13.36}$$

Example 13.2. *Laurent series:* We now expand the same function,

$$f(z) = \frac{1}{\sqrt{z^2 - 1}}, \tag{13.37}$$

in the region $|z| > 1$. We place the cutline outside our region of interest between the points -1 and 1 (Fig. 13.8). The outer boundary of the annular region in which $f(z)$ is analytic could be taken as a circle with infinite radius, while the inner boundary is a circle with radius infinitesimally larger than 1. We now write the Laurent series about $z = 0$

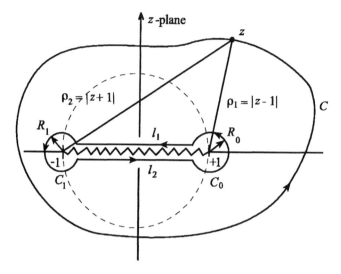

Fig. 13.8 For the $\frac{1}{\sqrt{z^2-1}}$ function we write the Laurent series in the region $|z| > 1$

as

$$f(z) = \sum_{n=-\infty}^{\infty} a_n z^n, \tag{13.38}$$

where the expansion coefficients are given as

$$a_n = \frac{1}{2\pi i} \oint_C \frac{1}{z'^{n+1}} \frac{dz'}{\sqrt{z'^2 - 1}}. \tag{13.39}$$

In this integral z' is a point on the contour C, which could be taken as any closed path inside the annular region where $f(z)$ is analytic. To evaluate the coefficients with $n \geq 0$, we first deform our contour so that it hugs the outer boundary of our annular region, which is a circle with infinite radius. For points on this contour we write

$$z' = Re^{i\theta} \tag{13.40}$$

and evaluate the coefficients a_n $(n \geq 0)$ in the limit as $R \to \infty$ we as

$$a_{n(n\geq 0)} = \lim_{R\to\infty} \frac{1}{2\pi i} \oint_C \frac{dz'}{z'^{n+1}\sqrt{z'^2 - 1}} \tag{13.41}$$

$$= \lim_{R\to\infty} \frac{1}{2\pi i} \int \frac{iRe^{i\theta}\,d\theta}{R^{n+2}e^{i(n+2)\theta}} \tag{13.42}$$

$$= 0. \tag{13.43}$$

To pick the coefficients with the negative values of n, we take our contour as a circle with radius infinitesimally larger than 1. Because $f(z)$ is analytic everywhere except the cutline, these coefficients can be evaluated by shrinking the contour to a bone-shape so that it hugs the cutline as shown in Figure 13.8; thus

$$a_{n(n<0)} = \frac{1}{2\pi i} \oint_{C_1 \, +\, \overleftarrow{l_1} \, +\, \overrightarrow{l_2} \, +\, C_0} \frac{1}{z'^{n+1}} \frac{dz'}{\sqrt{z'^2 - 1}}. \tag{13.44}$$

We evaluate the integrals over C_0 and C_1 in the limit as their radiuses go to zero. First, let us consider the integral over C_0 and take

$$z' - 1 = R_0 e^{i\theta_0}. \tag{13.45}$$

The contribution of this to a_n is zero:

$$\lim_{R_0 \to 0} \frac{1}{2\pi i} \int_{-\pi \atop C_0}^{\pi} \frac{(+1)^{|n|-1}}{\sqrt{2}} \frac{R_0 i e^{i\theta_0} d\theta_0}{\sqrt{R_0} e^{\frac{1}{2}\theta_0}} \to 0. \tag{13.46}$$

Similarly, the contribution of C_1 is also zero, thus leaving us with

$$a_{n(n<0)} = \frac{1}{2\pi i} \oint_{\overleftarrow{l_1} + \overrightarrow{l_2}} \frac{1}{z'^{n+1}} \frac{1}{\sqrt{z'^2 - 1}} dz'. \tag{13.47}$$

Integrals over l_1 and l_2 can be evaluated by defining the parameters

$$\begin{aligned} z' - 1 &= \rho_1 e^{i\theta_1}, \quad 0 \le \theta_1 < 2\pi, \\ z' + 1 &= \rho_2 e^{i\theta_2}, \quad 0 \le \theta_2 < 2\pi \end{aligned} \tag{13.48}$$

and writing

$$\begin{aligned} a_{n(n<0)} &= \frac{1}{2\pi i} \int_{L_1} \frac{z'^{|n|-1} dz'}{\sqrt{|z'-1||z'+1|}\, e^{i\theta_1} e^{i\theta_2}} \\ &+ \frac{1}{2\pi i} \int_{L_2} \frac{z'^{|n|-1} dz'}{\sqrt{|z'-1||z'+1|}\, e^{i\theta_1} e^{i\theta_2}}, \end{aligned} \tag{13.49}$$

$$\begin{aligned} a_{n(n<0)} &= \frac{1}{2\pi i} \int_1^{-1} \frac{x^{|n|-1} dx}{e^{i\frac{\pi}{2}} \sqrt{(1-x)(1+x)}} \\ &+ \frac{1}{2\pi i} \int_{-1}^1 \frac{x^{|n|-1} dx}{e^{i\frac{3\pi}{2}} \sqrt{(1-x)(1+x)}}, \end{aligned} \tag{13.50}$$

$$a_{n(n<0)} = -\frac{(-1)^{|n|}}{2\pi} \int_{-1}^1 \frac{x^{|n|-1} dx}{\sqrt{1-x^2}} + \frac{1}{2\pi} \int_{-1}^1 \frac{x^{|n|-1} dx}{\sqrt{1-x^2}}. \tag{13.51}$$

We finally obtain the coefficients as

$$a_{-|n|} = \frac{1}{2\pi} \left[1 - (-1)^{|n|} \right] \int_{-1}^{1} \frac{x^{|n|-1} dx}{\sqrt{1-x^2}}, \tag{13.52}$$

$$a_{-|n|} = \left\{ \begin{array}{ll} 0 & |n| = \text{even} \\ \frac{(|n|-1)!}{2^{|n|-1}[(|n|-1)/2!]^2} & |n| = \text{odd} \end{array} \right\}. \tag{13.53}$$

This gives us the Laurent expansion for the region $|z| > 1$ and about the origin as

$$\frac{1}{\sqrt{z^2-1}} = \frac{1}{z} + \frac{1}{2z^3} + \frac{3}{8}\frac{1}{z^5} + \frac{5}{16}\frac{1}{z^7} + \cdots . \tag{13.54}$$

Example 13.3. *Laurent series—a short cut:* In the previous example we found the Laurent expansion of the function

$$f(z) = \frac{1}{\sqrt{z^2-1}} \tag{13.55}$$

about the origin and in the region $|z| > 1$. We used the contour integral definition of the coefficients. However, using the uniqueness of power series and appropriate binomial expansions, we can also evaluate the same series. First, we write $f(z)$ as

$$f(z) = \frac{1}{\sqrt{z^2-1}} \tag{13.56}$$

$$= \frac{1}{(z+1)^{\frac{1}{2}}} \frac{1}{(z-1)^{\frac{1}{2}}} \tag{13.57}$$

$$= \frac{1}{z^{\frac{1}{2}}} \frac{1}{\left(1+\frac{1}{z}\right)^{\frac{1}{2}}} \frac{1}{z^{\frac{1}{2}}} \frac{1}{\left(1-\frac{1}{z}\right)^{\frac{1}{2}}} \tag{13.58}$$

$$= \frac{1}{z} \frac{1}{\left(1+\frac{1}{z}\right)^{\frac{1}{2}}} \frac{1}{\left(1-\frac{1}{z}\right)^{\frac{1}{2}}}. \tag{13.59}$$

Since for the region $|z| > 1$ the inequality $1/z < 1$ is satisfied, we can use the binomial formula for the factors $\left(1+\frac{1}{z}\right)^{-\frac{1}{2}}$ and $\left(1-\frac{1}{z}\right)^{-\frac{1}{2}}$ to write the Laurent expansion as

$$f(z) = \frac{1}{z} \left[1 + \frac{1}{2z} + \frac{3}{8z^2} + \frac{5}{16z^3} + \cdots \right] \left[1 - \frac{1}{2z} + \frac{3}{8z^2} - \frac{5}{16z^3} + \cdots \right] \tag{13.60}$$

$$f(z) = \frac{1}{z} + \frac{1}{2z^3} + \frac{3}{8z^3} + \cdots, \tag{13.61}$$

which is the same as our previous result [Eq. (13.54)].

13.4   CLASSIFICATION OF SINGULAR POINTS

Using Laurent the series we can classify singular points of a function.

Definition I

Isolated singular point: If a function is not analytic at z_0 but analytic at every other point in some neighborhood of z_0, then z_0 is called an isolated singular point.

Definition II

Essential singular point: In the Laurent series of a function:

$$f(z) = \sum_{n=-\infty}^{\infty} a_n (z - z_0)^n, \qquad (13.62)$$

if for $n < -|m| < 0$,

$$a_n = 0$$

and

$$a_{-|m|} \neq 0,$$

then z_0 is called a singular point of order m.

Definition III

Essential singular point: If m is infinity, then z_0 is called an essential singular point.

Definition IV

Simple Pole: In Definition II, if $m = 1$, then z_0 is called a simple pole.

Definition V

Entire function: When a function is analytic in the entire z-plane it is called an entire function.

13.5   RESIDUE THEOREM

If a function $f(z)$ is analytic in and on the closed contour C except for a finite number of isolated singular points (Fig. 13.9), then we can write the integral

$$\oint_C f(z)\, dz = 2\pi i \sum_{n=0}^{N} R_n, \qquad (13.63)$$

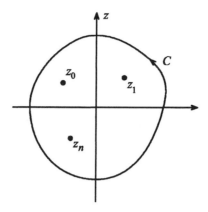

Fig. 13.9   Residue theorem

where R_n is the residue of $f(z)$ at the nth isolated singular point. The residue is defined as the coefficient of the $\dfrac{1}{(z-z_n)}$ term in the Laurent expansion of $f(z)$.

Proof: We change the contour C as shown in Figure 13.10 and use the Cauchy-Goursat theorem [Eq. (13.1)] to write

$$\oint_C f(z)\,dz$$

$$= \left[\sum_{n=0}^{N}\left[\oint_{c_n[\circlearrowleft]} + \oint_{l_n[\rightarrow]} + \oint_{l_n[\leftarrow]}\right] + \oint_{C'[\circlearrowleft]}\right] f(z)\,dz,$$

$$= 0. \tag{13.64}$$

Straight line segments of the integral cancel each other. Integrals over the small circles are evaluated clockwise and in the limit as their radius goes to zero. Since the integral over the closed path C' is equal to the integral over the closed path C, we write

$$\oint_C f(z)\,dz = \sum_{n=0}^{N}\oint_{c_n[\circlearrowleft]} f(z)\,dz, \tag{13.65}$$

where the integrals over c_n are now evaluated counterclockwise. Using the Laurent series expansion of $f(z)$ about z_0, the integral of the terms with the positive powers of $(z-z_0)$ gives

$$\oint_{c_0} (z-z_0)^n\,dz = 0, \quad n \geq 0. \tag{13.66}$$

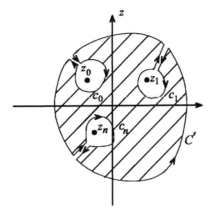

Fig. 13.10   Contour for the residue theorem

On the other hand, for the negative powers of $(z - z_0)$ we have

$$\oint_{c_n} \frac{dz}{(z - z_0)^n} = \lim_{R_0 \to 0} \int_0^{2\pi} \frac{i R_0 e^{i\theta_0} d\theta_0}{R_0^n e^{in\theta_0}}, \qquad n \geq 1, \qquad (13.67)$$

$$= \lim_{R_0 \to 0} \frac{i}{R_0^{n-1}} \int_0^{2\pi} e^{-i(n-1)\theta_0} d\theta_0, \qquad (13.68)$$

$$= \begin{cases} 0 & n = 2, 3, \dots \\ 2\pi i & n = 1. \end{cases} \qquad (13.69)$$

We repeat this for all the other poles to get

$$\oint_C f(z)\, dz = \sum_{n=0}^N \oint_{c_n} f(z)\, dz \qquad (13.70)$$

$$= 2\pi i \sum_{n=0}^N R_n. \qquad (13.71)$$

13.6   ANALYTIC CONTINUATION

When we discussed harmonic functions and mappings, we saw that analytic functions have very interesting properties. It is for this reason that it is very important to determine the region where a function is analytic and, if possible, to extend this region to other parts of the z-plane. This process is called analytic continuation. Sometimes functions like polynomials and trigonometric functions, which are defined on the real axis as

$$f(x) = a_0 + a_1 x + a_2 x^2 + \cdots + a_n x^n, \tag{13.72}$$
$$f(x) = \sin x, \tag{13.73}$$

can be analytically continued to the entire z-plane by simply replacing the real variable x with z, that is,

$$f(z) = a_0 + a_1 z + a_2 z^2 + \cdots + a_n z^n, \tag{13.74}$$
$$f(z) = \sin z. \tag{13.75}$$

However, analytic continuation is not always this easy. Let us now consider different series expansions of the function

$$f(z) = \frac{1}{1-z} + \frac{2}{2-z}. \tag{13.76}$$

This function has two isolated singular points at $z = 1$ and $z = 2$. We first make a Taylor series expansion about $z = 0$. We write

$$f(z) = \frac{1}{(1-z)} + \frac{1}{\left(1 - \dfrac{z}{2}\right)} \tag{13.77}$$

and use the binomial formula

$$(1-x)^{-1} = \sum_{n=0}^{\infty} x^n \tag{13.78}$$

to obtain

$$f(z) = \sum_{n=0}^{\infty} \left(1 + \frac{1}{2^n}\right) z^n, \qquad |z| < 1. \tag{13.79}$$

This expansion is naturally valid up to the nearest singular point at $z = 1$. Similarly, we can make another expansion, this time valid in the interval $1 < |z| < 2$ as

$$f(z) = -\left(\frac{1}{z}\right) \frac{1}{\left(1 - \dfrac{1}{z}\right)} + \frac{1}{\left(1 - \dfrac{z}{2}\right)} \tag{13.80}$$

$$= -\frac{1}{z} \sum_{n=0}^{\infty} \left(\frac{1}{z}\right)^n + \sum_{n=0}^{\infty} \left(\frac{z}{2}\right)^n \tag{13.81}$$

$$= 1 + \sum_{n=1}^{\infty} \left[\left(\frac{z}{2}\right)^n - \frac{1}{z^n}\right]. \tag{13.82}$$

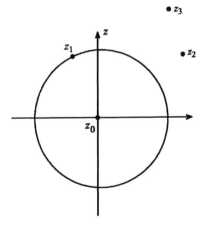

Fig. 13.11 Analytic continuation

Finally for $|z| > 2$, we obtain

$$f(z) = -\frac{1}{z}\left(\frac{1}{1-\frac{1}{z}}\right) - \left(\frac{2}{z}\right)\frac{1}{\left(1-\frac{2}{z}\right)}, \qquad (13.83)$$

$$= -\sum_{n=1}^{\infty}\left[\frac{1}{z^n} + \frac{2^n}{z^n}\right]. \qquad (13.84)$$

These three expansions of the same function [Eq. (13.77)] are valid for the intervals $|z| < 1$, $1 < |z| < 2$, and $|z| > 2$, respectively. Naturally it is not practical to use these series definitions, where each one is valid in a different part of the z-plane, when a closed expression like

$$f(z) = \frac{1}{1-z} + \frac{2}{2-z}$$

exists for the entire z-plane. However, it is not always possible to find a closed expression like this. Let us assume that we have a function with a finite number of isolated singular points at $z_1, z_2, ..., z_n$. Taylor series expansion of this function about a regular point z_0 will be valid only up to the nearest singular point z_1(Fig. 13.11). In such cases we can accomplish analytic continuation by successive Taylor series expansions, where each expansion is valid up to the nearest singular point (Fig. 13.12). We should make a note that during this process we are not making the function analytic at the points where it is not analytic.

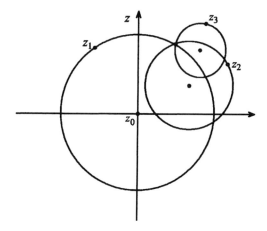

Fig. 13.12   How to accomplish analytic continuation

13.7   COMPLEX TECHNIQUES IN TAKING SOME DEFINITE INTEGRALS

Many of the definite integrals encountered in physics and engineering can be evaluated by using the complex integral theorems and analytic continuation:

I. Integrals of the form $I = \int_0^{2\pi} R\left(\cos\theta, \sin\theta\right) d\theta$.

In this integral R is a rational function of the form

$$R = \frac{a_1 \cos\theta + a_2 \sin\theta + a_3 \cos^2\theta + \cdots}{b_1 \cos\theta + b_2 \sin\theta + b_3 \cos^2\theta + b_4 \sin^2\theta + \cdots}. \qquad (13.85)$$

These integrals can be converted into a complex contour integral over the unit circle by the substitutions

$$\cos\theta = \frac{1}{2}\left(z + \frac{1}{z}\right), \quad \sin\theta = \frac{1}{2i}\left(z - \frac{1}{z}\right) \qquad (13.86)$$

and

$$z = e^{i\theta}, \quad d\theta = -i\left(\frac{dz}{z}\right) \qquad (13.87)$$

as

$$I = -i \oint_C R\left[\frac{1}{2}\left(z + \frac{1}{z}\right), \frac{1}{2i}\left(z - \frac{1}{z}\right)\right]\frac{dz}{z}. \qquad (13.88)$$

Example 13.4. *Complex contour integration technique:* Let us evaluate the integral

$$I = \int_0^{2\pi} \frac{d\theta}{a + \cos\theta} \; , \quad a > 1.$$

Using Equations (13.86) and (13.87) we can write this integral as

$$I = -i \oint_{|z|=1} \frac{dz}{\left(a + \dfrac{z}{2} + \dfrac{1}{2z}\right)z} \tag{13.89}$$

$$= -2i \oint_{|z|=1} \frac{dz}{z^2 + 2az + 1}. \tag{13.90}$$

The denominator can be factorized as

$$(z - \alpha)(z - \beta), \tag{13.91}$$

where

$$\alpha = -a + \left(a^2 - 1\right)^{\frac{1}{2}}, \tag{13.92}$$

$$\beta = -a - \left(a^2 - 1\right)^{\frac{1}{2}}. \tag{13.93}$$

For $a > 1$ we have $|\alpha| < 1$ and $|\beta| > 1$; thus only the root $z = \alpha$ is present inside the unit circle. We can now use the Cauchy integral theorem to find

$$I = -2i\left(2\pi i\right)\frac{1}{\alpha - \beta} \tag{13.94}$$

$$= \frac{2\pi}{\left(a^2 - 1\right)^{\frac{1}{2}}}. \tag{13.95}$$

Example 13.5. *Complex contour integral technique:* We now consider the integral

$$I = \frac{1}{2\pi} \int_0^{2\pi} \sin^{2l}\theta d\theta.$$

We can use Equations (13.86) and (13.87) to write I as a contour integral over the unit circle as

$$I = \frac{(-1)^l}{2\pi}\frac{(-i)}{2^{2l}} \oint \frac{dz}{z}\left(z - \frac{1}{z}\right)^{2l}. \tag{13.96}$$

We can now evaluate this integral by using the residue theorem as

$$I = \frac{(-1)^l}{2\pi}\frac{(-i)}{2^{2l}}2\pi i\left[\text{residue of } \frac{1}{z}\left(z - \frac{1}{z}\right)^{2l} \text{ at } z = 0\right]. \tag{13.97}$$

Using the binomial formula we can write

$$\frac{1}{z}\left(z - \frac{1}{z}\right)^{2l} = \frac{1}{z}\sum_{k=0}^{2l}\frac{(2l)!}{(2l-k)!k!}\left(z^{2l-k}\right)\left(-\frac{1}{z}\right)^{k}, \tag{13.98}$$

where the residue we need is the coefficient of the $1/z$ term. This can be easily found as

$$(-1)^{l}\frac{(2l)!}{(l!)^{2}}, \tag{13.99}$$

and the result of the definite integral I becomes

$$I = \frac{(2l)!}{2^{2l}(l!)^{2}}. \tag{13.100}$$

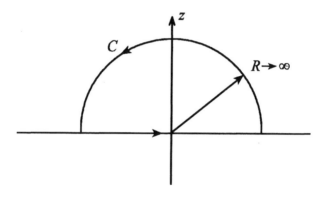

Fig. 13.13 Contour for the type II integrals

II. Integrals of the type $I = \int_{-\infty}^{\infty} dx R(x)$,

where $R(x)$ is a rational function of the form

$$R(x) = \frac{a_0 + a_1 x + a_2 x^2 + \cdots + a_n x^n}{b_0 + b_1 x + b_2 x^2 + \cdots + b_m x^m}, \tag{13.101}$$

a) With no singular points on the real axis,

b) $|R(z)|$ goes to zero at least as $\dfrac{1}{|z^2|}$ in the limit as $|z| \to \infty$.

Under these conditions I has the same value with the complex contour integral

$$I = \oint_{C} R(z)\, dz,$$

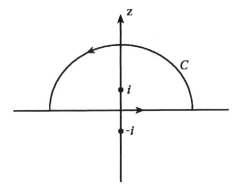

Fig. 13.14   Contour for Example 13.6

where C is a semicircle in the upper half of the z-plane considered in the limit as the radius goes to infinity (Fig. 13.13). Proof is fairly straightforward if we write I as

$$I = \oint_C R(z)\, dz = \int_{-\infty}^{\infty} R(x)dx + \oint_\cap R(z)dz \qquad (13.102)$$

and note that the integral over the semicircle vanishes in the limit as the radius goes to infinity. We can now evaluate I using the residue theorem.

Example 13.6. *Complex contour integral technique:*   Let us evaluate the integral

$$I = \int_{-\infty}^{\infty} \frac{dx}{(1+x^2)^n}\ , \quad n = 1, 2, \dots . \qquad (13.103)$$

Since the conditions of the above technique are satisfied, we write

$$I = \oint_C \frac{dz}{(z+i)^n (z-i)^n}. \qquad (13.104)$$

Only the singular point $z = i$ is inside the contour C (Fig. 13.14); thus we can write I as

$$I = 2\pi i \left[\text{residue of } \left(\frac{1}{(z+i)^n (z-i)^n} \right) \text{ at } z = i \right]. \qquad (13.105)$$

To find the residue we write

$$f(z) = \frac{1}{(z+i)^n} \qquad (13.106)$$

$$= \sum_{k=0}^{\infty} A_k (z-i)^k \qquad (13.107)$$

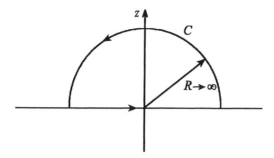

Fig. 13.15   Contour C in the limit $R \to \infty$ for type III integrals

and extract the A_{n-1} coefficient as

$$A_{n-1} = \frac{1}{(n-1)!} \frac{d^{n-1}f}{dz^{n-1}} \bigg|_{z=i} \tag{13.108}$$

$$= \frac{1}{(n-1)!} (-1)^{n-1} \frac{n(n+1)(n+2)\cdots(2n-2)}{(z+i)^{2n-1}} \bigg|_{z=i}.$$

This gives the value of the integral I as

$$I = \frac{2\pi i}{(n-1)!} \frac{(-1)^{n-1}}{2^{2n-1}} \frac{(2n-2)!i}{(n-1)!(-1)^n}. \tag{13.109}$$

III. Integrals of the type $I = \int_{-\infty}^{\infty} dx\, R(x)\, e^{i\kappa x}$,

where κ is a real parameter and $R(x)$ is a rational function with

a) No singular points on the real axis,

b) In the limit as $|z| \to \infty$, $|R(z)| \to 0$ independent of θ.

Under these conditions we can write the integral I as the contour integral

$$I = \oint_C R(z)\, e^{i\kappa z}\, dz, \tag{13.110}$$

where the contour C is shown in Figure 13.15. To show that this is true, we have to show the limit

$$I_A = \lim_{R\to\infty} \oint_\cap R(z)\, e^{i\kappa z}\, dz \to 0. \tag{13.111}$$

We start by taking the moduli of the quantities in the integrand to put an upper limit to this integral as

$$I_A \leq \int_0^\pi |R(z)| \left| e^{i\kappa(\rho\cos\theta + i\rho\sin\theta)} \right| \left| \rho i e^{i\theta} \right| d\theta. \tag{13.112}$$

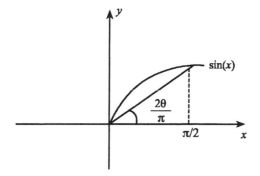

Fig. 13.16 Upper limit calculation

We now call the maximum value that $R(z)$ takes in the interval $[0, 2\pi]$

$$M\left(\rho\right) = \max\left|R\left(z\right)\right| \tag{13.113}$$

and improve this bound as

$$I_A \leq \rho M\left(\rho\right) \int_0^{\pi} e^{-\kappa\rho\sin\theta} d\theta\ , \tag{13.114}$$

$$I_A \leq 2\rho M\left(\rho\right) \int_0^{\frac{\pi}{2}} e^{-\kappa\rho\sin\theta} d\theta. \tag{13.115}$$

Since the straight line segment shown in Figure 13.16, in the interval $[0, \pi/2]$, is always less than the $\sin\theta$ function, we can also write Equation (13.115) as

$$I_A \leq 2\rho M\left(\rho\right) \int_0^{\frac{\pi}{2}} e^{-2\kappa\rho\frac{\theta}{\pi}} d\theta. \tag{13.116}$$

This integral can easily be taken to yield

$$I_A \leq 2\rho M\left(\rho\right) \frac{\pi}{2\kappa\rho} \left(1 - e^{-\kappa\rho}\right), \tag{13.117}$$

$$I_A \leq M\left(\rho\right) \frac{\pi}{\kappa} \left(1 - e^{-\kappa\rho}\right). \tag{13.118}$$

From here we see that in the limit as $\rho \to \infty$ the value of the integral I_A goes to zero, that is,

$$\lim_{\rho\to\infty} I_A \to 0.$$

This result is also called **Jordan's lemma**.

Example 13.7. _Complex contour integral technique:_ In calculating dispersion relations we frequently encounter integrals like

$$f(x) = \frac{1}{\sqrt{2\pi}} \int_{-\infty}^{\infty} dk g(k) e^{ikx}. \tag{13.119}$$

Let us consider a case where $g(k)$ is given as

$$g(k) = \frac{ik}{(k^2 + \mu^2)}. \tag{13.120}$$

i) For $x > 0$ we can write

$$f(x) = \frac{i}{\sqrt{2\pi}} \oint_C \frac{dk k e^{ikx}}{(k + i\mu)(k - i\mu)}. \tag{13.121}$$

In this integral k is now a point in the complex k-plane. Because we have a pole ($k = i\mu$) inside our contour, we use the Cauchy integral theorem [Eq. (13.6)] to find

$$f(x) = 2\pi i \frac{i}{\sqrt{2\pi}} \frac{i\mu}{2i\mu} e^{-\mu x} \tag{13.122}$$

$$= -\sqrt{\frac{\pi}{2}} e^{-\mu x}. \tag{13.123}$$

ii) For $x < 0$, we complete our contour C from below to find

$$f(x) = \frac{i}{2\pi} \oint \frac{k e^{-ik|x|} dk}{(k - i\mu)(k + i\mu)} \tag{13.124}$$

$$= -2\pi i \frac{i}{\sqrt{2\pi}} \left(\frac{-i\mu}{-2i\mu} \right) e^{-\mu|x|}$$

$$= \sqrt{\frac{\pi}{2}} e^{-\mu|x|}. \tag{13.125}$$

IV. Integrals of the type $I = \int_0^{\infty} dx x^{\lambda-1} R(x)$, where

a) $\lambda \neq$ integer,

b) $R(x)$ is a rational function with no poles on the positive real axis and the origin,

c) In the limit as $|z| \to 0$, $\left| z^{\lambda} R(z) \right| \to 0$ and

d) In the limit as $|z| \to \infty$, $\left| z^{\lambda} R(z) \right| \to 0$.

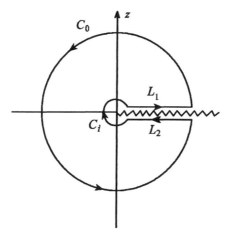

Fig. 13.17 Contour for the integrals of type IV

Under these conditions we can evaluate the integral I as

$$I = \int_0^\infty x^{\lambda-1} R(x)\, dx \tag{13.126}$$

$$= \frac{\pi(-1)^{\lambda-1}}{\sin \pi\lambda} \sum_{\text{inside } C} \text{residues of } \left[z^{\lambda-1} R(z)\right],$$

where C is the closed contour shown in Figure 13.17.

Proof: Let us write the integral I as a complex contour integral:

$$\oint_C z^{\lambda-1} R(z)\, dz. \tag{13.127}$$

In the limit as the radius of the small circle goes to zero the integral over the contour C_i goes to zero because of **c**. Similarly, in the limit as the radius of the large circle goes to infinity the integral over C_0 goes to zero because of **d**. This leaves us with

$$\oint_C z^{\lambda-1} R(z) dz = \oint_{\to L_1} z^{\lambda-1} R(z) dz$$
$$+ \oint_{\leftarrow L_2} z^{\lambda-1} R(z) dz.$$

We can now evaluate the integral on the left-hand side by using the residue theorem. On the other hand, the right-hand side can be written

as

$$\oint_{\to L_1 + \leftarrow L_2} z^{\lambda-1} R(z) dz$$

$$= \int_0^\infty x^{\lambda-1} R(x) \, dx \tag{13.128}$$

$$+ \int_\infty^0 x^{(\lambda-1)} e^{2i\pi(\lambda-1)} R(x) \, dx,$$

$$= \left(1 - e^{2\pi i(\lambda-1)}\right) \int_0^\infty dx x^{\lambda-1} R(x). \tag{13.129}$$

Thus we obtain

$$2\pi i \sum_{\text{inside } C} \text{residues of } \left[z^{\lambda-1} R(z)\right]$$

$$= -\frac{2i \sin \pi \lambda}{e^{-i\pi\lambda}} \int_0^\infty dx x^{\lambda-1} R(x). \tag{13.130}$$

Rearranging this, we write the final result as

$$\int_0^\infty x^{\lambda-1} R(x) \, dx = \frac{\pi (-1)^{\lambda-1}}{\sin \pi \lambda} \sum_{\text{inside } C} \text{residues of } \left[z^{\lambda-1} R(z)\right]. \tag{13.131}$$

13.8   GAMMA AND BETA FUNCTIONS

13.8.1   Gamma Function

For an important application of the type IV integrals we now introduce the gamma and beta functions, which are frequently encountered in applications. The gamma function is defined for all x values as

$$\Gamma(x) = \lim_{N \to \infty} \left\{ \frac{N! N^x}{x[x+1][x+2] \cdots [x+N]} \right\}. \tag{13.132}$$

Integral definition of the gamma function, even though restricted to $x > 0$, is also very useful:

$$\Gamma(x) = \int_0^\infty y^{x-1} \exp(-y) dy. \tag{13.133}$$

Using integration by parts we can write this as

$$\Gamma\left(x\right) = -\int_0^\infty d\left(e^{-y}\right) y^{x-1}, \tag{13.134}$$

$$\Gamma\left(x\right) = \left(x-1\right)\int_0^\infty dy e^{-y} y^{x-2}. \tag{13.135}$$

This gives us the formula

$$\Gamma\left(x\right) = \left(x-1\right)\Gamma\left(x-1\right), \tag{13.136}$$

which is one of the most important properties of the gamma function. For the positive integer values of x, this formula gives us

$$n = 1, \tag{13.137}$$

$$\Gamma\left(1\right) = 1, \tag{13.138}$$

$$\Gamma\left(n+1\right) = n!. \tag{13.139}$$

Besides, if we write

$$\Gamma\left(x-1\right) = \frac{\Gamma\left(x\right)}{\left(x-1\right)},$$

we can also define the gamma function for the negative integer arguments. Even though this expression gives infinity for the values of $\Gamma\left(0\right)$, $\Gamma\left(-1\right)$ and for all the other negative integer arguments, their ratios are finite:

$$\frac{\Gamma\left(-n\right)}{\Gamma\left(-N\right)} = \left[-N\right]\left[-N+1\right]\cdots\left[-N-2\right]\left[-N-1\right] \tag{13.140}$$

$$= \left[-1\right]^{N-n}\frac{N!}{n!}.$$

For some n values, the gamma function takes the values:

$\Gamma(-\tfrac{3}{2}) = \tfrac{4}{3}\sqrt{\pi}$	$\Gamma(1) = 1$
$\Gamma(-1) = \pm\infty$	$\Gamma(\tfrac{3}{2}) = \tfrac{1}{2}\sqrt{\pi}$
$\Gamma(-\tfrac{1}{2}) = -2\sqrt{\pi}$	$\Gamma(2) = 1$
$\Gamma(0) = \pm\infty$	$\Gamma(\tfrac{5}{2}) = \tfrac{3}{4}\sqrt{\pi}$
$\Gamma(\tfrac{1}{2}) = \sqrt{\pi}$	$\Gamma(3) = 2$

The inverse of the gamma function, $1/\Gamma\left(x\right)$, is single valued and always finite with the limit

$$\lim_{x\to\infty}\frac{1}{\Gamma\left(x\right)} = \frac{x^{\frac{1}{2}-x}}{\sqrt{2\pi}}\exp(x). \tag{13.141}$$

13.8.2 Beta Function

Let us write the multiplication of two gamma functions as

$$\Gamma(n+1)\Gamma(m+1) = \int_0^\infty e^{-u}u^n\,du \int_0^\infty e^{-v}v^m\,dv. \qquad (13.142)$$

Using the transformation

$$u = x^2 \text{ and } v = y^2, \qquad (13.143)$$

we can write

$$\Gamma(n+1)\Gamma(m+1)$$

$$= \left(2\int_0^\infty e^{-x^2}x^{2n+1}\,dx\right)\left(2\int_0^\infty e^{-y^2}y^{2m+1}\,dy\right)$$

$$= 4\int_0^\infty \int_0^\infty e^{-(x^2+y^2)}x^{2n+1}y^{2n+1}\,dx\,dy. \qquad (13.144)$$

In plane polar coordinates this becomes

$$\Gamma(n+1)\Gamma(m+1) \qquad (13.145)$$

$$= 4\int_0^\infty dr \int_0^{\frac{\pi}{2}} d\theta\, e^{-r^2}\, r^{2n+2m+2+1}\cos^{2n+1}\theta\sin^{2m+1}\theta$$

$$= 2\int_0^\infty \left[dr\,e^{-r^2}r^{2n+2m+2+1}\right]\left[2\int_0^{\frac{\pi}{2}} d\theta\sin^{2m+1}\theta\cos^{2n+1}\theta\right].$$

The first term on the right-hand side is $\Gamma(m+n+2)$ and the second term is called the beta function $B(m+1, n+1)$. The beta function is related to the gamma function through the relation

$$B(m+1, n+1) = \frac{\Gamma(n+1)\Gamma(m+1)}{\Gamma(m+n+2)}. \qquad (13.146)$$

Another definition of the beta function is obtained by the substitutions

$$\sin^2\theta = t \qquad (13.147)$$

and

$$t = \frac{x}{1-x} \qquad (13.148)$$

as

$$B(m+1, n+1) = \int_0^\infty \frac{x^m\,dx}{(1+x)^{m+n+2}}. \qquad (13.149)$$

Using the substitution

$$x = \frac{y}{1-y},$$ (13.150)

we can also write

$$B(p,q) = \int_0^1 y^{p-1}[1-y]^{q-1} dy, \qquad p > 0, \ q > 0.$$ (13.151)

To calculate the value of $B(\frac{1}{2}, \frac{1}{2})$ we have to evaluate the integral

$$B(\frac{1}{2}, \frac{1}{2}) = \int_0^\infty \frac{dx\, x^{\frac{1}{2}-1}}{(1+x)}.$$ (13.152)

Using formula (13.131) we can evaluate this as

$$B(\frac{1}{2}, \frac{1}{2}) = \frac{[\Gamma(\frac{1}{2})]^2}{\Gamma(1)} = [\Gamma(1/2)]^2$$

$$= -\pi \frac{(-1)^{-1/2}(-1)^{-1/2}}{\sin \pi/2}$$

$$= \pi.$$ (13.153)

From here we obtain

$$\Gamma(\frac{1}{2}) = \sqrt{\pi},$$ (13.154)

$$\Gamma(\frac{1}{2} + n) = \frac{(2n)!\sqrt{\pi}}{4^n n!},$$ (13.155)

$$\Gamma(\frac{1}{2} - n) = \frac{(-4)^n n!\sqrt{\pi}}{(2n)!}.$$ (13.156)

Another useful function related to the gamma function is given as

$$\Psi(x) = \frac{1}{\Gamma(x)} \frac{d\Gamma(x)}{dx}.$$ (13.157)

The function $\Psi(x)$ satisfies the recursion relation

$$\Psi(x+1) = \Psi(x) + x^{-1},$$ (13.158)

from which we obtain

$$\Psi(n+1) = \Psi(1) + \sum_{j=1}^n \frac{1}{j}.$$ (13.159)

The value of $\Psi(1)$ is given in terms of the Euler constant γ as

$$-\Psi(1) = \gamma = 0.5772157.$$ (13.160)

13.8.3   Useful Relations of the Gamma Functions

Among the useful relations of the gamma function we can write

$$\Gamma(-x) = \frac{-\pi \csc(\pi x)}{\Gamma(x+1)}, \tag{13.161}$$

$$\Gamma(2x) = \frac{4^x \Gamma(x)\Gamma(x+\frac{1}{2})}{2\sqrt{\pi}}, \tag{13.162}$$

$$\Gamma(nx) = \sqrt{\frac{2\pi}{n}} \left[\frac{n^x}{\sqrt{2\pi}} \right]^n \prod_{k=0}^{n-1} \Gamma(x + \frac{k}{n}). \tag{13.163}$$

In calculating ratios like

$$\frac{\Gamma(j-q)}{\Gamma(-q)} \quad \text{and} \quad \frac{\Gamma(j-q)}{\Gamma(j+1)}, \tag{13.164}$$

the ratio

$$\frac{\Gamma(j-q)}{\Gamma(-q)\Gamma(j+1)} = \frac{[-1]^j}{j!} \sum_{m=0}^{j} S_j^{(m)} q^m \tag{13.165}$$

is very useful. $S_j^{(m)}$ are the Stirling numbers of the first type:

$$S_{j+1}^{(m)} = S_j^{(m-1)} - j S_j^{(m)}, \ S_0^{(0)} = 1 \tag{13.166}$$

and for the others

$$S_0^{(m)} = S_j^{(0)} = 0. \tag{13.167}$$

In terms of the binomial coefficients this ratio can also be written as

$$\frac{\Gamma(j-q)}{\Gamma(-q)\Gamma(j+1)} = \left(\begin{array}{c} j-q-1 \\ j \end{array} \right) \tag{13.168}$$

$$= [-1]^j \left(\begin{array}{c} q \\ j \end{array} \right).$$

13.8.4   Incomplete Gamma and Beta Functions

Both the beta and the gamma functions have their incomplete forms. The definition of the incomplete beta function with respect to x is given as

$$B_x(p,q) = \int_0^x y^{p-1}[1-y]^{q-1} dy. \tag{13.169}$$

On the other hand, the incomplete gamma function is defined by

$$\gamma^*(c, x) = \frac{c^{-x}}{\Gamma(x)} \int_0^c y^{(x-1)} \exp(-y) dy \tag{13.170}$$

$$= \exp(-x) \sum_{j=0}^{\infty} \frac{x^j}{\Gamma(j + c + 1)}. \tag{13.171}$$

In this equation $\gamma^*(c, x)$ is a single-valued analytic function of c and x. Among the useful relations of $\gamma^*(c, x)$ we can give

$$\gamma^*(c - 1, x) = x\gamma^*(c, x) + \frac{\exp(-x)}{\Gamma(c)}, \tag{13.172}$$

$$\gamma^*(\frac{1}{2}, x) = \frac{\mathrm{erf}(\sqrt{x})}{\sqrt{x}}. \tag{13.173}$$

13.9   CAUCHY PRINCIPAL VALUE INTEGRAL

Sometimes we encounter integrals with poles on the real axis, such as the integral

$$I = \int_{-\infty}^{\infty} \frac{f(x)}{(x - a)} dx, \tag{13.174}$$

which is undefined (divergent) at $x = a$. However, because the problem is only at $x = a$, we can modify this integral by first integrating up to an infinitesimally close point, $(a - \delta)$, to a and then continue integration on the other side from an arbitrarily close point, $(a + \delta)$, to infinity, that is, define the integral I as

$$I = \lim_{\delta \to 0} \left[\int_{-\infty}^{a-\delta} \frac{f(x)\,dx}{(x - a)} + \int_{a+\delta}^{\infty} \frac{f(x)\,dx}{(x - a)} \right]$$

$$= P \int_{-\infty}^{\infty} \frac{f(x)}{(x - a)} dx. \tag{13.175}$$

This is called taking the **Cauchy principal value** of the integral, and it is shown as

$$\int_{-\infty}^{\infty} \frac{f(x)}{(x - a)} dx \rightarrow P \int_{-\infty}^{\infty} \frac{f(x)}{(x - a)} dx. \tag{13.176}$$

If $f(z)$ is analytic in the upper half z-plane, that is

$$\text{as } |z| \to \infty, \quad f(z) \to 0 \text{ for } y > 0,$$

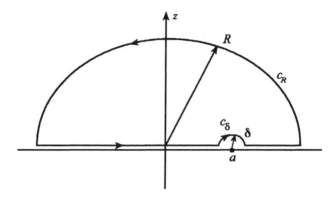

Fig. 13.18   Contour C for the Cauchy principal value integral

we can evaluate the Cauchy principal value of the integral (13.174) by using the contour in Figure 13.18. In this case we write

$$\oint_C \frac{f(z)\,dz}{(z-a)} \tag{13.177}$$

$$= \left[\oint_{\substack{c_R[\frown] \\ R\to\infty}} dz + \lim_{\substack{[R\to\infty] \\ \delta\to 0}} \int_{-R}^{a-\delta} dz + \oint_{\substack{c_\delta[\frown] \\ \delta\to 0}} dz + \lim_{\substack{[R\to\infty] \\ \delta\to 0}} \int_{a+\delta}^{R} dz \right] \frac{f(z)}{(z-a)}$$

and evaluate this integral by using the residue theorem as

$$\oint_C \frac{f(z)}{(z-a)}\,dz = 2\pi i \sum_{\text{inside } C} \text{residues of } \left[\frac{f(z)}{z-a} \right]. \tag{13.178}$$

If $f(z)/(z-a)$ does not have any isolated singular points inside the closed contour C [Fig. 13.18], the left-hand side of Equation (13.177) is zero, thus giving the Cauchy principal value of the integral (13.175) as

$$P\int_{-\infty}^{\infty} \frac{f(x)}{x-a}\,dx$$

$$= -\lim_{\delta\to 0} \oint_{c_\delta[\frown]} \frac{f(z)}{(z-a)}\,dz - \lim_{R\to\infty} \oint_{c_R[\frown]} \frac{f(z)}{(z-a)}\,dz. \tag{13.179}$$

From the condition $f(z) \to 0$ as $|z| \to \infty$ for $y > 0$, the second integral over c_R on the right-hand side is zero. To evaluate the integral over the small arc c_δ we write

$$z - a = \rho e^{i\theta}, \tag{13.180}$$

$$dz = i\,d\theta\,\rho e^{i\theta} \tag{13.181}$$

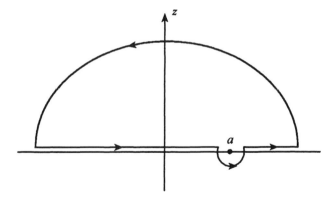

Fig. 13.19 Another path for the Cauchy principal value calculation

and find the Cauchy principal value as

$$P \int_{-\infty}^{\infty} \frac{f(x)\,dx}{(x-a)} = -i \int_{\pi}^{0} d\theta\, f(a) \qquad (13.182)$$

$$P \int_{-\infty}^{\infty} \frac{f(x)\,dx}{(x-a)} = i\pi f(a). \qquad (13.183)$$

Another contour that we can use to find the Cauchy principal value is given in Figure 13.19. In this case the pole at $x = a$ is inside our contour. Using the residue theorem we obtain

$$
\begin{aligned}
P \int_{-\infty}^{\infty} \frac{f(x)\,dx}{(x-a)} &= -i \int_{\pi}^{2\pi} d\theta\, f(a) + 2\pi i f(a) \\
&= -i\pi f(a) + 2\pi i f(a) \\
&= i\pi f(a).
\end{aligned}
\qquad (13.184)
$$

As expected, the Cauchy principal value is the same for both choices of detour about $z = a$.

If $f(z)$ is analytic in the lower half of the z-plane, that is,

$$f(z) \to 0 \quad \text{as} \quad |z| \to \infty \quad \text{for } y < 0,$$

then the Cauchy principal value is given as

$$P \int_{-\infty}^{\infty} \frac{f(x)\,dx}{(x-a)} = -i\pi f(a). \qquad (13.185)$$

In this case we again have two choices for the detour around the singular point on the real axis. Again the Cauchy principal value is $-i\pi f(a)$ for both choices.

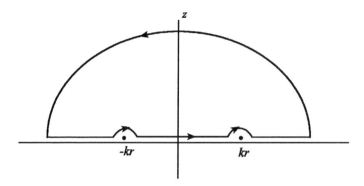

Fig. 13.20 Contour for I_1

Example 13.8. *Cauchy principal value integral:* Let us now evaluate the integral

$$I = \int_{-\infty}^{\infty} \frac{x \sin x \, dx}{[x^2 - k^2 r^2]}. \qquad (13.186)$$

We write I as

$$I = I_1 + I_2,$$

where

$$I_1 = \frac{1}{2i} \int_{-\infty}^{\infty} \frac{x e^{ix} dx}{(x - kr)(x + kr)}, \qquad (13.187)$$

$$I_2 = -\frac{1}{2i} \int_{-\infty}^{\infty} \frac{x e^{-ix} dx}{(x - kr)(x + kr)}. \qquad (13.188)$$

For I_1 we choose our path in the z-plane as in Figure 13.20 to obtain

$$I_1 = \frac{1}{2i} \left\{ i\pi \left[\frac{z e^{iz}}{z + kr} \right]_{z=kr} + i\pi \left[\left[\frac{z e^{ix}}{z - kr} \right]_{z=-kr} \right] \right\}$$

$$= \frac{\pi}{2} \cos kr. \qquad (13.189)$$

For the integral I_2 we use the path in Figure 13.21 to obtain

$$I_2 = -\frac{1}{2i} \left\{ -i\pi \left[\frac{z e^{-iz}}{z + kr} \right]_{z=kr} - i\pi \left[\left[\frac{z e^{-iz}}{z - kr} \right]_{z=-kr} \right] \right\}$$

$$= \frac{\pi}{2} \cos kr. \qquad (13.190)$$

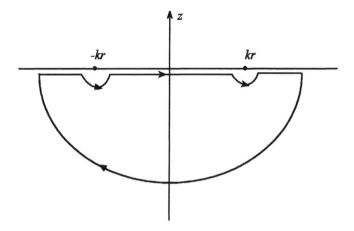

Fig. 13.21   Contour for I_2

Hence the divergent integral $\displaystyle\int_{-\infty}^{\infty} \frac{x \sin x\, dx}{[x^2 - k^2 r^2]}$ can now be replaced with its Cauchy principal value as

$$P \int_{-\infty}^{\infty} \frac{x \sin x\, dx}{[x^2 - k^2 r^2]} = \pi \cos kr.$$

13.10   CONTOUR INTEGRAL REPRESENTATIONS OF SOME SPECIAL FUNCTIONS

13.10.1   Legendre Polynomials

Let us write the Rodriguez formula for the Legendre polynomials:

$$P_l(x) = \frac{1}{2^l l!} \frac{d^l}{dx^l} \left(x^2 - 1\right)^l. \tag{13.191}$$

Using the Cauchy formula [Eq. (13.19)]:

$$\frac{d^l f(z)}{dz^l}\bigg|_{z_0} = \frac{l!}{2\pi i} \oint \frac{f(z)\, dz}{(z - z_0)^{l+1}}, \tag{13.192}$$

and taking $z_0 = x$ and $f(z) = \left(z^2 - 1\right)^l$ we obtain

$$\frac{d^l}{dz^l} \left(z^2 - 1\right)^l \bigg|_{z=x} = \frac{l!}{2\pi i} \oint \frac{\left(z'^2 - 1\right)^l dz'}{(z' - x)^{l+1}}. \tag{13.193}$$

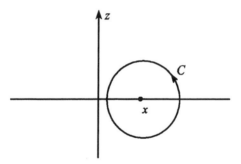

Fig. 13.22   Contour for the Schlöfli formula

This gives us the complex contour representation of the Legendre polynomials as

$$P_l(x) = \frac{1}{2\pi i}\frac{1}{2^l}\oint_C \frac{\left(z'^2 - 1\right)^l dz'}{\left(z' - x\right)^{l+1}}. \tag{13.194}$$

This is also called the **Schlöfli integral formula,** where the contour is given in Figure 13.22.   Using the Schlöfli formula [Eq. (13.194)] and the residue theorem, we can obtain the Legendre polynomials as

$$P_l(x) = \frac{1}{2^l}\left[\text{residue of }\left[\frac{\left(z^2 - 1\right)^l}{\left(z - x\right)^{l+1}}\right]\text{ at }x\right]. \tag{13.195}$$

We use the binomial formula to write $\left(z^2 - 1\right)^l$ in terms of powers of $(z - x)$ as

$$\left(z^2 - 1\right)^l = \sum_{k=0}^{l}\frac{l!\,(-1)^k}{k!\,(l-k)!}z^{2(l-k)}$$

$$= \sum_{k=0}^{l}\frac{l!\,(-1)^k}{k!\,(l-k)!}\left[z - x + x\right]^{2l-2k}$$

$$= \sum_{k=0}^{l}\frac{l!\,(-1)^k}{k!\,(l-k)!}\sum_{j=0}^{2l-2k}\frac{(2l-2k)!}{(2l-2k-j)!j!}\,(z - x)^j\,x^{2l-2k-j}. \tag{13.196}$$

For the residue we need the coefficient of $(z - x)^l$; hence we need the $j = l$ term in the above series, which is

$$\text{coefficient of }(z - x)^l = \sum_{k=0}^{\left[\frac{l}{2}\right]}\frac{l!\,(-1)^k}{k!\,(l-k)!}\frac{(2l-2k)!}{(l-2k)!l!}x^{l-2k}. \tag{13.197}$$

Using this in Equation (13.195) we finally obtain $P_l(x)$ as

$$P_l(x) = \sum_{k=0}^{[\frac{l}{2}]} \frac{(-1)^k}{k!\,(l-k)!} \frac{(2l-2k)!}{(l-2k)!} \frac{x^{l-2k}}{2^l}. \tag{13.198}$$

13.10.2   Laguerre Polynomials

The generating function of the Laguerre polynomials is defined as

$$\frac{e^{-\frac{xt}{1-t}}}{1-t} = \sum_{n=0}^{\infty} L_n(x)t^n. \tag{13.199}$$

The Taylor expansion in the complex t-plane of the function

$$f(t) = \frac{e^{-\frac{xt}{1-t}}}{1-t} \tag{13.200}$$

about the origin for a contour with unit radius is given as

$$f(t) = \sum_{n=0}^{\infty} \frac{1}{n!} f^{(n)}(0)t^n, \tag{13.201}$$

where

$$f^{(n)}(0) = \frac{n!}{2\pi i} \oint_C \frac{f(t)dt}{t^{n+1}} \tag{13.202}$$

$$= \frac{n!}{2\pi i} \oint_C \frac{e^{-\frac{xt}{1-t}}dt}{(1-t)t^{n+1}}.$$

Since $f(t)$ is analytic in and on the contour, where C includes the origin but excludes $t=1$, we use the above derivatives to write

$$f(t) = \sum_{n=0}^{\infty} \frac{1}{n!}\left[\frac{n!}{2\pi i} \oint_C \frac{e^{-\frac{xt}{1-t}}dt}{(1-t)t^{n+1}} \right] t^n \tag{13.203}$$

$$= \sum_{n=0}^{\infty} L_n(x)t^n, \tag{13.204}$$

to obtain

$$L_n(x) = \frac{1}{2\pi i} \oint_C \frac{e^{-\frac{xt}{1-t}}}{(1-t)t^{n+1}}dt. \tag{13.205}$$

Note that this is valid for a region enclosed by a circle centered at the origin with unit radius. To obtain $L_n(z)$ valid for the whole complex plane one might expand $f(t)$ about $t = 1$ in Laurent series.

Another contour integral representation of the Laguerre polynomials can be obtained by using the Rodriguez formula

$$L_n(x) = \frac{e^x}{n!} \frac{d^n}{dx^n}(x^n e^{-x}). \qquad (13.206)$$

Using the formula

$$\frac{d(x^n e^{-x})}{dx^n} = \frac{n!}{2\pi i} \oint_C \frac{f(z)dz}{(z - z_0)^{n+1}} \qquad (13.207)$$

and taking z_0 as a point on the real axis and

$$f(z) = z^n e^{-z} \qquad (13.208)$$

we can write

$$\frac{2\pi i}{n!} f^{(n)}(x) = \oint_C \frac{z^n e^{-z} dz}{(z - x)^{n+1}}, \qquad (13.209)$$

where C is a circle centered at some point $z = x$, thus obtaining

$$L_n(x) = \frac{1}{2\pi i} \oint_C \frac{z^n e^{x-z} dz}{(z - x)^{n+1}}. \qquad (13.210)$$

Problems

13.1   Use the contour integral representation of the Laguerre polynomials:

$$L_n(x) = \frac{1}{2\pi i} \oint \frac{z^n e^{x-z} dz}{(z-x)^{n+1}}$$

to obtain the coefficients C_k in the expansion

$$L_n(x) = \sum_{k=0}^{n} C_k x^{n-k}.$$

13.2   Establish the following contour integral representation for the Hermite polynomials:

$$H_n(x) = \frac{(-1)^n n!}{2\pi i} e^{x^2} \oint_C \frac{e^{-z^2} dz}{(z-x)^{n+1}},$$

where C encloses the point x, and use it to derive the series expansion

$$H_n(x) = \sum_{j=0}^{\left[\frac{n}{2}\right]} (-1)^j \frac{2^{n-2j} x^{n-2j} n!}{(n-2j)! j!}.$$

13.3   Using Taylor series prove the Cauchy-Goursat theorem

$$\oint_C f(z)\, dz = 0,$$

where $f(z)$ is an analytic function in and on the closed contour C in a simply connected domain.

13.4   Find the Laurent expansions of the function

$$f(z) = \frac{1}{1-z} + \frac{2}{(2-z)}$$

about the origin for the regions

$$|z| < 1, \quad |z| > 2, \text{ and } 1 < |z| < 2.$$

Use two different methods and show that the results agree with each other.

13.5   Using the path in Figure 13.23 evaluate the integral

$$\int_0^\infty e^{-ix^2} dx.$$

13.6   Evaluate the following integrals:

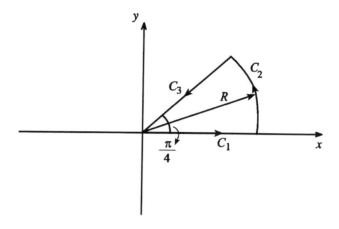

Fig. 13.23   Contour for problem 13.5

i)

$$\int_0^{2\pi} \frac{(\cos 3\theta)\, d\theta}{5 - 4\cos\theta}$$

ii)

$$\int_{-\infty}^{\infty} \frac{\sin x\, dx}{x^2 + 4x + 5}$$

iii)

$$\int_0^{\infty} \frac{\ln(x^2 + 1)\, dx}{x^2 + 1}$$

iv)

$$\int_0^{\infty} \frac{dx}{1 + x^5}$$

v)

$$\int_0^{\infty} \frac{\sin x\, dx}{x}$$

vi)

$$\int_0^{\infty} \frac{dx}{a^3 + x^3}$$

vii)

$$\int_{-\infty}^{\infty} \frac{x^2 \, dx}{(x+1)(x^2+2x+2)}$$

viii)

$$\int_{-\infty}^{\infty} \frac{x^{2a-1}}{b^2+x^2} \, dx$$

ix)

$$\int_{0}^{2\pi} \frac{\sin^2 \theta d\theta}{a+b\cos\theta}$$

x)

$$\int_{-\infty}^{\infty} \frac{\sin x dx}{x(a^2+x^2)}$$

13.7 Evaluate the following Cauchy principal value integral:

$$P\int_{0}^{\infty} \frac{dx}{[(x-x_0)^2+a^2](x-x_1)}, \quad x_1 > x_0.$$

13.8 Using the generating function for the polynomials $P_{nm}(x)$

$$\frac{e^{\frac{-xt}{(1-t)}}}{(1-t)^{m+1}} = \sum_{n=0}^{\infty} P_{nm}(x) t^n, \quad |t| < 1, \, m = \text{positive},$$

establish a contour integral representation in the complex t-plane. Use this representation to find $A(n, m, k)$ in

$$P_{nm}(x) = \sum_{k=0}^{n} A(n, m, k) x^k.$$

13.9 Use contour integral techniques to evaluate

$$\int_{-\infty}^{\infty} \frac{\sin^2 x dx}{x^2(1+x^2)}.$$

13.10 The Jacobi polynomials $P_n^{(a,b)}(\cos\theta)$, where $n = $ positive integer and a, b are arbitrary real numbers are defined by the Rodriguez formula

$$P_n^{(a,b)}(x) = \frac{(-1)^n}{2^n n!(1-x)^a(1+x)^b} \frac{d^n}{dx^n} \left[(1-x)^{n+a}(1+x)^{n+b}\right], \quad |x| < 1.$$

Find a contour integral representation for this polynomial valid for $|x| < 1$ and use this to show that the polynomial can be expanded as

$$P_n^{(a,b)}(\cos\theta) = \sum_{k=0}^{n} A(n, a, b, k)(\sin\frac{\theta}{2})^{2n-2k}(\cos\frac{\theta}{2})^{2k}.$$

Determine the coefficients $A(n, a, b, k)$ for the special case, where a and b are both integers.

13.11   For a function $F(z)$ analytic everywhere in the upper half plane and on the real axis with the property

$$F(z) \to b \text{ as } |z| \to \infty \text{ , } b \text{ is a real constant,}$$

show the following Cauchy principal value integrals:

$$F_R(z) = b + \frac{1}{\pi}P\int_{-\infty}^{\infty} \frac{F_I(x')dx'}{x' - x}$$

and

$$F_I(z) = -\frac{1}{\pi}P\int_{-\infty}^{\infty} \frac{F_R(x')dx'}{x' - x}.$$

13.12   Given the following contour integral definition of spherical Hankel function of the first kind

$$h_l^{(1)}(x) = \frac{(-1)^l 2^l l!}{\pi x^{l+1}} \oint_C \frac{e^{-ixz}dz}{(z^2 - 1)^{l+1}},$$

where the contour C encloses the point $x = -1$, show that $h_l^{(1)}(x)$ can be written as

$$h_l^{(1)}(x) = \sum_{k=0}^{\infty} A(k, l)\frac{e^{i[x-\beta(l,k)]}}{x^{k+1}},$$

that is,
    i) Show that this series breaks of at the $k = l$th term.
    ii) By using the contour integral definition given above, find explicitly the constants $A(k, l)$, and $\beta(l, k)$.

13.13   Another definition for the gamma function is given as

$$\Gamma(x) = x^{x-\frac{1}{2}}e^{-x}\sqrt{2\pi}e^{\theta(x)/12x} \text{ , } x > 0,$$

where $\theta(x)$ is a function satisfying $0 < \theta(x) < 1$. Using the above definition show the limit

$$\lim_{x\to\infty} \frac{\Gamma(x+1)}{x^{x+\frac{1}{2}}e^{-x}\sqrt{2\pi}} = 1.$$

When x is an integer this gives us the Stirling's approximation to $x!$ as $x \to \infty$:

$$x! \sim x^{x+\frac{1}{2}} e^{-x} \sqrt{2\pi}.$$

13.14   Show that

$$\Gamma(x+1) = x\Gamma(x), \text{ for } x > 0$$

and

$$\Gamma(n+1) = n! \text{ for } n = \text{integer} > 0.$$

13.15   Show that

$$(1+x)^{1/2} = \sum_{n=0}^{\infty} \binom{1/2}{n} x^n = 1 + \frac{x}{2} - \frac{x^2}{8} + \frac{x^3}{16} - \cdots$$

$$= 1 + \frac{x}{2} + \sum_{n=2}^{\infty} (-1)^{n-1} \frac{(2n-3)!!}{(2n)!!} x^n,$$

where the double factorial means

$$(2n+1)!! = 1 \cdot 3 \cdot 5 \cdots (2n+1),$$
$$(2n)!! = 2 \cdot 4 \cdot 6 \cdots (2n).$$

13.16   Use the factorization method (Chapter 9) to show that the spherical Hankel functions of the first kind,

$$h_l^{(1)} = j_l + i n_l,$$

can be expressed as

$$h_l^{(1)}(x) = (-1)^l x^l \left[\frac{1}{x} \frac{d}{dx} \right]^l h_0^{(1)}(x)$$

$$= (-1)^l x^l \left[\frac{1}{x} \frac{d}{dx} \right]^l \left(\frac{-ie^{ix}}{x} \right).$$

Hint. First define

$$u_l(x) = y_l(x)/x^{l+1}$$

in

$$y_l'' + \left[1 - \frac{l(l+1)}{x^2} \right] y_l = 0.$$

Using this result, define $h_l^{(1)}(x)$ as a contour integral in the complex j'-plane $(j' = t' + is')$, where

$$\frac{d}{dt} = \frac{1}{x}\frac{d}{dx}.$$

Indicate your contour by clearly showing the singularities that must be avoided.

13.17   If $f(z)$ is analytic in the lower half of the z-plane, that is,

$$f(z) \to 0 \quad \text{as} \quad |z| \to \infty \quad \text{for } y < 0,$$

then show that the Cauchy principal value is given as

$$P\int_{-\infty}^{\infty} \frac{f(x)\,dx}{(x-a)} = -i\pi f(a). \tag{13.211}$$

Identify your two choices for the detour around the singular point on the real axis and show that the Cauchy principal value is $-i\pi f(a)$ for both choices.

14

FRACTIONAL DERIVATIVES and INTEGRALS: "DIFFERINTEGRALS"

The diffusion equation in integral form is given as

$$\frac{\partial}{\partial t} \iiint_V c(\overrightarrow{r}, t)dv = - \oint_S \overrightarrow{J}(\overrightarrow{r}, t) \cdot d\overrightarrow{s}, \qquad (14.1)$$

where $c(\overrightarrow{r}, t)$ is the concentration of particles and $\overrightarrow{J}(\overrightarrow{r}, t)$ is the current density. The left-hand side gives the rate of change of the number of particles in volume V, and the right-hand side gives the number of particles flowing past the boundary S of this volume per unit time. In the absence of sources or sinks, these terms are naturally equal. Using the Gauss theorem we can write this equation as

$$\frac{\partial}{\partial t} \iiint_V c(\overrightarrow{r}, t)dv + \iiint_V \overrightarrow{\nabla} \cdot \overrightarrow{J}(\overrightarrow{r}, t)dv = \qquad (14.2)$$

$$\iiint_V \left[\frac{\partial}{\partial t}c(\overrightarrow{r}, t) + \overrightarrow{\nabla} \cdot \overrightarrow{J}(\overrightarrow{r}, t) \right] dv = 0. \qquad (14.3)$$

This gives us a partial differential equation to be solved for concentration as

$$\frac{\partial}{\partial t}c(\overrightarrow{r}, t) + \overrightarrow{\nabla} \cdot \overrightarrow{J}(\overrightarrow{r}, t) = 0. \qquad (14.4)$$

In order to solve this equation, we also need a relation between $c(\overrightarrow{r}, t)$ and $\overrightarrow{J}(\overrightarrow{r}, t)$. Because particles have a tendency to flow from regions of high to low concentration, as a first approximation we can assume a linear relation between the current density and the gradient of concentration as

$$\overrightarrow{J} = -k\overrightarrow{\nabla} c(\overrightarrow{r}, t). \qquad (14.5)$$

The proportionality constant k is called the diffusion constant. We can now write the diffusion equation as

$$\frac{\partial}{\partial t} c(\overrightarrow{r}, t) \; - k \overrightarrow{\nabla}^2 c(\overrightarrow{r}, t) = 0, \tag{14.6}$$

which is also called Fick's equation.

Einstein noticed that in a diffusion process concentration is also proportional to the probability, $P(\overrightarrow{r}, t)$, of finding a diffusing particle at position $\overrightarrow{r}$ and time t. Thus the probability distribution satisfies the same differential equation as the concentration. For a particle starting its motion from the origin, probability distribution can be found as

$$P(\overrightarrow{r}, t) = \frac{1}{(4\pi kt)^{\frac{3}{2}}} \exp\left(-\frac{r^2}{4kt}\right). \tag{14.7}$$

This means that even though the average displacement of a particle is zero $(<\overrightarrow{r}>= 0)$, mean square displacement is nonzero and is given as

$$< \overrightarrow{r}'^2 >=< \overrightarrow{r}'^2 > - < \overrightarrow{r} >^2$$

$$= \int r^2 P(\overrightarrow{r}, t) d^3 r$$

$$= 6kt. \tag{14.8}$$

What is important in this equation is the

$$< \overrightarrow{r}'^2 > \propto t \tag{14.9}$$

relation. For the particle to cover twice the distance, time must be increased by a factor of four. This scaling property results from the diffusion equation where the time derivative is of first and the space derivative is of second order. However, it has been experimentally determined that for some systems this relation goes as

$$< \overrightarrow{r}'^2 > \propto t^\alpha, \text{ where } \alpha \neq 1. \tag{14.10}$$

In terms of the diffusion equation this would imply

$$\frac{\partial^\alpha}{\partial t^\alpha} P(\overrightarrow{r}, t) - k_\alpha \overrightarrow{\nabla}^2 P(\overrightarrow{r}, t) = 0, \quad k_\alpha \neq 1. \tag{14.11}$$

However, what does this mean? Is a fractional derivative possible? If a fractional derivative is possible, can we also have a fractional integral? Actually, the geometric interpretation of derivative as the slope and integral as the area are so natural that most of us have not even thought of the possibility of fractional derivatives and integrals. On the other hand, the history of fractional calculus dates back as far as Leibniz (1695), and results have been accumulated over the past years in various branches of mathematics. The situation on the applied side of this branch of mathematics is now changing rapidly, and there are now a growing number of research areas in science and engineering that make use of fractional calculus. Chemical analysis of fluids, heat transfer, diffusion, the Schrödinger equation, and material science are some areas where fractional calculus is used. Interesting applications to economy, finance, and earthquake science should also be expected. It is well known that in the study of nonlinear situations and in the study of processes away from equilibrium fractal curves and surfaces are encountered, where ordinary mathematical techniques are not sufficient. In this regard the relation between fractional calculus and fractals is also being actively investigated.

Fractional calculus also offers us some useful mathematical techniques in evaluating definite integrals and finding sums of infinite series. In this chapter, we introduce some of the basic properties of fractional calculus along with some mathematical techniques and their applications.

14.1 UNIFIED EXPRESSION OF DERIVATIVES AND INTEGRALS

14.1.1 Notation and Definitions

In our notation we follow Oldham and Spanier, where a detailed treatment of the subject along with a survey of the history and various applications can be found. Unless otherwise specified we use n and N for positive integers, q and Q for any number. The nth derivative of a function is shown as

$$\frac{d^n f}{dx^n}. \tag{14.12}$$

Since an integral is the inverse of a derivative, we write

$$\frac{d^{-1} f}{d\left[x\right]^{-1}} = \int_0^x f(x_0)dx_0. \tag{14.13}$$

Successive integrations will be shown as

$$\frac{d^{-2} f}{d\left[x\right]^{-2}} = \int_0^x dx_1 \int_0^{x_1} f(x_0)dx_0 \tag{14.14}$$

$$\vdots$$

$$\frac{d^{-n} f}{d\left[x\right]^{-n}} = \int_0^x dx_{n-1} \int_0^{x_{n-1}} dx_{n-2} \cdots \int_0^{x_2} dx_1 \int_0^{x_1} f(x_0)dx_0. \tag{14.15}$$

When the lower limit differs from zero, we will write

$$\frac{d^{-1} f}{[d(x-a)]^{-1}} = \int_a^x f(x_0)dx_0 \tag{14.16}$$

$$\vdots$$

$$\frac{d^{-n}f}{[d(x-a)]^{-n}} = \int_a^x dx_{n-1} \int_a^{x_{n-1}} dx_{n-2} \cdots \int_a^{x_2} dx_1 \int_a^{x_1} f(x_0)dx_0. \quad (14.17)$$

We should remember that even though the equation

$$\frac{d^n}{[d(x-a)]^n} = \frac{d^n}{[dx]^n} \quad (14.18)$$

is true for derivatives, it is not true for integrals, that is,

$$\frac{d^{-n}}{[d(x-a)]^{-n}} \neq \frac{d^{-n}}{[dx]^{-n}}. \quad (14.19)$$

The nth derivative is frequently written as

$$f^{(n)}(x). \quad (14.20)$$

Hence for n successive integrals we will also use

$$f^{(-n)} = \int_{a_n}^x dx_{n-1} \int_{a_{n-2}}^{x_{n-1}} dx_{n-2} \cdots \int_{a_2}^{x_2} dx_1 \int_{a_1}^{x_1} f(x_0)dx_0. \quad (14.21)$$

When there is no room for confusion, we write

$$\frac{d^q f(x)}{[d(x-0)]^q} = \begin{cases} \frac{d^q f(x)}{[dx]^q} \\ \frac{d^q f(x)}{dx^q} \\ f^{(q)}(x) \end{cases}$$

The value of a differintegral at $x = b$ is shown as

$$\frac{d^q f(x)}{[d(x-a)]^q}\bigg|_{x=b} = \frac{d^q f}{[d(x-a)]^q}(b).$$

Other commonly used expressions for differintegrals are:

$$\frac{d^q f(x)}{[d(x-a)]^q} = \begin{cases} {}_aD_x^q f(x) \\ D_a^q f(x). \end{cases}$$

14.1.2   The nth Derivative of a Function

Before we introduce the differintegral, we derive a unified expression for the derivative and integral for integer orders. We first write the definition of a derivative as

$$\frac{d^1 f}{dx^1} = \frac{df(x)}{dx} = \lim_{\delta x \to 0}\{[\delta x]^{-1}[f(x) - f(x - \delta x)]\}. \quad (14.22)$$

Similarly, the second- and third-order derivatives can be written as

$$\frac{d^2 f}{dx^2} = \lim_{\delta x \to 0} \{ [\delta x]^{-2} [f(x) - 2f(x - \delta x) + f(x - 2\delta x)] \} \tag{14.23}$$

and

$$\frac{d^3 f}{dx^3} = \lim_{\delta x \to 0} \{ [\delta x]^{-3} [f(x) - 3f(x - \delta x) + 3f(x - 2\delta x) - f(x - 3\delta x)] \}. \tag{14.24}$$

Since the coefficients in these equations are the binomial coefficients, for the nth derivative we can write

$$\frac{d^n f}{dx^n} = \lim_{\delta x \to 0} \left\{ [\delta x]^{-n} \sum_{j=0}^{n} [-1]^j \begin{pmatrix} n \\ j \end{pmatrix} f(x - j\delta x) \right\}. \tag{14.25}$$

In these equations we have assumed that all the derivatives exist. In addition, we have assumed that $[\delta x]$ goes to zero continuously, that is, by taking all values on its way to zero. For a unified representation with the integral, we are going to need a restricted limit. For this we divide the interval $[x - a]$ into N equal segments;

$$\delta_N x = [x - a]/N, \qquad N = 1, 2, 3, \dots . \tag{14.26}$$

In this expression a is a number smaller than x. Thus Equation (14.25) becomes

$$\frac{d^n f}{[dx]^n} = \lim_{\delta_N x \to 0} \left\{ [\delta_N x]^{-n} \sum_{j=0}^{n} [-1]^j \begin{pmatrix} n \\ j \end{pmatrix} f(x - j\delta_N x) \right\}. \tag{14.27}$$

Since the binomial coefficients $\begin{pmatrix} n \\ j \end{pmatrix}$ are zero for the $j > n$ values, we can also write

$$\frac{d^n f}{[dx]^n} = \lim_{\delta_N x \to 0} \left\{ [\delta_N x]^{-n} \sum_{j=0}^{N-1} [-1]^j \begin{pmatrix} n \\ j \end{pmatrix} f(x - j\delta_N x) \right\}. \tag{14.28}$$

Now, assuming that this limit is also valid in the continuum limit, we write the nth derivative as

$$\frac{d^n f}{[dx]^n} = \lim_{N \to \infty} \left\{ \left[\frac{x - a}{N} \right]^{-n} \sum_{j=0}^{N-1} [-1]^j \begin{pmatrix} n \\ j \end{pmatrix} f \left(x - j \left[\frac{x - a}{N} \right] \right) \right\}. \tag{14.29}$$

14.1.3 Successive Integrals

We now concentrate on the expression for n successive integrations of $f(x)$. Because an integral of integer order is defined as area, we express it as a Riemann sum:

$$\frac{d^{-1}f}{[d(x-a)]^{-1}} = \int_a^x f(x_0)dx_0 \tag{14.30}$$

$$= \lim_{\delta_N x \to 0}\{\delta_N x[f(x) + f(x - \delta_N x) + f(x - 2\delta_N x) +$$

$$\cdots + f(a + \delta_N x)]\} \tag{14.31}$$

$$= \lim_{\delta_N x \to 0}\left\{\delta_N x \sum_{j=0}^{N-1} f(x - j\delta_N x)\right\}. \tag{14.32}$$

As above, we have taken $\delta_N x = [x - a]/N$. We also write the Riemann sum for the double integral as

$$\frac{d^{-2}f}{[d(x-a)]^{-2}} = \int_a^x dx_1 \int_a^{x_1} f(x_0)dx_0$$

$$= \lim_{\delta_N x \to 0}\{[\delta_N x]^2 [f(x) + 2f(x - \delta_N x) + 3f(x - 2\delta_N x) +$$

$$\cdots + Nf(a + \delta_N x)]\}$$

$$= \lim_{\delta_N x \to 0}\left\{[\delta_N x]^2 \sum_{j=0}^{N-1} [j + 1] f(x - j\delta_N x)\right\} \tag{14.33}$$

and for the triple integral as

$$\frac{d^{-3}f}{[d(x-a)]^{-3}} = \int_a^x dx_2 \int_a^{x_2} dx_1 \int_a^{x_1} f(x_0)dx_0 \tag{14.34}$$

$$= \lim_{\delta_N x \to 0}\{[\delta_N x]^3 \sum_{j=0}^{N-1} \frac{[j + 1][j + 2]}{2} f(x - j\delta_N x)\}.$$

Similarly for n successive integrals we write

$$\frac{d^{-n}f}{[d(x-a)]^{-n}} = \lim_{\delta_N x \to 0}\left\{[\delta_N x]^n \sum_{j=0}^{N-1} \binom{j+n-1}{j} f(x - j\delta_N x)\right\} \tag{14.35}$$

$$\frac{d^{-n}f}{[d(x-a)]^{-n}} = \lim_{N \to \infty}\left\{\left[\frac{x-a}{N}\right]^n \sum_{j=0}^{N-1} \binom{j+n-1}{j} f\left(x - j\left[\frac{x-a}{N}\right]\right)\right\}.$$

$$\tag{14.36}$$

Compared to Equation (14.29), the binomial coefficients in this equation are going as

$$\binom{j+n-1}{j}$$

and all the terms are positive.

14.1.4 Unification of Derivative and Integral Operations for Integer Orders

Using Equations (14.29) and (14.36) and also making use of the relation

$$[-1]^{j}\binom{n}{j} = \binom{j+n-1}{j} = \frac{\Gamma(j-n)}{\Gamma(-n)\Gamma(j+1)}, \tag{14.37}$$

we can write a single expression for both the derivative and integral of order n as

$$\frac{d^{n}f}{[d(x-a)]^{n}} = \lim_{N\to\infty}\left\{\left[\frac{x-a}{N}\right]^{-n}\frac{1}{\Gamma(-n)}\sum_{j=0}^{N-1}\frac{\Gamma(j-n)}{\Gamma(j+1)}f\left(x-j\frac{[x-a]}{N}\right)\right\}. \tag{14.38}$$

In this equation n takes integer values of both signs.

14.2 DIFFERINTEGRALS

14.2.1 Grünwald's Definition of Differintegrals

Considering that the gamma function in the above formula is valid for all n, we obtain the most general and basic definition of differintegral given by Grünwald as

$$\frac{d^{q}f}{[d(x-a)]^{q}} = \lim_{N\to\infty}\left\{\left[\frac{x-a}{N}\right]^{-q}\frac{1}{\Gamma(-q)}\sum_{j=0}^{N-1}\frac{\Gamma(j-q)}{\Gamma(j+1)}f\left(x-j\left[\frac{x-a}{N}\right]\right)\right\}. \tag{14.39}$$

In this expression q can take all values. A major advantage of this definition (also called the Grünwald-Letnikov definition) is that the differintegral is found by using only the values of the function without the need for its derivatives or integrals. On the other hand, evaluation of the infinite series could pose practical problems in applications. In this formula even though the gamma function $\Gamma(-q)$ is infinite for the positive values of q, their ratio $\dfrac{\Gamma(j-q)}{\Gamma(-q)}$ is finite.

We now show that for a positive integer n and for all q values the following relation is true:

$$\frac{d^n}{dx^n} \frac{d^q f}{[d(x-a)]^q} = \frac{d^{n+q} f}{[d(x-a)]^{n+q}}. \tag{14.40}$$

Using $\delta_N x = [x-a]/N$, we can write

$$\frac{d^q f}{[d(x-a)]^q} = \lim_{N \to \infty} \left\{ \frac{[\delta_N x]^{-q}}{\Gamma(-q)} \sum_{j=0}^{N-1} \frac{\Gamma(j-q)}{\Gamma(j+1)} f(x - j\delta_N x) \right\}. \tag{14.41}$$

If we further divide the interval $a \le x' \le x - \delta_N x$ into $N-1$ equal pieces we can write

$$\frac{d^q f}{[d(x-a)]^q}(x - \delta_N x)$$

$$= \lim_{N \to \infty} \left\{ \frac{[\delta_N x]^{-q}}{\Gamma(-q)} \sum_{j=0}^{N-2} \frac{\Gamma(j-q)}{\Gamma(j+1)} f(x - \delta_N x - j\delta_N x) \right\} \tag{14.42}$$

$$= \lim_{N \to \infty} \left\{ \frac{[\delta_N x]^{-q}}{\Gamma(-q)} \sum_{j=1}^{N-1} \frac{\Gamma(j-q-1)}{\Gamma(j)} f(x - j\delta_N x) \right\}. \tag{14.43}$$

Taking the derivative of Equation (14.41), and using Equation (14.43) gives us

$$\frac{d}{dx} \frac{d^q f}{[d(x-a)]^q}$$

$$= \lim_{N \to \infty} \left\{ [\delta_N x]^{-1} \left[\frac{d^q f}{[d(x-a)]^q}(x) - \frac{d^q f}{[d(x-a)]^q}(x - \delta_N x) \right] \right\} \tag{14.44}$$

$$= \lim_{N \to \infty} \left\{ \frac{[\delta_N x]^{-q-1}}{\Gamma(-q)} \left[\Gamma(-q)f(x) + \sum_{j=1}^{N-1} \left\{ \frac{\Gamma(j-q)}{\Gamma(j+1)} - \frac{\Gamma(j-q-1)}{\Gamma(j)} \right\} f(x - j\delta_N x) \right] \right.$$

We use the following relation among gamma functions:

$$\frac{\Gamma(j-q)}{\Gamma(j+1)} - \frac{\Gamma(j-q-1)}{\Gamma(j)} = \frac{\Gamma(-q)}{\Gamma(-q-1)} \frac{\Gamma(j-q-1)}{\Gamma(j+1)} \tag{14.45}$$

to write this as

$$\frac{d}{dx} \frac{d^q f}{[d(x-a)]^q} = \lim_{N \to \infty} \left\{ \frac{[\delta_N x]^{-q-1}}{\Gamma(-q-1)} \left[\sum_{j=0}^{N-1} \frac{\Gamma(j-q-1)}{\Gamma(j+1)} f(x - j\delta_N x) \right] \right\} \tag{14.46}$$

$$= \frac{d^{q+1} f}{[d(x-a)]^{q+1}}. \tag{14.47}$$

The general formula can be shown by assuming this to be true for $(n-1)$ and then showing it for n.

14.2.2   Riemann-Liouville Definition of Differintegrals

Another commonly used definition of the differintegral is given by Riemann and Liouville. Assume that the following integral is given:

$$I_n(x) = \int_a^x (x - \xi)^{n-1} f(\xi) d\xi, \tag{14.48}$$

where n is an integer greater than zero and a is a constant. Using the formula

$$\frac{d}{dx} \int_{A(x)}^{B(x)} F(x, \xi) d\xi$$

$$= \int_{A(x)}^{B(x)} \frac{\partial F(x, \xi)}{\partial x} d\xi + F(x, B(x)) \frac{dB(x)}{dx} - F(x, A(x)) \frac{dA(x)}{dx}, \tag{14.49}$$

we find the derivative of I_n as

$$\frac{dI_n}{dx} = (n-1) \int_a^x (x - \xi)^{n-2} f(\xi) d\xi + \left[(x - \xi)^{n-1} f(\xi) \right]_{\xi=x}. \tag{14.50}$$

For $n > 1$ this gives us

$$\frac{dI_n}{dx} = (n-1) I_{n-1} \tag{14.51}$$

and for $n = 1$

$$\frac{dI_1}{dx} = f(x). \tag{14.52}$$

Differentiating Equation (14.50) k times we find

$$\frac{d^k I_n}{dx^k} = (n-1)(n-2) \cdots (n-k) I_{n-k}, \tag{14.53}$$

which gives us

$$\frac{d^{n-1} I_n}{dx^{n-1}} = (n-1)! I_1(x) \tag{14.54}$$

or

$$\frac{d^n I_n}{dx^n} = (n-1)! f(x). \tag{14.55}$$

Using the fact that $I_n(a) = 0$ for $n \geq 1$, from Equations (14.54) and (14.55) we see that $I_n(x)$ and all of its $(n-1)$ derivatives evaluated at $x = a$ are zero. This gives us

$$I_1(x) = \int_a^x f(x_1) dx_1 \tag{14.56}$$

$$I_2(x) = \int_a^x I_1(x_2) dx_2 = \int_a^x \int_a^{x_2} f(x_1) dx_1 dx_2 \tag{14.57}$$

and in general

$$I_n(x) = (n-1)! \int_a^x \int_a^{x_n} \cdots \int_a^{x_3} \int_a^{x_2} f(x_1) dx_1 dx_2 \cdots dx_{n-1} dx_n. \quad (14.58)$$

From here we obtain a very useful formula also known as the **Cauchy formula:**

$$\int_a^x \int_a^{x_n} \cdots \int_a^{x_3} \int_a^{x_2} f(x_1) dx_1 dx_2 \cdots dx_{n-1} dx_n$$
$$= \frac{1}{(n-1)!} \int_a^x (x-\xi)^{n-1} f(\xi) d\xi. \quad (14.59)$$

To obtain the Riemann-Liouville definition of the differintegral, we write the above equation for all $q < 0$ as

$$\left[\frac{d^q f}{[d(x-a)]^q} \right]_{R-L} = \frac{1}{\Gamma(-q)} \int_a^x [x-x']^{-q-1} f(x') dx' \quad (q < 0). \quad (14.60)$$

However, this formula is valid only for the $q < 0$ values. In this definition, $[..]_{R-L}$ denotes the fact that differintegral is being evaluated by the Riemann-Liouville definition. Later, when we show that this definition agrees with the Grünwald definition for all q, we drop the subscript.

We first show that for $q < 0$ and for a finite function $f(x)$ in the interval $a \le x' \le x$, the two definitions agree. We now calculate the difference between the two definitions as

$$\Delta = \frac{d^q f}{[d(x-a)]^q} - \left[\frac{d^q f}{[d(x-a)]^q} \right]_{R-L}. \quad (14.61)$$

Using definitions (14.39) and (14.60), and changing the range of the integral (14.60), we write Δ as

$$\Delta = \lim_{N \to \infty} \left\{ \frac{[\delta_N x]^{-q}}{\Gamma(-q)} \sum_{j=0}^{N-1} \frac{\Gamma(j-q)}{\Gamma(j+1)} f(x - j\delta_N x) \right\} - \frac{1}{\Gamma(-q)} \int_0^{x-a} \frac{f(x-x')}{x'^{1+q}} dx'. \quad (14.62)$$

We write the integral in the second term as a Riemann sum to get

$$\Delta = \lim_{N \to \infty} \left\{ \frac{[\delta_N x]^{-q}}{\Gamma(-q)} \sum_{j=0}^{N-1} \frac{\Gamma(j-q)}{\Gamma(j+1)} f(x - j\delta_N x) \right\}$$
$$- \lim_{N \to \infty} \left\{ \sum_{j=0}^{N-1} \frac{f(x - j\delta_N x)\delta_N x}{\Gamma(-q)[j\delta_N x]^{1+q}} \right\}. \quad (14.63)$$

Taking $\delta_N x = (x-a)/N$, this becomes

$$\Delta = \lim_{N \to \infty} \left\{ \frac{[\delta_N x]^{-q}}{\Gamma(-q)} \sum_{j=0}^{N-1} f(x - j\delta_N x) \left[\frac{\Gamma(j-q)}{\Gamma(j+1)} - j^{-1-q} \right] \right\} \tag{14.64}$$

$$= \frac{[x-a]^{-q}}{\Gamma(-q)} \lim_{N \to \infty} \sum_{j=0}^{N-1} f\left(\frac{Nx - jx + ja}{N} \right) N^q \left[\frac{\Gamma(j-q)}{\Gamma(j+1)} - j^{-1-q} \right].$$

We now write the sum on the right-hand side as two terms, the first from 0 to $(j-1)$ and the other from j to $(n-1)$. Also, assuming that j is sufficiently large so that we can use the approximation $\Gamma(j-q)/\Gamma(j+1) \simeq j^{-1-q}\left[1 + q(q+1)/2j + 0(j^{-2})\right]$, we obtain

$$\Delta =$$

$$\frac{[x-a]^{-q}}{\Gamma(-q)} \lim_{N \to \infty} \left\{ \sum_{j=0}^{J-1} f\left(\frac{Nx - jx + ja}{N} \right) N^q \left[\frac{\Gamma(j-q)}{\Gamma(j+1)} - j^{-1-q} \right] \right\}$$

$$+ \frac{[x-a]^{-q}}{\Gamma(-q)} \lim_{N \to \infty} \left\{ \frac{1}{N} \sum_{j=J}^{N-1} f\left(\frac{Nx - jx + ja}{N} \right) \left[\frac{j}{N} \right]^{-2-q} \left[\frac{q(q+1)}{2N} + \frac{0(j^{-1})}{N} \right] \right\}.$$

$$\tag{14.65}$$

In the first sum, for $q < -1$, the quantity inside the parentheses is finite and in the limit as $N \to \infty$, because of the N^q factor it goes to zero. Similarly, for $q \le -2$, the second term also goes to zero as $N \to \infty$. Thus we have shown that in the interval $a \le x' \le x$, for a finite function f and for $q \le -2$, the two definitions agree:

$$\frac{d^q f}{[d(x-a)]^q} = \left[\frac{d^q f}{[d(x-a)]^q} \right]_{R-L}, \quad q \le -2. \tag{14.66}$$

To see that the Riemann-Liouville definition agrees with the Grünwald definition [Eq. (14.39)] for all q, as in the Grünwald definition we require the Riemann-Liouville definition to satisfy Equation (14.40):

$$\left[\frac{d^q f}{[d(x-a)]^q} \right]_{R-L} = \frac{d^n}{dx^n} \left[\frac{d^{q-n} f}{[d(x-a)]^{q-n}} \right]_{R-L}. \tag{14.67}$$

In the above formula, for a given q, if we choose n as $q - n \le -2$ and use Equation (14.66) to write

$$\left[\frac{d^q f}{[d(x-a)]^q} \right]_{R-L} = \frac{d^n}{dx^n} \left[\frac{d^{q-n} f}{[d(x-a)]^{q-n}} \right], \tag{14.68}$$

we see that the Grünwald definition and the Riemann-Liouville definition agree with each other for all q values:

$$\left[\frac{d^q f}{[d(x-a)]^q} \right]_{R-L} = \left[\frac{d^{q-n} f}{[d(x-a)]^{q-n}} \right]. \tag{14.69}$$

We can now drop the subscript *R-L*.

14.2.2.1 Riemann-Liouville Definition: We now summarize the Riemann-Liouville definition:

For $q < 0$ the differintegral is evaluated by using the formula

$$\left[\frac{d^q f}{[d(x-a)]^q}\right] = \frac{1}{\Gamma(-q)} \int_a^x [x-x']^{-q-1} f(x')dx', \qquad q < 0. \qquad (14.70)$$

For $q \geq 0$ we use

$$\left[\frac{d^q f}{[d(x-a)]^q}\right] = \frac{d^n}{dx^n}\left[\frac{1}{\Gamma(n-q)} \int_a^x [x-x']^{-(q-n)-1} f(x')dx'\right], \qquad q \geq 0,\ n > q,$$
$$(14.71)$$

where the integer n must be chosen such that $(q - n) < 0$. The Riemann-Liouville definition has found widespread application. In this definition the integral in Equation (14.70) is convergent only for the $q < 0$ values. However, for the $q \geq 0$ values the problem is circumvented by imposing the condition $n > q$ in Equation (14.71). The fact that we have to evaluate an n-fold derivative of an integral somewhat reduces the practicality of the Riemann-Liouville definition for the $q \geq 0$ values.

14.3   OTHER DEFINITIONS OF DIFFERINTEGRALS

The Grünwald and Riemann-Liouville definitions are the most basic definitions of differintegral, and they have been used widely. In addition to these, we can also define differintegral via the Cauchy integral formula and by using integral transforms. Even though these definitions are not as useful as the Grünwald and Riemann-Liouville definitions, they are worth discussing to show that other definitions are possible and when they are implemented properly they agree with the basic definitions. In the literature sometimes fractional derivatives and fractional integrals are treated separately. However, the unification of two approaches as the "differintegral" brings these two notions closer than one usually assumes and avoids confusion between different definitions.

14.3.1   Cauchy Integral Formula

We have seen that for a function $f(z)$ analytic on and inside a closed contour C, the nth derivative is given as

$$\frac{d^n f(z)}{dz^n} = \frac{n!}{2\pi i} \oint_C \frac{f(z')dz'}{(z'-z)^{n+1}}, \qquad n \geq 0 \text{ and an integer}, \qquad (14.72)$$

where z' denotes a point on the contour C and z is a point inside C (Fig.

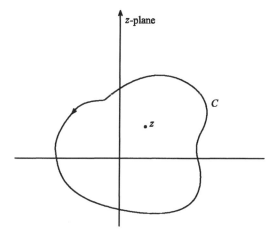

Fig. 14.1 Contour C for the Cauchy integral formula

14.1). We rewrite this formula for an arbitrary q and take z as a point on the real axis:

$$\frac{d^q f(x)}{dx^q} = \frac{\Gamma(q+1)}{2\pi i} \oint_C \frac{f(z')dz'}{(z'-x)^{q+1}}. \tag{14.73}$$

For the path shown in Figure 14.1 this formula is valid only for the positive integer values of q. For the negative integer values of q it is not defined because $\Gamma(q+1)$ diverges. However, it can still be used to define differintegrals for the negative but different than integer values of q. Now, x is a branch point; hence we have to be careful with the direction of the cut line. Thus our path is no longer as shown in Figure 14.1. We choose our cut line along the real axis and to the left of our branch point. We now modify the contour as shown in Figure 14.2 and write our definition of differintegral for the negative, noninteger values of q as

$$\frac{d^q f(x)}{dx^q} = \frac{\Gamma(q+1)}{2\pi i} \oint_C \frac{f(z')dz'}{(z'-x)^{q+1}}, \quad (q < 0 \text{ and } \neq \text{ integer}). \tag{14.74}$$

The integral is evaluated over the contour C in the limit as the radius goes to infinity.

Evaluating the integral in Equation (14.74), as it stands, is not easy. Thus we modify our contour to C' as shown in Figure 14.3. Since the function

$$\frac{f(z')}{(z'-x)^{q+1}} \tag{14.75}$$

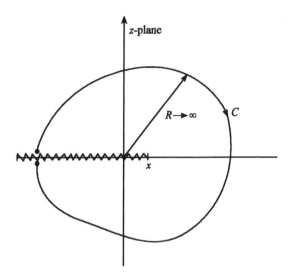

Fig. 14.2   Contour C in the differintegral formula

is analytic in and on the closed contour C', we can write

$$\oint_{C'} \frac{f(z')dz'}{(z'-x)^{q+1}} = 0, \qquad (14.76)$$

where the contour C' has the parts

$$\circlearrowright C' = \circlearrowright C + \circlearrowright C_0 + \leftarrow L_1 + \rightarrow L_2. \qquad (14.77)$$

We see that the integral we need to evaluate in Equation (14.74) is equal to the negative of the integral (Fig. 14.4)

$$\oint_{\circlearrowright C_0 + \leftarrow L_1 + \rightarrow L_2} \frac{f(z')dz'}{(z'-x)^{q+1}}.$$

Part of the integral over C_0 is taken in the limit as the radius goes to zero. For a point on the contour we write

$$z' - x = \delta_0 e^{i\theta}. \qquad (14.78)$$

Thus for $q < 0$ and noninteger, the integral $\oint_{C_0} \frac{f(z')dz'}{(z'-x)^{q+1}}$ becomes

$$\lim_{\delta_0 \to 0} \oint_{C_0} f(\xi) \frac{\delta_0 i e^{i\theta} d\theta}{\delta_0^{q+1} e^{i(q+1)\theta}} = \lim_{\delta_0 \to 0} f(x) i \delta_0^{-qi} \int_{-\pi}^{\pi} e^{-iq\theta} d\theta, \qquad (14.79)$$

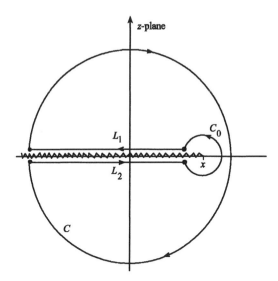

Fig. 14.3 Contour $C' = C + C_0 + L_1 + L_2$ in the differintegral formula

which goes to zero in the limit $\delta_0 \to 0$. For the C_0 integral to be zero in the limit $\delta_0 \to 0$, we have taken q as negative. Using this result we can write Equation (14.74) as

$$\oint_C \frac{f(z')dz'}{(z'-x)^{q+1}} = -\left[\oint_{\leftarrow L_1} \frac{f(z')dz'}{(z'-x)^{q+1}} + \oint_{\to L_2} \frac{f(z')dz'}{(z'-x)^{q+1}}\right]$$

$$= \left[\oint_{\to L_1} \frac{f(z')dz'}{(z'-x)^{q+1}} - \oint_{\to L_2} \frac{f(z')dz'}{(z'-x)^{q+1}}\right]. \qquad (14.80)$$

Now we have to evaluate the $\left[\oint_{\to L_1} - \oint_{\to L_2}\right]$ integral. We first evaluate the parts of the integral for $[-\infty, 0]$, which gives zero as

$$\int_{-\infty}^0 \frac{f(\delta e^{i(\pi-\varepsilon)})d\delta e^{i(\pi-\varepsilon)}}{(\delta e^{i(\pi-\varepsilon)} - x)^{q+1}} - \int_{-\infty}^0 \frac{f(\delta e^{i(\pi+\varepsilon)})d\delta e^{i(\pi+\varepsilon)}}{(\delta e^{i(\pi+\varepsilon)} - x)^{q+1}} \qquad (14.81)$$

$$= \left[e^{i(\pi-\varepsilon)} - e^{i(\pi+\varepsilon)}\right] \int_{-\infty}^0 \frac{f(\delta e^{i(\pi-\varepsilon)})d\delta}{(-\delta - x)^{q+1}}$$

$$= \lim_{\delta \to 0}\left[e^{i(\pi-\varepsilon)} - e^{i(\pi+\varepsilon)}\right] \int_{-\infty}^0 \frac{f(\delta e^{i(\pi-\varepsilon)})d\delta}{(-\delta - x)^{q+1}}$$

$$= 0.$$

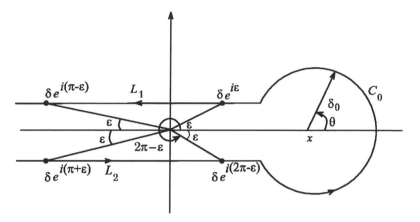

Fig. 14.4 Contours for the $\oint_{\leftarrow L_1}, \oint_{\rightarrow L_2}$, and $\oint_{C_0}$ integrals

Writing the remaining part of the $\oint_C dz$ integral we get

$$\oint_C \frac{f(z')dz'}{(z'-x)^{q+1}} \tag{14.82}$$

$$= \lim_{\delta_0,\varepsilon\to 0} \left[\int_0^x \frac{f(\delta e^{i\varepsilon})e^{i\varepsilon}d\delta}{(\delta-x)^{q+1}e^{i(q+1)\varepsilon}} - \int_0^x \frac{f(\delta e^{i(2\pi-\varepsilon)})e^{i(2\pi-\varepsilon)}d\delta}{(\delta-x)^{q+1}e^{i(q+1)(2\pi-\varepsilon)}} \right].$$

After taking the limit we substitute this into the definition [Eq. (14.74)] to obtain

$$\frac{d^q f(x)}{dx^q} = \frac{\Gamma(q+1)}{2\pi i} \left[1 - e^{-i(q+1)2\pi} \right] \int_0^x \frac{f(\delta)d\delta}{(\delta-x)^{q+1}}, \quad q < 0 \text{ and noninteger.} \tag{14.83}$$

Simplifying this we can write

$$\frac{d^q f(x)}{dx^q} = \frac{\Gamma(q+1)}{2\pi i} [2i\sin(\pi q)] (-1)^q \int_0^x \frac{f(\delta)d\delta}{(\delta-x)^{q+1}} \tag{14.84}$$

$$= -\frac{\Gamma(q+1)}{\pi} [\sin(\pi q)] \int_0^x \frac{f(\delta)d\delta}{(x-\delta)^{q+1}}. \tag{14.85}$$

To see that this agrees with the Riemann-Liouville definition we use the following relation of the gamma function:

$$\Gamma(-q) = \frac{-\pi \csc(\pi q)}{\Gamma(q+1)} \tag{14.86}$$

and write

$$\frac{d^q f(x)}{dx^q} = \frac{1}{\Gamma(-q)} \int_0^x \frac{f(\delta)d\delta}{(x-\delta)^{q+1}}, \quad q < 0 \text{ and noninteger.} \tag{14.87}$$

This is nothing but the Riemann-Liouville definition. Using Equation (14.71) we can extend this definition to positive values of q.

14.3.2 Riemann Formula

We now evaluate the differintegral of

$$f(x) = x^p, \tag{14.88}$$

which is very useful for finding differintegrals of functions the Taylor series of which can be given. Using formula (14.84) we write

$$\frac{d^q x^p}{dx^q} = \frac{\Gamma(q+1)}{\pi} \sin(\pi q)(-1)^q \int_0^x \frac{\delta^p \, d\delta}{(\delta - x)^{q+1}} \tag{14.89}$$

and

$$\frac{d^q x^p}{dx^q} = \frac{\Gamma(q+1)}{\pi} \sin(\pi q)(-1)^q \int_0^x \frac{\delta^p \, d\delta}{x^{q+1}(\frac{\delta}{x} - 1)^{q+1}}. \tag{14.90}$$

We define

$$\frac{\delta}{x} = s \tag{14.91}$$

so that Equation (14.90) becomes

$$\frac{d^q x^p}{dx^q} = -\frac{\Gamma(q+1)}{\pi} \sin(\pi q) x^{p-q} \int_0^1 \frac{s^p \, ds}{(1-s)^{q+1}}. \tag{14.92}$$

Remembering the definition of the beta function:

$$B(p, q) = \int_0^1 y^{p-1}[1 - y]^{q-1} dy, \quad p \text{ and } q > 0, \tag{14.93}$$

we can write Equation (14.92) as

$$\frac{d^q x^p}{dx^q} = -\frac{\Gamma(q+1)}{\pi} \sin(\pi q) x^{p-q} B(p+1, -q). \tag{14.94}$$

Also using the relation (14.86) and

$$B(p+1, -q) = \frac{\Gamma(p+1)\Gamma(-q)}{\Gamma(p+1-q)} \tag{14.95}$$

between the beta and the gamma functions, we obtain the result as

$$\frac{d^q x^p}{dx^q} = \frac{\Gamma(p+1)x^{p-q}}{\Gamma(p+1-q)}, \quad p > -1 \text{ and } q < 0. \tag{14.96}$$

Limits on the parameters p and q follow from the conditions of convergence for the beta integral.

For $q \geq 0$, as in the Riemann-Liouville definition, we write

$$\frac{d^q x^p}{dx^q} = \frac{d^n}{dx^n}\left[\frac{d^{q-n} x^p}{dx^{q-n}}\right] \tag{14.97}$$

and choose the integer n as $q - n < 0$. We now evaluate the differintegral inside the square brackets using formula (14.71) as

$$\frac{d^q x^p}{dx^q} = \frac{d^n}{dx^n}\left[\frac{\Gamma(p+1)x^{p-q+n}}{\Gamma(p-q+n+1)}\right] \tag{14.98}$$

$$= \frac{\Gamma(p+1)x^{p-q}}{\Gamma(p-q+1)}, \quad q \geq 0 .$$

Combining this with the results in Equations (14.96) and (14.98) we obtain a formula valid for all q as

$$\frac{d^q x^p}{dx^q} = \frac{\Gamma(p+1)x^{p-q}}{\Gamma(p-q+1)}, \quad p > -1 . \tag{14.99}$$

This formula is also known as the Riemann formula. It is a generalization of the formula

$$\frac{d^n x^m}{dx^n} = \frac{m!}{(m-n)!}x^{m-n}, \tag{14.100}$$

for $p > -1$, where m and n are positive integers. For $p \leq -1$ the beta function is divergent. Thus a generalization valid for all p values is yet to be found.

14.3.3   Differintegrals via Laplace Transforms

For the negative values of q we can define differintegrals by using Laplace transforms as

$$\frac{d^q f}{dx^q} = \mathcal{L}^{-1}[s^q \widetilde{f}(s)] , \quad q < 0, \tag{14.101}$$

where $\widetilde{f}(s)$ is the Laplace transform of $f(x)$. To see that this agrees with the Riemann-Liouville definition we make use of the convolution theorem

$$\mathcal{L}\int_0^x f(u)g(x-u)du = \widetilde{f}(s)\widetilde{g}(s). \tag{14.102}$$

In this equation we take $g(x)$ as

$$g(x) = \frac{1}{x^{q+1}}, \tag{14.103}$$

where its Laplace transform is

$$\widetilde{g}(s) = \int_0^\infty e^{-sx} \frac{dx}{x^{q+1}} \tag{14.104}$$

$$= \Gamma(-q)s^q, \tag{14.105}$$

and also write the Laplace transform of $f(x)$ as

$$\widetilde{f}(s) = \int_0^\infty e^{-sx} f(x) dx. \tag{14.106}$$

For $q < 0$ we obtain

$$\left[\frac{d^q f}{dx^q} \right]_L = \mathcal{L}^{-1}[s^q \widetilde{f}(s)] \ , \quad q < 0. \tag{14.107}$$

$$\left[\frac{d^q f}{dx^q} \right]_L = \frac{1}{\Gamma(-q)} \int_0^x \frac{f(\tau)d\tau}{(x-\tau)^{q+1}}$$

$$\left[\frac{d^q f}{dx^q} \right]_L = \left[\frac{d^q f}{dx^q} \right]_{R-L}, \quad q < 0, \tag{14.108}$$

The subscripts L and R-L denote the method used in evaluating the differintegral. Thus the two methods agree for $q < 0$.

For $q > 0$, the differintegral definition by the Laplace transforms is given as (Section 14.6.1)

$$\left[\frac{d^q f}{dx^q} \right]_L = \mathcal{L}^{-1} \left[s^q \widetilde{f}(s) - \sum_{k=0}^{n-1} s^k \frac{d^{q-1-k} f}{dx^{q-1-k}}(0) \right], \quad q > 0, \tag{14.109}$$

or

$$\left[\frac{d^q f}{dx^q} \right]_L = \mathcal{L}^{-1} \left[s^q \widetilde{f}(s) - \frac{d^{q-1} f}{dx^{q-1}}(0) - \cdots - s^{n-1} \frac{d^{q-n} f}{dx^{q-n}}(0) \right]. \tag{14.110}$$

In this definition $q > 0$ and the integer n must be chosen such that the inequality $n - 1 < q \leq n$ is satisfied. The differintegrals on the right-hand side are all evaluated via the L method. To show that the methods agree we write

$$A(x) = \frac{1}{\Gamma(n-q)} \int_0^x \frac{f(\tau)d\tau}{(x-\tau)^{q-n+1}}, \quad q < n, \tag{14.111}$$

and use the convolution theorem to find its Laplace transform as

$$\widetilde{A}(s) = \mathcal{L} \left[\frac{1}{\Gamma(q-n)} \int_0^x \frac{f(\tau)d\tau}{(x-\tau)^{q-n+1}} \right] \tag{14.112}$$

$$= s^{q-n} \widetilde{f}(s), \quad q - n < 0. \tag{14.113}$$

This gives us the $s^q \widetilde{f}(s) = s^n \widetilde{A}(s)$ relation. Using the Riemann-Liouville definition [Eqs. (14.70-71)] we can write

$$A(0) = \left[\frac{d^{(q-n)} f}{dx^{(q-n)}} (0) \right]_{R-L}. \tag{14.114}$$

Since $q - n < 0$ and because of Equation (14.108), we can write

$$A(0) = \left[\frac{d^{(q-n)} f}{dx^{(q-n)}} (0) \right]_{L}. \tag{14.115}$$

From the definition of $A(x)$ we can also write

$$A(x) = \frac{1}{\Gamma(n-q)} \int_0^x \frac{f(\tau) d\tau}{(x-\tau)^{q-n+1}}, \quad q - n < 0,$$
$$= \left[\frac{d^{(q-n)} f(x)}{dx^{(q-n)}} \right]_{R-L},$$
$$= \left[\frac{d^{(q-n)} f(x)}{dx^{(q-n)}} \right]_{L}. \tag{14.116}$$

As in the Grünwald and Riemann-Liouville definitions we assume that the $[..]_L$ definition also satisfies the relation [Eq. (14.40)]

$$\frac{d^n}{dx^n} \frac{d^q f(x)}{dx^q} = \frac{d^{n+q} f(x)}{dx^{n+q}}, \tag{14.117}$$

where n a is positive integer and q takes all values. We can now write

$$\frac{d^{n-1} A(x)}{dx^{n-1}} = \frac{d^{n-1}}{dx^{n-1}} \left[\frac{d^{q-n} f(x)}{dx^{q-n}} \right]_{L}, \tag{14.118}$$
$$= \left[\frac{d^{q-1} f(x)}{dx^{q-1}} \right]_{L},$$

which gives us

$$\left[\frac{d^{q-1} f(0)}{dx^{q-1}} \right]_{L} = \left[\frac{d^{n-1} A(0)}{dx^{n-1}} \right]. \tag{14.119}$$

Similarly we find the other terms in Equation (14.110) to write

$$\left[\frac{d^q f}{dx^q} \right]_{L} = \mathcal{L}^{-1} \left[s^n \widetilde{A}(s) - \frac{d^{n-1} A}{dx^{n-1}} (0) - \cdots - s^{n-1} A(0) \right],$$
$$= \mathcal{L}^{-1} \left[s^n \widetilde{A}(s) - s^{n-1} A(0) - \cdots - \frac{d^{n-1} A}{dx^{n-1}} (0) \right],$$
$$= \mathcal{L}^{-1} \left[\mathcal{L} \left[\frac{d^n A}{dx^n} \right]_{R-L} \right] = \left[\frac{d^n A}{dx^n} \right]_{R-L}. \tag{14.120}$$

Using Equation (14.111) we can now write

$$
\left[\frac{d^q f}{dx^q}\right]_L = \frac{d^n}{dx^n}\left[\frac{1}{\Gamma(n-q)}\int_0^x \frac{f(\tau)d\tau}{(x-\tau)^{q-n+1}}\right], \tag{14.121}
$$

$$
= \frac{1}{\Gamma(n-q)}\frac{d^n}{dx^n}\int_0^x \frac{f(\tau)d\tau}{(x-\tau)^{q-n+1}}, \quad n > q,
$$

$$
= \left[\frac{d^q f}{dx^q}\right]_{R-L}, \quad q > 0, \tag{14.122}
$$

which shows that for $q > 0$, too, both definitions agree.

In formula (14.110), if the function $f(x)$ satisfies the boundary conditions

$$
\frac{d^{q-1}f}{dx^{q-1}}(0) = \cdots = \frac{d^{q-n}f}{dx^{q-n}}(0) = 0, \quad q > 0, \tag{14.123}
$$

we can write a differintegral definition valid for all q values via the Laplace transform as

$$
\frac{d^q f}{dx^q} = \mathcal{L}^{-1}[s^q \widetilde{f}(s)]. \tag{14.124}
$$

However, because the boundary conditions (14.123) involve fractional derivatives this will create problems in interpretation and application. (See Problem 14.7 on the Caputo definition of fractional derivatives.)

14.4   PROPERTIES OF DIFFERINTEGRALS

In this section we see the basic properties of differintegrals. These properties are also useful in generating new differintegrals from the known ones.

14.4.1   Linearity

We express the linearity of differintegrals as

$$
\frac{d^q[f_1 + f_2]}{[d(x-a)]^q} = \frac{d^q f_1}{[d(x-a)]^q} + \frac{d^q f_2}{[d(x-a)]^q}. \tag{14.125}
$$

14.4.2   Homogeneity

Homogeneity of differintegrals is expressed as

$$
\frac{d^q(C_0 f)}{[d(x-a)]^q} = C_0\frac{d^q f}{[d(x-a)]^q}, \quad C_0 \text{ is any constant.} \tag{14.126}
$$

Both of these properties could easily be seen from the Grünwald definition [Eq. (14.39)].

14.4.3   Scale Transformation

We express the scale transformation of a function with respect to the lower limit a as

$$f(x) \rightarrow f(\gamma x - \gamma a + a), \tag{14.127}$$

where γ is a constant scale factor. If the lower limit is zero, this means that

$$f(x) \rightarrow f(\gamma x). \tag{14.128}$$

If the lower limit differs from zero, the scale change is given as

$$\frac{d^q f(\gamma X)}{[d(x-a)]^q} = \gamma^q \frac{d^q f(\gamma X)}{[d(\gamma X - a)]^q}, \quad X = x + [a - a\gamma]/\gamma \tag{14.129}$$

This formula is most useful when a is zero:

$$\frac{d^q f(\gamma x)}{[dx]^q} = \gamma^q \frac{d^q f(\gamma x)}{[d(\gamma x)]^q}. \tag{14.130}$$

14.4.4   Differintegral of a Series

Using the linearity of the differintegral operator we can find the differintegral of a uniformly convergent series for all q values as

$$\frac{d^q}{[d(x-a)]^q} \sum_{j=0}^{\infty} f_j(x) = \sum_{j=0}^{\infty} \frac{d^q f_j}{[d(x-a)]^q}. \tag{14.131}$$

Differintegrated series are also uniformly convergent in the same interval. For functions with power series expansions, using the Riemann formula we can write

$$\frac{d^q}{[d(x-a)]^q} \sum_{j=0}^{\infty} a_i [x-a]^{p+(j/n)} = \sum_{j=0}^{\infty} a_j \frac{\Gamma\left(\dfrac{pn+j+n}{n}\right)}{\Gamma\left(\dfrac{pn-qn+j+n}{n}\right)} [x-a]^{p-q+(j/n)}, \tag{14.132}$$

where q can take any value, but $p + (j/n) > -1$, $a_0 \neq 0$, and n is a positive integer.

14.4.5   Composition of Differintegrals

When working with differintegrals one always has to remember that operations like

$$d^q d^Q = d^Q d^q,$$
$$d^q d^Q = d^{q+Q} \quad \text{and} \tag{14.133}$$
$$d^q f = g \rightarrow f = d^{-q} g$$

are valid only under certain conditions. In these operations problems are not just restricted to the noninteger values of q and Q.

When n and N are positive integer numbers, from the properties of derivatives and integrals we can write

$$\frac{d^n}{[d(x-a)]^n}\left\{\frac{d^N f}{[d(x-a)]^N}\right\} = \frac{d^{n+N} f}{[d(x-a)]^{n+N}} \tag{14.134}$$

$$= \frac{d^N}{[d(x-a)]^N}\left\{\frac{d^n f}{[d(x-a)]^n}\right\}$$

and

$$\frac{d^{-n}}{[d(x-a)]^{-n}}\left\{\frac{d^{-N} f}{[d(x-a)]^{-N}}\right\} = \frac{d^{-n-N} f}{[d(x-a)]^{-n-N}} \tag{14.135}$$

$$= \frac{d^{-N}}{[d(x-a)]^{-N}}\left\{\frac{d^{-n} f}{[d(x-a)]^{-n}}\right\}.$$

However, if we look at the operation

$$\frac{d^{\pm n}}{[d(x-a)]^{\pm n}}\left\{\frac{d^{\mp N} f}{[d(x-a)]^{\mp N}}\right\}, \tag{14.136}$$

the result is not always

$$\frac{d^{\pm n \mp N} f}{[d(x-a)]^{\pm n \mp N}}. \tag{14.137}$$

Assume that the function $f(x)$ has continuous Nth-order derivative in the interval $[a, b]$ and let us take the integral of this Nth-order derivative as

$$\int_a^x f^{(N)}(x_1)dx_1 = f^{(N-1)}(x)\Big|_a^x = f^{(N-1)}(x) - f^{(N-1)}(a). \tag{14.138}$$

We integrate this once more:

$$\int_a^x \left(\int_a^{x_2} f^{(N)}(x_1)dx_1\right) dx_2 = \int_a^x \left[f^{(N-1)}(x) - f^{(N-1)}(a)\right] dx$$

$$= f^{(N-2)}(x) - f^{(N-2)}(a) - (x-a)f^{(N-1)}(a) \tag{14.139}$$

and repeat the process n times to get

$$\int_a^x \cdots \int_a^x f^{(N)}(x)(dx)^n = f^{(N-n)}(x) - f^{(N-n)}(a) - (x-a)f^{(N-n-1)}(a)$$

$$- \frac{(x-a)^2}{2!}f^{(N-n-2)}(a) \tag{14.140}$$

$$\cdots - \frac{(x-a)^{n-1}}{(n-1)!}f^{(N-1)}(a).$$

Since

$$\frac{d^{-n}}{[d(x-a)]^{-n}}\left\{\frac{d^N f}{[d(x-a)]^N}\right\} = \int_a^x \cdots \int_a^x f^{(N)}(x)(dx)^n, \qquad (14.141)$$

we write

$$\frac{d^{-n}}{[d(x-a)]^{-n}}\left\{\frac{d^N f}{[d(x-a)]^N}\right\} = f^{(N-n)}(x) - \sum_{k=0}^{n-1}\frac{[x-a]^k}{k!}f^{(N+k-n)}(a).$$
$$(14.142)$$

Writing Equation (14.142) for $N = 0$ gives us

$$\frac{d^{-n}f}{[d(x-a)]^{-n}} = f^{(-n)}(x) - \sum_{k=0}^{n-1}\frac{[x-a]^k}{k!}f^{(k-n)}(a). \qquad (14.143)$$

We differentiate this to get

$$\frac{d}{dx}\left\{\frac{d^{-n}f}{[d(x-a)]^{-n}}\right\} = f^{(1-n)}(x) - \sum_{k=1}^{n-1}\frac{[x-a]^{k-1}}{(k-1)!}f^{(k-n)}(a). \qquad (14.144)$$

After N-fold differentiation we obtain

$$\frac{d^N}{dx^N}\left\{\frac{d^{-n}f}{[d(x-a)]^{-n}}\right\} = f^{(N-n)}(x) - \sum_{k=N}^{n-1}\frac{[x-a]^{k-N}}{(k-N)!}f^{(k-n)}(a). \qquad (14.145)$$

For $N \geq n$, remembering that differentiation does not depend on the lower limit and also observing that in this case the summation in Equation (14.145) is empty, we write

$$\frac{d^N}{[d(x-a)]^N}\left\{\frac{d^{-n}f}{[d(x-a)]^{-n}}\right\} = \frac{d^{N-n}f}{[d(x-a)]^{N-n}} = f^{(N-n)}(x). \qquad (14.146)$$

On the other hand for $N < n$, we use Equation (14.143) to write

$$\frac{d^{N-n}f}{[d(x-a)]^{N-n}} = f^{(N-n)}(x) - \sum_{k=0}^{n-N-1}\frac{[x-a]^k}{k!}f^{(k+N-n)}(a). \qquad (14.147)$$

This equation also contains Equation (14.146). In Equation (14.145) we now make the transformation

$$k \rightarrow k + N \qquad (14.148)$$

to write

$$\frac{d^N}{[d(x-a)]^N}\left\{\frac{d^{-n}f}{[d(x-a)]^{-n}}\right\} = f^{(N-n)}(x) - \sum_{k=0}^{n-N-1}\frac{[x-a]^k}{k!}f^{(k+N-n)}(a).$$
$$(14.149)$$

Because the right-hand sides of Equations (14.149) and (14.147) are identical, we obtain the composition rule for n successive integrations followed by N differentiations as

$$\frac{d^N}{[d(x-a)]^N}\left\{\frac{d^{-n}f}{[d(x-a)]^{-n}}\right\} = \frac{d^{N-n}f}{[d(x-a)]^{N-n}}. \tag{14.150}$$

To find the composition rule for the cases where the differentiations are performed before the integrations, we turn to Equation (14.142) and write the sum in two pieces as

$$\frac{d^{-n}}{[d(x-a)]^{-n}}\left\{\frac{d^N f}{[d(x-a)]^N}\right\} = f^{(N-n)}(x) - \sum_{k=0}^{n-N-1}\frac{[x-a]^k}{k!}f^{(N+k-n)}(a)$$
$$- \sum_{k=n-N}^{n-1}\frac{[x-a]^k}{k!}f^{(N+k-n)}(a). \tag{14.151}$$

Comparing this with Equation (14.147), we now obtain the composition rule for the cases where N-fold differentiation is performed before n successive integrations as

$$\frac{d^{-n}}{[d(x-a)]^{-n}}\left\{\frac{d^N f}{[d(x-a)]^N}\right\} = \frac{d^{N-n}f}{[d(x-a)]^{N-n}} - \sum_{k=n-N}^{n-1}\frac{[x-a]^k}{k!}f^{(N+k-n)}(a). \tag{14.152}$$

Example 14.1. *Composition of differintegrals:* For the function $f(x) = e^{-3x}$, we first calculate

$$\frac{d}{[dx]}\left\{\frac{d^{-3}f(x)}{[dx]^{-3}}\right\}.$$

For this case we use Equations (14.150) and (14.143) to find

$$\frac{d}{[dx]}\left\{\frac{d^{-3}f(x)}{[dx]^{-3}}\right\} = \frac{d^{-2}f(x)}{[dx]^{-2}}$$
$$= \frac{e^{-3x}}{9} + \frac{x}{3} - \frac{1}{9}. \tag{14.153}$$

On the other hand, for

$$\frac{d^{-3}}{[dx]^{-3}}\left\{\frac{df(x)}{[dx]}\right\},$$

we have to use formula (14.152). Since $N = 1$ and $n = 3$, k takes only the value two, thus giving

$$\frac{d^{-3}}{[dx]^{-3}}\left\{\frac{df(x)}{[dx]}\right\} = \frac{e^{-3x}}{9} + \frac{x}{3} - \frac{1}{9} - \frac{x^2}{2}. \tag{14.154}$$

14.4.5.1 **Composition Rule for General q and Q:** When q and Q take any value, composition of differintegrals as

$$\frac{d^q}{[d(x-a)]^q}\left[\frac{d^Q f}{[d(x-a)]^Q}\right] = \frac{d^{q+Q}f}{[d(x-a)]^{q+Q}} \qquad (14.155)$$

is possible only under certain conditions. It is needless to say that we assume all the required differintegrals exist. Assuming that a series expansion for $f(x)$ can be given as

$$f(x) = \sum_{j=0}^{\infty} a_j[x-a]^{p+j}, \quad p \text{ is a noninteger such that } p+j > -1, \quad (14.156)$$

it can be shown that the composition rule [Eq. (14.155)] is valid only for functions satisfying the condition

$$f(x) - \frac{d^{-Q}}{[d(x-a)]^{-Q}}\left[\frac{d^Q f}{[d(x-a)]^Q}\right] = 0. \qquad (14.157)$$

In general, for functions that can be expanded as Equation (14.156) differintegrals are composed as (Oldham and Spanier)

$$\frac{d^q}{[d(x-a)]^q}\left[\frac{d^Q f}{[d(x-a)]^Q}\right]$$
$$= \frac{d^{q+Q}f}{[d(x-a)]^{q+Q}} - \frac{d^{q+Q}}{[d(x-a)]^{q+Q}}\left\{f - \frac{d^{-Q}}{[d(x-a)]^{-Q}}\left[\frac{d^Q f}{[d(x-a)]^Q}\right]\right\}. \qquad (14.158)$$

For such functions violation of condition (14.157) can be shown to result from the fact that $\frac{d^Q f}{[d(x-a)]^Q}$ vanishes even though $f(x)$ is different from zero. From here we see that, even though the operators $\frac{d^Q}{[d(x-a)]^Q}$ and $\frac{d^{-Q}}{[d(x-a)]^{-Q}}$ are in general inverses of each other, this is not always true.

In practice it is difficult to apply the composition rule as given in Equation (14.158). Because the violation of Equation (14.157) is equivalent to the vanishing of the derivative $\frac{d^Q f(x)}{[dx]^Q}$, let us first write the differintegral (for simplicity we set $a = 0$) of $f(x)$ as

$$\frac{d^Q f(x)}{[dx]^Q} = \sum_{j=0}^{\infty} a_j \frac{d^Q x^{p+j}}{[dx]^Q} = \sum_{j=0}^{\infty} a_j \frac{\Gamma(p+j+1)x^{p+j-Q}}{\Gamma(p+j-Q+1)}. \qquad (14.159)$$

Because the condition $p + j > -1$ (or $p > -1$), the gamma function in the numerator is always different from zero and finite. For the $Q < p+1$ values, gamma function in the denominator is always finite; thus condition (14.157) is satisfied. For the remaining cases condition (14.157) is violated. We now check the equivalent condition $\frac{d^Q f(x)}{[dx]^Q} = 0$ to identify the terms responsible for the violation of condition (14.157). For the derivative $\frac{d^Q f(x)}{[dx]^Q}$ to vanish, from Equation (14.159) it is seen that the gamma function in the denominator must diverge for all $a_j \neq 0$, that is,

$$p + j - Q + 1 = 0, -1, -2, \ldots .$$

For a given p (> -1) and positive Q, j will eventually make $(p - Q + j + 1)$ positive; therefore we can write

$$p + j = Q - 1, Q - 2, \ldots, Q - m , \qquad (14.160)$$

where m is an integer satisfying

$$0 < Q < m < Q + 1. \qquad (14.161)$$

For the j values that make $(p - Q + j + 1)$ positive, the gamma function in the denominator is finite, and the corresponding terms in the series satisfy condition (14.157). Thus the problem is located to the terms with the j values satisfying Equation (14.160). Now, in general for an arbitrary diffferintegrable function we can write the expression

$$f(x) - \frac{d^{-Q}}{[dx]^{-Q}} \left[\frac{d^Q f}{[dx]^Q} \right] = c_0 x^{Q-1} + c_1 x^{Q-2} + \cdots + c_m x^{Q-m}, \qquad (14.162)$$

where $c_1, c_2, \ldots, c_m$ are arbitrary constants. Note that the right-hand side of Equation (14.162) is exactly composed of the terms that vanish when $\frac{d^Q f(x)}{[dx]^Q} \neq 0$, that is, when Equation (14.157) is satisfied. This formula, which is very useful in finding solutions of extraordinary differential equations can now be used in Equation (14.158) to compose differintegrals.

Another useful formula is obtained when Q takes integer values N in Equation (14.158). We apply the composition rule [Eq. (14.158)] with Equation (14.142) written for $n = N$, and use the generalization of the Riemann formula:

$$\frac{d^q (x-a)^p}{[d(x-a)]^q} = \frac{\Gamma(p+1)(x-a)^{p-q}}{\Gamma(p-q+1)}, \qquad p > -1 \qquad (14.163)$$

to obtain

$$\frac{d^q}{[d(x-a)]^q} \left[\frac{d^N f}{[d(x-a)]^N} \right]$$

$$= \frac{d^{q+N} f}{[d(x-a)]^{q+N}} - \frac{d^{q+N}}{[d(x-a)]^{q+N}} \left\{ f - \frac{d^{-N}}{[d(x-a)]^{-N}} \left[\frac{d^N f}{[d(x-a)]^N} \right] \right\}$$

$$= \frac{d^{q+N} f}{[d(x-a)]^{q+N}} - \sum_{k=0}^{N-1} \frac{[x-a]^{k-q-N} f^{(k)}(a)}{\Gamma(k-q-N+1)}. \qquad (14.164)$$

Example 14.2. *Composition of differintegrals:* As another example we consider the function

$$f = x^{-1/2} \tag{14.165}$$

for the values $a = 0$, $Q = 1/2$, and $q = -1/2$. Since condition (14.157) is not satisfied, that is,

$$x^{-1/2} - \frac{d^{-\frac{1}{2}}}{[dx]^{-\frac{1}{2}}} \left[\frac{d^{\frac{1}{2}} x^{-1/2}}{[dx]^{\frac{1}{2}}} \right] = x^{-1/2} - \frac{d^{-\frac{1}{2}}}{[dx]^{-\frac{1}{2}}} \left[\frac{\Gamma(\frac{1}{2})}{\Gamma(0)} \right]$$

$$= x^{-1/2} - 0 \tag{14.166}$$

$$\neq 0, \tag{14.167}$$

we have to use Equation (14.158):

$$\frac{d^{-\frac{1}{2}}}{[dx]^{-\frac{1}{2}}} \frac{d^{\frac{1}{2}} x^{-1/2}}{[dx]^{\frac{1}{2}}} \tag{14.168}$$

$$= \frac{d^{-\frac{1}{2}+\frac{1}{2}} x^{-1/2}}{[dx]^{-\frac{1}{2}+\frac{1}{2}}} - \frac{d^{-\frac{1}{2}+\frac{1}{2}}}{[dx]^{-\frac{1}{2}+\frac{1}{2}}} \left\{ x^{-1/2} - \frac{d^{-\frac{1}{2}}}{[dx]^{-\frac{1}{2}}} \left[\frac{d^{\frac{1}{2}} x^{-1/2}}{[dx]^{\frac{1}{2}}} \right] \right\}.$$

Since

$$\frac{d^{\frac{1}{2}} x^{-1/2}}{[dx]^{\frac{1}{2}}} = 0, \tag{14.169}$$

we have

$$\frac{d^{-\frac{1}{2}}}{[dx]^{-\frac{1}{2}}} \left\{ \frac{d^{\frac{1}{2}} x^{-1/2}}{[dx]^{\frac{1}{2}}} \right\} = 0, \tag{14.170}$$

which leads to

$$\frac{d^{-\frac{1}{2}}}{[dx]^{-\frac{1}{2}}} \frac{d^{\frac{1}{2}} x^{-1/2}}{[dx]^{\frac{1}{2}}} \tag{14.171}$$

$$= \frac{d^{-\frac{1}{2}+\frac{1}{2}} x^{-1/2}}{[dx]^{-\frac{1}{2}+\frac{1}{2}}} - \frac{d^{-\frac{1}{2}+\frac{1}{2}}}{[dx]^{-\frac{1}{2}+\frac{1}{2}}} \left(x^{-1/2} - 0 \right)$$

$$= x^{-1/2} - x^{-1/2} \tag{14.172}$$

$$= 0. \tag{14.173}$$

Contrary to what we expect $\frac{d^{-\frac{1}{2}}}{[dx]^{-\frac{1}{2}}}$ is not the inverse of $\frac{d^{\frac{1}{2}}}{[dx]^{\frac{1}{2}}}$ for $x^{-1/2}$.

Example 14.3. *Inverse of differintegrals:* We now consider the function

$$f = x \tag{14.174}$$

for the values $Q = 2$ and $a = 0$. Since

$$\frac{d^2 x}{[dx]^2} = 0 \tag{14.175}$$

is true, contrary to our expectations we find

$$\frac{d^{-2}}{[dx]^{-2}} \frac{d^2 x}{[dx]^2} = 0. \tag{14.176}$$

The problem is again that the function $f = x$ does not satisfy condition (14.157).

14.4.6   Leibniz's Rule

The differintegral of the qth order of the multiplication of two functions f and g is given by the formula

$$\frac{d^q [fg]}{[d(x-a)]^q} = \sum_{j=0}^{\infty} \binom{q}{j} \frac{d^{q-j} f}{[d(x-a)]^{q-j}} \frac{d^j g}{[d(x-a)]^j}, \tag{14.177}$$

where the binomial coefficients are to be calculated by replacing the factorials with the corresponding gamma functions.

14.4.7   Right- and Left-Handed Differintegrals

The Riemann-Liouville definition of differintegral was given as

$$\frac{d^q f(t)}{[d(t-a)]^q} = \frac{1}{\Gamma(k-q)} \left(\frac{d}{dt} \right)^k \int_a^t (t-\tau)^{k-q-1} f(\tau) d\tau, \tag{14.178}$$

where k is an integer satisfying

$$\begin{aligned} k &= 0 && \text{for} \quad q < 0 \\ k-1 &< q < k && \text{for} \quad q \geq 0. \end{aligned} \tag{14.179}$$

This is also called the right-handed Riemann-Liouville definition. If $f(t)$ is a function representing a dynamic process, in general t is a timelike variable. The principle of causality justifies the usage of the right-handed derivative because the present value of a differintegral is determined from the past values of $f(t)$ starting from an initial time $t = a$. Similar to the advanced potentials, it is also possible to define a left-handed Riemann-Liouville differintegral as

$$\frac{d^q f(t)}{[d(b-t)]^q} = \frac{1}{\Gamma(k-q)} \left(-\frac{d}{dt} \right)^k \int_t^b (\tau-t)^{k-q-1} f(\tau) d\tau, \tag{14.180}$$

where k is again an integer satisfying Equation (14.179). Even though for dynamic processes it is difficult to interpret the left-handed definition, in general the boundary or the initial conditions determine which definition is to be used. It is also possible to give a left-handed version of the Grünwald definition. In this chapter we confine ourselves to the right-handed definition.

14.4.8   Dependence on the Lower Limit

We now discuss the dependence of

$$\frac{d^q f(x)}{[d(x-a)]^q} \tag{14.181}$$

on the lower limit. For $q < 0$, using Equation (14.178) we write the difference

$$\delta = \frac{d^q f(x)}{[d(x-a)]^q} - \frac{d^q f(x)}{[d(x-b)]^q}$$

as

$$
\begin{aligned}
\delta &= \frac{1}{\Gamma(-q)} \int_a^x (x-\tau)^{-q-1} f(\tau) d\tau - \frac{1}{\Gamma(-q)} \int_b^x (x-\tau)^{-q-1} f(\tau) d\tau \\
&= \frac{1}{\Gamma(-q)} \int_a^b (x-\tau)^{-q-1} f(\tau) d\tau \tag{14.182} \\
&= \frac{1}{\Gamma(-q)} \int_a^b (x-b+b-\tau)^{-q-1} f(\tau) d\tau \\
&= \frac{1}{\Gamma(-q)} \int_a^b \left[\sum_{l=0}^\infty \begin{pmatrix} -l-q \\ l \end{pmatrix} (x-b)^{-q-1-l}(b-\tau)^l \right] f(\tau) d\tau. \tag{14.183}
\end{aligned}
$$

For the binomial coefficients we write

$$\begin{pmatrix} -l-q \\ l \end{pmatrix} = \frac{\Gamma(-q)}{\Gamma(-q-l)\Gamma(l+1)} \tag{14.184}$$

to obtain

$$
\begin{aligned}
\delta &= \int_a^b \left[\sum_{l=0}^\infty \frac{(x-b)^{-q-1-l}(b-\tau)^l}{\Gamma(-q-l)\Gamma(l+1)} \right] f(\tau) d\tau \\
&= \sum_{l=0}^\infty \left[\frac{(x-b)^{-q-1-l}}{\Gamma(-q-l)} \right] \left[\int_a^b \frac{(b-\tau)^l f(\tau) d\tau}{\Gamma(l+1)} \right] \\
&= \sum_{l=0}^\infty \frac{d^{q+l+1}[1]}{[d(x-b)]^{q+l+1}} \frac{d^{-l-1} f(b)}{[d(b-a)]^{-l-1}}, \quad (\text{ see Section 14.5.1}) \\
&= \sum_{l=1}^\infty \frac{d^{q+l}[1]}{[d(x-b)]^{q+l}} \frac{d^{-l} f(b)}{[d(b-a)]^{-l}}. \tag{14.185}
\end{aligned}
$$

Even though we have obtained this expression for $q < 0$, it is also valid for all q (Oldham and Spanier, Section 3.2). For $q = 0, 1, 2, ...$, that is, for ordinary derivatives, we have

$$\delta = 0 \tag{14.186}$$

as expected. For $q = -1$ the above equation simplifies to

$$\delta = \frac{d^{-1}f(b)}{[d(b-a)]^{-1}} = \int_a^b f(\tau)d\tau. \tag{14.187}$$

For all other values of q, δ not only depends on a and b but also on x. This is due to the fact that the differintegral, except when it reduces to an ordinary derivative, is a global operator and requires a knowledge of f over the entire space. This is apparent from the Riemann-Liouville definition [Eq. (14.178)], which is given as an integral, and the Grünwald definition [Eq. (14.39)] which is given as an infinite series.

14.5 DIFFERINTEGRALS OF SOME FUNCTIONS

In this section we discuss differintegrals of some selected functions. For an extensive list and discussion of the differintegrals of functions of mathematical physics we refer the reader to Oldham and Spanier.

14.5.1   Differintegral of a Constant

First we take the number one and find its differintegral using the Grünwald definition [Eq. (14.39)] as

$$\frac{d^q[1]}{[d(x-a)]^q} = \lim_{N \to \infty} \left\{ \left[\frac{N}{x-a} \right]^q \sum_{j=0}^{N-1} \frac{\Gamma(j-q)}{\Gamma(j+1)\Gamma(-q)} \right\}. \tag{14.188}$$

Using the properties of gamma functions; $\sum_{j=0}^{N-1} \Gamma(j-q)/\Gamma(-q)\Gamma(j+1) = \Gamma(N-q)/\Gamma(1-q)\Gamma(N)$, and $\lim_{N \to \infty}[N^q\Gamma(N-q)/\Gamma(N)] = 1$, we find

$$\frac{d^q[1]}{[d(x-a)]^q} = \frac{[x-a]^{-q}}{\Gamma(1-q)}. \tag{14.189}$$

When q takes integer values, this reduces to the expected result. For an arbitrary constant C, including zero, the differintegral is (see Problem 14.7)

$$\frac{d^q[C]}{[d(x-a)]^q} = C\frac{d^q[1]}{[d(x-a)]^q} = C\frac{[x-a]^{-q}}{\Gamma(1-q)}. \tag{14.190}$$

14.5.2 Differintegral of $[x - a]$

For the differintegral of $[x - a]$, we again use Equation (14.39) and write

$$\frac{d^q[x - a]}{[d(x - a)]^q} = \lim_{N \to \infty} \left\{ \left[\frac{N}{x - a}\right]^q \sum_{j=0}^{N-1} \frac{\Gamma(j - q)}{\Gamma(-q)\Gamma(j + 1)} \left[\frac{Nx - jx + ja}{N} - a\right]\right\}$$

(14.191)

$$= [x - a]^{1-q} \left[\lim_{N \to \infty} \left\{ N^q \sum_{j=0}^{N-1} \frac{\Gamma(j - q)}{\Gamma(-q)\Gamma(j + 1)}\right\} \right.$$

$$\left. - \lim_{N \to \infty} \left\{ [N]^{q-1} \sum_{j=0}^{N-1} \frac{j\Gamma(j - q)}{\Gamma(-q)\Gamma(j + 1)}\right\}\right].$$

(14.192)

In addition to the properties used in Section 14.5.1, we also use the following relation between the gamma functions: $\sum_{j=0}^{N-1} \Gamma(j - q)/\Gamma(-q)\Gamma(j) = (-q)\Gamma(N - q)/\Gamma(2 - q)\Gamma(N - 1)$, to obtain

$$\frac{d^q[x - a]}{[d(x - a)]^q} = [x - a]^{1-q}\left[\frac{1}{\Gamma(1 - q)} + \frac{q}{\Gamma(2 - q)}\right]$$

(14.193)

$$\frac{d^q[x - a]}{[d(x - a)]^q} = \frac{[x - a]^{1-q}}{\Gamma(2 - q)}.$$

(14.194)

We now use the Riemann-Liouville formula to find the same differintegral. We first write

$$\frac{d^q[x - a]}{[d(x - a)]^q} = \frac{1}{\Gamma(-q)} \int_a^x \frac{[x' - a]dx'}{[x - x']^{q+1}}.$$

(14.195)

For $q < 0$ values we make the transformation $y = x - x'$ and write

$$\frac{d^q[x - a]}{[d(x - a)]^q} = \frac{1}{\Gamma(-q)} \int_a^{x-a} \frac{[x - a - y]dy}{y^{q+1}}$$

(14.196)

$$= \frac{1}{\Gamma(-q)} \left[\int_a^{x-a} \frac{[x - a]dy}{y^{q+1}} + \int_a^{x-a} \frac{dy}{y^q}\right]$$

(14.197)

$$= \frac{1}{\Gamma(-q)} \left[\frac{[x - a]^{1-q}}{-q} + \frac{[x - a]^{1-q}}{1 - q}\right]$$

(14.198)

$$= \frac{[x - a]^{1-q}}{[-q][1 - q]\Gamma(-q)},$$

(14.199)

which leads us to

$$\frac{d^q[x - a]}{[d(x - a)]^q} = \frac{[x - a]^{1-q}}{\Gamma(2 - q)}, \qquad q < 0.$$

(14.200)

For the other values of q we use formula (14.40) to write

$$\frac{d^q[x-a]}{[d(x-a)]^q} = \frac{d^n}{dx^n}\left[\frac{d^{q-n}[x-a]}{[d(x-a)]^{q-n}}\right] \tag{14.201}$$

and choose n such that $q - n < 0$ is satisfied. We use the Riemann formula [Eq. (14.99)] to write

$$\frac{d^{q-n}[x-a]}{[d(x-a)]^{q-n}} = \frac{\Gamma(2)[x-a]^{1-q+n}}{\Gamma(2-q+n)}, \tag{14.202}$$

which leads to the following result:

$$\frac{d^q[x-a]}{[d(x-a)]^q} = \frac{d^n}{[d(x-a)]^n}\left[\frac{[x-a]^{1-q+n}}{\Gamma(2-q+n)}\right] \tag{14.203}$$

$$= \frac{\Gamma(2-q+n)}{\Gamma(2-q)}\frac{[x-a]^{1-q}}{\Gamma(2-q+n)} \tag{14.204}$$

$$= \frac{[x-a]^{1-q}}{\Gamma(2-q)}. \tag{14.205}$$

This is now valid for all q.

14.5.3   Differintegral of $[x-a]^p$ $(p > -1)$

Here there is no restriction on p other than $p > -1$. We start with the Riemann-Liouville definition and write

$$\frac{d^q[x-a]^p}{[d(x-a)]^q} = \frac{1}{\Gamma(-q)}\int_a^x \frac{[x'-a]^p dx'}{[x-x']^{q+1}}, \quad q < 0. \tag{14.206}$$

When we use the transformation $x' - a = v$, this becomes

$$\frac{d^q[x-a]^p}{[d(x-a)]^q} = \frac{1}{\Gamma(-q)}\int_0^{x-a} \frac{v^p dv}{[x-a-v]^{q+1}}. \tag{14.207}$$

Now we make the transformation $v = (x-a)u$ to write

$$\frac{d^q[x-a]^p}{[d(x-a)]^q} = \frac{(x-a)^{p-q}}{\Gamma(-q)}\int_0^1 u^p(1-u)^{-q-1}du, \quad q < 0. \tag{14.208}$$

Using the definition of the beta function [Eq. (13.151)] and its relation with the gamma functions, we finally obtain

$$\frac{d^q[x-a]^p}{[d(x-a)]^q} = \frac{[x-a]^{p-q}}{\Gamma(-q)}B(p+1,-q) \tag{14.209}$$

$$= \frac{\Gamma(p+1)[x-a]^{p-q}}{\Gamma(p-q+1)}, \tag{14.210}$$

where $q < 0$ and $p > -1$. Actually, we could remove the restriction on q and use Equation (14.210) for all q (see the derivation of the Riemann formula with the substitution $x \to x - a$).

14.5.4 Differintegral of $[1 - x]^p$

To find a formula valid for all p and q values we write

$$1 - x = 1 - a - (x - a) \qquad (14.211)$$

and use the binomial formula to write

$$(1 - x)^p = \sum_{j=0}^{\infty} \frac{\Gamma(p + 1)}{\Gamma(j + 1)\Gamma(p - j + 1)}(-1)^j(1 - a)^{p-j}(x - a)^j. \qquad (14.212)$$

We now use Equation (14.132) and the Riemann formula (14.99), along with the properties of gamma and the beta functions to find

$$\frac{d^q[1 - x]^p}{[d(x - a)]^q} = \frac{(1 - x)^{p-q}}{\Gamma(-q)}B_x(-q, q - p), \quad |x| < 1, \qquad (14.213)$$

where B_x is the incomplete beta function.

14.5.5 Differintegral of $\exp(\pm x)$

We first write the Taylor series of the exponential function as

$$\exp(\pm x) = \sum_{j=0}^{\infty} \frac{[\pm x]^j}{\Gamma(j + 1)} \qquad (14.214)$$

and use the Riemann formula (14.99) to obtain

$$\frac{d^q \exp(\pm x)}{dx^q} = \frac{\exp(\pm x)}{x^q}\gamma^*(-q, \pm x), \qquad (14.215)$$

where γ^* is the incomplete gamma function.

14.5.6 Differintegral of $\ln(x)$

For all values of q the differintegral of $\ln(x)$ is given as

$$\frac{d^q \ln(x)}{dx^q} = \frac{x^{-q}}{\Gamma(1 - q)}[\ln(x) - \gamma - \psi(1 - q)], \qquad (14.216)$$

where γ is the Euler constant, the value of which is 0.5772157, and the $\psi(x)$ function is defined as

$$\psi(x) = \frac{1}{\Gamma(x)}\frac{d\Gamma(x)}{dx}. \qquad (14.217)$$

14.5.7 Some Semiderivatives and Semi-integrals

We conclude this section with a table of the frequently used semiderivatives and semi-integrals of some functions:

f	$d^{\frac{1}{2}}f/[dx]^{\frac{1}{2}}$	$d^{-\frac{1}{2}}f/[dx]^{-\frac{1}{2}}$
C	$C/\sqrt{\pi x}$	$2C\sqrt{x/\pi}$
$1/\sqrt{x}$	0	$\sqrt{\pi}$
$\sqrt{x}$	$\sqrt{\pi}/2$	$x\sqrt{\pi}/2$
x	$2\sqrt{x/\pi}$	$\frac{4}{3}x^{3/2}\sqrt{\pi}$
$x^{\mu} \ (\mu > -1)$	$[\Gamma(\mu+1)/\Gamma(\mu+1/2)]x^{\mu-1/2}$	$\Gamma(\mu+1)/\Gamma(\mu+3/2)]x^{\mu+1/2}$
$\exp(x)$	$1/\sqrt{\pi x} + \exp(x)\operatorname{erf}(\sqrt{x})$	$\exp(x)\operatorname{erf}(\sqrt{x})$
$\ln x$	$\ln(4x)/\sqrt{\pi x}$	$2\sqrt{\pi/x}[\ln(4x) - 2]$
$\exp(x)\operatorname{erf}(\sqrt{x})$	$\exp(x)$	$\exp(x) - 1$

14.6 MATHEMATICAL TECHNIQUES WITH DIFFERINTEGRALS

14.6.1 Laplace Transform of Differintegrals

The Laplace transform of a differintegral is defined as

$$\mathcal{L}\left\{\frac{d^q f}{dx^q}\right\} = \int_0^\infty \exp(-sx)\frac{d^q f}{dx^q}dx. \qquad (14.218)$$

When q takes integer values, the Laplace transforms of derivatives and integrals are given as

$$\mathcal{L}\left\{\frac{d^q f}{dx^q}\right\} = s^q\mathcal{L}\{f\} - \sum_{k=0}^{q-1} s^{q-1-k}\frac{d^q f}{dx^q}(0), \ q = 1, 2, 3..., \qquad (14.219)$$

$$\mathcal{L}\left\{\frac{d^q f}{dx^q}\right\} = s^q\mathcal{L}\{f\}, \ q = 0, -1, -2, ... \ . \qquad (14.220)$$

We can unify these equations as

$$\mathcal{L}\left\{\frac{d^n f}{dx^n}\right\} = s^n\mathcal{L}\{f\} - \sum_{k=0}^{n-1} s^k\frac{d^{n-1-k}f}{dx^{q-1-k}}(0), \quad n = 0, \pm1, \pm2, \pm3, ... \ .$$

$$(14.221)$$

In this equation we can replace the upper limit in the sum by any number greater than $n-1$. We are now going to show that this expression is generalized for all q values as

$$\mathcal{L}\left\{\frac{d^q f}{dx^q}\right\} = s^q \mathcal{L}\{f\} - \sum_{k=0}^{n-1} s^k \frac{d^{q-1-k} f}{dx^{q-1-k}}(0), \qquad (14.222)$$

where n is an integer satisfying the inequality $n-1 < q \leq n$.

We first consider the $q < 0$ case. We write the Riemann-Liouville definition as

$$\frac{d^q f}{dx^q} = \frac{1}{\Gamma(-q)} \int_0^x \frac{f(x')dx'}{[x-x']^{q+1}}, \qquad q < 0, \qquad (14.223)$$

and use the convolution theorem

$$\mathcal{L}\left\{\int_0^\infty f_1(x-x')f_2(x')\right\} = \mathcal{L}\{f_1\}\mathcal{L}\{ f_2\}, \qquad (14.224)$$

where we take

$$f_1(x) = x^{-q-1} \quad \text{and} \quad f_2(x) = f(x)$$

to write

$$\mathcal{L}\left\{\frac{d^q f}{dx^q}\right\} = \frac{1}{\Gamma(-q)}\mathcal{L}\{x^{-1-q}\}\mathcal{L}\{ f\} \qquad (14.225)$$
$$= s^q \mathcal{L}\{f\}.$$

For the $q < 0$ values the sum in Equation (14.222) is empty. Thus we see that the expression in Equation (14.222) is valid for all $q < 0$ values.

For the $q > 0$ case we write the condition [Eq. (14.40)] that the Grünwald and Riemann-Liouville definitions satisfy as

$$\frac{d^n}{dx^n}\frac{d^{q-n} f}{dx^{q-n}} = \frac{d^q f}{dx^q}, \qquad (14.226)$$

where n is positive integer, and choose n as

$$n-1 < q < n. \qquad (14.227)$$

We now take the Laplace transform of Equation (14.226) to find

$$\mathcal{L}\left\{\frac{d^q f}{dx^q}\right\} = \mathcal{L}\left\{\frac{d^n}{dx^n}\left[\frac{d^{q-n} f}{dx^{q-n}}\right]\right\}$$

$$= s^n \mathcal{L}\left\{\frac{d^{q-n} f}{dx^{q-n}}\right\} - \sum_{k=0}^{n-1} s^k \frac{d^{n-1-k}}{dx^{n-1-k}}\left[\frac{d^{q-n} f}{dx^{q-n}}\right](0), \qquad q-n < 0. \qquad (14.228)$$

Since $q - n < 0$, from Equations (14.223-225) the first term on the right-hand side becomes $s^q \mathcal{L}\{f\}$. When $n - 1 - k$ takes integer values, the term,

$$\frac{d^{n-1-k}}{dx^{n-1-k}}\left[\frac{d^{q-n}f}{dx^{q-n}}\right](0),$$

under the summation sign, with the $q - n < 0$ condition and the composition formula [Eq. (14.226)], can be written as

$$\frac{d^{q-1-k}f}{dx^{q-1-k}}(0),$$

which leads us to

$$\mathcal{L}\left\{\frac{d^q f}{dx^q}\right\} = s^q \mathcal{L}\{f\} - \sum_{k=0}^{n-1} s^k \frac{d^{q-1-k}f}{dx^{q-1-k}}(0), \quad 0 < q \neq 1, 2, 3... \ . \quad (14.229)$$

We could satisfy this equation for the integer values of q by taking the condition $n - 1 < q \leq n$ instead of Equation (14.227).

Example 14.4. *Heat transfer equation:* We consider the heat transfer equation for a semi-infinite slab:

$$\frac{\partial T(x,t)}{\partial t} = K\frac{\partial^2 T(x,t)}{\partial x^2}, \ t \in [0, \infty], \ x \in [0, \infty]. \quad (14.230)$$

K is the heat transfer coefficient, which depends on conductivity, density, and the specific heat of the slab. We take $T(x,t)$ as the difference of the local temperature from the ambient temperature; t is the time, and x is the distance from the surface of interest. As the boundary conditions we take

$$T(x,0) = 0 \quad (14.231)$$

and

$$T(\infty, t) = 0. \quad (14.232)$$

Taking the Laplace transform of Equation (14.230) with respect to t we get

$$\mathcal{L}\left\{\frac{\partial T(x,t)}{\partial t}\right\} = K\mathcal{L}\left\{\frac{\partial^2 T(x,t)}{\partial x^2}\right\}, \quad (14.233)$$

$$s\widetilde{T}(x,s) - T(x,0) = K\frac{\partial^2 \widetilde{T}(x,s)}{\partial x^2}, \quad (14.234)$$

$$s\widetilde{T}(x,s) = K\frac{\partial^2 \widetilde{T}(x,s)}{\partial x^2}. \tag{14.235}$$

Using the boundary condition [Eq. (14.232)] we can immediately write the solution, which is finite for all x as

$$\widetilde{T}(x,s) = F(s)e^{-x\sqrt{s/K}}, \tag{14.236}$$

In this solution $F(s)$ is the Laplace transform of the boundary condition $T(0,t)$:

$$F(s) = \mathcal{L}\left\{T(0,t)\right\}, \tag{14.237}$$

which remains unspecified. In most of the engineering applications we are interested in the heat flux, which is given as

$$J(x,t) = -k\frac{\partial T(x,t)}{\partial x}, \tag{14.238}$$

where k is the conductivity. In particular, the surface flux given by

$$J(0,t) = -k\left.\frac{\partial T(x,t)}{\partial x}\right|_{x=0}. \tag{14.239}$$

For the surface flux we differentiate Equation (14.236) with respect to x as

$$\frac{\partial \widetilde{T}(x,s)}{\partial x} = -\sqrt{s/K}F(s)e^{-x\sqrt{s}} \tag{14.240}$$

and eliminate $F(s)$ by using Equation (14.236) to get

$$\frac{\partial \widetilde{T}(x,s)}{\partial x} = -\sqrt{s/K}\widetilde{T}(x,s). \tag{14.241}$$

We now use Equation (14.229) and choose $n = 1$:

$$\mathcal{L}\left\{\frac{d^{1/2}T(x,t)}{dt^{1/2}}\right\} = s^{1/2}\mathcal{L}\{T(x,t)\} - \frac{d^{-1/2}T(x,0)}{dt^{-1/2}}. \tag{14.242}$$

Using the other boundary condition [Eq. (14.231)] the second term on the right-hand side is zero; thus we write

$$\mathcal{L}\left\{\frac{d^{1/2}T(x,t)}{dt^{1/2}}\right\} = s^{1/2}\mathcal{L}\{T(x,t)\} \tag{14.243}$$

$$= s^{1/2}\widetilde{T}(x,s).$$

Substituting Equation (14.241) into this equation and taking the inverse Laplace transform we get

$$\frac{d^{1/2}T(x,t)}{dt^{1/2}} = -\sqrt{K}\frac{\partial T(x,t)}{\partial x}. \tag{14.244}$$

Using this in the surface heat flux expression we finally obtain

$$J(0,t) = -k \left. \frac{\partial T(x,t)}{\partial x} \right|_{x=0} \tag{14.245}$$

$$= \frac{k}{\sqrt{K}} \frac{d^{1/2}T(0,t)}{dt^{1/2}}. \tag{14.246}$$

The importance of this result is that the surface heat flux is given in terms of the surface temperature distribution, that is $T(0,t)$, which is experimentally easier to measure.

14.6.2   Extraordinary Differential Equations

An equation composed of the differintegrals of an unknown function is called an extraordinary differential equation. Naturally, solutions of such equations involve some constants and integrals. A simple example of such an equation can be given as

$$\frac{d^Q f}{dx^Q} = F(x). \tag{14.247}$$

Here Q is any number, $F(x)$ is a given function, and $f(x)$ is the unknown function. For simplicity we have taken the lower limit a as zero. We would like to write the solution of this equation simply as

$$f(x) = \frac{d^{-Q}F}{dx^{-Q}}. \tag{14.248}$$

However, we have seen that the operators $\frac{d^{-Q}}{dx^{-Q}}$ and $\frac{d^Q}{dx^Q}$ are not the inverses of each other, unless condition

$$f - \frac{d^{-Q}}{dx^{-Q}} \frac{d^Q f}{dx^Q} = 0 \tag{14.249}$$

is satisfied. It is for this reason that extraordinary differential equations are in general much more difficult to solve.

A commonly encountered equation in science is

$$\frac{dx(t)}{dt} = -\alpha x^n. \tag{14.250}$$

For $n = 1$ the solution is given as an exponential function

$$x(t) = x_0 \exp(-\alpha t). \tag{14.251}$$

For $n \neq 1$ solutions are given with a power dependence as

$$x(t)^{1-n} = \alpha(n-1)(t-t_0). \tag{14.252}$$

Solutions of

$$\frac{d^n x}{dt^n} = (\mp \alpha)^n x, \ \ n = 1, 2, \dots \tag{14.253}$$

are the Mittag-Leffler functions

$$x(t) = x_0 E_n[(\mp \alpha t)^n], \tag{14.254}$$

which correspond to extrapolations between exponential and power dependence. A fractional generalization of Equation (14.253) as

$$\frac{d^{-q}N(t)}{[d(t)]^{-q}} = -\alpha^{-q}[N(t) - N_0], \ \ q > 0,$$

is frequently encountered in kinetic theory, and its solutions are given in terms of the Mittag-Leffler functions as

$$N(t) = N_0 E_q(-\alpha^q t^q).$$

14.6.3 Mittag-Leffler Functions

Mittag-Leffler functions are encountered in many different branches of science such as; biology, chemistry, kinetic theory, and Brownian motion. They are defined by the series (Fig. 14.5)

$$E_\alpha(x) = \sum_{k=0}^{\infty} \frac{x^k}{\Gamma(\alpha k + 1)}, \ \ \alpha > 0. \tag{14.255}$$

For some α values Mittag-Leffler functions are given as

$$
\begin{aligned}
E_0(x) &= \frac{1}{1-x} \\
E_1(x) &= \exp(x) \\
E_2(x) &= \cosh(\sqrt{x}) \\
E_3(x) &= \frac{1}{3}[\exp(\sqrt[3]{x}) + 2\exp(-\sqrt[3]{x}/2)\cos(\frac{\sqrt{3}}{2}\sqrt[3]{x})] \\
E_4(x) &= \frac{1}{2}[\cos(\sqrt[4]{x}) + \cosh(\sqrt[4]{x})]
\end{aligned}
\tag{14.256}
$$

A frequently encountered Mittag-Leffler function is given for the $q = 1/2$ value and can be written in terms of the error function as

$$E_{1/2}(\pm\sqrt{x}) = \exp(x)\left[1 + \text{erf}(\pm\sqrt{x})\right], \ \ x > 0. \tag{14.257}$$

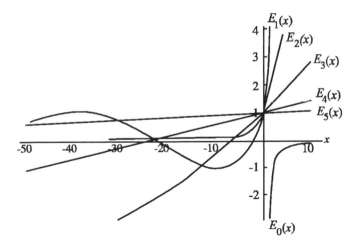

Fig. 14.5   Mittag-Leffler functions

14.6.4   Semidifferential Equations

In applications we frequently encounter extraordinary differential equations like

$$\frac{d^3 f}{dx^3} + \sin(x)\frac{d^{3/2} f}{dx^{3/2}} = \exp(2x), \tag{14.258}$$

$$\frac{d^{-1/2} f}{dx^{-1/2}} - 4\frac{d^{3/2} f}{dx^{3/2}} + 5f = x, \tag{14.259}$$

which involves semiderivatives of the unknown function. However, an equation

$$\frac{d^4 f}{dx^4} - \frac{d^3 f}{dx^3} + 5f = \frac{d^{3/2} F}{dx^{3/2}}, \tag{14.260}$$

where $F(x)$ is a known function is not considered to be a semidifferential equation.

Example 14.5. *Semidifferential equation solution:* Consider the following semidifferential equation:

$$\frac{d^{1/2}}{dx^{1/2}} f + af = 0, \quad a = \text{constant}. \tag{14.261}$$

Applying the $\dfrac{d^{1/2}}{dx^{1/2}}$ operator to this equation and using the composition rule [Eq. (14.158)] with Equation (14.162), and with m taken as one we get

$$\frac{df}{dx} - C_1 x^{-3/2} + a\frac{d^{1/2}}{dx^{1/2}}f = 0. \qquad (14.262)$$

Using Equation (14.261) again we find

$$\frac{df}{dx} - a^2 f = C_1 x^{-3/2}. \qquad (14.263)$$

This is a first-order ordinary differential equation the solution of which is given as

$$f(x) = C \exp(a^2 x) + C_1 \exp(a^2 x) \int_0^x \exp(-a^2 x') x'^{-3/2} dx'. \qquad (14.264)$$

What is new is that the solution involves two integration constants and a divergent integral. However, this integral can be defined by using the incomplete gamma function $\gamma^*(c, x)$, which is defined as

$$\gamma^*(c, x) = \frac{c^{-x}}{\Gamma(x)} \int_0^c x'^{x-1} \exp(-x') dx' \qquad (14.265)$$

$$= \exp(-x) \sum_{j=0}^\infty \frac{x^j}{\Gamma(j + c + 1)}, \qquad (14.266)$$

where $\gamma^*(c, x)$ is a single-valued and analytic function of c and x. Using the relations

$$\gamma^*(c - 1, x) = x\gamma^*(c, x) + \frac{\exp(-x)}{\Gamma(c)} \qquad (14.267)$$

and

$$\gamma^*(\frac{1}{2}, x) = \frac{\mathrm{erf}(\sqrt{x})}{\sqrt{x}}, \qquad (14.268)$$

we can determine the value of the divergent integral:

$$I = \int_0^x \exp(-a^2 x') x'^{-3/2} dx' \qquad (14.269)$$

to be

$$I = -2a\sqrt{\pi}\,\mathrm{erf}(\sqrt{a^2 x}) - 2a\frac{\exp(-a^2 x)}{\sqrt{a^2 x}}. \qquad (14.270)$$

Substituting this into Equation (14.264), we find the solution

$$f(x) = C \exp(a^2 x) + C_1 \exp(a^2 x)\left[-2a\sqrt{\pi}\,\mathrm{erf}(\sqrt{a^2 x}) - \frac{2a\exp(-a^2 x)}{\sqrt{a^2 x}}\right] \qquad (14.271)$$

$$= C \exp(a^2 x) - 2a\sqrt{\pi}C_1 \exp(a^2 x)\,\mathrm{erf}(\sqrt{a^2 x}) - \frac{2aC_1}{\sqrt{a^2 x}}. \qquad (14.272)$$

This solution still contains two arbitrary constants. To check that it satisfies the semidifferential Equation (14.261), we first find its semiderivative as

$$\frac{d^{1/2}f}{dx^{1/2}} = a\left[\frac{C}{\sqrt{\pi a^2 x}} + C\exp(a^2 x)\ \mathrm{erf}(\sqrt{a^2 x}) - 2a\sqrt{\pi}C_1\exp(a^2 x)\right],$$

where we have used the scale transformation formula [Eq. 14.130)] and the semiderivative given in Section 14.5.7. Substituting the above equation into Equation (14.261) gives

$$a\left[\frac{C}{\sqrt{\pi a^2 x}} + C\exp(a^2 x)\ \mathrm{erf}(\sqrt{a^2 x}) - 2a\sqrt{\pi}C_1\exp(a^2 x)\right] \qquad (14.273)$$
$$= -a\left[-\frac{2aC_1}{\sqrt{a^2 x}} - 2a\sqrt{\pi}C_1\exp(a^2 x)\ \mathrm{erf}(\sqrt{a^2 x}) + C\exp(a^2 x)\right],$$

thus we obtain a relation between C and C_1 as

$$\frac{C}{\sqrt{\pi}} = 2aC_1. \qquad (14.274)$$

Now the final solution is obtained as

$$f(x) = C\exp(a^2 x)\ \left[1 - \mathrm{erf}(a\sqrt{x})\right] - \frac{C}{a^2\sqrt{\pi x}}. \qquad (14.275)$$

14.6.5 Evaluating Definite Integrals by Differintegrals

We have seen how analytic continuation and complex integral theorems can be used to evaluate some definite integrals. Fractional calculus can also be used for evaluating some definite integrals. Using the transformation

$$x' = x - x\lambda \qquad (14.276)$$

and the Riemann-Liouville definition Equations (14.70–71), we can write the differintegral of the function x^q as

$$\frac{d^q x^q}{dx^q} = \frac{1}{\Gamma(-q)}\int_0^1 \frac{[1-\lambda]^q d\lambda}{\lambda^{q+1}} \qquad (14.277)$$
$$= \Gamma(q+1),$$

where $-1 < q < 0$ and we have used Equation (14.210) to write $d^q x^q/dx^q = \Gamma(q+1)$. Making one more transformation,

$$t = -\ln(\lambda), \qquad (14.278)$$

we obtain the following definite integral:

$$\int_0^\infty \frac{dt}{[\exp(t)-1]^{-q}} \qquad (14.279)$$
$$= \Gamma(-q)\Gamma(q+1),$$
$$= -\pi\csc(q\pi), \quad -1 < q < 0. \qquad (14.280)$$

We can also use the transformation [Eq. (14.276)] in the Riemann-Liouville definition for an arbitrary function to write (Oldham and Spanier)

$$\int_0^1 f(x - x\lambda)d(\lambda^{-q}) = \Gamma(1 - q)x^q \frac{d^q f(x)}{dx^q}, \quad q < 0. \tag{14.281}$$

If we also make the replacements $\lambda^{-q} \to t$ and $-1/q \to p$ (p is positive but does not have to be an integer) to obtain the formula

$$\int_0^1 f(x - xt^p)dt = \Gamma\left(\frac{p+1}{p}\right) x^{-1/p} \frac{d^{-1/p} f(x)}{dx^{-1/p}}, \quad p > 0, \tag{14.282}$$

this is very useful in the evaluation of some definite integrals. As a special case we may choose $x = 1$ to write

$$\int_0^1 f(1 - t^p)dt = \Gamma\left(\frac{p+1}{p}\right) \frac{d^{-1/p} f(x)}{dx^{-1/p}} \bigg|_{x=1}. \tag{14.283}$$

Example 14.6. *Evaluation of some definite integrals by differintegrals:*
Using differintegrals we evaluate

$$\int_0^1 \exp(2 - 2t^{2/3})dt. \tag{14.284}$$

Using formula (14.281) with $x = 2$ and $p = 2/3$ along with Equation (14.215) we find

$$\int_0^1 \exp(2 - 2t^{2/3})dt = \Gamma\left(\frac{5}{2}\right) 2^{-3/2} \frac{d^{-3/2}(\exp x)}{dx^{-3/2}} \bigg|_{x=2}$$

$$= \Gamma\left(\frac{5}{2}\right) 2^{-3/2} \left[\frac{\exp(x)}{x^{-3/2}} \gamma^*(\frac{3}{2}, x)\right]_{x=2}$$

$$= \left(3\sqrt{\pi}e^2/4\right) \gamma^*(\frac{3}{2}, 2). \tag{14.285}$$

In 1972 Osler gave the integral version of the Leibniz rule [Eq. (14.177)], which can be useful in evaluating some definite integrals as

$$\frac{d^q[fg]}{dx^q} = \int_{-\infty}^\infty \left(\begin{array}{c} q \\ \lambda + \gamma \end{array}\right) \frac{d^{q-\gamma-\lambda}f}{dx^{q-\gamma-\lambda}} \frac{d^{\gamma+\lambda}g}{dx^{\gamma+\lambda}} d\lambda, \tag{14.286}$$

where γ is any constant.

Example 14.7. *Evaluation of definite integrals by differintegrals:*
In the Osler formula we may choose $f = x^\alpha$, $g = x^\beta$, $\gamma = 0$ to write

$$\frac{d^q[x^{\alpha+\beta}]}{dx^q} = \int_{-\infty}^\infty \frac{\Gamma(q+1)\Gamma(\alpha+1)\Gamma(\beta+1)x^{\alpha-q+\lambda+\beta-\lambda}d\lambda}{\Gamma(q-\lambda+1)\Gamma(\lambda+1)\Gamma(\alpha-q+\lambda+1)\Gamma(\beta-\lambda+1)}.$$

Using the derivative [Eq. (14.210)]

$$\frac{d^q[x^{\alpha+\beta}]}{dx^q} = \frac{\Gamma(\alpha+\beta+1)x^{\alpha+\beta-q}}{\Gamma(\alpha+\beta-q+1)},$$

and after simplification, we obtain the definite integral

$$\int_{-\infty}^{\infty} \frac{\Gamma(q+1)\Gamma(\alpha+1)\Gamma(\beta+1)d\lambda}{\Gamma(q-\lambda+1)\Gamma(\lambda+1)\Gamma(\alpha-q+\lambda+1)\Gamma(\beta-\lambda+1)}$$
$$= \frac{\Gamma(\alpha+\beta+1)}{\Gamma(\alpha+\beta-q+1)}. \tag{14.287}$$

Furthermore, if we set $\beta = 0$ and $\alpha - q + 1 = 1$ so that we can use the relation

$$\Gamma(\lambda+1)\Gamma(1-\lambda) = \frac{\lambda\pi}{\sin\lambda\pi}$$

to obtain the following useful result:

$$\int_{-\infty}^{\infty} \frac{\sin(\pi\lambda)d\lambda}{\lambda\Gamma(\lambda+1)\Gamma(q+1-\lambda)} = \frac{\pi}{\Gamma(q+1)}. \tag{14.288}$$

14.6.6   Evaluation of Sums of Series by Differintegrals

In 1970 Osler gave the summation version of the Leibniz rule, which is very useful in finding sums of infinite series:

$$\frac{d^q[u(x)v(x)]}{dx^q} = \sum_{n=-\infty}^{\infty} \binom{q}{n+\gamma} \frac{d^{q-n-\gamma}u}{dx^{q-n-\gamma}} \frac{d^{\gamma+n}v}{dx^{\gamma+n}} \tag{14.289}$$
$$= \sum_{n=-\infty}^{\infty} \frac{\Gamma(q+1)}{\Gamma(q-\gamma-n+1)\Gamma(\gamma+n+1)} \frac{d^{q-\gamma-n}u}{dx^{q-\gamma-n}} \frac{d^{\gamma+n}v}{dx^{\gamma+n}},$$

where γ is any constant..

Example 14.8. *Evaluation of sums of series by differintegrals:* In the above formula, we choose $u = x^\alpha$, $v = x^\beta$, $\gamma = 0$ and use Equation (14.210) to obtain the sum

$$\sum_{n=-\infty}^{\infty} \frac{\Gamma(q+1)\Gamma(\alpha+1)\Gamma(\beta+1)}{\Gamma(q-n+1)\Gamma(n+1)\Gamma(\alpha-q+n+1)\Gamma(\beta-n+1)}$$
$$= \frac{\Gamma(\alpha+\beta+1)}{\Gamma(\alpha+\beta-q+1)}. \tag{14.290}$$

Furthermore, we set $\alpha = -1/2$, $\beta = 1/2$, $q = -1/2$ to get

$$\sum_{n=0}^{\infty} \frac{[(2n)!]^2}{2^{4n}(n!)^4(1-2n)} = \frac{2}{\pi}.$$

14.6.7   Special Functions Expressed as Differintegrals

Using (14.289) we can also express hypergeometric functions and some special functions as differintegrals (Osler 1970):

Hypergeometric Functions :

$$F(\alpha, \beta, \gamma, x) = \frac{\Gamma(\gamma)x^{1-\gamma}}{\Gamma(\beta)} \frac{d^{\beta-\gamma}}{dx^{\beta-\gamma}} \left(\frac{x^{\beta-1}}{(1-x)^{\alpha}} \right)$$

Confluent hypergeometric Functions :

$$M(\alpha, \gamma, x) = \frac{\Gamma(\gamma)x^{1-\gamma}}{\Gamma(\alpha)} \frac{d^{\alpha-\gamma}}{dx^{\alpha-\gamma}} (e^x x^{\alpha-1})$$

Bessel Functions :

$$J_{\nu}(x) = \frac{x^{-\nu}}{2^{\nu}\sqrt{\pi}} \frac{d^{-\nu-1/2}}{d(x^2)^{-\nu-1/2}} \left(\frac{\cos x}{x} \right)$$

Legendre Polynomials :

$$P_{\nu}(x) = \frac{1}{\Gamma(\nu+1)2^{\nu}} \frac{d^{\nu}}{d(1-x)^{\nu}} (1-x^2)^{\nu}$$

Incomplete gamma Function :

$$\gamma^*(\alpha, x) = \Gamma(\alpha)e^{-x} \frac{d^{-\alpha}e^x}{dx^{-\alpha}}$$

14.7   APPLICATIONS OF DIFFERINTEGRALS IN SCIENCE AND ENGINEERING

14.7.1   Continuous Time Random Walk (CTRW)

We have seen that the diffusion equation is given as

$$\frac{\partial c(\overrightarrow{r}, t)}{\partial t} = -\overrightarrow{\nabla} \cdot \overrightarrow{J}(\overrightarrow{r}, t), \tag{14.291}$$

where $\overrightarrow{J}(\overrightarrow{r}, t)$ represents the current density and $c(\overrightarrow{r}, t)$ is the concentration. As a first approximation we can assume a linear relation between $\overrightarrow{J}$ and the gradient of the concentration as

$$\overrightarrow{J} = -k\overrightarrow{\nabla}c, \quad k \text{ is the diffusion constant.} \tag{14.292}$$

This gives the partial differential equation to be solved for $c(\overrightarrow{r}, t)$ as

$$\frac{\partial c(\overrightarrow{r}, t)}{\partial t} = k\overrightarrow{\nabla}^2 c(\overrightarrow{r}, t). \tag{14.293}$$

To prove the molecular structure of matter, Einstein studied the random motion of particles in suspension in a fluid. This motion is also known as Brownian motion and results from the random collisions of the fluid molecules with the particles in suspension. Diffusion is basically many particles undergoing Brownian motion at the same time. Hence, division of the concentration $c(r, t)$ by the total number of particles gives us the probability of finding a particle at position $\overrightarrow{r}$ and time t as

$$P(\overrightarrow{r}, t) - \frac{1}{N}\rho(\overrightarrow{r}, t).$$

Thus $P(\overrightarrow{r}, t)$ satisfies the same differential equation as the concentration:

$$\frac{\partial P(\overrightarrow{r}, t)}{\partial t} = k\overrightarrow{\nabla}^2 P(\overrightarrow{r}, t). \tag{14.294}$$

In d dimensions and for a particle initially at the origin, the solution of Equation (14.294) is a Gaussian:

$$P(\overrightarrow{r}, t) = \frac{1}{(4\pi kt)^{d/2}} \exp\left(-\frac{r^2}{4kt}\right). \tag{14.295}$$

In Brownian motion, even though the mean distance covered is zero

$$< \overrightarrow{r}(t) >= \int \overrightarrow{r} P(\overrightarrow{r}, t) d\overrightarrow{r} = 0, \tag{14.296}$$

the mean square distance is given as

$$< r^2(t) >= \int r^2 P(\overrightarrow{r}, t) d\overrightarrow{r} = 2kdt. \tag{14.297}$$

This equation sets the scale of the process as

$$< r^2(t) > \propto t. \tag{14.298}$$

Hence the root mean square of the distance covered by a particle is

$$\sqrt{< r^2(t) >} \propto t^{1/2}. \tag{14.299}$$

In Figure 14.6, the first figure shows the distance covered by a Brown particle. In Brownian motion or Einstein random walk, even though the particles are hit by the fluid particles symmetrically, they slowly drift away from the origin with the relation (14.299).

In Einstein's theory of random walk steps are taken with equal intervals. Recently theories in which steps are taken according to a waiting distribution $\Psi(t)$ have been developed. This distribution function essentially carries information about the delays and the traps present in the system. Thus, in a way, memory effects are introduced to the random walk process. These theories are called continuous time random walk (CTRW) theories. In CTRW, if the integral

$$\tau = \int t\Psi(t)dt, \tag{14.300}$$

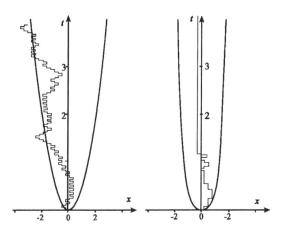

Fig. 14.6 Random walk and CTRW

that gives the average waiting time of the system is finite, we can study the problem by taking the diffusion constant in Equation (14.294) as $a^2/2\tau$. If the average waiting time is divergent, as in

$$\Psi(t) \propto \frac{1}{(1 + t/\tau)^{1+\alpha}}, \quad 0 < \alpha < 1, \tag{14.301}$$

the situation changes dramatically. In CTRW theories $< r^2 >$ in general grows as

$$< r^2 > \propto t^\alpha. \tag{14.302}$$

Cases with $\alpha < 1$ are called subdiffusive and as shown in the second figure in Figure 14.6 the distance covered is less than Einstein's theory. On the other hand, $\alpha > 1$ cases are called superdiffusive and more distance is covered. In CTRW cases, waiting times between steps changes (Sokolov, Klafter, and Blumen, 2002). This is reminiscent of stock markets or earthquakes, where there could be long waiting times before the systems make the next move. For the $\alpha = 1/2$ value, the root mean square distance covered is given by

$$\sqrt{< r^2 >} \propto t^{1/4} \tag{14.303}$$

and the probability distribution $P(\vec{r}, t)$ behaves like the second curve in Figure 14.7, which has a cusp compared to a Gaussian.

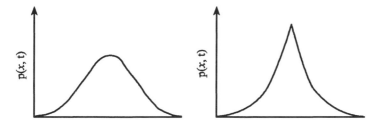

Fig. 14.7 Probability distribution in random walk and CTRW

An important area of application for fractional derivatives is that the extraordinary diffusion phenomenon, studied in CTRW, can also be studied by the differintegral form of Equation (14.294) as

$$\frac{\partial^\alpha P(\overrightarrow{r}, t)}{[\partial(t)]^\alpha} = k_\alpha \overrightarrow{\nabla}^2 P(\overrightarrow{r}, t), \ \alpha \text{ is a real number.} \tag{14.304}$$

Another advantage of this approach is that known solutions for simple cases can be used as seeds to generate solutions for more complicated cases.

14.7.2 Fractional Fokker-Planck Equations

In standard diffusion problems particles move because of their random collisions with the molecules. However, there could also exist a deterministic force due to some external agent like gravity, external electromagnetic fields, etc. Effects of such forces can be included by taking the current density as

$$\overrightarrow{J}(\overrightarrow{r}, t) = -k\overrightarrow{\nabla} P(\overrightarrow{r}, t) + \mu f(\overrightarrow{r}, t) P(\overrightarrow{r}, t). \tag{14.305}$$

The diffusion equation now becomes

$$\frac{\partial P(\overrightarrow{r}, t)}{\partial t} = \overrightarrow{\nabla} \cdot \left[k\overrightarrow{\nabla} P(\overrightarrow{r}, t) - \mu f(\overrightarrow{r}, t) P(\overrightarrow{r}, t) \right], \tag{14.306}$$

which is also called the Fokker-Planck equation.

If we consider particles moving under the influence of a harmonic oscillator potential $U = \frac{1}{2} b x^2$, the probability distribution for particles initially concentrated at some point x_0 is given as shown in Figure 14.8 by the thin curves. When we study the same phenomenon using the fractional Fokker-Planck equation

$$\frac{\partial^\alpha P(\overrightarrow{r}, t)}{[\partial(t)]^\alpha} = \overrightarrow{\nabla} \cdot \left[k\overrightarrow{\nabla} P(\overrightarrow{r}, t) - \mu f(\overrightarrow{r}, t) P(\overrightarrow{r}, t) \right], \tag{14.307}$$

with $\alpha = 1/2$, the general behavior of the probability distribution looks like the thick curves in Figure 14.8. Both distributions become Gaussian for large

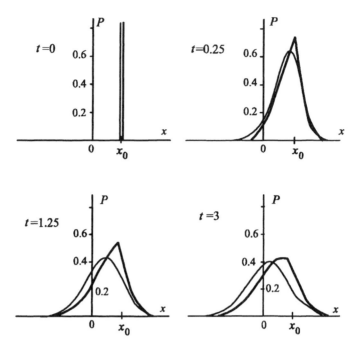

Fig. 14.8   Evolution of probability distribution with harmonic oscillator potential

times. However, for the fractional Fokker-Planck case it not only takes longer but also initially it is very different from a Gaussian and shows CTRW characteristics (Sokolov, Klafter, and Blumen, 2002). For the standard diffusion case the distribution is always a Gaussian.

For the cases known as superdiffusive ($\alpha > 1$), use of fractional derivatives in the Fokker-Planck equation is not restricted to time derivatives, either. Chaotic diffusion and Levy processes, which relate far away points and regions, are also active areas of research where the use of fractional space derivatives is being investigated.

Problems

14.1   Show that the following differintegral is valid for all q values:

$$\frac{d^q[x-a]^p}{[d(x-a)]^q} = \frac{\Gamma(p+1)[x-a]^{p-q}}{\Gamma(p-q+1)}, \qquad p > -1.$$

14.2   Derive the formula [Eq. (14.213)]

$$\frac{d^q[1-x]^p}{[d(x-a)]^q} = \frac{(1-x)^{p-q}}{\Gamma(-q)} B_x(-q, q-p), \quad -1 < x < 1.$$

14.3   Show that the differintegral of an exponential function is given as

$$\frac{d^q \exp(\pm x)}{dx^q} = \frac{\exp(\pm x)}{x^q} \gamma^*(-q, \pm x).$$

14.4   Show that the upper limit $(q-1)$ in the summation

$$\mathcal{L}\left\{\frac{d^q f}{dx^q}\right\} = s^q \mathcal{L}\{f\} - \sum_{k=0}^{q-1} s^k \frac{d^{q-1-k} f}{dx^{q-1-k}}(0), \qquad q = 0, \pm 1, \pm 2, \pm 3...,$$

can be replaced by any number.

14.5   Show that the solution of the extraordinary differential equation

$$\frac{df}{dx} + \frac{d^{1/2} f}{dx^{1/2}} - 2f = 0$$

is given as

$$f(x) = \frac{C}{3}(2\exp(4x)\,\mathrm{erf}\,c(2\sqrt{x}) + \exp(x)\,\mathrm{erf}\,c(-\sqrt{x})),$$

where

$$\mathrm{erf}\,c(x) = 1 - \mathrm{erf}(x).$$

14.6   Show the integral

$$\int_0^1 \sin(\sqrt{1-t^2})\,dt = 0.69123$$

by using differintegrals.

14.7   **Caputo fractional derivative:** Another definition for the fractional derivative was given in the late 1960s by Caputo, for modeling dissipation effects in linear viscoelasticity problems as

$$\left(\frac{d^q f(t)}{dt^q}\right)_C = \frac{1}{\Gamma(1-q)} \int_0^t \frac{df(\tau)}{d\tau} \frac{d\tau}{(t-\tau)^q}, \qquad 0 < q < 1,$$

where C stands for the Caputo derivative.

i) As in the Riemann-Liouville and Grünwald definitions, impose the condition

$$\frac{d}{dt}\left[\frac{d^q f}{dt^q}\right] = \frac{d^{1+q} f}{dt^{1+q}}$$

to show that the Riemann-Liouville and Caputo derivatives are related by

$$\left[\frac{d^q f}{dt^q}\right]_C = \left[\frac{d^q f}{dt^q}\right]_{R-L} - \frac{t^{-q} f(0)}{\Gamma(1-q)}, \quad 0 < q < 1.$$

ii) Using the above result, show that with the Caputo definition the fractional derivative of a constant is zero.

iii) Show that the Laplace transform of the Caputo derivative is

$$\mathcal{L}\left\{\left[\frac{d^q f}{dt^q}\right]_C\right\} = s^q \widetilde{f}(s) - s^{q-1} f(0^+),$$

where $\widetilde{f}(s)$ stands for the Laplace transform of $f(t)$.

iv) Also show that the Laplace transform of the Grünwald (or the Riemann-Liouville) definition of differintegral is

$$\mathcal{L}\left\{\left[\frac{d^q f}{dt^q}\right]_{R-L}\right\} = s^q \widetilde{f}(s) - \frac{d^{q-1} f}{dx^{q-1}}(0^+), \quad 0 < q < 1.$$

Compare your result with the Laplace transform of the Caputo derivative found above. Because of the difficulty in experimentally defining the value of the initial condition $\frac{d^{q-1} f}{dx^q}(0^+)$, the Caputo definition has also found some use in the literature.

14.8   Show the differintegral

$$\frac{d^q \exp(c_0 - c_1 x)}{[d(x-a)]^q} = \frac{\exp(c_0 - c_1 x)}{[x-a]^q} \gamma^*(-q, -c_1(x-a)).$$

15

INFINITE SERIES

In physics and engineering applications sometimes physical properties are expressed in terms of infinite sums. We also frequently encounter differential or integral equations, which can only be solved by the method of infinite series. Given an infinite series, we first need to check its convergence. In this chapter we start by introducing convergence tests for series of numbers and then extend our discussion to series of functions and in particular to power series. We then introduce some analytic techniques for finding infinite sums. We also discuss asymptotic series and infinite products. In conjunction with the Casimir effect, we show how finite and meaningful results can be obtained from some divergent series in physics by the methods of regularization and renormalization.

15.1 CONVERGENCE OF INFINITE SERIES

We write an infinite series with the general term a_n as

$$\sum_{n=1}^{\infty} a_n = a_1 + a_2 + \cdots . \tag{15.1}$$

Summation of the first N terms is called the Nth partial sum of the series. If the Nth partial sum of a series has the limit

$$\lim_{N \to \infty} \left(\sum_{n=1}^{N} a_n \right) \to S, \tag{15.2}$$

we say the series is convergent and write the infinite series as

$$\sum_{n=1}^{\infty} a_n = S. \tag{15.3}$$

When S is infinity we say the series is divergent. When a series is not convergent, it is divergent. The nth term of a convergent series always satisfies the limit

$$\lim_{n \to \infty} a_n \to 0. \tag{15.4}$$

However, the converse is not true.

Example 15.1. *Harmonic series:* Even though the nth term of the harmonic series,

$$\sum_{n=1}^{\infty} a_n, \quad a_n = \frac{1}{n}, \tag{15.5}$$

goes to zero as $n \to \infty$, the series diverges.

15.2 ABSOLUTE CONVERGENCE

If a series constructed by taking the absolute values of the terms of a given series as

$$\sum_{n=1}^{\infty} |a_n| \tag{15.6}$$

is convergent, we say the series is absolutely convergent. An absolutely convergent series is also convergent, but the converse is not true. Series that are convergent but not absolutely convergent are called **conditionally convergent** series. In working with series absolute convergence is a very important property.

Example 15.2. *Conditionally convergent series:* The series

$$1 - \frac{1}{2} + \frac{1}{3} - \cdots = \sum_{n=1}^{\infty} (-1)^{n+1} \frac{1}{n} \tag{15.7}$$

converges to $\ln 2$. However, since it is not absolutely convergent it is only conditionally convergent.

15.3   CONVERGENCE TESTS

There exist a number of tests for checking the convergence of a given series. In what follows we give some of the most commonly used tests for convergence. The tests are ordered in increasing level of complexity. In practice one starts with the simplest test and, if the test fails, moves on to the next one. In the following tests we either consider series with positive terms or take the absolute value of the terms; hence we check for absolute convergence.

15.3.1   Comparison Test

The simplest test for convergence is the comparison test. We compare a given series term by term with another series convergence or divergence of which has been established. Let two series with the general terms a_n and b_n be given. For all $n \geq 1$ if $|a_n| \leq |b_n|$ is true and if the series $\sum_{n=1}^{\infty} |b_n|$ is convergent, then the series $\sum_{n=1}^{\infty} a_n$ is also convergent. Similarly, if $\sum_{n=1}^{\infty} a_n$ is divergent, then the series $\sum_{n=1}^{\infty} |b_n|$ is also divergent.

Example 15.3. *Comparison test:* Consider the series with the general term $a_n = n^{-p}$ where $p = 0.999$. We compare this series with the harmonic series which has the general term $b_n = n^{-1}$. Since for $n \geq 1$ we can write $n^{-1} < n^{-0.999}$ and since the harmonic series is divergent, we also conclude that the series $\sum_{n=1}^{\infty} n^{-p}$ is divergent.

15.3.2   Ratio Test

For the series $\sum_{n=1}^{\infty} a_n$, let $a_n \neq 0$ for all $n \geq 1$. When we find the limit

$$\lim_{n \to \infty} \left| \frac{a_{n+1}}{a_n} \right| = r, \tag{15.8}$$

for $r < 1$ the series is convergent, for $r > 1$ the series is divergent, and for $r = 1$ the test is inconclusive.

15.3.3   Cauchy Root Test

For the series $\sum_{n=1}^{\infty} a_n$, when we find the limit

$$\lim_{n \to \infty} \sqrt[n]{|a_n|} = l, \tag{15.9}$$

for $l < 1$ the series is convergent, for $l > 1$ the series is divergent, and for $l = 1$ the test is inconclusive.

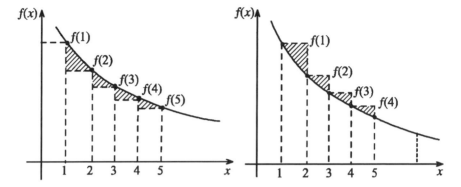

Fig. 15.1 Integral test

15.3.4 Integral Test

Let $a_n = f(n)$ be the general term of a given series with positive terms. If for $n > 1$, $f(n)$ is continuous and a monotonic decreasing function, that is, $f(n+1) < f(n)$, then the series converges or diverges with the integral $\int_1^\infty f(x)\,dx$.

Proof

As shown in Figure 15.1, we can put a lower and an upper bound to the series $\sum_{n=1}^\infty a_n$ as

$$\int_1^{N+1} f(x)\,dx < \sum_{n=1}^N a_n, \tag{15.10}$$

$$\sum_{n=1}^N a_n < \int_1^N f(x)\,dx + a_1. \tag{15.11}$$

From here it is apparent that in the limit as $N \to \infty$, if the integral $\int_1^N f(x)\,dx$ is finite, then the series $\sum_{n=1}^\infty a_n$ is convergent. If the integral diverges, then the series also diverges.

Example 15.4. *Integral test:* Let us consider the Riemann zeta function

$$\xi(s) = 1 + \frac{1}{2^s} + \frac{1}{3^s} + \cdots. \tag{15.12}$$

To use the ratio test we make use of the binomial formula and write

$$\frac{a_{n+1}}{a_n} = \left(\frac{n}{n+1}\right)^s = \left(1 + \frac{1}{n}\right)^{-s},$$

$$\simeq 1 - \frac{s}{n} + \cdots. \tag{15.13}$$

In the limit as $n \to \infty$ this gives $\dfrac{a_{n+1}}{a_n} \to 1$; thus the ratio test fails. However, using the integral test we find

$$\int_1^\infty \frac{dx}{x^s} = \frac{x^{-s+1}}{-s+1}\Big|_1^\infty = \begin{cases} \frac{1}{s-1} & s > 1 \quad \Rightarrow \quad \text{series is convergent} \\ \\ \infty & s < 1 \quad \Rightarrow \quad \text{series is divergent} \end{cases} \tag{15.14}$$

15.3.5 Raabe Test

For a series with positive terms, $a_n > 0$, when we find the limit

$$\lim_{n \to \infty} n\left(\frac{a_n}{a_{n+1}} - 1\right) = m, \tag{15.15}$$

for $m > 1$ the series is convergent and for $m < 1$ the series is divergent. For $m = 1$ the Raabe test is inconclusive.

The Raabe test can also be expressed as follows: Let N be a positive integer independent of n. For all $n \geq N$, if $n\left(\frac{a_n}{a_{n+1}} - 1\right) \geq P > 1$ is true, then the series is convergent and if $n\left(\frac{a_n}{a_{n+1}} - 1\right) \leq 1$ is true, then the series is divergent.

Example 15.5. *Raabe test:* For the series $\sum_{n=1}^\infty \frac{1}{n^2}$ the ratio test is inconclusive. However, using the Raabe test we see that it converges:

$$\lim_{n \to \infty} n\left(\frac{a_n}{a_{n+1}} - 1\right) = \lim_{n \to \infty} n\left(\frac{(n+1)^2}{n^2} - 1\right) \tag{15.16}$$

$$= \lim_{n \to \infty} \left(2 + \frac{1}{n}\right) = 2 > 1.$$

Example 15.6. *Raabe test:* The second form of the Raabe test shows that the harmonic series $\sum_{n=1}^\infty \frac{1}{n}$ is divergent. This follows from the fact that for all n values,

$$n\left(\frac{a_n}{a_{n+1}} - 1\right) = n\left(\frac{n+1}{n} - 1\right) = 1. \tag{15.17}$$

When the available tests fail, we can also use theorems like the Cauchy theorem.

15.3.6 Cauchy Theorem

A given series $\sum_{n=1}^\infty a_n$ with positive decreasing terms $(a_n \geq a_{n+1} \geq \cdots \geq 0)$ converges or diverges with the series

$$\sum_{n=1}^\infty c^n a_{c^n} = ca_c + c^2 a_{c^2} + c^3 a_{c^3} + \cdots \quad (c \text{ an integer}). \tag{15.18}$$

Example 15.7. *Cauchy theorem:* Let us check the convergence of the series

$$\frac{1}{2\ln^\alpha 2} + \frac{1}{3\ln^\alpha 3} + \frac{1}{4\ln^\alpha 4} + \cdots = \sum_{n=2}^{\infty} \frac{1}{n\ln^\alpha n}, \qquad (15.19)$$

by using the Cauchy theorem for $\alpha \geq 0$. Choosing the value of c as two, we construct the series $\sum_{n=1}^{\infty} 2^n a_{2^n} = 2a_2 + 4a_4 + 8a_8 + \cdots$, where the general term is given as

$$2^k a_{2^k} = 2^k \frac{1}{2^k \ln^\alpha 2^k} = \left(\frac{1}{\ln^\alpha 2}\right)\frac{1}{k^\alpha}. \qquad (15.20)$$

Since the series $\frac{1}{\ln^\alpha 2} \sum_{n=1}^{\infty} \frac{1}{n^\alpha}$ converges for $\alpha > 1$, our series is also convergent for $\alpha > 1$. On the other hand, for $\alpha \leq 1$, both series are divergent.

15.3.7 Gauss Test and Legendre Series

Legendre series are given as

$$\sum_{n=0}^{\infty} a_{2n}x^{2n} \quad \text{and} \quad \sum_{n=0}^{\infty} a_{2n+1}x^{2n+1}, \qquad x \in [-1, 1]. \qquad (15.21)$$

Both series have the same recursion relation

$$a_{n+2} = a_n \frac{(n - l)(l + n + 1)}{(n + 1)(n + 2)}, \quad n = 0, 1, \dots . \qquad (15.22)$$

For $|x| < 1$, convergence of both series can be established by using the ratio test. For the even series the general term is given as $u_n = a_{2n}x^{2n}$; hence we write

$$\frac{u_{n+1}}{u_n} = \frac{a_{2n+1}x^{2n+1}}{a_{2n}x^{2n}} = \frac{(2n - l)(2n + l + 1)x^2}{(2n + 1)(2n + 2)}, \qquad (15.23)$$

$$\lim_{n \to \infty} \left|\frac{u_{n+1}}{u_n}\right| = x^2. \qquad (15.24)$$

Using the ratio test we conclude that the Legendre series with the even terms is convergent for the interval $x \in (-1, 1)$. The argument and the conclusion for the other series are exactly the same. However, at the end points the ratio test fails. For these points we can use the Gauss test:

Gauss test:

Let $\sum_{n=0}^{\infty} u_n$ be a series with positive terms. If for $n \geq N$ (N is a given constant) we can write

$$\frac{u_n}{u_{n+1}} \simeq 1 + \frac{\mu}{n} + 0\left(\frac{1}{n^i}\right), \quad i > 0, \qquad (15.25)$$

where $0\left(\frac{1}{n^i}\right)$ means that for a given function $f(n)$ the limit $\lim_{n\to\infty}\{f(n)/\frac{1}{n^i}\}$ is finite, then the $\sum_{n=0}^{\infty} u_n$ series converges for $\mu > 1$ and diverges for $\mu \leq 1$. Note that there is no case here where the test fails.

Example 15.8. *Legendre series:* We now investigate the convergence of the Legendre series at the end points, $x = \pm 1$, by using the Gauss test. We find the required ratio as

$$\frac{u_n}{u_{n+1}} = \frac{(2n+1)(2n+2)}{(2n-l)(2n+l+1)} = \frac{4n^2+6n+2}{4n^2+2n-l(l+1)}, \qquad (15.26)$$

$$\frac{u_n}{u_{n+1}} \simeq 1 + \frac{1}{n} + \frac{l(l+1)(1+n)}{[4n^2+2n-l(l+1)]n}. \qquad (15.27)$$

From the limit

$$\lim_{n\to\infty} \frac{l(l+1)(1+n)}{[4n^2+2n-l(l+1)]n}/(\frac{1}{n^2}) = \frac{l(l+1)}{4}, \qquad (15.28)$$

we see that this ratio is constant and goes as $O(\frac{1}{n^2})$. Since $\mu = 1$ in $\frac{u_n}{u_{n+1}}$, we conclude that the Legendre series (both the even and the odd series) diverge at the end points.

Example 15.9. *Chebyshev series:* The Chebyshev equation is given as

$$(1-x^2)\frac{d^2y}{dx^2} - x\frac{dy}{dx} + n^2y = 0. \qquad (15.29)$$

Let us find finite solutions of this equation in the interval $x \in [-1, 1]$ by using the Frobenius method. We substitute the following series and its derivatives into the Chebyshev equation:

$$y = \sum_{k=0}^{\infty} a_k x^{k+\alpha}, \qquad (15.30)$$

$$y' = \sum_{k=0}^{\infty} a_k(k+\alpha)x^{k+\alpha-1}, \qquad (15.31)$$

$$y'' = \sum_{k=0}^{\infty} a_k(k+\alpha)(k+\alpha-1)x^{k+\alpha-2} \qquad (15.32)$$

to get

$$\sum_{k=0}^{\infty} a_k(k+\alpha)(k+\alpha-1)x^{k+\alpha-2} - \sum_{k=0}^{\infty} a_k(k+\alpha)(k+\alpha-1)x^{k+\alpha}$$

$$- \sum_{k=0}^{\infty} a_k(k+\alpha)x^{k+\alpha} + n^2 \sum_{k=0}^{\infty} a_k x^{k+\alpha} = 0. \qquad (15.33)$$

After rearranging we first get

$$a_0\alpha\,(\alpha-1)\,x^{\alpha-2} + a_1\alpha\,(\alpha+1)\,x^{\alpha-1} + \sum_{k=2}^{\infty} a_k(k+\alpha)(k+\alpha-1)x^{k+\alpha-2}$$

$$+ \sum_{k=0}^{\infty} a_k x^{k+\alpha}\left[n^2 - (k+\alpha)^2\right] = 0 \qquad (15.34)$$

and then

$$a_0\alpha\,(\alpha-1)\,x^{\alpha-2} + a_1\alpha\,(\alpha+1)\,x^{\alpha-1} + \sum_{k=0}^{\infty} a_{k+2}(k+\alpha+2)(k+\alpha+1)x^{k+\alpha}$$

$$+ \sum_{k=0}^{\infty} a_k x^{k+\alpha}\left[n^2 - (k+\alpha)^2\right] = 0. \qquad (15.35)$$

This gives the indicial equation as

$$a_0\alpha\,(\alpha-1) = 0, \ a_0 \neq 0. \qquad (15.36)$$

The remaining coefficients are given by

$$a_1\alpha\,(\alpha+1) = 0 \qquad (15.37)$$

and the recursion relation

$$a_{k+2} = \frac{(k+\alpha)^2 - n^2}{(k+\alpha+2)(k+\alpha+1)}a_k, \qquad k = 0, 1, 2, \dots. \qquad (15.38)$$

Since $a_0 \neq 0$, roots of the indicial equation are 0 and 1. Choosing the smaller root gives the general solution with the recursion relation

$$a_{k+2} = \frac{k^2 - n^2}{(k+2)(k+1)}a_k \qquad (15.39)$$

and the series solution of the Chebyshev equation is obtained as

$$y(x) = a_0 \left(1 - \frac{n^2}{2}x^2 - \frac{n^2\left(2^2 - n^2\right)}{4 \cdot 3 \cdot 2}x^4 - \cdots\right)$$

$$+ a_1 \left(x + \frac{(1 - n^2)}{3 \cdot 2}x^3 + \frac{\left(3^2 - n^2\right)\left(1 - n^2\right)}{5 \cdot 4 \cdot 3 \cdot 2}x^5 + \cdots\right). \qquad (15.40)$$

We now investigate the convergence of these series. Since the argument for both series is the same, we study the series with the general term $u_k = a_{2k}x^{2k}$ and write

$$\left|\frac{u_{k+1}}{u_k}\right| = \left|\frac{a_{2k+2}x^{2k+2}}{a_{2k}x^{2k}}\right| = \left|\frac{a_{2k+2}}{a_{2k}}\right| x^2. \qquad (15.41)$$

This gives us the limit

$$\lim_{k \to \infty} \left| \frac{a_{2k+2}}{a_{2k}} \right| x^2 = x^2. \tag{15.42}$$

Using the ratio test it is clear that this series converges for the interval $(-1, 1)$. However, at the end points the ratio test fails, where we now use the Raabe test. We first evaluate the ratio

$$\lim_{k \to \infty} k \left[\frac{a_{2k}}{a_{2k+2}} - 1 \right] = \lim_{k \to \infty} k \left[\frac{(2k+2)(2k+1)}{(2k)^2 - n^2} - 1 \right] \tag{15.43}$$

$$= \lim_{k \to \infty} k \left[\frac{6k + 2 + n^2}{(2k)^2 - n^2} \right] = \frac{3}{2} > 1 , \tag{15.44}$$

which indicates that the series is convergent at the end points as well. This means that for the polynomial solutions of the Chebyshev equation, restricting n to integer values is an additional assumption, which is not required by the finite solution condition at the end points. The same conclusion is valid for the series with the odd powers.

15.3.8   Alternating Series

For a given series of the form $\sum_{n=1}^{\infty} (-1)^{n+1} a_n$, if a_n is positive for all n, then the series is called an alternating series. In an alternating series for sufficiently large values of n, if a_n is monotonic decreasing or constant and the limit

$$\lim_{n \to \infty} a_n = 0 \tag{15.45}$$

is true, then the series is convergent. This is also known as the **Leibniz rule**.

Example 15.10. *Leibniz rule:* In the alternating series

$$\sum_{n=1}^{\infty} (-1)^{n+1} \frac{1}{n} = 1 - \frac{1}{2} + \frac{1}{3} - \frac{1}{4} + \cdots, \tag{15.46}$$

since $\frac{1}{n} > 0$ and $\frac{1}{n} \to 0$ as $n \to \infty$, the series is convergent.

15.4   ALGEBRA OF SERIES

Absolute convergence is very important in working with series. It is only for absolutely convergent series that ordinary algebraic manipulations (addition, subtraction, multiplication, etc.) can be done without problems:

1. An absolutely convergent series can be rearranged without affecting the sum.

2. Two absolutely convergent series can be multiplied with each other. The result is another absolutely convergent series, which converges to the multiplication of the individual series sums.

All these operations look very natural; however, when applied to conditionally convergent series they may lead to erroneous results.

Example 15.11. *Conditionally convergent series:* The following conditionally convergent series:

$$\sum_{n=1}^{\infty} (-1)^{n+1} \frac{1}{n} = 1 - \frac{1}{2} + \frac{1}{3} - \frac{1}{4} + \cdots$$

$$= 1 - (\frac{1}{2} - \frac{1}{3}) - (\frac{1}{4} - \frac{1}{5}) - \cdots \qquad (15.47)$$

$$= 1 - 0.167 - 0.05 - \cdots , \qquad (15.48)$$

obviously converges to some number less than one, actually to $\ln 2 = 0.693$. We now rearrange this sum as

$$(1 + \frac{1}{3} + \frac{1}{5}) - (\frac{1}{2}) + (\frac{1}{7} + \frac{1}{9} + \frac{1}{11} + \frac{1}{13} + \frac{1}{15}) - (\frac{1}{4})$$

$$+ (\frac{1}{17} + \cdots + \frac{1}{25}) - (\frac{1}{6}) + (\frac{1}{27} + \cdots + \frac{1}{35}) - (\frac{1}{8}) + \cdots , \qquad (15.49)$$

and consider each term in parenthesis as the terms of a new series. Partial sums of this new series are

$$
\begin{array}{ll}
s_1 = 1.5333, & s_2 = 1.0333, \\
s_3 = 1.5218, & s_4 = 1.2718, \\
s_5 = 1.5143, & s_6 = 1.3476, \quad \cdots . \\
s_7 = 1.5103, & s_8 = 1.3853, \\
s_9 = 1.5078, & s_{10} = 1.4078,
\end{array}
\qquad (15.50)
$$

It is now seen that this alternating series added in this order converges to $3/2$. What we have done is very simple. First we added positive terms until the partial sum was equal or just above $3/2$ and then subtracted negative terms until the partial sum fell just below $3/2$. In this process we have neither added nor subtracted anything from the series; we have simply added its terms in a different order.

By a suitable arrangement of its terms a conditionally convergent series can be made to converge to any desired value or even to diverge. This result is also known as the **Riemann theorem**.

15.4.1 Rearrangement of Series

Let us write the partial sum of a double series as

$$s_{nm} = \sum_{i=1}^{n} \sum_{j=1}^{m} a_{ij}. \qquad (15.51)$$

If the limit

$$\lim_{\substack{n\to\infty \\ m\to\infty}} s_{nm} = s \tag{15.52}$$

exists, then we can write

$$\sum_{i=1}^{\infty}\sum_{j=1}^{\infty} a_{ij} = s \tag{15.53}$$

and say that the double series $\sum_{i,j=1}^{\infty} a_{ij}$ is convergent and has the sum s.
When a double sum

$$\sum_{i=0}^{\infty}\sum_{j=0}^{\infty} a_{ij} \tag{15.54}$$

converges absolutely, that is, when $\sum_{i=0}^{\infty}\sum_{j=0}^{\infty} |a_{ij}|$ is convergent, then we can rearrange its terms without affecting the sum.
Let us define new dummy variables q and p as

$$i = q \geq 0 \text{ and } j = p - q \geq 0. \tag{15.55}$$

Now the sum $\sum_{i=0}^{\infty}\sum_{j=0}^{\infty} a_{ij}$ becomes

$$\sum_{i=0}^{\infty}\sum_{j=0}^{\infty} a_{ij} = \sum_{p=0}^{\infty}\sum_{q=0}^{p} a_{q(p-q)}. \tag{15.56}$$

Writing both sums explicitly we get

$$
\begin{aligned}
& a_{00} + a_{01} + a_{02} + \cdots \\
& + a_{10} + a_{11} + a_{12} + \cdots \\
& + a_{20} + a_{21} + a_{22} + \cdots \\
& \vdots \\
& = a_{00} \\
& + a_{01} + a_{10} \\
& + a_{02} + a_{11} + a_{20} \\
& + a_{03} + a_{12} + a_{21} + a_{30} \\
& \vdots
\end{aligned}
\tag{15.57}
$$

Another rearrangement can be obtained by the definitions

$$i = s \geq 0 \text{ and } j = r - 2s \geq 0 , \quad (s \leq \frac{r}{2}), \tag{15.58}$$

as

$$\sum_{i=0}^{\infty} \sum_{j=0}^{\infty} a_{ij} = \sum_{r=0}^{\infty} \sum_{s=0}^{[\frac{r}{2}]} a_{s,r-2s} \tag{15.59}$$

$$= a_{00} + a_{01} + a_{02} + + a_{10} + a_{03} + a_{11} + \cdots . \tag{15.60}$$

15.5   USEFUL INEQUALITIES ABOUT SERIES

Let $\frac{1}{p} + \frac{1}{q} = 1$; then we can state the following useful inequalities about series:

Hölder's Inequality: If $a_n \geq 0$, $b_n \geq 0$, $p > 1$, then

$$\sum_{n=1}^{\infty} a_n b_n \leq \left(\sum_{n=1}^{\infty} a_n^p \right)^{1/p} \cdot \left(\sum_{n=1}^{\infty} b_n^q \right)^{1/q} . \tag{15.61}$$

Minkowski's Inequality: If $a_n \geq 0$, $b_n \geq 0$ and $p \geq 1$, then

$$\left[\sum_{n=1}^{\infty} (a_n + b_n)^p \right]^{1/p} \leq \left(\sum_{n=1}^{\infty} a_n^p \right)^{1/p} + \left(\sum_{n=1}^{\infty} b_n^p \right)^{1/p} . \tag{15.62}$$

Schwarz-Cauchy Inequality: If $a_n \geq 0$, and $b_n \geq 0$, then

$$\left(\sum_{n=1}^{\infty} a_n b_n \right)^2 \leq \left(\sum_{n=1}^{\infty} a_n^2 \right) \cdot \left(\sum_{n=1}^{\infty} b_n^2 \right) . \tag{15.63}$$

Thus, if the series $\sum_{n=1}^{\infty} a_n^2$ and $\sum_{n=1}^{\infty} b_n^2$ converges, then the series $\sum_{n=1}^{\infty} a_n b_n$ also converges.

15.6   SERIES OF FUNCTIONS

We can also define series of functions with the general term $u_n = u_n(x)$. In this case the partial sums S_n are also functions of x :

$$S_n(x) = u_1(x) + u_2(x) + \cdots + u_n(x). \tag{15.64}$$

If $\lim_{n \to \infty} S_n(x) \to S(x)$ is true, then we can write

$$\sum_{n=1}^{\infty} u_n(x) = S(x). \tag{15.65}$$

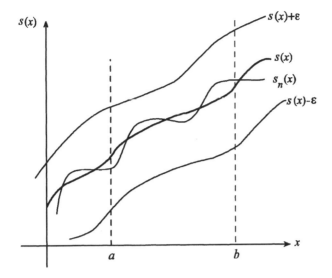

Fig. 15.2 Uniform convergence is very important

In studying the properties of series of functions we need a new concept called the **uniform convergence**.

15.6.1   Uniform Convergence

For a given positive small number ε, if there exists a number N independent of x for $x \in [a, b]$, and if for all $n \geq N$ we can say the inequality

$$|s(x) - s_n(x)| < \varepsilon \tag{15.66}$$

is true, then the series with the general term $u_n(x)$ is uniformly convergent in the interval $[a, b]$. This also means that for a uniformly convergent series and for a given error margin ε, we can always find N number of terms independent of x such that when added the remainder of the series, that is

$$\left| \sum_{i=N+1}^{\infty} u_i(x) \right|, \tag{15.67}$$

is always less than ε for all x in the interval $[a, b]$. Uniform convergence can also be shown as in Figure 15.2.

15.6.2   Weierstrass M-Test

For uniform convergence the most commonly used test is the Weierstrass M or in short the M-test: Let us say that we found a series of numbers

$\sum_{i=1}^{\infty} M_i$, such that for all x in $[a, b]$ the inequality $M_i \geq |u_i(x)|$ is true. Then the uniform convergence of the series of functions $\sum_{i=1}^{\infty} u_i(x)$, in the interval $[a, b]$, follows from the convergence of the series of numbers $\sum_{i=1}^{\infty} M_i$. Note that because the absolute values of $u_i(x)$ are taken, the M-test also checks absolute convergence. However, it should be noted that absolute convergence and uniform convergence are two independent concepts and neither of them implies the other.

Example 15.12 *M-test:* The following series are uniformly convergent, but not absolutely convergent:

$$\sum_{n=1}^{\infty} \frac{(-1)^n}{n + x^4} \quad , \quad -\infty < x < \infty \quad \text{and} \qquad (15.68)$$

$$\sum_{n=1}^{\infty} (-1)^{n-1} \frac{x^n}{n} = \ln(1 + x) \quad , \quad 0 \leq x \leq 1,$$

while the series (the so-called Riemann zeta function)

$$\zeta(x) = \frac{1}{1^x} + \frac{1}{2^x} + \cdots + \frac{1}{n^x} + \cdots \qquad (15.69)$$

converges uniformly and absolutely in the interval $[a, \infty)$, where a is any number greater than one. Because the M-test checks for uniform and absolute convergence together, for conditionally convergent series we can use the Abel test.

15.6.3   Abel Test

Let a series with the general term $u_n(x) = a_n f_n(x)$ be given. If the series of numbers $\sum a_n = A$ is convergent and if the functions $f_n(x)$ are bounded, $0 \leq f_n(x) \leq M$, and monotonic decreasing, $f_{n+1}(x) \leq f_n(x)$, in the interval $[a, b]$, then the series $\sum u_n(x)$ is uniformly convergent in $[a, b]$.

Example 15.13 *Uniform convergence:*   The series

$$\sum_{n=0}^{\infty} (1 - x)x^n \begin{aligned} &= 1, & 0 \leq x < 1, \\ &= 0, & x = 1 \end{aligned} \qquad (15.70)$$

is absolutely convergent but not uniformly convergent in $[0, 1]$.

From the definition of uniform convergence it is clear that any series

$$f(x) = \sum_{n=1}^{\infty} u_n(x), \qquad (15.71)$$

where all $u_n(x)$ are continuous functions, cannot be uniformly convergent in any interval containing a discontinuity of $f(x)$.

15.6.4 Properties of Uniformly Convergent Series

For a uniformly convergent series the following are true:

1. If $u_n(x)$ for all n are continuous, then the series

$$f(x) = \sum_{n=1}^{\infty} u_n(x) \tag{15.72}$$

   is also continuous.

2. Provided $u_n(x)$ are continuous for all n in $[a, b]$, then the series can be integrated as

$$\int_a^b f(x)dx = \int_a^b dx \left[\sum_{n=1}^{\infty} u_n(x) \right] = \sum_{n=1}^{\infty} \int_a^b u_n(x)dx, \tag{15.73}$$

   where the integral sign can be interchanged with the summation sign.

3. If for all n in the interval $[a, b]$, $u_n(x)$ and $\frac{d}{dx}u_n(x)$ are continuous, and the series $\sum_{n=1}^{\infty} \frac{d}{dx}u_n(x)$ is uniformly convergent, then we can differentiate the series term by term as

$$\frac{d}{dx}f(x) = \frac{d}{dx}\left[\sum_{n=1}^{\infty} u_n(x) \right] \tag{15.74}$$

$$= \sum_{n=1}^{\infty} \frac{d}{dx}u_n(x). \tag{15.75}$$

15.7 TAYLOR SERIES

Let us assume that a function has a continuous nth derivative, $f^{(n)}(x)$, in the interval $[a, b]$. Integrating this derivative we get

$$\int_a^x f^{(n)}(x_1)dx_1 = f^{(n-1)}(x_1)\Big|_a^x = f^{(n-1)}(x) - f^{(n-1)}(a). \tag{15.76}$$

Integrating again,

$$\int_a^x \left(\int_a^{x_2} f^{(n)}(x_1)dx_1 \right) dx_2 = \int_a^x \left[f^{(n-1)}(x_2) - f^{(n-1)}(a) \right] dx_2$$

$$= f^{(n-2)}(x) - f^{(n-2)}(a) - (x-a)f^{(n-1)}(a) \tag{15.77}$$

and after n-fold integrations we get

$$\int_a^x \cdots \int_a^x f^{(n)}(x)(dx)^n = f(x) - f(a) - (x-a)f'(a) \tag{15.78}$$
$$- \frac{(x-a)^2}{2!}f''(a)$$
$$\cdots - \frac{(x-a)^{n-1}}{(n-1)!}f^{(n-1)}(a).$$

We now solve this equation for $f(x)$ to write

$$f(x) = f(a) + (x-a)f'(a) + \frac{(x-a)^2}{2!}f''(a) + \cdots$$
$$+ \frac{(x-a)^{n-1}}{(n-1)!}f^{(n-1)}(a) + \int_a^x \cdots \int_a^x f^{(n)}(x)(dx)^n. \tag{15.79}$$

In this equation

$$R_n = \int_a^x \cdots \int_a^x f^{(n)}(x)(dx)^n \tag{15.80}$$

is called the remainder, and it can also be written as

$$R_n = \frac{(x-a)^n}{n!}f^{(n)}(\xi), \quad a \le \xi \le x. \tag{15.81}$$

Note that Equation (15.79) is exact. When the limit

$$\lim_{n \to \infty} R_n = 0 \tag{15.82}$$

is true, then we have a series expansion of the function $f(x)$ in terms of the positive powers of $(x-a)$ as

$$f(x) = \sum_{n=0}^{\infty} \frac{(x-a)^n}{n!} f^{(n)}(a). \tag{15.83}$$

This is called the Taylor series expansion of $f(x)$ about the point $x = a$.

15.7.1 Maclaurin Theorem

In the Taylor series if we take the point of expansion as the origin, we obtain the Maclaurin series:

$$f(x) = \sum_{n=0}^{\infty} \frac{f^{(n)}(0)}{n!} x^n. \tag{15.84}$$

15.7.2 Binomial Theorem

We now write the Taylor series for the function

$$f(x) = (1+x)^m \tag{15.85}$$

about $x = 0$ as

$$(1+x)^m = 1 + mx + \frac{m(m-1)}{2!}x^2 + \cdots + R_n \tag{15.86}$$

with the remainder term

$$R_n = \frac{x^n}{n!}(1+\xi)^{m-n} m(m-1)\cdots(m-n+1), \tag{15.87}$$

where m can be negative and noninteger and ξ is a point such that $0 \leq \xi < x$. Since for $n > m$ the function $(1+\xi)^{m-n}$ takes its maximum value for $\xi = 0$, we can write the following upper bound for the remainder term:

$$R_n \leq \frac{x^n}{n!}m(m-1)\cdots(m-n+1). \tag{15.88}$$

From Equation (15.88) it is seen that in the interval $0 \leq x < 1$ the remainder goes to zero as $n \to \infty$; thus we obtain the binomial formula as

$$(1+x)^m = \sum_{n=0}^{\infty} \frac{m!}{n!(m-n)!}x^n = \sum_{n=0}^{\infty} \binom{m}{n}x^n. \tag{15.89}$$

It can be easily shown that this series is convergent in the interval $-1 < x < 1$. Note that for $m = n$ (integer) the sum automatically terminates after a finite number of terms, where the quantity $\binom{m}{n} = m!/n!(m-n)!$ is called the binomial coefficient.

Example 15.14. *Relativistic kinetic energy:* The binomial formula is probably one of the most widely used formulas in science and engineering. An important application of the binomial formula was given by Einstein in his celebrated paper where he announced his famous formula for the energy of a freely moving particle of mass m as

$$E = mc^2. \tag{15.90}$$

In this equation c is the speed of light and m is the mass of the moving particle, which is related to the rest mass m_0 by

$$m = \frac{m_0}{\left(1 - \dfrac{v^2}{c^2}\right)^{1/2}}, \tag{15.91}$$

where v is the velocity. Relativistic kinetic energy can be defined by subtracting the rest energy from the energy in motion:

$$K.E. = mc^2 - m_0c^2. \tag{15.92}$$

Since $v < c$, we can use the binomial formula to write the kinetic energy as

$$K.E. = m_0c^2 + \frac{1}{2}m_0v^2 + \frac{3}{8}m_0v^2(\frac{v^2}{c^2}) + \frac{5}{16}m_0v^2(\frac{v^2}{c^2})^2 + \cdots - m_0c^2 \tag{15.93}$$

and after simplifying we obtain

$$K.E. = \frac{1}{2}m_0v^2 + \frac{3}{8}m_0v^2(\frac{v^2}{c^2}) + \frac{5}{16}m_0v^2(\frac{v^2}{c^2})^2 + \cdots. \tag{15.94}$$

From here we see that in the nonrelativistic limit, that is, $v/c << 1$ or when $c \to \infty$, the above formula reduces to the well-known classical expression for kinetic energy:

$$K.E. \cong \frac{1}{2}m_0v^2. \tag{15.95}$$

15.7.3   Taylor Series for Functions with Multiple Variables

For a function with two independent variables, $f(x,y)$, the Taylor series is given as

$$f(x,y) = f(a,b) + (x-a)\frac{\partial f}{\partial x} + (x-b)\frac{\partial f}{\partial y}$$

$$+\frac{1}{2!}\left[(x-a)^2\frac{\partial^2 f}{\partial x^2} + 2(x-a)(y-b)\frac{\partial^2 f}{\partial x \partial y} + (y-b)^2\frac{\partial^2 f}{\partial y^2}\right]$$

$$+\frac{1}{3!}\left[(x-a)^3\frac{\partial^3 f}{\partial x^3} + 3(x-a)^2(y-b)\frac{\partial^3 f}{\partial x^2 \partial y}\right.$$

$$\left.+3(x-a)(y-b)^2\frac{\partial^3 f}{\partial x \partial y^2} + (y-b)^3\frac{\partial^3 f}{\partial y^3}\right] + \cdots \tag{15.96}$$

All the derivatives are evaluated at the point (a,b). In the presence of m independent variables Taylor series becomes

$$f(x_1, x_2, ..., x_m) = \sum_{n=o}^{\infty} \frac{1}{n!}\left\{\left[\sum_{i=1}^{m}(x_j - x_{j0})\frac{\partial}{\partial x_i}\right]^n f(x_1, x_2, ..., x_m)\right\}_{x_{10}, x_{20}, ..., x_{m0}}. \tag{15.97}$$

15.8   POWER SERIES

Series with their general term given as $u_n(x) = a_n x^n$ are called power series:

$$f(x) = a_0 + a_1 x + a_2 x^2 + \cdots = \sum_{n=0}^{\infty} a_n x^n, \qquad (15.98)$$

where the coefficients a_n are independent of x. To use the ratio test we write

$$\left| \frac{u_{n+1}}{u_n} \right| = \left| \frac{a_{n+1} x^{n+1}}{a_n x^n} \right| = \left| \frac{a_{n+1}}{a_n} \right| |x| \qquad (15.99)$$

and find the limit

$$\lim_{n \to \infty} \left| \frac{a_{n+1}}{a_n} \right| = \frac{1}{R}. \qquad (15.100)$$

Hence the condition for the convergence of a power series is obtained as

$$|x| < R \Longrightarrow -R < x < R, \qquad (15.101)$$

where R is called the radius of convergence. At the end points the ratio test fails; hence these points must be analyzed separately.

Example 15.14. *Power series:* For the power series

$$1 + x + \frac{x^2}{2} + \frac{x^3}{3} + \cdots + \frac{x^n}{n} + \cdots, \qquad (15.102)$$

the radius of convergence R is one; thus the series converges in the interval $-1 < x < 1$. On the other hand, at the end point $x = 1$ it is divergent, while at the other end point, $x = -1$, it is convergent.

Example 15.15. *Power series:* The radius of convergence can also be zero. For the series

$$1 + x + 2!x^2 + 3!x^3 + \cdots + n!x^n + \cdots, \qquad (15.103)$$

the ratio

$$\frac{a_{n+1}}{a_n} = \frac{(n+1)!}{n!} = n + 1 \qquad (15.104)$$

gives

$$\lim_{n \to \infty} (n+1) = \frac{1}{R} \to \infty. \qquad (15.105)$$

Thus the radius of convergence is zero. Note that this series converges only for $x = 0$.

Example 15.16. *Power series:* For the power series

$$1 + x + \frac{x^2}{2!} + \frac{x^3}{3!} + \cdots + \frac{x^n}{n!} + \cdots \qquad (15.106)$$

we find

$$\frac{a_{n+1}}{a_n} = \frac{n!}{(n+1)!} = \frac{1}{n+1} \qquad (15.107)$$

and

$$\lim_{n\to\infty} \frac{1}{n+1} = \frac{1}{R} \to 0. \qquad (15.108)$$

Hence the radius of convergence is infinity. This series converges for all x values.

15.8.1   Convergence of Power Series

If a power series is convergent in the interval $-R < x < R$, then it is uniformly and absolutely convergent in any subinterval S:

$$-S \leq x \leq S, \quad \text{where} \quad 0 < S < R. \qquad (15.109)$$

This can be seen by taking $M_i = |a_i| \, S^i$ in the M-test.

15.8.2   Continuity

In a power series, since every term, that is, $u_n(x) = a_n x^n$, is a continuous function and since in the interval $-S \leq x \leq S$ the series $f(x) = \sum a_n x^n$ is uniformly convergent, $f(x)$ is a continuous function. Considering that in Fourier series even though the $u_n(x)$ functions are continuous we expand discontinuous functions shaped like a saw tooth, this is an important property.

15.8.3   Differentiation and Integration of Power Series

In the interval of uniform convergence a power series can be differentiated and integrated as often as desired. These operations do not change the radius of convergence.

15.8.4   Uniqueness Theorem

Let us assume that a function has two power series expansions about the origin with overlapping radii of convergence, that is,

$$f(x) = \sum_{n=0}^{\infty} a_n x^n \Rightarrow -R_a < x < R_a \qquad (15.110)$$

$$= \sum_{n=0}^{\infty} b_n x^n \Rightarrow -R_b < x < R_b,$$

then $b_n = a_n$ is true for all n. Hence the power series is unique.

Proof: Let us write

$$\sum_{n=0}^{\infty} a_n x^n = \sum_{n=0}^{\infty} b_n x^n \Rightarrow -R < x < R,$$

where R is equal to the smaller of the two radii R_a and R_b. If we set $x = 0$ in this equation we find $a_0 = b_0$. Using the fact that a power series can be differentiated as often as desired, we differentiate the above equation once to write

$$\sum_{n=1}^{\infty} a_n n x^{n-1} = \sum_{n=1}^{\infty} b_n n x^{n-1}.$$

We again set $x = 0$, this time to find $a_1 = b_1$. Similarly, by repeating this process we show that $a_n = b_n$ for all n.

15.8.5   Inversion of Power Series

Consider the power series expansion of the function $y(x) - y_0$ in powers of $(x - x_0)$ as

$$y - y_0 = a_1(x - x_0) + a_2(x - x_0)^2 + \cdots, \qquad (15.111)$$

that is,

$$y - y_0 = \sum_{n=1}^{\infty} a_n(x - x_0)^n. \qquad (15.112)$$

Sometimes it is desirable to express this series as

$$x - x_0 = \sum_{n=1}^{\infty} b_n(y - y_0)^n. \qquad (15.113)$$

For this we can substitute Equation (15.113) into Equation (15.112) and compare equal powers of $(y - y_0)$ to get the new coefficients b_n as

$$b_1 = \frac{1}{a_1},$$

$$b_2 = -\frac{a_2}{a_1^3},$$

$$b_3 = \frac{1}{a_1^5}(2a_2^2 - a_1 a_3), \tag{15.114}$$

$$b_4 = \frac{1}{a_1^7}(5a_1 a_2 a_3 - a_1^2 a_4 - 5a_2^3),$$

$$\vdots$$

A closed expression for these coefficients can be found by using the residue theorem as

$$b_n = \frac{1}{n!}\left[\frac{d^{n-1}}{dt^{n-1}}\left(\frac{t}{w(t)}\right)^n\right]_{t=0}, \tag{15.115}$$

where $w(t) = \sum_{n=1}^{\infty} a_n t^n$.

15.9 SUMMATION OF INFINITE SERIES

After we conclude that a given series is convergent, the next and most important thing we need in applications is the value or the function that it converges to. For uniformly convergent series it is sometimes possible to identify an unknown series as the derivatives or the integrals of a known series. In this section we introduce some analytic techniques to evaluate the sums of infinite series. We start with the Euler-Maclaurin sum formula, which has important applications in quantum field theory and Green's function calculations. Next we discuss how some infinite series can be summed by using the residue theorem. Finally, we show that differintegrals can also be used to sum infinite series.

15.9.1 Bernoulli Polynomials and Their Properties

In deriving the Euler-Maclaurin sum formula we make use of the properties of the Bernoulli polynomials, $B_s(x)$, where their generating function definition is given as

$$\frac{te^{xt}}{e^t - 1} = \sum_{s=0}^{\infty} B_s(x)\frac{t^s}{s!}, \quad |t| < 2\pi. \tag{15.116}$$

Some of the Bernoulli polynomials can be written as

$$
\begin{array}{llll}
B_0(x) = & 1 & B_1(x) = & x - \frac{1}{2} \\
B_2(x) = & x^2 - x + \frac{1}{6} & B_3(x) = & x(x - \frac{1}{2})(x - 1) \\
B_4(x) = & x^4 - 2x^3 + x^2 - \frac{1}{30} & B_5(x) = & x(x - \frac{1}{2})(x - 1)(x^2 - x - \frac{1}{3}) \\
\end{array}
$$

$$
B_6(x) = \quad x^6 - 3x^5 + \frac{5}{2}x^4 - \frac{1}{2}x^2 + \frac{1}{42} \qquad \vdots \qquad\qquad \vdots
$$

$$\tag{15.117}$$

Values of the Bernoulli polynomials at $x = 0$ are known as the Bernoulli numbers,

$$B_s = B_s(0), \tag{15.118}$$

where the first nine of them are given as

$$B_0 = 1 \qquad B_1 = -\frac{1}{2} \qquad B_2 = \frac{1}{6} \qquad B_3 = 0 \qquad B_4 = -\frac{1}{30}$$

$$B_5 = 0 \qquad B_6 = \frac{1}{42} \qquad B_7 = 0 \qquad B_8 = -\frac{1}{30} \qquad B_9 = 0 \quad \cdots .$$

$$\tag{15.119}$$

Some of the important properties of the Bernoulli polynomials can be listed as follows:

1.

$$B_s(x) = \sum_{j=0}^{s} \binom{s}{j} B_{s-j} x^j. \tag{15.120}$$

2.

$$B_s'(x) = s B_{s-1}(x), \quad \int_0^1 B_s(x)\,dx = 0, \quad (s \geq 1). \tag{15.121}$$

3.

$$B_s(1 - x) = (-1)^s B_s(x). \tag{15.122}$$

Note that when we write $B_s(1 - x)$ we mean the Bernoulli polynomial with the argument $(1 - x)$. We show the Bernoulli numbers as B_s.

4.

$$\sum_{j=1}^{n} j^s = \frac{1}{(s+1)} \{B_{s+1}(n+1) - B_{s+1}\}, \quad (s \geq 1). \tag{15.123}$$

5.

$$\sum_{j=1}^{\infty} \frac{1}{j^{2s}} = (-1)^{s-1} (2\pi)^{2s} \frac{B_{2s}}{2(2s)!}, \quad (s \geq 1). \tag{15.124}$$

6.

$$\int_0^{\infty} \frac{x^{2s-1}}{e^{2\pi x} - 1} dx = (-1)^{s-1} \frac{B_{2s}}{4s}, \quad (s \geq 1). \tag{15.125}$$

7. In the interval $[0, 1]$ and for $s \geq 1$, the only zeroes of $B_{2s+1}(x)$ are $0, \frac{1}{2}$, and 1. In the same interval 0 and 1 are the only zeroes of $(B_{2s}(x) - B_{2s})$. Bernoulli polynomials also satisfy the inequality

$$|B_{2s}(x)| \leq |B_{2s}|, \quad 0 \leq x \leq 1. \tag{15.126}$$

8. The Bernoulli periodic function, which is continuous and has the period one, is defined as

$$P_s(x) = B_s(x - [x]), \tag{15.127}$$

where $[x]$ means the greatest integer in the interval $(x - 1, x]$. The Bernoulli periodic function also satisfies the relations

$$P_s'(x) = s P_{s-1}(x), \quad s = 1, 2, 3, \ldots \tag{15.128}$$

and

$$P_s(1) = (-1)^s P_s(0), \quad s = 0, 1, 2, 3, \ldots . \tag{15.129}$$

15.9.2   Euler-Maclaurin Sum Formula

We first write the integral

$$\int_0^1 f(x) dx \tag{15.130}$$

by using the properties of the Bernoulli polynomials. Since the first Bernoulli polynomial is

$$B_0(x) = 1, \tag{15.131}$$

we can write this integral as

$$\int_0^1 f(x)dx = \int_0^1 f(x)B_0(x)dx. \qquad (15.132)$$

Using the following property of Bernoulli polynomials:

$$B_1'(x) = B_0(x) = 1, \qquad (15.133)$$

we can write Equation (15.132) as

$$\int_0^1 f(x)dx = \int_0^1 f(x)B_0(x)dx = \int_0^1 f(x)B_1'(x)dx. \qquad (15.134)$$

Integrating this by parts we obtain

$$\int_0^1 f(x)dx = f(x)B_1(x)|_0^1 - \int_0^1 f'(x)B_1(x)dx \qquad (15.135)$$

$$= \frac{1}{2}[f(1) + f(0)] - \int_0^1 f'(x)B_1(x)dx, \qquad (15.136)$$

where we have used $B_1(1) = \frac{1}{2}$ and $B_1(0) = -\frac{1}{2}$. In the above integral we now use

$$B_1(x) = \frac{1}{2}B_2'(x) \qquad (15.137)$$

and integrate by parts again to obtain

$$\int_0^1 f(x)dx = \frac{1}{2}[f(1) + f(0)] - \frac{1}{2!}[f'(1)B_2(1) - f'(0)B_2(0)]$$

$$+ \frac{1}{2!}\int_0^1 f''(x)B_2(x)dx. \qquad (15.138)$$

Using the values

$$B_{2n}(1) = \quad B_{2n}(0) = B_{2n} \qquad n = 0, 1, 2, ...,$$

$$\qquad (15.139)$$

$$B_{2n+1}(1) = \quad B_{2n+1}(0) = 0 \qquad n = 1, 2, 3, ...,$$

and continuing like this we obtain

$$\int_0^1 f(x)dx = \frac{1}{2}[f(1) + f(0)] - \sum_{p=1}^{q}\frac{B_{2p}}{(2p)!}\left[f^{(2p-1)}(1) - f^{(2p-1)}(0)\right]$$

$$+ \frac{1}{(2q)!}\int_0^1 f^{(2q)}(x)B_{2q}(x)dx. \qquad (15.140)$$

This equation is called the Euler-Maclaurin sum formula. We have assumed that all the necessary derivatives of $f(x)$ exist and q is an integer greater than one.

We now change the limits of the integral from 0 to 1 to 1 to 2. We write

$$\int_1^2 f(x)dx = \int_0^1 f(y+1)dy \tag{15.141}$$

and repeat the same steps in the derivation of the Euler-Maclaurin sum formula for $f(y+1)$ to obtain

$$\int_0^1 f(y+1)dy = \frac{1}{2}[f(2)+f(1)] \tag{15.142}$$

$$- \sum_{p=1}^q \frac{B_{2p}}{(2p)!}\left[f^{(2p-1)}(2) - f^{(2p-1)}(1)\right]$$

$$+ \frac{1}{(2q)!}\int_0^1 f^{(2q)}(y+1)P_{2q}(y)dy,$$

We have now used the Bernoulli periodic function [Eq. (15.127)]. Making the transformation

$$y+1 = x, \tag{15.143}$$

we write

$$\int_1^2 f(x)dx = \int_0^1 f(y+1)dy \tag{15.144}$$

$$= \frac{1}{2}[f(2)+f(1)] \tag{15.145}$$

$$- \sum_{p=1}^q \frac{B_{2p}}{(2p)!}\left[f^{(2p-1)}(2) - f^{(2p-1)}(1)\right]$$

$$+ \frac{1}{(2q)!}\int_0^1 f^{(2q)}(x)P_{2q}(x-1)dx.$$

Repeating this for the interval $[2,3]$ we can write

$$\int_2^3 f(x)dx = \frac{1}{2}[f(3)+f(2)] \tag{15.146}$$

$$- \sum_{p=1}^q \frac{B_{2p}}{(2p)!}\left[f^{(2p-1)}(3) - f^{(2p-1)}(2)\right]$$

$$+ \frac{1}{(2q)!}\int_2^3 f^{(2q)}(x)P_{2q}(x-2)dx.$$

Integrals for the other intervals can be written similarly. Since the integral for the interval $[0, n]$ can be written as

$$\int_0^n f(x)dx = \int_0^1 f(x)dx + \int_1^2 f(x)dx + \cdots + \int_{n-1}^n f(x)dx, \qquad (15.147)$$

we substitute the formulas found above in the right-hand side to obtain

$$\int_0^n f(x)dx = \frac{1}{2}f(0) + f(1) + f(2) + \cdots + \frac{1}{2}f(n) \qquad (15.148)$$

$$- \sum_{p=1}^q \frac{B_{2p}}{(2p)!} \left[f^{(2p-1)}(n) - f^{(2p-1)}(0) \right]$$

$$+ \frac{1}{(2q)!} \int_0^n f^{(2q)}(x)P_{2q}(x)dx.$$

We have used the fact that the function $P_{2q}(x)$ is periodic with the period one. Rearranging this, we write

$$\sum_{j=0}^n f(j) = \int_0^n f(x)dx + \frac{1}{2}[f(0) + f(n)]$$

$$+ \sum_{p=1}^{q-1} \frac{B_{2p}}{(2p)!} \left[f^{(2p-1)}(n) - f^{(2p-1)}(0) \right] \qquad (15.149)$$

$$+ \frac{B_{2q}}{(2q)!} \left[f^{(2q-1)}(n) - f^{(2q-1)}(0) \right] - \frac{1}{(2q)!} \int_0^n f^{(2q)}(x)P_{2q}(x)dx.$$

The last two terms on the right-hand side can be written under the same integral sign, which gives us the final form of the Euler-Maclaurin sum formula as

$$\sum_{j=0}^n f(j) = \int_0^n f(x)dx + \frac{1}{2}[f(0) + f(n)]$$

$$+ \sum_{p=1}^{q-1} \frac{B_{2p}}{(2p)!} \left[f^{(2p-1)}(n) - f^{(2p-1)}(0) \right] \qquad (15.150)$$

$$+ \int_0^n \frac{[B_{2q} - B_{2q}(x - [x])]}{(2q)!} f^{(2q)}(x)dx.$$

In this derivation we have assumed that $f(x)$ is continuous and has all the required derivatives. This is a very versatile formula that can be used in several ways. When q is chosen as a finite number it allows us to evaluate a given series as an integral plus some correction terms. When q is chosen as infinity it could allow us to replace a slowly converging series with a rapidly converging one. If we take the integral to the left-hand side, it can be used for numerical evaluation of integrals.

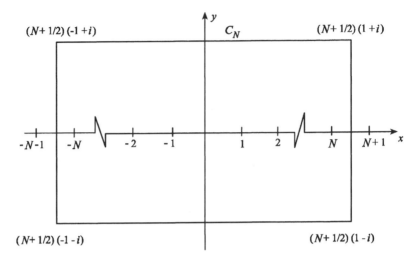

Fig. 15.3 Contour for finding series sums using the residue theorem

15.9.3 Using Residue Theorem to Sum Infinite Series

Some of the infinite series can be summed by using the residue theorem. First we take a rectangular contour C_N in the z-plane with the corners as shown in Figure 15.3. We now prove a property that will be useful to us shortly.

Lemma: On the contour C_N the inequality $|\cot \pi z| < A$ is always satisfied, where A is a constant independent of N.

We prove this by considering the parts of C_N with $y > \frac{1}{2}$, $-\frac{1}{2} \leq y \leq \frac{1}{2}$, and $y < -\frac{1}{2}$ separately.

1. **Case:** $(y > \frac{1}{2})$: We write a complex number as $z = x + iy$; thus

$$|\cot \pi z| = \left| \frac{e^{i\pi z} + e^{-i\pi z}}{e^{i\pi z} - e^{-i\pi z}} \right| = \left| \frac{e^{i\pi x - \pi y} + e^{-i\pi x + \pi y}}{e^{i\pi x - \pi y} - e^{-i\pi x + \pi y}} \right|. \tag{15.151}$$

Using the triangle inequality

$$|z_1| - |z_2| \leq |z_1 + z_2| \leq |z_1| + |z_2| \tag{15.152}$$

and considering that $y > \frac{1}{2}$ we find

$$
\begin{aligned}
|\cot \pi z| &\leq \frac{\left|e^{i\pi x - \pi y}\right| + \left|e^{-i\pi x + \pi y}\right|}{\left|e^{-i\pi x + \pi y}\right| - \left|e^{i\pi x - \pi y}\right|} \\
&\leq \frac{e^{-\pi y} + e^{\pi y}}{e^{\pi y} - e^{-\pi y}} = \frac{1 + e^{-2\pi y}}{1 - e^{-2\pi y}} \\
&\leq \frac{1 + e^{-\pi}}{1 - e^{-\pi}} = A_1.
\end{aligned}
\tag{15.153}
$$

2. Case: $(y < -\frac{1}{2})$: Following the procedure of the first case we find

$$|\cot \pi z| \leq \frac{\left|e^{i\pi x - \pi y}\right| + \left|e^{-i\pi x + \pi y}\right|}{\left|e^{i\pi x - \pi y}\right| - \left|e^{-i\pi x + \pi y}\right|} = \frac{e^{-\pi y} + e^{\pi y}}{e^{-\pi y} - e^{\pi y}}$$

$$\leq \frac{1 + e^{2\pi y}}{1 - e^{2\pi y}} \leq \frac{1 + e^{-\pi}}{1 - e^{-\pi}} = A_1. \tag{15.154}$$

3. Case: $(-\frac{1}{2} \leq y \leq \frac{1}{2})$: A point on the right-hand side of C_N can be written as $z = N + \frac{1}{2} + iy$; thus for $-\frac{1}{2} \leq y \leq \frac{1}{2}$ we obtain

$$|\cot \pi z| = \left|\cot \pi \left(N + \frac{1}{2} + iy\right)\right| = \left|\cot \left(\frac{\pi}{2} + i\pi y\right)\right|$$

$$= |\tanh \pi y| \leq \tanh \left(\frac{\pi}{2}\right) = A_2. \tag{15.155}$$

Similarly, a point on the left-hand side of C_N is written as $z = -N - \frac{1}{2} + iy$, which gives us

$$|\cot \pi z| = \left|\cot \pi \left(-N - \frac{1}{2} + iy\right)\right| = \left|\cot \left(\frac{\pi}{2} + i\pi y\right)\right|$$

$$= |\tanh \pi y| \leq \tanh \left(\frac{\pi}{2}\right) = A_2. \tag{15.156}$$

Thus choosing the greater of the A_1 and A_2 and calling it A proves that on C_N the inequality $|\cot \pi z| < A$ is satisfied. We also note that A is a constant independent of N. Actually, since $A_2 < A_1$ we could also write

$$|\cot \pi z| \leq A_1 = \cot \frac{\pi}{2}. \tag{15.157}$$

We now state the following useful theorem:

Theorem: If a function $f(z)$ satisfies the inequality $|f(z)| \leq \dfrac{M}{|z|^k}$ on the contour C_N, where $k > 1$ and M is a constant independent of N, then the sum of the series $\sum_{j=-\infty}^{\infty} f(j)$ is given as

$$\sum_{j=-\infty}^{\infty} f(j) = -\{\text{residues of the function } \pi \cot \pi z f(z) \text{ at} \atop \text{the isolated singular points of } f(z)\}. \tag{15.158}$$

Proof:

1. Case: Let us assume that $f(z)$ has a finite number of isolated singular points. In this case we choose the number N such that the closed contour C_N encloses all of the singular points of $f(z)$. On the other hand, $\cot \pi z$ has poles at the points

$$z = n, \quad n = 0, \pm 1, \pm 2, ...,$$

where the residues of the function $\pi \cot \pi z f(z)$ at these points are given as

$$\lim_{z \to n} (z - n) \pi \cot \pi z f(z) = \lim_{z \to n} \pi \frac{(z - n) f(z) \cos \pi z}{\sin z \pi} = f(n).$$

$$(15.159)$$

For this result we have used the L'Hopital's rule and assumed that $f(z)$ has no poles at the points $z = n$. Using the residue theorem we can now write

$$\oint_{C_N} \pi \cot \pi z f(z) dz = \sum_{n=-N}^{N} f(n) + S,$$

$$(15.160)$$

where S represents the finite number of residues of $\pi \cot \pi z f(z)$ at the poles of $f(z)$. We can put an upper bound to the integral on the left as

$$\oint_{C_N} \pi \cot \pi z f(z) dz \leq \left| \oint_{C_N} \pi \cot \pi z f(z) dz \right| \leq \oint_{C_N} \pi \left| \cot \pi z \right| \left| f(z) \right| \left| dz \right|.$$

Since $f(z)$ satisfies $|f(z)| \leq \dfrac{M}{|z|^k}$ on the contour, this becomes

$$\left| \oint_{C_N} \pi \cot \pi z f(z) dz \right| \leq \frac{\pi A M}{N^k} (8N + 4),$$

$$(15.161)$$

where $(8N + 4)$ is the length of C_N. From here we see that as $N \to \infty$ value of the integral in Equation (15.160) goes to zero:

$$\lim_{N \to \infty} \oint_{C_N} \pi \cot \pi z f(z) dz = 0.$$

$$(15.162)$$

We can now write

$$\sum_{n=-\infty}^{\infty} f(n) = -S.$$

$$(15.163)$$

2. **Case:** If $f(z)$ has infinitely many singular points the result can be obtained similarly by an appropriate limiting process.

Example 15.17. *Series sum by the residue theorem:* In quantum field theory and in Green's function calculations we occasionally encounter series like

$$\sum_{n=0}^{\infty} \frac{1}{n^2 + a^2} , \quad a > 0 ,$$

$$(15.164)$$

where

$$f(z) = \frac{1}{z^2 + a^2} \tag{15.165}$$

has two isolated singular points located at $z = \pm ia$ and satisfies the conditions of the above theorem. The residue of

$$\frac{\pi \cot \pi z}{z^2 + a^2} \tag{15.166}$$

at $z = ia$ is found as

$$\lim_{z \to ia} (z - ia) \frac{\pi \cot \pi z}{(z + ia)(z - ia)} = \frac{\pi \cot \pi ia}{2ia} \tag{15.167}$$

$$= -\frac{\pi}{2a} \coth \pi a.$$

Similarly, the residue at $z = -ia$ is $-\dfrac{\pi}{2a} \coth \pi a$; thus using the conclusion of the above theorem [Eq. (15.158)] we can write

$$\sum_{n=-\infty}^{\infty} \frac{1}{n^2 + a^2} = \frac{\pi}{a} \coth \pi a. \tag{15.168}$$

From this result we obtain the sum $\sum_{n=0}^{\infty} \dfrac{1}{n^2 + a^2}$ as

$$\sum_{n=-\infty}^{\infty} \frac{1}{n^2 + a^2} = \sum_{n=-\infty}^{-1} \frac{1}{n^2 + a^2} + \frac{1}{a^2} + \sum_{n=1}^{\infty} \frac{1}{n^2 + a^2},$$

$$2\sum_{n=1}^{\infty} \frac{1}{n^2 + a^2} + \frac{1}{a^2} = \frac{\pi}{a} \coth \pi a,$$

$$\sum_{n=0}^{\infty} \frac{1}{n^2 + a^2} = \frac{\pi}{2a} \coth \pi a + \frac{1}{2a^2}. \tag{15.169}$$

15.9.4 Evaluating Sums of Series by Differintegrals

In 1970 Osler gave the following summation version of the Leibniz rule, which is very useful in finding sums of infinite series:

$$\frac{d^q [u(x)v(x)]}{dx^q} = \sum_{n=-\infty}^{\infty} \binom{q}{n + \gamma} \frac{d^{q-\gamma-n} u}{dx^{q-\gamma-n}} \frac{d^{\gamma+n} v}{dx^{\gamma+n}}$$

$$= \sum_{n=-\infty}^{\infty} \frac{\Gamma(q+1)}{\Gamma(q - \gamma - n + 1)\Gamma(\gamma + n + 1)} \frac{d^{q-\gamma-n} u}{dx^{q-\gamma-n}} \frac{d^{\gamma+n} v}{dx^{\gamma+n}}, \tag{15.170}$$

where γ is any constant.

Example 15.18. *Evaluating sums of series by differintegrals:* In the above formula if we choose $u = x^a$, $v = x^b$, and $\gamma = 0$, and use the differintegral $\dfrac{d^q x^p}{dx^q} = \dfrac{\Gamma(p+1)x^{p-q}}{\Gamma(p-q+1)}$, where $p > -1$, we find the sum

$$\frac{\Gamma(a+b+1)}{\Gamma(a+b-q+1)} =$$

$$\sum_{n=0}^{\infty} \frac{\Gamma(q+1)\Gamma(a+1)\Gamma(b+1)}{\Gamma(q-n+1)\Gamma(n+1)\Gamma(a-q+n+1)\Gamma(b-n+1)}. \tag{15.171}$$

15.9.5 Asymptotic Series

Asymptotic series are frequently encountered in applications. They are generally used in numerical evaluation of certain functions approximately. Two typical functions where asymptotic series are used for their evaluation are given as $I_1(x)$ and $I_2(x)$:

$$I_1(x) = \int_x^{\infty} e^{-u} f(u) du, \tag{15.172}$$

$$I_2(x) = \int_0^{\infty} e^{-u} f\left(\frac{u}{x}\right) du. \tag{15.173}$$

In astrophysics we frequently work on gasses obeying the Maxwell-Boltzman distribution, where we encounter gamma functions defined as

$$I(x,p) = \int_x^{\infty} e^{-u} u^{-p} du = \Gamma(1-p,x), \quad (x,p > 0). \tag{15.174}$$

We now calculate $I(x,p)$ for large values of x. We first start by integrating the above integral by parts twice to get

$$I(x,p) = \frac{e^{-x}}{x^p} - p \int_x^{\infty} e^{-u} u^{-p-1} du \tag{15.175}$$

and then

$$I(x,p) = \frac{e^{-x}}{x^p} - \frac{pe^{-x}}{x^{p+1}} + p(p+1) \int_x^{\infty} e^{-u} u^{-p-2} du. \tag{15.176}$$

We keep on integrating by parts to obtain the series

$$I(x,p) = e^{-x} \left\{ \frac{1}{x^p} - \frac{p}{x^{p+1}} + \frac{p(p+1)}{x^{p+2}} - \cdots \right.$$
$$+ (-1)^{n-1} \frac{(p+n-2)!}{(p-1)! x^{p+n-1}} \right\}$$
$$+ (-1)^n \frac{(p+n-1)!}{(p-1)!} \int_x^\infty e^{-u} u^{-p-n} du. \tag{15.177}$$

This is a rather interesting series, if we apply the ratio test we find

$$\lim_{n\to\infty} \frac{|u_{n+1}|}{|u_n|} = \lim_{n\to\infty} \frac{(p+n)!}{(P+n-1)!} \cdot \frac{1}{x}$$
$$= \lim_{n\to\infty} \frac{p+n}{x} \to \infty. \tag{15.178}$$

Thus the series diverges for all finite values of x. Before we discard this series as useless in calculating the values of the function $I(x,p)$, let us write the absolute value of the difference of $I(x,p)$ and the nth partial sum as

$$|I(x,p) - S_n(x,p)| \leq \left| (-1)^{n+1} \frac{(p+n)!}{(p-1)!} \int_x^\infty e^{-u} u^{-p-n-1} du \right| = |R_n(x,p)|. \tag{15.179}$$

Using the transformation $u = v + x$ we can write the above integral as

$$\int_x^\infty e^{-u} u^{-p-u-1} du = e^{-x} \int_0^\infty e^{-v} (v+x)^{-p-n-1} dv \tag{15.180}$$

$$= \frac{e^{-x}}{x^{p+n+1}} \int_0^\infty e^{-v} \left(1 + \frac{v}{x} \right)^{-p-n-1} dv. \tag{15.181}$$

For the large values of x we find the limit

$$\lim_{x\to\infty} \int_0^\infty e^{-v} \left(1 + \frac{v}{x} \right)^{-p-n-1} dv \to 1 \tag{15.182}$$

and the remainder term R_n is

$$|R_n| = |I(x,p) - S_n(x,p)| \approx \frac{(p+n)!}{(p-1)!} \frac{e^{-x}}{x^{p+n+1}}, \tag{15.183}$$

which shows that for sufficiently large values of x we can use S_n for evaluating the values of the $I(x,p)$ function to sufficient accuracy. Naturally, the R_n value of the partial sum depends on the desired accuracy. For this reason such series are sometimes called **asymptotic** or **semi-convergent** series.

Example 15.19. *Asymptotic expansions:* We now consider the integral

$$I = \int_0^x e^{-t^2} dt. \tag{15.184}$$

Using the expansion

$$e^{-t^2} = 1 - \frac{t^2}{1!} + \frac{t^4}{2!} - \frac{t^6}{3!} + \cdots, \ (r = \infty), \tag{15.185}$$

where r is the radius of convergence we write

$$I = \int_0^x e^{-t^2} dt = x - \frac{x^3}{3.1!} + \frac{x^5}{5.2!} - \frac{x^7}{7.3!} + \cdots, \ (r = \infty). \tag{15.186}$$

For small values of x this series can be used to evaluate I to any desired level of accuracy. However, even though this series is convergent for all x, it is not practical to use for large x. For the large values of x we can use the method of asymptotic expansions. Writing

$$I = \int_0^x e^{-t^2} dt \tag{15.187}$$

$$= \int_0^\infty e^{-t^2} dt - \int_x^\infty e^{-t^2} dt$$

$$= \frac{\sqrt{\pi}}{2} - \left[\int_x^\infty (\frac{-1}{2t}) d(e^{-t^2}) \right]$$

and integrating by parts we obtain

$$I = \frac{\sqrt{\pi}}{2} - \left[\frac{e^{-x^2}}{2x} - \frac{1}{2} \int_x^\infty \frac{e^{-t^2}}{t^2} dt \right]. \tag{15.188}$$

Repeated application of integration by parts, after n times, yields

$$\int_x^\infty e^{-t^2} dt \tag{15.189}$$

$$= \frac{e^{-x^2}}{2x} \left[1 - \frac{1}{2x^2} + \frac{1.3}{(2x^2)^2} - \cdots + (-1)^{n-1} \frac{1 \cdot 3 \cdot 5 \cdots (2n-3)}{(2x^2)^{n-1}} \right] + R_n,$$

where

$$R_n = (-1)^n \frac{1 \cdot 3 \cdot 5 \cdots (2n-1)}{2^n} \int_x^\infty \frac{e^{-t^2}}{t^{2n}} dt. \tag{15.190}$$

As $n \to \infty$ this series diverges for all x. However, using the inequalities

$$\int_x^\infty \frac{e^{-t^2}}{t^{2n}} dt < \frac{1}{x^{2n}} \int_x^\infty e^{-t^2} dt \qquad (15.191)$$

and

$$\int_x^\infty e^{-t^2} dt < \frac{e^{-x^2}}{2x}, \qquad (15.192)$$

we see that the remainder satisfies

$$|R_n| < \frac{1 \cdot 3 \cdot 5 \cdots (2n-1)}{2^{n+1} x^{2n+1}} e^{-x^2}. \qquad (15.193)$$

Hence $|R_n|$ can be made sufficiently small by choosing n sufficiently large, thus the series (15.189) can be used to evaluate I as

$$I = \frac{\sqrt{\pi}}{2} - \frac{e^{-x^2}}{2x} \left[1 - \frac{1}{2x^2} + \frac{1.3}{(2x^2)^2} - \cdots + (-1)^{n-1} \frac{1 \cdot 3 \cdot 5 \cdots (2n-3)}{(2x^2)^{n-1}} \right] - R_n. \tag{15.194}$$

For $x = 5$ if we choose $n = 13$, we have $|R_n| < 10^{-20}$.

15.10   DIVERGENT SERIES IN PHYSICS

So far we have seen how to test a series for convergence and introduced some techniques for evaluating infinite sums. In quantum field theory we occasionally encounter divergent series corresponding to physical properties like energy, mass, etc. These divergences are naturally due to some pathologies in our theory, which are expected not to exist in the next generation of field theories. However, even within the existing theories it is sometimes possible to obtain meaningful results, which agree with experiments to an incredibly high degree of accuracy. The process of obtaining finite and meaningful results from divergent series is accomplished in two steps. The first step is called **regularization**, where the divergent pieces are written out explicitly, and the second step is called **renormalization**, where the divergent pieces identified in the first part are subtracted by suitable physical arguments. Whether a given theory is renormalizable or not is very important. In 1999 Gerardus 't Hooft and J. G. Martinus Veltman received the Nobel Prize for showing that Salam and Weinberg's theory of unified electromagnetic and weak interactions is renormalizable. On the other hand, quantum gravity is nonrenormalizable because it contains infinitely many divergent pieces.

15.10.1   Casimir Effect and Renormalization

To demonstrate the regularization and renormalization techniques we consider a massless (conformal) scalar field in a one-dimensional box with length L .

Using the periodic boundary conditions we can find the eigenfrequencies as

$$\omega_n = \frac{2\pi cn}{L} \qquad n = 0, 1, 2, ...,$$ (15.195)

where each frequency is two-fold degenerate. In quantum field theory vacuum energy is a divergent expression given by

$$\overline{E}_0 = \sum_n g_n \frac{\hbar \omega_n}{2},$$ (15.196)

where g_n stands for the degeneracy of the nth eigenstate. For the one-dimensional box problem this gives

$$\overline{E}_0 = \frac{2\pi c\hbar}{L} \sum_{n=0}^{\infty} n.$$ (15.197)

Because the high frequencies are reason for the divergence of the vacuum energy, we have to suppress them for a finite result. Let us multiply Equation (15.197) with a regularization (cutoff) function like $e^{-\alpha \omega_n}$ and write

$$\overline{E}_0 = \frac{2\pi c\hbar}{L} \sum_{n=0}^{\infty} n e^{-2\pi cn\alpha/L},$$ (15.198)

where α is a cutoff parameter. This sum is now convergent and can be evaluated easily by using the geometric series as

$$\overline{E}_0 = \frac{2\pi}{L} e^{-2\pi \alpha/L} [1 - e^{-2\pi \alpha/L}]^{-2}, \quad \text{we set } \hbar = c = 1.$$ (15.199)

The final and finite result is naturally going to be obtained in the limit where the effects of the regularization function disappear, that is, when $\alpha \to 0$. We now expand Equation (15.199) in terms of the cutoff parameter α to write

$$\overline{E}_0 = \frac{L}{2\pi \alpha^2} - \frac{\pi}{6L} + (\text{terms in positive powers of } \alpha).$$ (15.200)

Note that in the limit as $\alpha \to 0$ vacuum energy is still divergent; however, the regularization function has helped us to identify the divergent piece explicitly as

$$\frac{L}{2\pi \alpha^2}.$$ (15.201)

The second term in $\overline{E}_0$ is finite and independent of α, while the remaining terms are all proportional to the positive powers of α, which disappears in the limit $\alpha \to 0$.

The second part of the process is renormalization, which is subtracting the divergent piece by a physical argument. We now look at the case where the walls are absent, or taken to infinity. In this case the eigenfrequencies are continuous; hence we write the vacuum energy in terms of an integral as

$$\overline{E}_0 \to \tilde{E}_0 = \frac{L}{2\pi} \int_0^\infty \omega d\omega. \tag{15.202}$$

This integral is also divergent. We regularize it with the same function and evaluate its value as

$$\tilde{E}_0 = \frac{L}{2\pi} \int_0^\infty \omega e^{-\alpha\omega} d\omega \tag{15.203}$$

$$= \frac{L}{2\pi\alpha^2}, \tag{15.204}$$

which is identical to the divergent term in the series (15.200).

To be consistent with our expectations, we now argue that in the absence of walls the quantum vacuum energy should be zero, thus we define the **renormalized quantum vacuum energy,** E_0, by subtracting the divergent piece [Eq. (15.204)] from the unrenormalized energy, $\overline{E}_0$, and then by taking the limit $\alpha \to 0$ as

$$E_0 = \lim_{\alpha \to 0} [\overline{E}_0 - \tilde{E}_0]. \tag{15.205}$$

For the renormalized quantum vacuum energy between the walls this prescription gives

$$E_0 = \lim_{\alpha \to 0} [\frac{L}{2\pi\alpha^2} - \frac{\pi}{6L} + (\text{terms in positive powers of } \alpha) - \frac{L}{2\pi\alpha^2}]$$

$$= -\frac{\pi}{6L}. \tag{15.206}$$

The minus sign means that the force between the walls is attractive. In three dimensions this method gives the renormalized electromagnetic vacuum energy between two perfectly conducting neutral plates held at absolute zero and separated by a distance L as

$$E_0 = -\frac{c\hbar\pi^2}{720L^3}S, \tag{15.207}$$

where S is the surface area of the plates and we have inserted c and $\hbar$. This gives the attractive force per unit area between the plates as

$$F_0 = -\frac{\partial E_0}{\partial L} = -\frac{\pi^2 c\hbar}{240L^4}.$$

In quantum field theory this interesting effect is known as the Casimir effect, and it has been verified experimentally. The Casimir effect has also been calculated for plates with different geometries and also in curved background spacetimes. More powerful and covariant techniques like the point splitting method, which are independent of cutoff functions, have confirmed the results obtained by the simple mode sum method used here. Note that in the classical limit, that is, as $\hbar \to 0$, the Casimir effect disappears, that is, it is a pure quantum effect.

One should keep in mind that in the regularization and renormalization process we have not cured the divergence problem of the quantum vacuum energy. In the absence of gravity only the energy differences are observable; hence all we have done is to define the renormalized quantum vacuum energy in the absence of plates as zero and then scaled all the other energies with respect to it.

15.10.2   Casimir Effect and MEMS

The Casimir force between two neutral metal plates is very small and goes as A/d^4. For plates with a surface area of 1 cm^2 and a separation of 1 μm, the Casimir force is around $10^{-7} N$. This is roughly the weight of a water drop. When we reduce the separation to 1 nm, roughly 100 times the size of a typical atom, pressure on the plates becomes 1 atm. The Casimir effect plays an important role in microelectromechanical devices, in short, MEMS. These are systems with moving mechanical parts embedded in silicon chips at micro- and submicroscales. Examples of MEMS are microrefrigerators, actuators, sensors, and switches. In the production of MEMS, the Casimir effect can sometimes produce unwanted effects like sticking between parts, but it can also be used to produce mechanical effects like bending and twisting. A practical use for the Casimir effect in our everyday lives is the pressure sensors of airbags in cars. Casimir energy is bound to make significant changes in our concept of vacuum.

15.11   INFINITE PRODUCTS

Infinite products are closely related to infinite series. Most of the known functions can be written as infinite products, which are also useful in calculating some of the transcendental numbers. We define the Nth partial product of an infinite product of positive terms as

$$P_N = f_1 \cdot f_2 \cdot f_3 \cdots f_N = \prod_{n=1}^{N} f_n. \tag{15.208}$$

If the limit

$$\lim_{N \to \infty} \prod_{n=1}^{N} f_n \to P \tag{15.209}$$

exists, then we say the infinite product is convergent and write

$$\prod_{n=1}^{\infty} f_n = P. \tag{15.210}$$

Infinite products satisfying the condition $\lim_{n\to\infty} f_n > 1$ are divergent. When the condition $0 < \lim_{n\to\infty} f_n < 1$ is satisfied it is advantageous to write the product as

$$\prod_{n=1}^{\infty} (1 + a_n). \tag{15.211}$$

The condition $a_n \to 0$ as $n \to \infty$ is necessary, but not sufficient, for convergence. Using the ln function we can write an infinite product as an infinite sum as

$$\ln \prod_{n=1}^{\infty} (1 + a_n) = \sum_{n=1}^{\infty} \ln(1 + a_n). \tag{15.212}$$

Theorem: When the inequality $0 \le a_n < 1$ is true, then the infinite products $\prod_{n=1}^{\infty}(1+a_n)$ and $\prod_{n=1}^{\infty}(1-a_n)$ converge or diverge with the infinite series

$$\sum_{n=1}^{\infty} a_n. \tag{15.213}$$

Proof: Since $1 + a_n \le e^{a_n}$ is true, we can write

$$e^{a_n} = 1 + a_n + \frac{a_n^2}{2!} + \cdots . \tag{15.214}$$

Thus the inequality

$$P_N = \prod_{n=1}^{N} (1 + a_n) \le \prod_{n=1}^{N} e^{a_n} = \exp \left\{ \sum_{n=1}^{N} a_n \right\} = e^{S_N} \tag{15.215}$$

is also true. Since in the limit as $N \to \infty$ we can write

$$\prod_{n=1}^{\infty} (1 + a_n) \le \exp \left\{ \sum_{n=1}^{\infty} a_n \right\}, \tag{15.216}$$

we obtain an upper bound to the infinite product. For a lower bound we write the Nth partial sum as

$$P_N = 1 + \sum_{i=1}^{N} a_i + \sum_{i=1}^{N} \sum_{j=1}^{N} a_i a_j + \cdots . \tag{15.217}$$

Since $a_i \geq 0$, we obtain the lower bound as

$$\prod_{n=1}^{\infty}(1+a_n) \geq \sum_{n=1}^{\infty} a_n. \tag{15.218}$$

Both the upper and the lower bounds to the infinite product $\Pi_{n=1}^{\infty}(1 + a_n)$ depend on the series $\sum_{n=1}^{\infty} a_n$; thus both of them converge or diverge together. Proof for the product $\Pi_{n=1}^{\infty}(1 - a_n)$ is done similarly.

15.11.1   Sine, Cosine, and the Gamma Functions

An nth order polynomial with n real roots can be written as a product:

$$P_n(x) = (x - x_1)(x - x_2)\cdots(x - x_n) = \prod_{i=1}^{n}(x - x_i). \tag{15.219}$$

Naturally a function with infinitely many roots can be expressed as an infinite product. We can find the infinite product representations of the sine and cosine functions using complex analysis:

In the z-plane a function, $h(z)$, with simple poles at $z = a_n$ $(0 < |a_1| < |a_2| < \cdots)$ can be written as

$$h(z) = h(0) + \sum_{n=1}^{\infty} b_n \left[\frac{1}{(z - a_n)} + \frac{1}{a_n} \right], \tag{15.220}$$

where b_n is the residue of the function at the pole a_n. This is also known as the Mittag-Leffler theorem. We have seen that a function analytic on the entire z-plane is called an entire function. For such a function its logarithmic derivative, f'/f, has poles and its Laurent expansion must be given about the poles. If an entire function $f(z)$ has a simple zero at $z = a_n$, then we can write

$$f(z) = (z - a_n)g(z). \tag{15.221}$$

$g(z)$ is again an analytic function satisfying $g(z) \neq g(a_n)$. Using the above equation we can write

$$\frac{f'}{f} = \frac{1}{(z - a_n)} + \frac{g'(z)}{g(z)}. \tag{15.222}$$

Since a_n is a simple pole of f'/f, we can take $b_n = 1$ and $h(z) = f'/f$ in Equation (15.220) to write

$$\frac{f'(z)}{f(z)} = \frac{f'(0)}{f(0)} + \sum_{n=1}^{\infty} \left[\frac{1}{(z - a_n)} + \frac{1}{a_n} \right]. \tag{15.223}$$

Integrating Equation (15.223) gives

$$\ln \frac{f(z)}{f(0)} = z\frac{f'(0)}{f(0)} + \sum_{n=1}^{\infty}\left[\ln(z - a_n) - \ln(-a_n) + \frac{z}{a_n}\right], \qquad (15.224)$$

and finally the general expression is obtained as

$$f(z) = f(0)\exp\left[z\frac{f'(0)}{f(0)}\right]\prod_{n=1}^{\infty}\left(1 - \frac{z}{a_n}\right)\exp(\frac{z}{a_n}). \qquad (15.225)$$

Applying this formula with $z = x$ to the sine and cosine functions we obtain

$$\sin x = x\prod_{n=1}^{\infty}\left(1 - \frac{x^2}{n^2\pi^2}\right) \qquad (15.226)$$

and

$$\cos x = \prod_{n=1}^{\infty}\left(1 - \frac{4x^2}{(2n - 1)^2\pi^2}\right). \qquad (15.227)$$

These products are finite for all the finite values of x. For $\sin x$ this can easily be seen by taking $a_n = x^2/n^2\pi^2$. Since the series $\sum_{n=1}^{\infty} a_n$ is convergent, the infinite product is also convergent:

$$\sum_{n=1}^{\infty} a_n = \frac{x^2}{\pi^2}\sum_{n=1}^{\infty}\frac{1}{n^2} = \frac{x^2}{\pi^2}\zeta(2) = \frac{x^2}{6}. \qquad (15.228)$$

In the $\sin x$ expression if we take $x = \frac{\pi}{2}$ we obtain

$$1 = \frac{\pi}{2}\prod_{n=1}^{\infty}\left(1 - \frac{1}{(2n)^2}\right) = \frac{\pi}{2}\prod_{n=1}^{\infty}\left(\frac{(2n)^2 - 1}{(2n)^2}\right). \qquad (15.229)$$

Writing this as

$$\frac{\pi}{2} = \prod_{n=1}^{\infty}\left(\frac{(2n)^2}{(2n - 1)(2n + 1)}\right) = \frac{2\cdot 2\,4\cdot 4\,6\cdot 6}{1\cdot 3\,3\cdot 5\,5\cdot 7}\cdots, \qquad (15.230)$$

we obtain Wallis' famous formula for $\pi/2$.

Infinite products can also be used to write the Γ function as

$$\Gamma(x) = \left[xe^{\gamma x}\prod_{r=1}^{\infty}(1 + \frac{x}{r})e^{-x/r}\right]^{-1}, \qquad (15.231)$$

where γ is the Euler-Masheroni constant

$$\gamma = 0.577216\ldots\,. \qquad (15.232)$$

Using Equation (15.231) we can write

$$\Gamma(-x)\Gamma(x) \tag{15.233}$$

$$= \left[-xe^{-\gamma x} \prod_{r=1}^{\infty} (1 - \frac{x}{r})e^{x/r} \right]^{-1} \cdot \left[xe^{\gamma x} \prod_{r=1}^{\infty} (1 + \frac{x}{r})e^{-x/r} \right]^{-1}$$

$$= -\left[x^2 \prod_{r=1}^{\infty} (1 - \frac{x^2}{r^2}) \right]^{-1},$$

which is also equal to

$$\Gamma(x)\Gamma(-x) = -\frac{\pi}{x \sin \pi x}. \tag{15.234}$$

Problems

15.1   Show that the sum of the series

$$\overline{E}_0 = \frac{2\pi}{L} \sum_{n=0}^{\infty} ne^{-2\pi n\alpha/L}$$

is given as

$$\overline{E}_0 = \frac{2\pi}{L} e^{-2\pi\alpha/L} [1 - e^{-2\pi\alpha/L}]^{-2}.$$

Expand the result in powers of α to obtain

$$\overline{E}_0 = \frac{L}{2\pi\alpha^2} - \frac{\pi}{6L} + (\text{terms in positive powers of } \alpha).$$

15.2   Using the Euler-Maclaurin sum formula find the sum of the series given in Problem 15.1, that is,

$$\overline{E}_0 = \frac{2\pi}{L} \sum_{n=0}^{\infty} ne^{-2\pi n\alpha/L},$$

and then show that it agrees with the expansion given in the same problem.

15.3   Find the Casimir energy for the massless conformal scalar field on the surface of a sphere (S-2) with constant radius R_0. The divergent vacuum energy is given as (we set $c=\hbar=1$)

$$\overline{E}_0 = \frac{1}{2} \sum_{l=0}^{\infty} g_l \omega_l,$$

where the degeneracy, g_n, and the eigenfrequencies, w_n, are given as

$$g_l = (2l+1), \quad w_l = \frac{(l+\frac{1}{2})}{R_0}.$$

Note: Interested students can obtain the eigenfrequencies and the degeneracy by solving the wave equation for the massless conformal scalar field:

$$\Box\Phi(\vec{r},t) + \frac{1}{4}\frac{(n-2)}{(n-1)}R\Phi(\vec{r},t) = 0,$$

where n is the dimension of spacetime, R is the scalar curvature, and $\Box$ is the d'Alembert (wave) operator

$$\Box = g_{\mu\nu}\partial^\mu\partial_\nu, \tag{15.235}$$

where ∂_ν stands for the covariant derivative. Use the separation of variables method and impose the boundary condition Φ = finite on the sphere. For this problem spacetime dimension n is 3 and for a sphere of constant radius, R_0, the curvature scalar is $2/R_0^2$.

15.4   Using asymptotic series evaluate the logarithmic integral

$$I = \int_0^x \frac{dt}{\ln t}, \quad 0 < x < 1.$$

Hint: Use the substitutions $t = e^{-u}$ and $a = -\ln x$, $a > 0$, and integrate by parts successively to write the series

$$I = -x\left[\frac{1}{a} - \frac{1!}{a^2} + \frac{2!}{a^3} - \cdots + (-1)^{n-1}\frac{(n-1)!}{a^n}\right] + R_n,$$

where

$$R_n = (-1)^{n+1}n!\int_a^\infty \frac{e^{-t}}{t^{n+1}}dt,$$

so that

$$|R_n| < \frac{n!e^{-a}}{a^{n+1}}.$$

15.5   In a closed Einstein universe the renormalized energy density of a massless conformal scalar field with thermal spectrum can be written as

$$\langle\rho\rangle_{\text{ren.}} = \frac{1}{2\pi^2 R_0^3}\left[\frac{\hbar c}{R_0}\sum_{n=1}^\infty \frac{n^3}{\exp\left(\frac{n\hbar c}{kR_0T}\right) - 1} + \frac{\hbar c}{240R_0}\right],$$

where R_0 is the constant radius of the universe, T is the temperature of the radiation, and $(2\pi^2 R_0^3)$ is the volume of the universe. The second term $(\dfrac{\hbar c}{240R_0})$ inside the square brackets is the well-known renormalized quantum vacuum energy, that is, the Casimir energy for the Einstein universe.

First find the high and low temperature limits of $\langle \rho \rangle_{\mathrm{ren.}}$ and then obtain the flat spacetime limit $R_0 \to \infty$.

15.6 Without using a calculator evaluate the following sum to five decimal places:

$$\sum_{n=6}^{\infty} \frac{1}{n^2}.$$

How many terms did you have to add?

15.7 Check the convergence of the series

(a) $\sum_{n=1}^{\infty} \dfrac{(\ln n)^2}{n}$ (b) $\sum_{n=1}^{\infty} n^2 \exp(-n^2)$

(c) $\sum_{n=1}^{\infty} \ln(1 + \dfrac{1}{n})$ (d) $\sum_{n=1}^{\infty} \dfrac{(-1)^n n^2}{(2n+1)^2}$

(e) $\sum_{n=1}^{\infty} \dfrac{\sqrt{n}}{\sqrt{n^4 + 1}}$ (f) $\sum_{n=1}^{\infty} \dfrac{\sin^2(nx)}{n^4}$

15.8 Find the interval of convergence for the series

(a) $\sum_{n=1}^{\infty} \dfrac{x^n}{\ln(n+2)}$ (b) $\sum_{n=1}^{\infty} \dfrac{(x+1)^n}{\sqrt{n}}$

15.9 Evaluate the sums

(a) $\sum_{n=0}^{\infty} a^n \cos n\theta$

(b) $\sum_{n=0}^{\infty} a^n \sin n\theta,$ a is a constant

Hint: Try using complex variables.

15.10 Verify the following Taylor series:

$$e^x = \sum_{n=0}^{\infty} \frac{x^n}{n!} \quad \text{for all } x$$

and

$$\frac{1}{1+x^2} = 1 - x^2 + \cdots + (-1)^n x^{2n} + \cdots \quad \text{for } |x| < 1.$$

15.11   Find the first three nonzero terms of the following Taylor series:

a)

$$f(x) = x^3 + 2x + 2 \quad \text{about } x = 2$$

b)

$$f(x) = e^{2x} \cos x \quad \text{about } x = 0$$

15.12   Another important consequence of the Lorentz transformation is the formula for the addition of velocities, where the velocities measured in the K and $\overline{K}$ frames are related by the formula

$$u^1 = \frac{\overline{u}^1 + v}{1 + \overline{u}^1 v / c^2},$$

where $u^1 = \dfrac{dx^1}{dt}$ and $\overline{u}^1 = \dfrac{d\overline{x}^1}{d\overline{t}}$ are the velocities measured in the K and $\overline{K}$ frames, respectively, and $\overline{K}$ is moving with respect to K with velocity v along the common direction of the x- and $\overline{x}$-axes. Using the binomial formula find an appropriate expansion of the above formula and show that in the limit of small velocities this formula reduces to the well-known Galilean result

$$u^1 = \overline{u}^1 + v.$$

15.13   In Chapter 10 we have obtained the formulas for the Doppler shift as

$$\omega = \gamma \omega'(1 - \beta \cos \theta)$$
$$\tan \theta' = \sin \theta / \gamma (\cos \theta - \beta),$$

where θ, θ' are the angles of the wave vectors $\overrightarrow{k}$ and $\overrightarrow{\overline{k}}$ with respect to the relative velocity $\overrightarrow{v}$ of the source and the observer. Find the nonrelativistic limit of these equations and interpret your results.

15.14   Given a power series

$$g(x) = \sum_{n=0}^{\infty} a_n x^n, \qquad |x| < R,$$

show that the differentiated and integrated series will have the same radius of convergence.

15.15   Expand

$$h(x) = \tanh x - \frac{1}{x}$$

as a power series of x.

15.16   Find the sum

$$g(x) = \frac{1}{1 \cdot 2} + \frac{x}{2 \cdot 3} + \frac{x^2}{3 \cdot 4} + \frac{x^4}{4 \cdot 5} + \cdots$$

Hint: First try to convert into geometric series.

Answer: $\left[g(x) = \frac{1}{x} + \frac{1-x}{x^2} \ln(1-x) \right]$.

15.17   Using the geometric series evaluate the sum

$$\sum_{n=1}^{\infty} n^3 x^n$$

exactly for the interval $|x| < 1$, then expand your answer in powers of x.

15.18   By using the Euler-Maclaurin sum formula evaluate the sum

$$\sum_{n=1}^{\infty} n^3 x^n.$$

Show that it agrees with the expansion found in Problem 15.17.

16

INTEGRAL
TRANSFORMS

Integral transforms are among the most versatile mathematical tools. Their applications range from solution of differential equations to evaluation of definite integrals and from solution of systems of coupled differential equations to integral equations. They can even be used for defining differintegrals, that is, fractional derivatives and integrals (Chapter 14). In this chapter, after a general introduction we mainly discuss two of the most frequently used integral transforms, the Fourier and the Laplace transforms, their properties, and techniques.

Commonly encountered integral transforms allow us to relate two functions through the integral

$$g(\alpha) = \int_a^b \kappa(\alpha, t) f(t) dt, \tag{16.1}$$

where $g(\alpha)$ is called the integral transform of $f(t)$ with respect to the kernel $\kappa(\alpha, t)$. These transformations are also linear, that is, if the transforms

$$g_1(\alpha) = \int_a^b f_1(t)\kappa(\alpha, t)dt \quad \text{and} \quad g_2(\alpha) = \int_a^b f_2(t)\kappa(\alpha, t)dt \tag{16.2}$$

exist, then one can write

$$g_1(\alpha) + g_2(\alpha) = \int_a^b [f_1(t) + f_2(t)] \kappa(\alpha, t)dt \tag{16.3}$$

and

$$cg_1(\alpha) = \int_a^b [cf_1(t)]\kappa(\alpha, t)dt, \qquad (16.4)$$

where c is a constant. Integral transforms can also be shown as an operator:

$$g(\alpha) = \pounds(\alpha, t)f(t), \qquad (16.5)$$

where the operator $\pounds(\alpha, t)$ is defined as

$$\pounds = \int_a^b dt\kappa(\alpha, t). \qquad (16.6)$$

We can now show the inverse transform as

$$f(t) = \pounds^{-1}(t, \alpha)g(\alpha). \qquad (16.7)$$

16.1   SOME COMMONLY ENCOUNTERED INTEGRAL TRANSFORMS

Fourier transforms are among the most commonly encountered integral transforms. They are defined as

$$g(\alpha) = \frac{1}{\sqrt{2\pi}} \int_{-\infty}^{\infty} f(t)e^{i\alpha t}dt. \qquad (16.8)$$

Because the kernel of the Fourier transform is also used in defining waves, they are generally used in the study of wave phenomena. Scattering of X-rays from atoms is a typical example. The Fourier transform of the amplitude of the scattered waves gives the electron distribution. Fourier cosine and sine transforms are defined as

$$g(\alpha) = \sqrt{\frac{2}{\pi}} \int_0^{\infty} f(t) \cos(\alpha t)dt, \qquad (16.9)$$

$$g(\alpha) = \sqrt{\frac{2}{\pi}} \int_0^{\infty} f(t) \sin(\alpha t)dt. \qquad (16.10)$$

Other frequently used kernels are

$$e^{-\alpha t}, \quad tJ_n(\alpha t), \text{ and } t^{\alpha-1}. \qquad (16.11)$$

The Laplace transform is defined as

$$g(\alpha) = \int_0^{\infty} f(t)e^{-\alpha t}dt \qquad (16.12)$$

and it is very useful in finding solutions of systems of ordinary differential equations by converting them into a system of algebraic equations. The Hankel or Fourier-Bessel transform is defined as

$$g(\alpha) = \int_0^\infty f(t) t J_n(\alpha t) dt, \tag{16.13}$$

and it is usually encountered in potential energy calculations in cylindrical coordinates. Another useful integral transform is the Mellin transform:

$$g(\alpha) = \int_0^\infty f(t) t^{\alpha-1} dt. \tag{16.14}$$

The Mellin transform is useful in the reconstruction of "Weierstrass-type functions" from power series expansions. Weierstrass function is defined as

$$f_W(x) = \sum_{n=0}^\infty b^n \cos[a^n \pi x], \tag{16.15}$$

where a and b are constants. It has been proven that, provided $0 < b < 1$, $a > 1$, and $ab > 1$, the Weierstrass function has the interesting property of being continuous but nowhere differentiable. These interesting functions have found widespread use in the study of earthquakes, rupture, financial crashes, etc. (Gluzman and Sornette).

16.2   DERIVATION OF THE FOURIER INTEGRAL

16.2.1   Fourier Series

Fourier series are very useful in representing a function in a finite interval, like $[0, 2\pi]$ or $[-L, L]$, or a periodic function in the infinite interval $(-\infty, \infty)$. We now consider a nonperiodic function in the infinite interval $(-\infty, \infty)$. Physically this corresponds to expressing an arbitrary signal in terms of sine and cosine waves. We first consider the trigonometric Fourier expansion of a sufficiently smooth function in the finite interval $[-L, L]$ as

$$f(x) = \frac{1}{2L} \int_{-L}^L f(t) dt + \sum_{n=1}^\infty a_n \cos\frac{n\pi x}{L} + \sum_{n=1}^\infty b_n \sin\frac{n\pi x}{L}. \tag{16.16}$$

Fourier expansion coefficients a_n and b_n are given as

$$a_n = \frac{1}{L} \int_{-L}^L f(t) \cos\frac{n\pi t}{L} dt, \tag{16.17}$$

$$b_n = \frac{1}{L} \int_{-L}^L f(t) \sin\frac{n\pi t}{L} dt. \tag{16.18}$$

Substituting a_n and b_n explicitly into the Fourier series and using the trigonometric identity

$$\cos(a - b) = \cos a \cos b + \sin a \sin b, \tag{16.19}$$

we get

$$f(x) = \frac{1}{2L} \int_{-L}^{L} f(t)dt + \frac{1}{L} \sum_{n=1}^{\infty} \int_{-L}^{L} f(t) \cos\left[\frac{n\pi}{L}(t - x)\right] dt. \tag{16.20}$$

Since the eigenfrequencies are given as $\omega = \frac{n\pi}{L}$, where $n = 0, 1, 2, \ldots$, the distance between two neighboring eigenfrequencies is

$$\Delta \omega = \frac{\pi}{L}. \tag{16.21}$$

Using Equation (16.21) we can write

$$f(x) = \frac{1}{2L} \int_{-L}^{L} f(t)dt + \frac{1}{\pi} \sum_{n=1}^{\infty} \Delta \omega \int_{-\infty}^{\infty} f(t) \cos \omega(t - x)dt. \tag{16.22}$$

We now take the continuum limit, $L \to \infty$, where we can make the replacement

$$\sum_{n=1}^{\infty} \Delta \omega \to \int_{0}^{\infty} d\omega. \tag{16.23}$$

Thus the Fourier integral is obtained as

$$f(x) = \frac{1}{\pi} \int_{0}^{\infty} d\omega \int_{-\infty}^{\infty} f(t) \cos \omega(t - x)dt. \tag{16.24}$$

In this expression we have assumed the existence of the integral

$$\int_{-\infty}^{\infty} f(t)dt. \tag{16.25}$$

For the Fourier integral of a function to exist, it is sufficient for the integral $\int_{-\infty}^{\infty} |f(t)|\, dt$ to be convergent.

We can also write the Fourier integral in exponential form. Using the fact that $\sin \omega(t - x)$ is an odd function with respect to ω, we can write

$$\frac{1}{2\pi} \int_{-\infty}^{\infty} d\omega \int_{-\infty}^{\infty} f(t) \sin \omega(t - x)dt = 0. \tag{16.26}$$

Also since $\cos \omega(t - x)$ is an even function with respect to ω, we can extend the range of the ω integral in the Fourier integral to $(-\infty, \infty)$ as

$$f(x) = \frac{1}{2\pi} \int_{-\infty}^{\infty} d\omega \int_{-\infty}^{\infty} f(t) \cos \omega(t - x)dt. \tag{16.27}$$

We now multiply Equation (16.26) by i and then add Equation (16.27) to obtain the exponential form of the Fourier integral as

$$f(x) = \frac{1}{2\pi} \int_{-\infty}^{\infty} d\omega\, e^{-i\omega x} \int_{-\infty}^{\infty} f(t) e^{i\omega t} dt, \qquad (16.28)$$

where ω is a parameter; however, in applications to waves it is the angular frequency.

16.2.2   Dirac-Delta Function

Let us now write the Fourier integral as

$$f(x) = \int_{-\infty}^{\infty} f(t) \left\{ \frac{1}{2\pi} \int_{-\infty}^{\infty} e^{i\omega(t-x)} d\omega \right\} dt, \qquad (16.29)$$

where we have interchanged the order of integration. The expression inside the curly brackets is nothing but the Dirac-delta function:

$$\delta(t - x) = \frac{1}{2\pi} \int_{-\infty}^{\infty} e^{i\omega(t-x)} d\omega, \qquad (16.30)$$

which has the following properties:

$$\delta(x - a) = 0, \quad (x \neq a), \qquad (16.31)$$

$$\int_{-\infty}^{\infty} \delta(x - a) dx = 1, \qquad (16.32)$$

$$\int_{-\infty}^{\infty} \delta(x - a) f(x) dx = f(a), \qquad (16.33)$$

where $f(x)$ is continuous at $x = a$.

16.3   FOURIER AND INVERSE FOURIER TRANSFORMS

We write the Fourier integral theorem [Eq. (16.28)] as

$$f(t) = \frac{1}{\sqrt{2\pi}} \int_{-\infty}^{\infty} d\omega\, e^{-i\omega t} \left[\frac{1}{\sqrt{2\pi}} \int_{-\infty}^{\infty} f(t') e^{i\omega t'} dt' \right].$$

We now define the Fourier transform of $f(t)$ as

$$g(\omega) = \frac{1}{\sqrt{2\pi}} \int_{-\infty}^{\infty} f(t) e^{i\omega t} dt, \qquad (16.34)$$

where the inverse Fourier transform is defined as

$$f(t) = \frac{1}{\sqrt{2\pi}} \int_{-\infty}^{\infty} g(\omega) e^{-i\omega t} d\omega. \qquad (16.35)$$

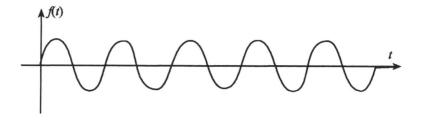

Fig. 16.1   Wave train with $N = 5$

16.3.1   Fourier Sine and Cosine Transforms

When $f(t)$ is an even function we can write

$$f_c(-x) = f_c(x). \tag{16.36}$$

Using the identity

$$e^{i\omega t} = \cos \omega t + i \sin \omega t, \tag{16.37}$$

we can also write

$$g_c(\omega) = \frac{1}{\sqrt{2\pi}} \int_{-\infty}^{\infty} f_c(t) \left(\cos \omega t + i \sin \omega t \right) dt. \tag{16.38}$$

Considering that $\sin \omega t$ is an odd function with respect to t, the Fourier cosine transform is obtained as

$$g_c(\omega) = \sqrt{\frac{2}{\pi}} \int_0^{\infty} f_c(t) \cos \omega t dt. \tag{16.39}$$

The inverse Fourier cosine transform is given as

$$f_c(t) = \sqrt{\frac{2}{\pi}} \int_0^{\infty} g_c(\omega) \cos \omega t d\omega. \tag{16.40}$$

Similarly, for an odd function we can write

$$f_s(-x) = -f_s(x). \tag{16.41}$$

From the Fourier integral we obtain its Fourier sine transform as

$$g_s(\omega) = \sqrt{\frac{2}{\pi}} \int_0^{\infty} f_s(t) \sin \omega t dt, \tag{16.42}$$

and its inverse Fourier sine transform is

$$f_s(x) = \sqrt{\frac{2}{\pi}} \int_0^{\infty} g_s(\omega) \sin \omega x d\omega. \tag{16.43}$$

Example 16.1. *Fourier analysis of finite wave train:* We now find the Fourier transform of a finite wave train, which is given as

$$
f(t) = \begin{cases} \sin \omega_0 t & |t| < \dfrac{N\pi}{\omega_0} \\ 0 & |t| > \dfrac{N\pi}{\omega_0}. \end{cases}
\tag{16.44}
$$

For $N = 5$ this wave train is shown in Figure 16.1.

Since $f(t)$ is an odd function we find its Fourier sine transform as

$$
g_s(\omega) = \sqrt{\frac{2}{\pi}} \left[\frac{\sin(\omega_0 - \omega)\frac{N\pi}{\omega_0}}{2(\omega_0 - \omega)} - \frac{\sin(\omega_0 + \omega)\frac{N\pi}{\omega_0}}{2(\omega_0 + \omega)} \right].
\tag{16.45}
$$

For frequencies $\omega \sim \omega_0$ only the first term in Equation (16.45) dominates. Thus $g_s(\omega)$ is given as in Figure 16.2.
This is the diffraction pattern for a single slit. It has zeroes at

$$
\frac{\omega_0 - \omega}{\omega_0} = \frac{\Delta\omega}{\omega_0} = \pm\frac{1}{N}, \pm\frac{2}{N}, \dots .
\tag{16.46}
$$

Because the contribution coming from the central maximum dominates the others, to form the wave train [Eq. (16.44)] it is sufficient to take waves with the spread in their frequency distribution as

$$
\Delta\omega = \frac{\omega_0}{N}.
\tag{16.47}
$$

For a longer wave train, naturally the spread in frequency is less.

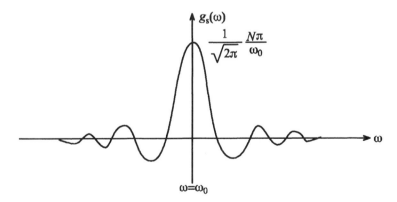

Fig. 16.2 $g_s(\omega)$ function

16.3.2   Fourier Transform of a Derivative

First we find the Fourier transform of $\dfrac{df(x)}{dx}$ as

$$g_1(\omega) = \frac{1}{\sqrt{2\pi}} \int_{-\infty}^{\infty} \frac{df(x)}{dx} e^{i\omega x} dx. \tag{16.48}$$

Using integration by parts we write

$$g_1(\omega) = \frac{e^{i\omega x}}{\sqrt{2\pi}} f(x)\Big|_{-\infty}^{\infty} - \frac{i\omega}{\sqrt{2\pi}} \int_{-\infty}^{\infty} f(x) e^{i\omega x} dx. \tag{16.49}$$

Assuming that $f(x) \to 0$ as $x \to \pm\infty$ we obtain the Fourier transform of the first derivative as

$$g_1(\omega) = -i\omega g(\omega). \tag{16.50}$$

Assuming that all the derivatives

$$f^{n-1}(x), \ f^{n-2}(x), \ f^{n-3}(x), \dots, \ f(x) \tag{16.51}$$

go to zero as $x \to \pm\infty$, we write the Fourier transform of the nth derivative as

$$g_n(\omega) = (-i\omega)^n g(\omega). \tag{16.52}$$

Example 16.2. *Partial differential equations and Fourier transforms:*
One of the many uses of integral transforms is solving partial differential equations. Consider vibrations of an infinitely long wire. The equation to be solved is the wave equation, which is given as

$$\frac{\partial^2 y(x,t)}{\partial x^2} = \frac{1}{v^2} \frac{\partial^2 y(x,t)}{\partial t^2}, \tag{16.53}$$

where v is the velocity of the wave and $y(x,t)$ is the displacement of the wire from its equilibrium position as a function of position and time. As our initial condition we take the shape of the wire at $t = 0$ as

$$y(x,0) = f(x). \tag{16.54}$$

We now take the Fourier transform of the wave equation with respect to x as

$$\int_{-\infty}^{\infty} \frac{d^2 y(x,t)}{dx^2} e^{i\alpha x} dx = \frac{1}{v^2} \int_{-\infty}^{\infty} \frac{d^2 y(x,t)}{dt^2} e^{i\alpha x} dx, \tag{16.55}$$

$$(-i\alpha)^2 Y(\alpha,t) = \frac{1}{v^2} \frac{d^2 Y(\alpha,t)}{dt^2}, \tag{16.56}$$

where $Y(\alpha, t)$ represents the Fourier transform of $y(x, t)$:

$$Y(\alpha, t) = \frac{1}{\sqrt{2\pi}} \int_{-\infty}^{\infty} y(x, t) e^{i\alpha x} dx. \tag{16.57}$$

From Equation (16.56) we see that the effect of the integral transform on the partial differential equation is to reduce the number of independent variables. Thus the differential equation to be solved for $Y(a, t)$ is now an ordinary differential equation, solution of which can be written easily as

$$Y(\alpha, t) = F(\alpha) e^{\pm ivat}. \tag{16.58}$$

$F(\alpha)$ is the Fourier transform of the initial condition

$$y(x, 0) = f(x),$$

which gives

$$F(\alpha) = Y(\alpha, 0) \tag{16.59}$$

$$= \frac{1}{\sqrt{2\pi}} \int_{-\infty}^{\infty} f(x) e^{i\alpha x} dx. \tag{16.60}$$

To be able to interpret this solution we must go back to $y(x, t)$ by taking the inverse Fourier transform of $Y(a, t)$ as

$$y(x, t) = \frac{1}{\sqrt{2\pi}} \int_{-\infty}^{\infty} Y(\alpha, t) e^{-i\alpha x} d\alpha, \tag{16.61}$$

$$= \frac{1}{\sqrt{2\pi}} \int_{-\infty}^{\infty} F(\alpha) e^{-i\alpha(x \mp vt)} d\alpha.$$

Because the last expression is nothing but the inverse Fourier transform of $F(\alpha)$, we can write the final solution as

$$y(x, t) = f(x \mp vt). \tag{16.62}$$

This represents a wave moving to the right or left with the velocity v and with its shape unchanged.

16.3.3   Convolution Theorem

Let $F(t)$ and $G(t)$ be the Fourier transforms of two functions $f(x)$ and $g(x)$, respectively. Convolution of $f(x)$ and $g(x)$ is defined as

$$f * g = \frac{1}{\sqrt{2\pi}} \int_{-\infty}^{\infty} dy g(y) f(x - y). \tag{16.63}$$

Using the definition of Fourier transforms we can write the right-hand side as

$$\int_{-\infty}^{\infty} g(y)f(x-y)dy = \frac{1}{\sqrt{2\pi}} \int_{-\infty}^{\infty} dy g(y) \left[\int_{-\infty}^{\infty} F(t)e^{-it(x-y)} dt \right],$$

$$= \frac{1}{\sqrt{2\pi}} \int_{-\infty}^{\infty} dt F(t)e^{-itx} \int_{-\infty}^{\infty} dy g(y)e^{ity},$$

$$= \int_{-\infty}^{\infty} dt F(t)G(t)e^{-itx}, \qquad (16.64)$$

which is nothing but the convolution of $f(x)$ and $g(x)$. For the special case with $x = 0$ we get

$$\int_{-\infty}^{\infty} F(t)G(t)dt = \int_{-\infty}^{\infty} g(y)f(-y)dy. \qquad (16.65)$$

16.3.4   Existence of Fourier Transforms

We can show the Fourier transform of $f(x)$ in terms of an integral operator $\Im$ as

$$F(k) = \Im\{f(x)\}, \qquad (16.66)$$

$$\Im = \int_{-\infty}^{+\infty} dx e^{ikx}.$$

For the existence of the Fourier transform of $f(x)$, a sufficient but not necessary condition is the convergence of the integral

$$\int_{-\infty}^{\infty} |f(x)| \, dx. \qquad (16.67)$$

16.3.5   Fourier Transforms in Three Dimensions

Fourier transforms can also be defined in three dimensions as

$$\phi(\overrightarrow{k}) = \frac{1}{\sqrt[3]{2\pi}} \int_{-\infty}^{\infty} \int_{-\infty}^{\infty} \int_{-\infty}^{\infty} d^3\overrightarrow{r} f(\overrightarrow{r})e^{i\overrightarrow{k}\cdot\overrightarrow{r}}, \qquad (16.68)$$

$$f(\overrightarrow{r}) = \frac{1}{\sqrt[3]{2\pi}} \int_{-\infty}^{\infty} \int_{-\infty}^{\infty} \int_{-\infty}^{\infty} d^3k \phi(\overrightarrow{k})e^{-i\overrightarrow{k}\cdot\overrightarrow{r}}. \qquad (16.69)$$

Substituting Equation (16.68) back in Equation (16.69) and interchanging the order of integration we obtain the three dimensional Dirac-delta function as

$$\delta(\overrightarrow{r} - \overrightarrow{r}') = \frac{1}{(2\pi)^3} \int_{-\infty}^{\infty} \int_{-\infty}^{\infty} \int_{-\infty}^{\infty} d^3k e^{i\overrightarrow{k}\cdot(\overrightarrow{r} - \overrightarrow{r}')}. \qquad (16.70)$$

These formulas can easily be extended to n dimensions.

16.4 SOME THEOREMS ON FOURIER TRANSFORMS

Parseval Theorem I

$$\int_{-\infty}^{\infty} |F(k)|^2 \, dk = \int_{-\infty}^{\infty} |f(x)|^2 \, dx, \tag{16.71}$$

Parseval Theorem II

$$\int_{-\infty}^{\infty} F(k)G(-k) dk = \int_{-\infty}^{\infty} g(x)f(x) dx, \tag{16.72}$$

where $F(k)$ and $G(k)$ are the Fourier transforms of $f(x)$ and $g(x)$, respectively.

Proof: To prove these theorems we make the $k \to -k$ change in the Fourier transform of $g(x)$:

$$G(-k) = \frac{1}{\sqrt{2\pi}} \int_{-\infty}^{\infty} g(x)e^{-ikx} dx. \tag{16.73}$$

Multiplying the integral in Equation (16.73) with $F(k)$ and integrating it over k in the interval $(-\infty, \infty)$ we get

$$\int_{-\infty}^{\infty} dk F(k)G(-k) = \int_{-\infty}^{\infty} dk F(k)\frac{1}{\sqrt{2\pi}} \int_{-\infty}^{\infty} dx g(x)e^{-ikx}. \tag{16.74}$$

Assuming that the integrals

$$\int_{-\infty}^{\infty} |f(x)| \, dx \text{ and } \int_{-\infty}^{\infty} |g(x)| \, dx \tag{16.75}$$

converge, we can change the order of the k and x integrals as

$$\int_{-\infty}^{\infty} F(k)G(-k) dk = \int_{-\infty}^{\infty} dx g(x)\frac{1}{\sqrt{2\pi}} \int_{-\infty}^{\infty} dk F(k)e^{-ikx}. \tag{16.76}$$

Assuming that the inverse Fourier transform of $F(k)$ exists, the second Parceval theorem is proven.

If we take

$$f(x) = g(x) \tag{16.77}$$

in Equation (16.72), remembering that

$$G(-k) = G(k)^*, \tag{16.78}$$

we can write

$$\int_{-\infty}^{\infty} |G(k)|^2 \, dk = \int_{-\infty}^{\infty} |g(x)|^2 \, dx, \tag{16.79}$$

which is the first Parceval theorem. From this proof it is seen that pointwise existence of the inverse Fourier transform is not necessary; that is, as long as the value of the integral

$$\int_{-\infty}^{\infty} g(x)f(x)dx \tag{16.80}$$

does not change, the value of the integral

$$\frac{1}{\sqrt{2\pi}} \int_{-\infty}^{\infty} F(k)dke^{-ikx} \tag{16.81}$$

can be different from the value of $f(x)$ at some isolated singular points. In quantum mechanics wave functions in position and momentum spaces are each others' Fourier transforms. The significance of Parseval's theorems is that normalization in one space ensures normalization in the other.

Example 16.3. *Diffusion problem in one dimension:* Let us consider a long thin pipe filled with water and with M g of salt located at x_0. We would like to find the concentration of salt as a function of position and time. Because we have a thin pipe, we can neglect the change in concentration across the width of the pipe. The density (concentration×mass)

$$\rho = \rho(x,t) \text{ g/cm}^3 \tag{16.82}$$

satisfies the diffusion equation:

$$\frac{\partial \rho}{\partial t} = D\frac{\partial^2 \rho}{\partial x^2}. \tag{16.83}$$

Because at $t = 0$, the density is zero everywhere except at x_0, we take our initial condition as

$$\rho(x,0) = \left(\frac{M}{A}\right)\delta(x - x_0). \tag{16.84}$$

In addition, for all times the limits

$$\lim_{x \to \pm\infty} \rho(x,t) = \rho(\pm\infty, t) = 0 \tag{16.85}$$

must be satisfied. Because we have an infinite pipe and the density vanishes at the end points, we have to use Fourier transforms rather than the Fourier series. Because the total amount of salt is conserved, we have

$$\int_{-\infty}^{\infty} \rho(x,t)Adx = m, \tag{16.86}$$

which is sufficient for the existence of the Fourier transform. Taking the Fourier transform of the diffusion equation with respect to x we get

$$\frac{dR(k,t)}{dt} = -Dk^2 R(k,t).$$

(16.87)

This is an ordinary differential equation, where $R(k,t)$ is the Fourier transform of the density. The initial condition for $R(k,t)$ is the Fourier transform of the initial condition for the density, that is,

$$R(k,0) = \Im\left\{\left(\frac{M}{A}\right)\delta\left(x - x_0\right)\right\} = \frac{M}{A}\frac{1}{\sqrt{2\pi}}e^{ikx_0}.$$

(16.88)

The solution of Equation (16.87) can easily be written as

$$R(k,t) = R(k,0)e^{-Dk^2 t}.$$

(16.89)

To find the density we have to find the inverse Fourier transform of

$$R(k,t) = \frac{M}{A\sqrt{2\pi}}e^{ikx_0}e^{-Dk^2 t}.$$

(16.90)

After some rearrangement this gives

$$
\begin{aligned}
\rho(x,t) &= \frac{M}{A2\pi}\int_{-\infty}^{\infty} e^{ikx_0}e^{-Dk^2 t}e^{-ikx}\,dk \\
&= \frac{M}{A2\pi}\int_{-\infty}^{\infty} e^{-Dk^2 t - ik(x-x_0)}\,dk \\
&= \frac{M}{A2\pi}\int_{-\infty}^{\infty} e^{-Dt(k^2 + ik\frac{(x-x_0)}{Dt})}\,dk \\
&= \frac{M}{A2\pi}\int_{-\infty}^{\infty} e^{-Dt\left[k^2 + ik\frac{(x-x_0)}{Dt} - \left(\frac{(x-x_0)}{2Dt}\right)^2 + \left(\frac{(x-x_0)}{2Dt}\right)^2\right]}\,dk \\
&= \frac{M}{A2\pi}\int_{-\infty}^{\infty} e^{-Dt\left[k + \frac{i(x-x_0)}{2Dt}\right]^2}e^{-\frac{(x-x_0)^2}{4Dt}}\,dk \\
&= \frac{M}{A2\pi}e^{-\frac{(x-x_0)^2}{4Dt}}\int_{-\infty}^{\infty} e^{-Dt\left[k + \frac{i(x-x_0)}{2Dt}\right]^2}\,dk \\
&= \frac{M}{A2\pi}e^{-\frac{(x-x_0)^2}{4Dt}}\int_{-\infty}^{\infty} e^{-Dtu^2}\,du \\
&= \frac{M}{A2\pi}e^{-\frac{(x-x_0)^2}{4Dt}}\sqrt{\frac{\pi}{Dt}}.
\end{aligned}
$$

(16.91)

Hence the density is finally obtained as

$$\rho(x,t) = \frac{M}{A}\frac{1}{\sqrt{4\pi Dt}}e^{-\frac{(x-x_0)^2}{4Dt}}.$$

(16.92)

Check that this solution satisfies the diffusion equation with the initial condition

$$\lim_{t \to 0} \rho(x, t) = \left(\frac{M}{A}\right) \delta\left(x - x_0\right). \tag{16.93}$$

16.5 LAPLACE TRANSFORMS

The Laplace transform of a function is defined as

$$
\begin{aligned}
f(s) &= \mathcal{L}\left\{F(t)\right\} \\
&= \lim_{a \to \infty} \int_0^a e^{-st} F(t) dt \\
&= \int_0^\infty e^{-st} F(t) dt. \quad s > 0 \text{ and real.}
\end{aligned}
\tag{16.94}
$$

For this transformation to exist we do not need the existence of the integral

$$\int_0^\infty F(t) dt. \tag{16.95}$$

In other words, $F(t)$ could diverge exponentially for large values of t. However, if there exists a constant s_0 and a positive constant C, such that for sufficiently large t, that is $t > t_0$, the inequality

$$\left|e^{-s_0 t} F(t)\right| \leq C \tag{16.96}$$

is satisfied, then the Laplace transform of this function exists for $s > s_0$. An example is the

$$F(t) = e^{2t^2} \tag{16.97}$$

function. For this function we cannot find a suitable s_0 and a C value that satisfies Equation (16.96); hence its Laplace transform does not exist. The Laplace transform may also fail to exist if the function $F(t)$ has a sufficiently strong singularity as $t \to 0$. The Laplace transform of t^n

$$\mathcal{L}\left\{t^n\right\} = \int_0^\infty e^{-st} t^n dt \tag{16.98}$$

does not exist for $n \leq -1$ values, because it has a singular point at the origin. On the other hand, for $s > 0$ and $n > -1$, the Laplace transform is given as

$$\mathcal{L}\left\{t^n\right\} = \frac{n!}{s^{n+1}}. \tag{16.99}$$

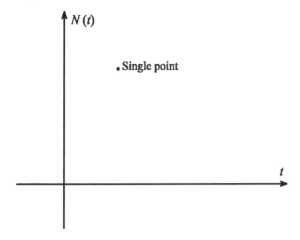

Fig. 16.3   Null function

16.6   INVERSE LAPLACE TRANSFORMS

Using operator language we can show the Laplace transform of a function as

$$f(s) = \mathcal{L}\{F(t)\}. \tag{16.100}$$

The inverse transform of $f(s)$ is now shown with $\mathcal{L}^{-1}$ as

$$\mathcal{L}^{-1}\{f(s)\} = F(t). \tag{16.101}$$

In principle, the inverse transform is not unique. Two functions $F_1(t)$ and $F_2(t)$ could have the same Laplace transform; however, in such cases the difference of these functions is

$$F_1(t) - F_2(t) = N(t), \tag{16.102}$$

where for all t_0 values $N(t)$ satisfies

$$\int_0^{t_0} N(t)dt = 0. \tag{16.103}$$

$N(t)$ is called a null function, and this result is also known as the **Lerch theorem**. In practice we can take $N(t)$ as zero, thus making the inverse Laplace transform unique. In Figure 16.3 we show a null function.

16.6.1 Bromwich Integral

A formal expression for the inverse Laplace transform is given in terms of the Bromwich integral as

$$F(t) = \lim_{\alpha \to \infty} \frac{1}{2\pi i} \int_{\gamma - i\alpha}^{\gamma + i\alpha} e^{st} f(s) ds, \qquad (16.104)$$

where γ is real and s is now a complex variable. The contour for the above integral is an infinite straight line passing through the point γ and parallel to the imaginary axis in the complex s-plane. γ is chosen such that all the singularities of $e^{st} f(s)$ are to the left of the straight line. For $t > 0$ we can close the contour with an infinite semicircle to the left-hand side of the line. The above integral can now be evaluated by using the residue theorem to find the inverse Laplace transform.

The Bromwich integral is a powerful tool for inverting complicated Laplace transforms when other means prove inadequate. However, in practice using the fact that Laplace transforms are linear and with the help of some basic theorems we can generate many of the inverses needed from a list of elementary Laplace transforms.

16.6.2 Elementary Laplace Transforms

1. Many of the discontinuous functions can be expressed in terms of the Heavyside step function (Fig. 16.4)

$$U(t - a) = \begin{cases} 0 & t < a \\ 1 & t > a \end{cases}, \qquad (16.105)$$

   the Laplace transform of which is given as

$$\mathcal{L}\left\{U(t - a)\right\} = \frac{e^{-as}}{s}, \quad s > 0. \qquad (16.106)$$

2. For $F(t) = 1$, the Laplace transform is given as

$$\mathcal{L}\left\{1\right\} = \int_0^\infty e^{-st} dt = \frac{1}{s}, \quad s > 0. \qquad (16.107)$$

3. The Laplace transform of $F(t) = e^{kt}$ for $t > 0$ is

$$\mathcal{L}\left\{e^{kt}\right\} = \int_0^\infty e^{kt} e^{-st} dt = \frac{1}{s - k}, \quad s > k. \qquad (16.108)$$

4. Laplace transforms of hyperbolic functions

$$F(t) = \cosh kt = \frac{1}{2}\left(e^{kt} + e^{-kt}\right)$$

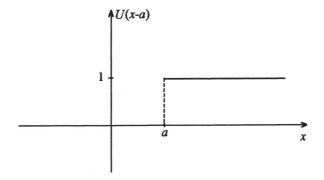

Fig. 16.4   Heavyside step function

and

$$F(t) = \sinh kt = \frac{1}{2} \left(e^{kt} - e^{-kt} \right)$$

can be found by using the fact that $\mathcal{L}$ is a linear operator as

$$\mathcal{L} \left\{ \cosh kt \right\} = \frac{1}{2} \left(\frac{1}{s - k} + \frac{1}{s + k} \right) = \frac{s}{s^2 - k^2}, \qquad (16.109)$$

$$\mathcal{L} \left\{ \sinh kt \right\} = \frac{k}{s^2 - k^2}, \qquad (16.110)$$

where $s > k$ for both.

5. Using the relations

$$\cos kt = \cosh ikt \qquad \text{and} \qquad \sin kt = -i \sinh kt,$$

we can find the Laplace transforms of the cosine and the sine functions as

$$\mathcal{L} \left\{ \cos kt \right\} = \frac{s}{s^2 + k^2}, \quad s > 0, \qquad (16.111)$$

$$\mathcal{L} \left\{ \sin kt \right\} = \frac{k}{s^2 + k^2}, \quad s > 0. \qquad (16.112)$$

6. For $F(t) = t^n$ we have

$$\mathcal{L}\{t^n\} = \frac{n!}{s^{n+1}}, \quad s > 0, \; n > -1. \qquad (16.113)$$

16.6.3   Theorems About Laplace Transforms

By using the entries in a list of transforms, the following theorems are very useful in finding inverses of unknown transforms:

Theorem I: First Translation Theorem.

If the function $f(t)$ has the Laplace transform

$$\mathcal{L}\left\{f(t)\right\} = F(s), \tag{16.114}$$

then the Laplace transform of $e^{at}f(t)$ is given as

$$\mathcal{L}\left\{e^{at}f(t)\right\} = F(s-a). \tag{16.115}$$

Similarly, if $\mathcal{L}^{-1}\left\{F(s)\right\} = f(t)$ is true, then we can write

$$\mathcal{L}^{-1}\left\{\ F(s-a)\right\} = e^{at}f(t). \tag{16.116}$$

Proof:

$$\mathcal{L}\left\{e^{at}f(t)\right\} = \int_0^\infty e^{at}f(t)e^{-st}dt \tag{16.117}$$

$$= \int_0^\infty e^{-(s-a)t}f(t)dt$$

$$= F(s-a).$$

Theorem II: Second Translation Theorem.

If $F(s)$ is the Laplace transform of $f(t)$ and the Heavyside step function is shown as $U(t-a)$, we can write

$$\mathcal{L}\left\{U(t-a)f(t-a)\right\} = e^{-as}F(s). \tag{16.118}$$

Similarly, if $\mathcal{L}^{-1}\left\{F(s)\right\} = f(t)$ is true, then we can write

$$\mathcal{L}^{-1}\left\{\ e^{-as}F(s)\right\} = U(t-a)f(t-a). \tag{16.119}$$

Proof: Since the Heavyside step function is defined as

$$U(t-a) = \left\{ \begin{array}{ll} 0 & t < a \\ 1 & t > a \end{array} \right. \tag{16.120}$$

we can write

$$\mathcal{L}\left\{U(t-a)f(t-a)\right\} = \int_0^\infty e^{-st}U(t-a)f(t-a)dt$$

$$= \int_a^\infty e^{-st}f(t-a)dt. \tag{16.121}$$

Changing the integration variable to $v = t - a$ we obtain

$$\mathcal{L}\{U(t-a)f(t-a)\} = \int_0^\infty e^{-s(v+a)}f(v)dv \qquad (16.122)$$

$$= e^{-as}\int_0^\infty e^{-sv}f(v)dv$$

$$= e^{-as}F'(s).$$

Theorem III: If $\mathcal{L}\{f(t)\} = F(s)$ is true, then we can write

$$\mathcal{L}\{f(at)\} = \frac{1}{a}F(\frac{s}{a}). \qquad (16.123)$$

If $\mathcal{L}^{-1}\{F(s)\} = f(t)$ is true, then we can write the inverse as

$$\mathcal{L}^{-1}\left\{F(\frac{s}{a})\right\} = af(at). \qquad (16.124)$$

Proof: Using the definition of the Laplace transform we write

$$\mathcal{L}\{f(at)\} = \int_0^\infty e^{-st}f(at)dt. \qquad (16.125)$$

Changing the integration variable to $v = at$ we find

$$\mathcal{L}\{f(at)\} = \frac{1}{a}\int_0^\infty e^{-sv/a}f(v)dv \qquad (16.126)$$

$$= \frac{1}{a}F(\frac{s}{a}).$$

Theorem IV: Derivative of a Laplace Transform.

If the Laplace transform of $f(t)$ is $F(s)$, then the Laplace transform of $t^n f(t)$ is given as

$$\mathcal{L}\{t^n f(t)\} = (-1)^n \frac{d^n F(s)}{dt^n} = (-1)^n F^{(n)}(s), \qquad (16.127)$$

where $n = 0, 1, 2, 3... $.

Similarly, if $\mathcal{L}^{-1}\{F(s)\} = f(t)$ is true, then we can write

$$\mathcal{L}^{-1}\left\{F^{(n)}(s)\right\} = (-1)^n t^n f(t). \qquad (16.128)$$

Proof: Since $\mathcal{L}\{f(t)\} = F(s)$, we write

$$F(s) = \int_0^\infty e^{-st}f(t)dt. \qquad (16.129)$$

Taking the derivative of both sides with respect to s we get

$$\int_0^\infty e^{-st} t f(t) dt = -F'(s).$$
(16.130)

If we keep on differentiating we find

$$\int_0^\infty e^{-st} t^2 f(t) dt = F''(s),$$
(16.131)

$$\int_0^\infty e^{-st} t^3 f(t) dt = -F'''(s)$$
(16.132)

and eventually the nth derivative as

$$\int_0^\infty e^{-st} t^n f(t) dt = (-1)^n F^{(n)}(s).$$
(16.133)

Theorem V: Laplace Transform of Periodic Functions.

If $f(t)$ is a periodic function with the period $p > 0$, that is, $f(t + p) = f(t)$, then we can write

$$\pounds\{f(t)\} = \frac{\int_0^p e^{-st} f(t) dt}{1 - e^{-sp}}.$$
(16.134)

On the other hand, if $\pounds^{-1}\{F(s)\} = f(t)$ is true, then we can write

$$\pounds^{-1}\left\{\frac{\int_0^p e^{-st} f(t) dt}{1 - e^{-sp}}\right\} = f(t).$$
(16.135)

Proof: We first write

$$\pounds\{f(t)\} = \int_0^\infty e^{-st} f(t) dt,$$
(16.136)

$$= \int_0^p e^{-st} f(t) dt + \int_p^{2p} e^{-st} f(t) dt$$

$$+ \int_{2p}^{3p} e^{-st} f(t) dt + \cdots,$$
(16.137)

$$= \int_0^p e^{-st} f(t) dt + \int_0^p e^{-s(v+p)} f(v + p) dv$$

$$+ \int_0^p e^{-s(v+2p)} f(v + 2p) dv + \cdots.$$
(16.138)

Making the variable change $v \to t$ and using the fact that $f(t)$ is periodic we get

$$\int_0^\infty e^{-st} f(t) dt \tag{16.139}$$

$$= \int_0^p e^{-st} f(t) dt + e^{-sp} \int_0^p e^{-st} f(t) dt$$

$$+ e^{-s2p} \int_0^p e^{-st} f(t) dt + \cdots \tag{16.140}$$

$$= \left(1 + e^{-sp} + e^{-s2p} + e^{-s3p} + \cdots\right) \int_0^p e^{-st} f(t) dt \tag{16.141}$$

$$= \frac{\int_0^p e^{-st} f(t) dt}{1 - e^{-sp}}, \quad s > 0. \tag{16.142}$$

Theorem VI: Laplace Transform of an Integral.

If the Laplace transform of $f(t)$ is $F(s)$, then we can write

$$£\left\{\int_0^t f(u) du\right\} = \frac{F(s)}{s}. \tag{16.143}$$

Similarly, if $£^{-1}\{F(s)\} = f(t)$ is true, then the inverse will be given as

$$£^{-1}\left\{\frac{F(s)}{s}\right\} = \int_0^t f(u) du. \tag{16.144}$$

Proof: Let us define the $G(t)$ function as

$$G(t) = \int_0^t f(u) du. \tag{16.145}$$

Now we have $G'(t) = f(t)$ and $G(0) = 0$; thus we can write

$$£\{G'(t)\} = s£\{G(t)\} - G(0) \tag{16.146}$$

$$= s£\{G(t)\}, \tag{16.147}$$

which gives

$$£\{G(t)\} = £\left\{\int_0^t f(u) du\right\} = \frac{1}{s}£\{f(t)\} = \frac{F(s)}{s}. \tag{16.148}$$

Theorem VII: If the limit $\lim_{t \to 0} \frac{f(t)}{t}$ exists and if the Laplace transform of $f(t)$ is $F(s)$, then we can write

$$£\left\{\frac{f(t)}{t}\right\} = \int_s^\infty F(u) du. \tag{16.149}$$

Similarly, if $\mathcal{L}^{-1}\{F(s)\} = f(t)$ is true, then we can write

$$\mathcal{L}^{-1}\left\{\int_s^\infty F(u)du\right\} = \frac{f(t)}{t}. \tag{16.150}$$

Proof: If we write $g(t) = \frac{f(t)}{t}$, we can take $f(t) = tg(t)$. Hence we can write

$$F(s) = \mathcal{L}\{f(t)\} \tag{16.151}$$

$$= \mathcal{L}\{tg(t)\} = -\frac{d}{ds}\mathcal{L}\{g(t)\} = -\frac{dG(s)}{ds}, \tag{16.152}$$

where we have used theorem VI. Thus we can write

$$G(s) = -\int_c^s F(u)du. \tag{16.153}$$

From the limit $\lim_{s\to\infty} G(s) = 0$ we conclude that $c = \infty$. We finally obtain

$$G(s) = \mathcal{L}\left\{\frac{f(t)}{t}\right\} = \int_s^\infty F(u)du \tag{16.154}$$

Theorem VIII: Convolution Theorem.

If the Laplace transforms of $f(t)$ and $g(t)$ are given as $F(s)$ and $G(s)$, respectively, we can write

$$\mathcal{L}\left\{\int_0^t f(u)g(t-u)du\right\} = F(s)G(s). \tag{16.155}$$

Similarly, if the inverses $\mathcal{L}^{-1}\{F(s)\} = f(t)$ and $\mathcal{L}^{-1}\{G(s)\} = g(t)$ exist, then we can write

$$\mathcal{L}^{-1}\{F(s)G(s)\} = \int_0^t f(u)g(t-u)du. \tag{16.156}$$

The above integral is called the convolution of $f(t)$ and $g(t)$, and it is shown as $f * g$:

$$f * g = \int_0^t f(u)g(t-u)du. \tag{16.157}$$

The convolution operation has the following properties:

$$\begin{aligned} f * g &= g * f, \\ f * (g+h) &= f * g + f * h, \\ f * (g * h) &= (f * g) * h. \end{aligned} \tag{16.158}$$

Proof: We first form the convolution of the following transforms:

$$F(s)G(s) = \left[\int_0^\infty e^{-su} f(u) du \right] \left[\int_0^\infty e^{-sv} g(v) dv \right] \tag{16.159}$$

$$= \int_0^\infty \int_0^\infty e^{-s(v+u)} g(v) f(u) du dv. \tag{16.160}$$

Using the transformation $v = t - u$, we obtain

$$F(s)G(s) = \int_{t=0}^\infty \int_{u=0}^t e^{-st} g(t-u) f(u) du dt \tag{16.161}$$

$$= \int_{t=0}^\infty e^{-st} \left[\int_{u=0}^t g(t-u) f(u) du \right] dt \tag{16.162}$$

$$= \mathcal{L} \left\{ \int_0^t f(u) g(t-u) du \right\}. \tag{16.163}$$

Note that with the $t = u + v$ transformation, we have gone from the uv-plane to the ut-plane.

Example 16.4. *Inverse Laplace transforms:* .

1. We now find the function with the Laplace transform

$$\frac{se^{-2s}}{s^2 + 16}. \tag{16.164}$$

Since $\mathcal{L} \left\{ \frac{s}{s^2+16} \right\} = \cos 4t$ we can use theorem II, which says

$$\mathcal{L} \left\{ U(t-a) f(t-a) \right\} = e^{-as} F(s). \tag{16.165}$$

Using the inverse

$$\mathcal{L}^{-1} \left\{ e^{-as} F(s) \right\} = U(t-a) f(t-a), \tag{16.166}$$

we find

$$\mathcal{L}^{-1} \left\{ e^{-2s} \left(\frac{s}{s^2+16} \right) \right\} = U(t-2) \cos 4(t-2) \tag{16.167}$$

$$= \begin{cases} 0, & t < 2, \\ \cos 4(t-2), & t > 2. \end{cases} \tag{16.168}$$

2. To find the inverse Laplace transform of $\ln \left(1 + \frac{1}{s} \right)$ we write

$$F(s) = \ln \left(1 + \frac{1}{s} \right) \tag{16.169}$$

and find its derivative as

$$F'(s) = \frac{1}{s+1} - \frac{1}{s}. \tag{16.170}$$

Using theorem IV we write

$$\mathcal{L}^{-1}\left\{F^{(n)}(s)\right\} = (-1)^n t^n f(t). \tag{16.171}$$

Applying this to our case we find

$$\mathcal{L}^{-1}\left\{F'(s)\right\} = -t\mathcal{L}^{-1}\left\{\ln\left(1+\frac{1}{s}\right)\right\}, \tag{16.172}$$

$$\mathcal{L}^{-1}\left\{\ln\left(1+\frac{1}{s}\right)\right\} = -\frac{1}{t}\mathcal{L}^{-1}\left\{\frac{1}{s+1} - \frac{1}{s}\right\}, \tag{16.173}$$

$$= \frac{1-e^{-t}}{t}. \tag{16.174}$$

3. The inverse Laplace transform of

$$\frac{1}{s\sqrt{s+1}} \tag{16.175}$$

can be found by making use of theorem VI. Since

$$\mathcal{L}^{-1}\left\{\frac{1}{\sqrt{s+1}}\right\} = \frac{t^{-\frac{1}{2}}e^{-t}}{\sqrt{\pi}}, \tag{16.176}$$

theorem VI allows us to write

$$\mathcal{L}^{-1}\left\{\frac{F(s)}{s}\right\} = \int_0^t f(u)du, \tag{16.177}$$

$$\mathcal{L}^{-1}\left\{\frac{1}{s\sqrt{s+1}}\right\} = \int_0^t \frac{u^{-\frac{1}{2}}e^{-u}}{\sqrt{\pi}}du, \tag{16.178}$$

$$= \frac{2}{\sqrt{\pi}}\int_0^{\sqrt{t}} e^{-v^2}dv. \tag{16.179}$$

We now make the transformation $u = v^2$ to write the result in terms of the error function as

$$\mathcal{L}^{-1}\left\{\frac{1}{s\sqrt{s+1}}\right\} = \mathrm{erf}(\sqrt{t}). \tag{16.180}$$

16.6.4 Method of Partial Fractions

We frequently encounter Laplace transforms, which are expressed in terms of rational functions as

$$F(s) = g(s)/h(s), \tag{16.181}$$

where $g(s)$ and $h(s)$ are two polynomials with no common factor and the order of $g(s)$ is less than $h(s)$.

We have the following cases:

i) When all the factors of $h(s)$ are linear and distinct we can write $f(s)$ as

$$f(s) = \frac{c_1}{s - a_1} + \frac{c_2}{s - a_2} + \cdots + \frac{c_n}{s - a_n}, \tag{16.182}$$

where c_i are constants independent of s.

ii) When one of the roots of $h(s)$ is mth order we write $f(s)$ as

$$f(s) = \frac{c_{1,m}}{(s - a_1)^m} + \frac{c_{1,m-1}}{(s - a_1)^{m-1}} + \cdots + \frac{c_{1,1}}{(s - a_1)}$$

$$+ \sum_{i=2}^{n} \frac{c_i}{s - a_i}. \tag{16.183}$$

iii) When one of the factors of $h(s)$ is quadratic like $\left(s^2 + ps + q\right)$, we add a term to the partial fractions with two constants as

$$\frac{as + b}{(s^2 + ps + q)}. \tag{16.184}$$

To find the constants we usually compare equal powers of s. In the first case, we can also use the limit

$$\lim_{s \to a_i} (s - a_i) f(s) = c_i \tag{16.185}$$

to evaluate the constants.

Example 16.5. *Method of partial fractions:* We use the method of partial fractions to find the inverse Laplace transform of

$$f(s) = \frac{k^2}{(s + 2)(s^2 + 2k^2)}. \tag{16.186}$$

We write $f(s)$ as

$$f(s) = \frac{c}{s + 2} + \frac{as + b}{s^2 + 2k^2} \tag{16.187}$$

and equate both expressions as

$$\frac{k^2}{(s + 2)(s^2 + 2k^2)} = \frac{c\left(s^2 + 2k^2\right) + (s + 2)(as + b)}{(s + 2)(s^2 + 2k^2)}. \tag{16.188}$$

Comparing equal powers of s, we obtain three equations to be solved for $a, b,$ and c as

$$
\left.\begin{array}{r}
c + a = 0 \\[2mm]
b + 2a = 0 \\[2mm]
2b + 2ck^2 = k^2
\end{array}\right\}
\begin{array}{l}
\text{coefficient of } s^2 \\[2mm]
\text{coefficient of } s \quad, \\[2mm]
\text{coefficient of } s^0
\end{array}
\qquad (16.189)
$$

which gives

$$
c = -a, \quad b = -2a, \quad a = -k^2/(2k^2 + 4). \qquad (16.190)
$$

We now have

$$
f(s) = -\frac{2}{(s+2)} + \frac{a(s-2)}{s^2 + 2k^2}, \qquad (16.191)
$$

inverse Laplace transform of which can be found easily as

$$
\mathcal{L}^{-1}\{f(s)\} = -a\left[e^{-2t} + \cos\sqrt{2}kt - \frac{\sqrt{2}}{k}\sin\sqrt{2}kt\right], \quad a = -k^2/((2k^2 + 4)). \qquad (16.192)
$$

Example 16.6. *Definite integrals and Laplace transforms:* We can also use integral transforms to evaluate some definite integrals. Let us consider

$$
F(t) = \int_0^\infty \frac{\sin tx}{x} dx. \qquad (16.193)
$$

Taking the Laplace transform of both sides we get

$$
\mathcal{L}\left\{\int_0^\infty \frac{\sin tx}{x} dx\right\} = \int_0^\infty e^{-st} \int_0^\infty \frac{\sin tx}{x} dx\, dt \qquad (16.194)
$$

$$
= \int_0^\infty \frac{1}{x}\left[\int_0^\infty dt\, e^{-st}\sin(tx)\right] dx.
$$

The quantity inside the square brackets is the Laplace transform of $\sin tx$. Thus we find

$$
\mathcal{L}\left\{\int_0^\infty \frac{\sin tx}{x} dx\right\} = \int_0^\infty \frac{1}{s^2 + x^2} dx \qquad (16.195)
$$

$$
\mathcal{L}\{F(t)\} = \frac{1}{s}\tan^{-1}\left(\frac{x}{s}\right)\Big|_0^\infty \qquad (16.196)
$$

$$
\mathcal{L}\{F(t)\} = \frac{\pi}{2s}. \qquad (16.197)
$$

Finding the inverse Laplace transform of this gives us the value of the definite integral as

$$F(t) = \frac{\pi}{2}, \quad t > 0. \tag{16.198}$$

16.7   LAPLACE TRANSFORM OF A DERIVATIVE

One of the main applications of Laplace transforms is to differential equations. In particular, systems of ordinary linear differential equations with constant coefficients can be converted into systems of linear algebraic equations, which are a lot easier to solve both analytically and numerically. The Laplace transform of a derivative is given as

$$\mathcal{L}\left\{F'(t)\right\} = \int_0^\infty e^{-st} \frac{dF(t)}{dt} dt \tag{16.199}$$

$$= e^{-st} F(t) \big|_0^\infty + s \int_0^\infty e^{-st} F(t) dt \tag{16.200}$$

$$= s\mathcal{L}\left\{F(t)\right\} - F(0). \tag{16.201}$$

To be precise, we mean that $F(0) = F(+0)$ and $dF(t)/dt$ is piecewise continuous in the interval $0 \leq t < \infty$. Similarly, the Laplace transform of higher-order derivatives can be written as

$$\mathcal{L}\left\{F^{(2)}(t)\right\} = s^2 \mathcal{L}\left\{F(t)\right\} - sF(+0) - F'(+0), \tag{16.202}$$

$$\mathcal{L}\left\{F^{(n)}(t)\right\} = s^n \mathcal{L}\left\{F(t)\right\} - s^{n-1} F(+0) - s^{n-2} F'(+0) -$$
$$\cdots - F^{(n-1)}(+0). \tag{16.203}$$

Example 16.7. *Laplace transforms and differential equations:* We start with a simple example; simple harmonic oscillator with the equation of motion given as

$$m\frac{d^2 x(t)}{dt^2} + kx(t) = 0. \tag{16.204}$$

Let us find a solution satisfying the boundary conditions

$$x(0) = x_0 \quad \text{and} \quad \frac{dx}{dt}\bigg|_0 = 0. \tag{16.205}$$

Taking the Laplace transform of the equation of motion we obtain the Laplace transform of the solution as

$$m\mathcal{L}\left\{\frac{d^2 x(t)}{dt^2}\right\} + k\mathcal{L}\left\{x(t)\right\} = 0,$$

$$ms^2 X(s) - msx_0 + kX(s) = 0,$$

$$X(s) = x_0 \frac{s}{s^2 + \omega_0^2}. \tag{16.206}$$

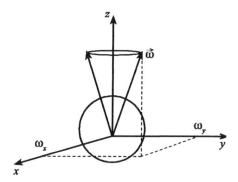

Fig. 16.5   Nutation of Earth

We have written $\omega_0^2 = k/m$. We now take the inverse Laplace transform to obtain the solution as

$$x(t) = \mathcal{L}^{-1}\left\{x_0\frac{s}{s^2+\omega_0^2}\right\} = x_0\mathcal{L}^{-1}\left\{\frac{s}{s^2+\omega_0^2}\right\} = x_0\cos\omega_0 t. \quad (16.207)$$

Example 16.8. *Nutation of Earth:* For the force-free rotation of Earth (Fig. 16.5), the Euler equations are given as

$$\begin{cases} \dfrac{dX}{dt} = -aY \\[2mm] \dfrac{dY}{dt} = aX. \end{cases} \quad (16.208)$$

This is a system of coupled ordinary differential equations with constant coefficients where we have defined a as

$$a = \left[\frac{I_z - I_x}{I_z}\right]\omega_z \quad (16.209)$$

and X and Y as

$$X = \omega_x, \quad Y = \omega_y. \quad (16.210)$$

I_z is the moment of inertia about the z-axis, and because of axial symmetry we have set $I_x = I_y$. Taking the Laplace transform of this system we obtain a set of coupled algebraic equations:

$$sx(s) - X(0) = -ay(s) \tag{16.211}$$

and

$$sy(s) - Y(0) = ax(s). \tag{16.212}$$

The solution for $x(s)$ can be obtained easily as

$$x(s) = X(0)\frac{s}{s^2 + a^2} - Y(0)\frac{a}{s^2 + a^2}. \tag{16.213}$$

Taking the inverse Laplace transform we find the solution as

$$X(t) = X(0)\cos at - Y(0)\sin at. \tag{16.214}$$

Similarly, the $Y(t)$ solution is found as

$$Y(t) = X(0)\sin at + Y(0)\cos at. \tag{16.215}$$

Example 16.9. *Damped harmonic oscillator:* Equation of motion for the damped harmonic oscillator is given as

$$m\ddot{x}(t) + b\dot{x}(t) + kx(t) = 0. \tag{16.216}$$

Let us solve this equation with the initial conditions $x(0) = x_0$ and $\dot{x}(0) = 0$. Taking the Laplace transform of the equation of motion, we obtain the Laplace transform of the solution as

$$m\left[s^2 X(s) - sx_0\right] + b\left[sX(s) - x_0\right] + kX(s) = 0, \tag{16.217}$$

$$X(s) = x_0\frac{ms + b}{ms^2 + bs + k}. \tag{16.218}$$

Completing the square in the denominator we write

$$s^2 + \frac{b}{m}s + \frac{k}{m} = \left(s + \frac{b}{2m}\right)^2 + \left(\frac{k}{m} - \frac{b^2}{4m^2}\right). \tag{16.219}$$

For weak damping, $b^2 < 4km$, the last term is positive. Calling this ω_1^2, we find

$$X(s) = x_0 \frac{s + \dfrac{b}{m}}{\left(s + \dfrac{b}{2m}\right)^2 + \omega_1^2} \qquad (16.220)$$

$$= x_0 \frac{s + \dfrac{b}{2m} + \dfrac{b}{2m}}{\left(s + \dfrac{b}{2m}\right)^2 + \omega_1^2}$$

$$= x_0 \frac{s + \dfrac{b}{2m}}{\left(s + \dfrac{b}{2m}\right)^2 + \omega_1^2} + x_0 \frac{\dfrac{b\omega_1}{2m\omega_1}}{\left(s + \dfrac{b}{2m}\right)^2 + \omega_1^2}.$$

Taking the inverse Laplace transform of $X(s)$ we find the final solution as

$$x(t) = x_0 e^{-\left(\frac{b}{2m}\right)t} \left[\cos \omega_1 t + \frac{b}{2m\omega_1} \sin \omega_1 t\right]. \qquad (16.221)$$

Check that this solution satisfies the given initial conditions.

Example 16.10. *Laplace transform of the te^{kt} function:* Using the elementary Laplace transform

$$\mathcal{L}\left\{e^{kt}\right\} = \int_0^\infty e^{-st} e^{kt} dt = \frac{1}{s-k}, \quad s > k, \qquad (16.222)$$

and theorem IV, we can obtain the desired transform by differentiation with respect to s as

$$\mathcal{L}\left\{te^{kt}\right\} = \frac{1}{(s-k)^2}, \quad s > k. \qquad (16.223)$$

Example 16.11. *Electromagnetic waves:* For a transverse electromagnetic wave propagating along the x-axis,

$$E = E_x \text{ or } E_y$$

satisfies the wave equation

$$\frac{\partial^2 E(x,t)}{\partial x^2} - \frac{1}{v^2} \frac{\partial^2 E(x,t)}{\partial t^2} = 0. \qquad (16.224)$$

We take the initial conditions as

$$E(x,0) = 0 \quad \text{and} \quad \left.\frac{\partial E(x,t)}{\partial t}\right|_{t=0} = 0. \qquad (16.225)$$

Taking the Laplace transform of the wave equation with respect to t we obtain

$$\frac{\partial^2}{\partial x^2} \mathcal{L}\left\{E(x,t)\right\} - \frac{s^2}{v^2} \mathcal{L}\left\{E(x,t)\right\} + \frac{s}{v^2} E(x,0) + \frac{1}{v^2} \left.\frac{\partial E(x,t)}{\partial t}\right|_{t=0} = 0.$$
(16.226)

Using the initial conditions this becomes

$$\frac{d^2}{dx^2} \mathcal{L}\left\{E(x,t)\right\} = \frac{s^2}{v^2} \mathcal{L}\left\{E(x,t)\right\},$$
(16.227)

which is an ordinary differential equation for $\mathcal{L}\left\{E(x,t)\right\}$ and can be solved easily as

$$\mathcal{L}\left\{E(x,t)\right\} = c_1 e^{-(s/v)x} + c_2 e^{(s/v)x},$$
(16.228)

where c_1 and c_2 are constants independent of x but could depend on s. In the limit as $x \to \infty$, we expect the wave to be finite; hence we choose c_2 as zero. If we are also given the initial shape of the wave as

$$E(0,t) = F(t),$$
(16.229)

we determine c_1 as

$$\mathcal{L}\left\{E(0,t)\right\} = c_1 = f(s).$$
(16.230)

Thus, with the given initial conditions, the Laplace transform of the solution is given as

$$\mathcal{L}\left\{E(x,t)\right\} = e^{-\left(\frac{s}{v}\right)x} f(s).$$
(16.231)

Using theorem II we can find the inverse Laplace transform, and the final solution is obtained as

$$E(x,t) = \begin{cases} F\left(t - \dfrac{x}{v}\right), & t \geq \dfrac{x}{v}, \\ \\ 0, & t < \dfrac{x}{v}. \end{cases}$$
(16.232)

This is a wave moving along the positive x-axis with velocity v and without distortion. Note that the wave still has not reached the region $x > vt$.

Example 16.12. *Bessel's equation:* We now consider Bessel's equation, which is an ordinary differential equation with variable coefficients:

$$x^2 y''(x) + xy'(x) + x^2 y(x) = 0.$$
(16.233)

Dividing this by x we get

$$xy''(x) + y'(x) + xy(x) = 0. \tag{16.234}$$

Using Laplace transforms we can find a solution satisfying the boundary condition

$$y(0) = 1. \tag{16.235}$$

From the differential equation (16.234), this also means that $y'(+0) = 0$. Assuming that the Laplace and the inverse Laplace transforms of the unknown function exist, that is,

$$\mathcal{L}\{y(x)\} = f(s), \quad \mathcal{L}^{-1}\{f(s)\} = y(x), \tag{16.236}$$

we write the Laplace transform of Equation (16.234) as

$$-\frac{d}{ds}\left[s^2 f(s) - s\right] + sf(s) - 1 - \frac{d}{ds}f(s) = 0 \tag{16.237}$$

$$(s^2 + 1)f'(s) + sf(s) = 0$$

$$\frac{df}{f} = -\frac{s}{s^2 + 1}ds.$$

After integration, we find $f(s)$ as

$$\ln\frac{f(s)}{c} = -\frac{1}{2}\ln(s^2 + 1), \tag{16.238}$$

$$f(s) = \frac{c}{\sqrt{s^2 + 1}}. \tag{16.239}$$

To find the inverse we write the binomial expansion of $f(s)$ as

$$f(s) = \frac{c}{s\sqrt{1 + \frac{1}{s^2}}}, \tag{16.240}$$

$$= \frac{c}{s}\left[1 - \frac{1}{2s^2} + \frac{1.3}{2^2 2! s^4} - \cdots\right.$$

$$\left.\cdots + \frac{(-1)^n (2n)!}{(2^n n!)^2 s^{2n}} + \cdots\right]. \tag{16.241}$$

Using the fact that Laplace transforms are linear, we find the inverse as

$$y(x) = c\sum_{n=0}^{\infty}\frac{(-1)^n x^{2n}}{(2^n n!)^2}. \tag{16.242}$$

Using the condition $y(0) = 1$, we determine the constant as c as one. This solution is nothing but the zeroth-order Bessel function $J_0(x)$. Thus we have determined the Laplace transform of $J_0(x)$ as

$$\mathcal{L}\{J_0(x)\} = \frac{1}{\sqrt{s^2 + 1}}. \tag{16.243}$$

In general one can show

$$\mathcal{L}\left\{J_n(ax)\right\} = \frac{a^{-n}\left(\sqrt{s^2+a^2}-s\right)^n}{\sqrt{s^2+a^2}}. \tag{16.244}$$

In this example we see that the Laplace transforms can also be used for finding solutions of ordinary differential equations with variable coefficients, however, there is no guarantee that they will work in general.

Example 16.13. *Solution of* $y'' + (1/2)y = (a_0/2)\sin t - (1/2)y^{(iv)}$: This could be interpreted as a harmonic oscillator with a driving force depending on the fourth derivative of displacement as $(a_0/2)\sin t - (1/2)y^{(iv)}$. We write this equation as

$$y^{(iv)} + 2y'' + y = a_0 \sin t \tag{16.245}$$

and use the following boundary conditions (a_0 is a constant):

$$y(0) = 1, \; y'(0) = -2 \,, \; y''(0) = 3, \; y'''(0) = 0.$$

Taking the Laplace transform and using partial fractions we write

$$\left[s^4 Y - s^3(1) - s^2(-2) - s(3) - 0\right] \tag{16.246}$$
$$+ 2\left[s^2 Y - s(1) - (-2)\right] + Y = \frac{a_0}{s^2+1},$$

where $Y(s)$ is the Laplace transform of $y(x)$. We solve this for Y to obtain

$$\left(s^4 + 2s^2 + 1\right)Y = \frac{a_0}{s^2+1} + s^3 - 2s^2 + 5s - 4, \tag{16.247}$$

$$Y = \frac{a_0}{\left(s^2+1\right)^3} + \frac{s^3 - 2s^2 + 5s - 4}{\left(s^2+1\right)^2}, \tag{16.248}$$

$$= \frac{a_0}{\left(s^2+1\right)^3} + \frac{\left(s^3+s\right) - 2\left(s^2+1\right) + 4s - 2}{\left(s^2+1\right)^2}, \tag{16.249}$$

$$= \frac{a_0}{\left(s^2+1\right)^3} + \frac{s}{\left(s^2+1\right)} - \frac{2}{\left(s^2+1\right)} + \frac{4s-2}{\left(s^2+1\right)^2}. \tag{16.250}$$

Using the theorems we have introduced the following inverses can be found:

$$\mathcal{L}^{-1}\left\{\frac{a_0}{\left(s^2+1\right)^3}\right\} = a_0\left[\frac{3}{8}\sin t - \frac{3}{8}t\cos t - \frac{t^2}{8}\sin t\right], \tag{16.251}$$

$$\mathcal{L}^{-1}\left\{\frac{4s-2}{\left(s^2+1\right)^2}\right\} = 2t\sin t - \sin t + t\cos t. \tag{16.252}$$

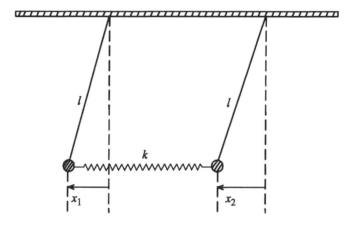

Fig. 16.6   Pendulums connected by a spring

Finally, the solution is obtained as

$$y(t) = \left[t(1 - \frac{3}{8}a_0) + 1\right] \cos t + \left[\frac{a_0}{8}(3 - t^2) - 3 + 2t\right] \sin t. \quad (16.253)$$

Note that the solution is still oscillatory but the amplitude changes with time.

Example 16.14. *Two Pendlums interacting through a spring:* Consider two pendulums connected by a vertical spring as shown in Figure 16.6. We investigate small oscillations of this system.

As our initial conditions we take

$$x_1(0) = x_2(0) = 0, \quad \dot{x}_1\big|_0 = v, \quad \dot{x}_2\big|_0 = 0. \quad (16.254)$$

For this system and for small oscillations, equations of motion are given as

$$m\ddot{x}_1 = -\frac{mg}{l}x_1 + k(x_2 - x_1), \quad (16.255)$$

$$m\ddot{x}_2 = -\frac{mg}{l}x_2 + k(x_1 - x_2). \quad (16.256)$$

We show the Laplace transforms of x_1 and x_2 as

$$\mathcal{L}\{x_i(t)\} = X_i(s), \quad i = 1, 2 \quad (16.257)$$

and take the Laplace transform of both equations as

$$m(s^2 X_1 - v) = -\frac{mg}{l}X_1 + k(X_2 - X_1), \quad (16.258)$$

$$ms^2 X_2 = -\frac{mg}{l} X_2 + k(X_1 - X_2). \tag{16.259}$$

This gives us two coupled algebraic equations. We first solve them for $X_1(s)$ to get

$$X_1(s) = \frac{v}{2} \left[(s^2 + g/l + 2k/m)^{-1} + (s^2 + g/l)^{-1} \right]. \tag{16 260}$$

Taking the inverse Laplace transform of $X_1(s)$ gives us $x_1(t)$ as

$$x_1(t) = \frac{v}{2} \left[\frac{\sin \sqrt{(\frac{g}{l} + 2\frac{k}{m})} t}{\sqrt{(\frac{g}{l} + 2\frac{k}{m})}} + \frac{\sin \sqrt{(\frac{g}{l})} t}{\sqrt{(\frac{g}{l})}} \right]. \tag{16.261}$$

In this solution

$$\sqrt{(\frac{g}{l} + 2\frac{k}{m})} \text{ and } \sqrt{(\frac{g}{l})} \tag{16.262}$$

are the normal modes of the system. The solution for $x_2(t)$ can be obtained similarly.

16.7.1 Laplace Transforms in n Dimensions

Laplace transforms are defined in two dimensions as

$$F(u, v) = \int_0^\infty \int_0^\infty f(x, y) e^{-ux - vy} dx dy. \tag{16.263}$$

This can also be generalized to n dimensions:

$$F(u_1, u_2, \dots, u_n) \tag{16.264}$$
$$= \int_0^\infty \int_0^\infty \cdots \int_0^\infty f(x_1, x_2, \dots, x_n) e^{-u_1 x_1 - u_2 x_2 - \cdots - u_n x_n} dx_1 dx_2 \dots dx_n.$$

16.8 RELATION BETWEEN LAPLACE AND FOURIER TRANSFORMS

The Laplace transform of a function is defined as

$$F(p) = \mathcal{L}\{f(x)\} = \int_0^\infty f(x) e^{-px} dx. \tag{16.265}$$

We now use $f(x)$ to define another function:

$$f_+(x) = \begin{cases} f(x), & x > 0, \\ 0, & x < 0. \end{cases} \tag{16.266}$$

The Fourier transform of this function is given as

$$F_+(k) = \frac{1}{\sqrt{2\pi}} \int_0^\infty f(x)e^{ikx}\,dx. \tag{16.267}$$

Thus we can write the relation between the Fourier and Laplace transforms as

$$F(p) = \sqrt{2\pi}F_+(ip). \tag{16.268}$$

16.9   MELLIN TRANSFORMS

Another frequently encountered integral transform is the Mellin transform:

$$F_m(s) = \int_0^\infty f(x)x^{s-1}\,dx. \tag{16.269}$$

The Mellin transform of $\exp(-x)$ is the gamma function. We write

$$x = e^z \tag{16.270}$$

in the Mellin transform to get

$$F_m(s) = \int_{-\infty}^\infty f(e^z)e^{sz}\,dz \tag{16.271}$$

$$= \int_{-\infty}^\infty g(z)e^{sz}\,dz, \quad g(z) = f(e^z).$$

Comparing this with

$$G(k) = \frac{1}{\sqrt{2\pi}} \int_{-\infty}^\infty g(z)e^{ikz}\,dz, \tag{16.272}$$

We get the relation between the Fourier and Mellin transforms as

$$F_m(s) = \sqrt{2\pi}G(-is). \tag{16.273}$$

Now all the properties we have discussed for the Fourier transforms can also be adopted to the Mellin transforms.

Problems

16.1   Show that the Fourier transform of a Gaussian,

$$f(\overrightarrow{r}) = (\frac{2}{\pi a^2})^{3/4}e^{-r^2/a^2},$$

is again a Gaussian.

16.2   Show that the Fourier transform of

$$f(t) = \begin{cases} \sin \omega_0 t & |t| < \dfrac{N\pi}{\omega_0} \\[2mm] 0 & |t| > \dfrac{N\pi}{\omega_0} \end{cases}$$

is given as

$$g_s(\omega) = \sqrt{\frac{2}{\pi}} \left[\frac{\dfrac{N\pi}{\omega_0} \sin(\omega_0 - \omega)}{2(\omega_0 - \omega)} - \frac{\dfrac{N\pi}{\omega_0} \sin(\omega_0 + \omega)}{2(\omega_0 + \omega)} \right].$$

16.3   Using the Laplace transform technique, find the solution of the following second-order inhomogeneous differential equation:

$$y'' - 3y' + 2y = 2e^{-t},$$

with the boundary conditions

$$y(0) = 2 \text{ and } y'(0) = -1.$$

16.4   Solve the following system of differential equations:

$$2x(t) - y(t) - y'(t) = 4(1 - \exp(-t))$$
$$2x'(t) + y(t) = 2(1 + 3\exp(-2t))$$

with the boundary conditions

$$x(0) = y(0) = 0.$$

16.5   One end of an insulated semi-infinite rod is held at temperature

$$T(t,0) = T_0,$$

with the initial conditions

$$T(0,x) = 0 \text{ and } T(t,\infty) = 0.$$

Solve the heat transfer equation

$$\frac{\partial T(t,x)}{\partial t} = (k/c\rho)\frac{\partial^2 T(t,x)}{\partial x^2}, \quad k > 0,$$

where k is the thermal conductivity, c is the heat capacity, and ρ is the density.

Hint: The solution is given in terms of erf c as

$$T(t,x) = T_0 \operatorname{erf} c \left[\frac{x}{2} \frac{\sqrt{c\rho/k}}{\sqrt{t}} \right],$$

where the erf c is defined in terms of erf as

$$\operatorname{erf} c(x) = 1 - \operatorname{erf} x$$

$$= 1 - \frac{2}{\sqrt{\pi}} \int_0^x e^{-u^2} du = \frac{2}{\sqrt{\pi}} \int_x^\infty e^{-u^2} du.$$

16.6 Find the current, I, for the IR circuit represented by differential equation

$$L\frac{dI}{dt} + RI = E,$$

with the initial condition

$$I(0) = 0.$$

E is the electromotive force and L, R, and E are constants.

16.7 Using the Laplace transforms find the solution of the following system of differential equations

$$\frac{dx}{dt} + y = 3e^{2t}$$

$$\frac{dy}{dt} + x = 0,$$

subject to the initial conditions

$$x(0) = 2, \ y(0) = 0.$$

16.8 Using the Fourier sine transform show the integral

$$e^{-x}\cos x = \frac{2}{\pi} \int_0^\infty \frac{s^3 \sin sx}{s^4 + 4} ds, \ x > 0.$$

16.9 Using the Fourier cosine transform show the integral

$$e^{-x}\cos x = \frac{2}{\pi} \int_0^\infty \frac{s^2 + 2}{s^4 + 4}(\cos sx)ds, \ x \geq 0.$$

16.10 Let a semi-infinite string be extended along the positive x-axis with the end at the origin fixed. The shape of the string at $t = 0$ is given as

$$y(x,0) = f(x),$$

where $y(x, t)$ represents the displacement of the string perpendicular to the x-axis and satisfies the wave equation

$$\frac{\partial^2 y}{\partial t^2} = a^2 \frac{\partial^2 y}{\partial x^2}, \quad a \text{ is a constant.}$$

Show that the solution is given as

$$y(x, t) = \frac{2}{\pi} \int_0^\infty ds \cos(sat) \sin(sx) \int_0^\infty d\xi f(\xi) \sin(s\xi).$$

16.11   Establish the Fourier sine integral representation

$$\frac{x}{x^2 + k^2} = \frac{2}{\pi} \int_0^\infty dy \sin(xy) \int_0^\infty dz \frac{z \sin(yz)}{z^2 + k^2}.$$

Hint: First show that

$$e^{-ky} = \frac{2}{\pi} \int_0^\infty \frac{z \sin(yz)}{z^2 + k^2} dz, \quad x > 0, \ k > 0.$$

16.12   Show that the Fourier sine transform of

$$xe^{-ax}$$

is given as

$$\sqrt{\frac{2}{\pi}} \frac{a^2 - k^2}{(a^2 + k^2)^2}.$$

16.13   Establish the result

$$\mathcal{L}\left\{\frac{1 - \cos at}{t}\right\} = \frac{1}{2} \log\left(1 + \frac{a^2}{s^2}\right).$$

16.14   Use the convolution theorem to show that

$$\mathcal{L}^{-1}\left\{\frac{s^2}{(s^2 + b^2)^2}\right\} = \frac{t}{2} \cos bt + \frac{1}{2b} \sin bt.$$

16.15   Use Laplace transforms to find the solution of the following system of differential equations:

$$\frac{dy_1}{dx} = -\alpha_1 y_1$$

$$\frac{dy_2}{dx} = -\alpha_1 y_1 - \alpha_2 y_2$$

$$\frac{dy_3}{dx} = -\alpha_2 y_2 - \alpha_3 y_3,$$

with the boundary conditions $y_1(0) = C_0$, $y_2(0) = y_3(0) = 0$.

16.16   Laguerre polynomials satisfy

$$xL_n'' + (1-t)L_n' + nL_n(x) = 0.$$

Show that $\mathcal{L}\{L_n(ax)\} = (s-a)^n/s^{n+1}$, $s > 0$.

17

VARIATIONAL ANALYSIS

Variational analysis is basically the study of changes. We are often interested in finding how a system reacts to small changes in its parameters. It is for this reason that variational analysis has found a wide range of applications not just in physics and engineering but also in finance and economics. In applications we frequently encounter cases where a physical property is represented by an integral, the extremum of which is desired. Compared to ordinary calculus, where we deal with functions of numbers, these integrals are functions of some unknown function and its derivatives; thus, they are called functionals. Search for the extremum of a function yields the point at which the function is extremum. In the case of functionals, variational analysis gives us a differential equation, which is to be solved for the extremizing function.

After Newton's formulation of mechanics Lagrange developed a new approach, where the equations of motion are obtained from a variational integral called action. This new formulation makes applications of Newton's theory to many particles and continuous systems possible. Today in making the transition to quantum mechanics and to quantum field theories Lagrangian formulation is a must.

Geodesics are the shortest paths between two points in a given geometry and constitute one of the main applications of variational analysis. In Einstein's theory of gravitation geodesics play a central role as the paths of freely moving particles in curved spacetime. Variational techniques also form the mathematical basis for the finite elements method, which is a powerful tool for solving complex boundary value problems, and stability analysis. Variational analysis and the Rayleigh-Ritz method also allows us to find approximate eigenvalues and eigenfunctions of a Sturm-Liouville system.

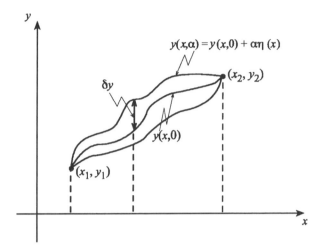

Fig. 17.1   Variation of paths

17.1   PRESENCE OF ONE DEPENDENT AND ONE INDEPENDENT VARIABLE

17.1.1   Euler Equation

A majority of the variational problems encountered in physics and engineering are expressed in terms of an integral:

$$J\left[y\left(x\right)\right] = \int_{x_1}^{x_2} f\left(y, y_x, x\right) dx, \tag{17.1}$$

where $y(x)$ is the desired function and f is a known function depending on y, its derivative with respect to x, that is y_x, and x. Because the unknown of this problem is a function, J is called a functional and we write it as

$$J\left[y\left(x\right)\right]. \tag{17.2}$$

Usually the purpose of these problems is to find a function, which is a path in the xy-plane between the points (x_1, y_1) and (x_2, y_2), which makes the functional $J\left[y\left(x\right)\right]$ an extremum. In Figure 17.1 we have shown two potentially possible paths; actually, there are infinitely many such paths. The difference of these paths from the desired path is called the variation of y, and we show it as δy. Because δy depends on position, we use $\eta(x)$ for its position dependence and use a scalar parameter α as a measure of its magnitude. Paths close to the desired path can now be parametrized in terms of α as

$$y(x, \alpha) = y(x, 0) + \alpha\eta(x) + 0(\alpha^2), \tag{17.3}$$

where $y(x, \alpha = 0)$ is the desired path, which extremizes the functional $J[y(x)]$. We can now express δy as

$$\delta y = y(x, \alpha) - y(x, 0) = \alpha \eta(x) \tag{17.4}$$

and write J as a function of α as

$$J(\alpha) = \int_{x_1}^{x_2} f[y(x, \alpha), y_x(x, \alpha), x] \, dx. \tag{17.5}$$

Now the extremum of J can be found as in ordinary calculus by imposing the condition

$$\left| \frac{\partial J(\alpha)}{\partial \alpha} \right|_{\alpha=0} = 0. \tag{17.6}$$

In this analysis we assume that $\eta(x)$ is a differentiable function and the variations at the end points are zero, that is,

$$\eta(x_1) = \eta(x_2) = 0. \tag{17.7}$$

Now the derivative of J with respect to α is

$$\frac{\partial J(\alpha)}{\partial \alpha} = \int_{x_1}^{x_2} \left[\frac{\partial f}{\partial y} \frac{\partial y}{\partial \alpha} + \frac{\partial f}{\partial y_x} \frac{\partial y_x}{\partial \alpha} \right] dx. \tag{17.8}$$

Using equation (17.3) we can write

$$\frac{\partial y(x, \alpha)}{\partial \alpha} = \eta(x) \tag{17.9}$$

and

$$\frac{\partial y_x(x, \alpha)}{\partial \alpha} = \frac{d\eta(x)}{dx}. \tag{17.10}$$

Substituting these in Equation (17.8) we obtain

$$\left| \frac{\partial J(\alpha)}{\partial \alpha} \right|_{\alpha=0} = \int_{x_1}^{x_2} \left[\frac{\partial f}{\partial y} \eta(x) + \frac{\partial f}{\partial y_x} \frac{d\eta(x)}{dx} \right] dx. \tag{17.11}$$

Integrating the second term by parts gives

$$\int_{x_1}^{x_2} \frac{\partial f}{\partial y_x} \frac{d\eta(x)}{dx} dx = \left| \frac{\partial f}{\partial y_x} \eta(x) \right|_{x_1}^{x_2} - \int_{x_1}^{x_2} \eta(x) \left(\frac{d}{dx} \frac{\partial f}{\partial y_x} \right) dx. \tag{17.12}$$

Using the fact that the variation at the end points are zero, we can write Equation (17.11) as

$$\int_{x_1}^{x_2} \left(\frac{\partial f}{\partial y} - \frac{d}{dx} \frac{\partial f}{\partial y_x} \right) \eta(x) dx = 0. \tag{17.13}$$

Because the variation $\eta(x)$ is arbitrary, the only way to satisfy this equation is by setting the expression inside the brackets to zero, that is,

$$\frac{\partial f}{\partial y} - \frac{d}{dx}\frac{\partial f}{\partial y_x} = 0. \tag{17.14}$$

In conclusion, variational analysis has given us a second-order differential equation to be solved for the path that extremizes the functional $J[y(x)]$. This differential equation is called the **Euler equation**.

17.1.2   Another Form of the Euler Equation

To find another version of the Euler equation we write the total derivative of the function $f(y, y_x, x)$ as

$$\frac{df}{dx} = \frac{\partial f}{\partial y}y_x + \frac{\partial f}{\partial y_x}\frac{dy_x}{dx} + \frac{\partial f}{\partial x}. \tag{17.15}$$

Using the Euler equation [Eq. (17.14)] we write

$$\frac{\partial f}{\partial y} = \frac{d}{dx}\frac{\partial f}{\partial y_x} \tag{17.16}$$

and substitute into Equation (17.15) to get

$$\frac{df}{dx} = y_x\frac{d}{dx}\frac{\partial f}{\partial y_x} + \frac{\partial f}{\partial y_x}\frac{dy_x}{dx} + \frac{\partial f}{\partial x}. \tag{17.17}$$

This can also be written as

$$\frac{\partial f}{\partial x} - \frac{d}{dx}\left[f - y_x\frac{\partial f}{\partial y_x}\right] = 0. \tag{17.18}$$

This is another version of the Euler equation, which is extremely useful when $f(y, y_x, x)$ does not depend on the independent variable x explicitly. In such cases we can immediately write the first integral as

$$f - y_x\frac{\partial f}{\partial y_x} = \text{constant}, \tag{17.19}$$

which reduces the problem to the solution of a first-order differential equation.

17.1.3   Applications of the Euler Equation

Example 17.1. *Shortest path between two points:* To find the shortest path between two points in two dimensions we write the line element as

$$ds = \left[(dx)^2 + (dy)^2\right]^{\frac{1}{2}} \tag{17.20}$$
$$= dx\left[1 + y_x^2\right]^{\frac{1}{2}}.$$

The distance between two points is now given as a functional of the path and in terms of the integral

$$J[y(x)] = \int_{(x_1,y_1)}^{(x_2,y_2)} ds \tag{17.21}$$

$$= \int_{x_1}^{x_2} \left[1+y_x^2\right]^{\frac{1}{2}} dx. \tag{17.22}$$

To find the shortest path we must solve the Euler equation for

$$f(y,y_x,x) = \left[1+y_x^2\right]^{\frac{1}{2}}.$$

Because $f(y,y_x,x)$ does not depend on the independent variable explicitly, we use the second form of the Euler equation [Eq. (17.19)] to write

$$\frac{1}{\left[1+y_x^2\right]^{\frac{1}{2}}} = c, \tag{17.23}$$

where c is a constant. This is a first-order differential equation for $y(x)$, and its solution can easily be found as

$$y = ax + b. \tag{17.24}$$

This is the equation of a straight line, where the integration constants a and b are to be determined from the coordinates of the end points. The shortest paths between two points in a given geometry are called geodesics. Geodesics in spacetime play a crucial role in Einstein's theory of gravitation as the paths of free particles.

Example 17.2. *Shape of a soap film between two rings:* Let us find the shape of a soap film between two rings separated by a distance of $2x_0$. Rings pass through the points (x_1,y_1) and (x_2,y_2) as shown in Figure 17.2. Ignoring gravitation, the shape of the film is a surface of revolution; thus it is sufficient to find the equation of a curve, $y(x)$, between two points (x_1,y_1) and (x_2,y_2). Because the energy of a soap film is proportional to its surface area, $y(x)$ should be the one that makes the area a minimum.

We write the infinitesimal area element of the soap film as

$$dA = 2\pi y ds \tag{17.25}$$

$$= 2\pi y \left[1+y_x^2\right]^{\frac{1}{2}} dx. \tag{17.26}$$

The area, aside from a factor of 2π, is given by the integral

$$J = \int_{x_1}^{x_2} y \left[1+y_x^2\right]^{\frac{1}{2}} dx. \tag{17.27}$$

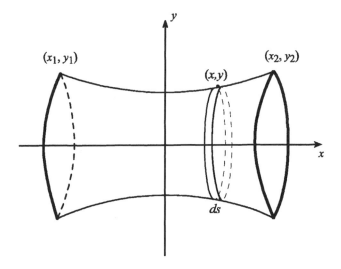

Fig. 17.2   Soap film between two circles

Since $f(y, y_x, x)$ is given as

$$f(y, y_x, x) = y \left[1 + y_x^2\right]^{\frac{1}{2}}, \tag{17.28}$$

which does not depend on x explicitly, we write the Euler equation as

$$\frac{y}{\left[1 + y_x^2\right]^{\frac{1}{2}}} = c_1, \tag{17.29}$$

where c_1 is a constant. Taking the square of both sides we write

$$\frac{y^2}{\left[1 + y_x^2\right]} = c_1^2. \tag{17.30}$$

This leads us to the first-order differential equation

$$(y_x)^{-1} = \frac{dx}{dy} = \frac{c_1}{\sqrt{y^2 - c_1^2}}, \quad c_1^2 \le y_{\min}^2, \tag{17.31}$$

which on integration gives

$$x = c_1 \cosh^{-1} \frac{y}{c_1} + c_2. \tag{17.32}$$

Thus the function $y(x)$ is determined as

$$y(x) = c_1 \cosh\left(\frac{x - c_2}{c_1}\right). \tag{17.33}$$

Integration constants c_1 and c_2 are to be determined so that $y(x)$ passes through the points (x_1, y_1) and (x_2, y_2). Symmetry of the problem gives $c_2 = 0$. For two rings with unit radius and $x_0 = 1/2$ we obtain

$$1 = c_1 \cosh\left(\frac{1}{2c_1}\right) \tag{17.34}$$

as the equation to be solved for c_1. This equation has two solutions. One of these is $c_1 = 0.2350$, which is known as the "deep curve," and the other one is $c_1 = 0.8483$, which is known as the "flat curve." To find the correct shape we have to check which one of these makes the area, and hence the energy, a minimum. Using Equations (17.27) and (17.29) we write the surface area as

$$A = \frac{4\pi}{c_1} \int_0^{x_0} y^2 \, dx. \tag{17.35}$$

Substituting the solution (17.33) into Equation (17.35) we get

$$A = \pi c_1^2 \left[\sinh\left(\frac{2x_0}{c_1}\right) + \frac{2x_0}{c_1} \right]. \tag{17.36}$$

For $x_0 = \frac{1}{2}$ we obtain

$$\left. \begin{matrix} c_1 = 0.2350 \\ \\ c_1 = 0.8483 \end{matrix} \right\} \rightarrow \left\{ \begin{matrix} A = 6.8456 \\ \\ A = 5.9917 \end{matrix} \right. . \tag{17.37}$$

This means that the correct value of c_1 is 0.8483. If we increase the separation between the rings beyond a certain point we expect the film to break. In fact, the transcendental equation

$$1 = c_1 \cosh\left(\frac{x_0}{c_1}\right) \tag{17.38}$$

does not have a solution for

$$x_0 \geq 1.$$

17.2   PRESENCE OF MORE THAN ONE DEPENDENT VARIABLE

In the variational integral if the function f depends on more than one dependent variable

$$y_1(x), y_2(x), y_3(x), ..., y_n(x) \tag{17.39}$$

and one independent variable x, then the functional J is written as

$$J = \int_{x_1}^{x_2} f\left[y_1(x), y_2(x), ..., y_n(x), y_{1x}(x), y_{2x}(x), ..., y_{nx}(x), x\right] dx, \tag{17.40}$$

where $y_{ix} = \partial y_i / \partial x$, $i = 1, 2, .., n$. We can now write small deviations from the desired paths, $y_i(x, 0)$, which makes the functional J an extremum as

$$y_i(x, \alpha) = y_i(x, 0) + \alpha \eta_i(x) + 0(\alpha^2), \quad i = 1, 2, ..., n, \tag{17.41}$$

where α is again a small parameter and the functions $\eta_i(x)$ are independent of each other. We again take the variation at the end points as zero:

$$\eta_i(x_1) = \eta_i(x_2) = 0. \tag{17.42}$$

Taking the derivative of $J(\alpha)$ with respect to α and setting α to zero we get

$$\int_{x_1}^{x_2} \sum_i \left(\frac{\partial f}{\partial y_i} \eta_i(x) + \frac{\partial f}{\partial y_{ix}} \frac{d\eta_i(x)}{dx} \right) dx = 0. \tag{17.43}$$

Integrating the second term by parts and using the fact that at the end points variations are zero, we write Equation (17.43) as

$$\int_{x_1}^{x_2} \sum_i \left(\frac{\partial f}{\partial y_i} - \frac{d}{dx} \frac{\partial f}{\partial y_{ix}} \right) \eta_i(x) dx = 0. \tag{17.44}$$

Because the variations $\eta_i(x)$ are independent, this equation can only be satisfied if all the coefficients of $\eta_i(x)$ vanish simultaneously, that is,

$$\frac{\partial f}{\partial y_i} - \frac{d}{dx} \frac{\partial f}{\partial y_{ix}} = 0, \quad i = 1, 2, ..., n. \tag{17.45}$$

We now have a system of n Euler equations to be solved simultaneously for the n dependent variables. An important example for this type of variational problems is the Lagrangian formulation of classical mechanics.

17.3   PRESENCE OF MORE THAN ONE INDEPENDENT VARIABLE

Sometimes the unknown function u and f in the functional J depend on more than one independent variable. For example, in three-dimensional problems J may be given as

$$J = \int \int \int_V f[u, u_x, u_y, u_z, x, y, z] \, dxdydz, \tag{17.46}$$

where u_x denotes the partial derivative with respect to x. We now have to find a function $u(x, y, z)$ such that J is an extremum. We again let $u(x, y, z, \alpha = 0)$ be the function that extremizes J and write the variation about this function as

$$u(x, y, z, \alpha) = u(x, y, z, \alpha = 0) + \alpha \eta(x, y, z) + O(\alpha^2), \tag{17.47}$$

where $\eta(x, y, z)$ is a differentiable function. We take the derivative of Equation (17.46) with respect to α and set $\alpha = 0$, that is,

$$\left(\frac{\partial J}{\partial \alpha}\right)_{\alpha=0} = 0. \tag{17.48}$$

We then integrate terms like $\dfrac{\partial f}{\partial u_x}\eta_x$ by parts and use the fact that variation at the end points are zero to write

$$\int\int\int \left(\frac{\partial f}{\partial u} - \frac{\partial}{\partial x}\frac{\partial f}{\partial u_x} - \frac{\partial}{\partial y}\frac{\partial f}{\partial u_y} - \frac{\partial}{\partial z}\frac{\partial f}{\partial u_z}\right)\eta(x, y, z)dxdydz = 0. \tag{17.49}$$

Because the variation $\eta(x, y, z)$ is arbitrary, the expression inside the parentheses must be zero; thus yielding

$$\frac{\partial f}{\partial u} - \frac{\partial}{\partial x}\frac{\partial f}{\partial u_x} - \frac{\partial}{\partial y}\frac{\partial f}{\partial u_y} - \frac{\partial}{\partial z}\frac{\partial f}{\partial u_z} = 0. \tag{17.50}$$

This is the Euler equation for one dependent and three independent variables.

Example 17.3. Laplace equation: In electrostatics energy density is given as

$$\rho = \frac{1}{2}\varepsilon E^2, \tag{17.51}$$

where E is the magnitude of the electric field. Because the electric field can be obtained from a scalar potential as

$$\vec{E} = -\vec{\nabla}\Phi, \tag{17.52}$$

we can also write

$$\rho = \frac{1}{2}\varepsilon\left(\vec{\nabla}\Phi\right)^2. \tag{17.53}$$

Ignoring the $\frac{1}{2}\varepsilon$ factor, let us find the Euler equation for the functional

$$J = \int\int\int_V \left(\vec{\nabla}\Phi\right)^2 dxdydz. \tag{17.54}$$

Since

$$\left(\vec{\nabla}\Phi\right)^2 = \Phi_x^2 + \Phi_y^2 + \Phi_z^2, \tag{17.55}$$

f is given as

$$f\left[\Phi, \Phi_x, \Phi_y, \Phi_z, x, y, z\right] = \Phi_x^2 + \Phi_y^2 + \Phi_z^2. \tag{17.56}$$

Writing the Euler equation [Eq. (17.50)] for this f, we obtain

$$-2\left(\Phi_{xx} + \Phi_{yy} + \Phi_{zz}\right) = 0, \tag{17.57}$$

which is the Laplace equation

$$\vec{\nabla}^2\Phi(x,y,z) = 0. \tag{17.58}$$

A detailed investigation will show that this extremum is actually a minimum.

17.4   PRESENCE OF MORE THAN ONE DEPENDENT AND INDEPENDENT VARIABLES

In general, if the f function depends on three dependent and three independent variables as

$$f = f\left[p, p_x, p_y, p_z, q, q_x, q_y, q_z, r, r_x, r_y, r_z, x, y, z\right], \tag{17.59}$$

we can parametrize the variation in terms of three scalar parameters α, β, and γ as

$$p(x,y,z;\alpha) = p(x,y,z,\alpha = 0) + \alpha\xi(x,y,z) + 0(\alpha^2)$$
$$q(x,y,z;\beta) = q(x,y,z,\beta = 0) + \beta\eta(x,y,z) + 0(\beta^2) \tag{17.60}$$
$$r(x,y,z;\gamma) = r(x,y,z,\gamma = 0) + \gamma\psi(x,y,z) + 0(\gamma^2).$$

Now, the p, q, and the r functions that extremize $J = \int\int\int f\,dx\,dy\,dz$ will be obtained from the solutions of the following system of three Euler equations:

$$\frac{\partial f}{\partial p} - \frac{\partial}{\partial x}\frac{\partial f}{\partial p_x} - \frac{\partial}{\partial y}\frac{\partial f}{\partial p_y} - \frac{\partial}{\partial z}\frac{\partial f}{\partial p_z} = 0 \tag{17.61}$$

$$\frac{\partial f}{\partial q} - \frac{\partial}{\partial x}\frac{\partial f}{\partial q_x} - \frac{\partial}{\partial y}\frac{\partial f}{\partial q_y} - \frac{\partial}{\partial z}\frac{\partial f}{\partial q_z} = 0 \tag{17.62}$$

$$\frac{\partial f}{\partial r} - \frac{\partial}{\partial x}\frac{\partial f}{\partial r_x} - \frac{\partial}{\partial y}\frac{\partial f}{\partial r_y} - \frac{\partial}{\partial z}\frac{\partial f}{\partial r_z} = 0. \tag{17.63}$$

If we have more than three dependent and three independent variables, we can use y_i to denote the dependent variables and x_j for the independent variables and write the Euler equations as

$$\frac{\partial f}{\partial y_i} - \sum_j \frac{\partial}{\partial x_j}\frac{\partial f}{\partial y_{ij}} = 0, \quad i = 1, 2, ..., \tag{17.64}$$

where

$$y_{ij} \equiv \frac{\partial y_i}{\partial x_j}. \tag{17.65}$$

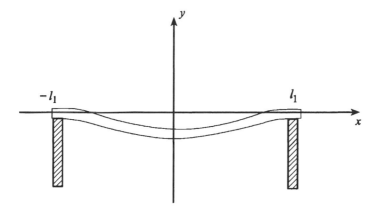

Fig. 17.3 Deformation of an elastic beam

17.5 PRESENCE OF HIGHER-ORDER DERIVATIVES

Sometimes in engineering problems we encounter functionals given as

$$J[y(x)] = \int_a^b F\left(x, y, y', ..., y^{(n)}\right) dx, \qquad (17.66)$$

where $y^{(n)}$ stands for the nth order derivative, the independent variable x takes values in the closed interval $[a, b]$, and the dependent variable $y(x)$ satisfies the boundary conditions

$$\begin{aligned} y(a) &= y_0, \quad y'(a) = y_0', \quad \ldots, \quad y^{(n-1)}(a) = y_0^{(n-1)}, \\ y(b) &= y_1, \quad y'(b) = y_1', \quad \ldots, \quad y^{(n-1)}(b) = y_1^{(n-1)}. \end{aligned} \qquad (17.67)$$

Using the same method that we have used for the other cases, we can show that the Euler equation that $y(x)$ satisfies is

$$F_y - \frac{d}{dx} F_{y'} + \frac{d^2}{dx^2} F_{y''} - \cdots + (-1)^n \frac{d^n}{dx^n} F_{y^{(n)}} = 0, \quad F_{y^{(n)}} = \frac{\partial F}{\partial y^{(n)}}. \qquad (17.68)$$

This equation is also known as the **Euler-Poisson** equation.

Example 17.4. *Deformation of an elastic beam:* Let us consider a homogeneous elastic beam supported from its end points at $(-l_1, 0)$ and $(0, l_1)$ as shown in Figure 17.3. Let us find the shape of the centerline of this beam.

From elasticity theory the potential energy E of the beam is given as

$$E = \int_{-l_1}^{l_1} \left[\frac{1}{2} \mu \frac{(y'')^2}{(1 + y'^2)} + \rho y \sqrt{1 + y'^2} \right] dx, \qquad (17.69)$$

where μ and ρ are parameters that characterize the physical properties of the beam. Assuming that the deformation is small, we can take

$$1 + y'^2 \approx 1. \tag{17.70}$$

Now the energy becomes

$$E = \int_{-l_1}^{l_1} \left[\frac{1}{2}\mu(y'')^2 + \rho y \right] dx. \tag{17.71}$$

For stable equilibrium the energy of the beam must be a minimum. Thus we have to minimize the energy integral with the conditions

$$y(l_1) = y(-l_1) = 0 \ \text{ and } \ y'(l_1) = y'(-l_1) = 0. \tag{17.72}$$

Using

$$F(x, y, y', y'') = \frac{1}{2}\mu \left(y'' \right)^2 + \rho y, \tag{17.73}$$

we write the Euler-Poisson equation as

$$\mu y^{(4)} + \rho = 0. \tag{17.74}$$

Solution of the Euler-Poisson equation is easily obtained as

$$y = \alpha x^3 + \beta x^2 + \gamma x + \delta - \frac{\rho}{24\mu}x^4. \tag{17.75}$$

Using the boundary conditions given in Equation (17.72) we can determine $\alpha, \beta, \gamma, \delta$ and find $y(x)$ as

$$y = \frac{\rho}{24\mu} \left[-x^4 + 2l_1^2 x^2 - l_1^4 \right]. \tag{17.76}$$

For the cases where there are m dependent variables we can generalize the variational problem in Equation (17.66) as

$$I\left(y_1, ..., y_m, x\right) =$$
$$\int F(x, y_1, y_1', ..., y_1^{(n_1)},$$
$$y_2, y_2', ..., y_2^{(n_2)}, ..., y_m, y_m', ..., y_m^{(n_m)})dx. \tag{17.77}$$

The boundary conditions are now given as

$$y_i^{(k)}(a) = y_{i0}^{(k)}, \qquad y_i^{(k)}(b) = y_{i1}^{(k)}, \\ k = 0, 1, ..., n_i - 1, \quad i = 1, 2, ..., m, \tag{17.78}$$

and the Euler-Poisson equations become

$$\sum_{k=0}^{n_i} (-1)^k \frac{d^k}{dx^k} F_{y_i}(k) = 0, \qquad i = 1, 2, ..., m. \tag{17.79}$$

17.6   ISOPERIMETRIC PROBLEMS AND THE PRESENCE OF CONSTRAINTS

In some applications we search for a function that not only extremizes a given functional

$$I = \int_{x_A}^{x_B} f(x, y, y')\, dx \qquad (17.80)$$

but also keeps another functional

$$J = \int_{x_A}^{x_B} g(x, y, y')\, dx \qquad (17.81)$$

at a fixed value. To find the Euler equation for such a function satisfying the boundary conditions

$$y(x_A) = y_A \ \text{ and } y(x_B) = y_B \qquad (17.82)$$

we parametrize the possible paths in terms of two parameters ε_1 and ε_2 as $\bar{y}(x, \varepsilon_1, \varepsilon_2)$. These paths also have the following properties:
 i) For all values of ε_1 and ε_2 they satisfy the boundary conditions

$$\bar{y}(x_A, \varepsilon_1, \varepsilon_2) = y_A \ \text{ and } \ \bar{y}(x_B, \varepsilon_1, \varepsilon_2) = y_B. \qquad (17.83)$$

 ii) $\bar{y}(x, 0, 0) = y(x)$ is the desired path.
 iii) $\bar{y}(x, \varepsilon_1, \varepsilon_2)$ has continuous derivatives with respect to all variables to second order.
 We now substitute these paths into Equations (17.80) and (17.81) to get two integrals depending on two parameters ε_1 and ε_2 as

$$I(\varepsilon_1, \varepsilon_2) = \int_{x_A}^{x_B} f(x, \bar{y}, \bar{y}')\, dx, \qquad (17.84)$$

$$J(\varepsilon_1, \varepsilon_2) = \int_{x_A}^{x_B} g(x, \bar{y}, \bar{y}')\, dx. \qquad (17.85)$$

While we are extremizing $I(\varepsilon_1, \varepsilon_2)$ with respect to ε_1 and ε_2, we are also going to ensure that $J(\varepsilon_1, \varepsilon_2)$ takes a fixed value; thus ε_1 and ε_2 cannot be independent. Using Lagrange undetermined multiplier λ we form

$$K(\varepsilon_1, \varepsilon_2) = I(\varepsilon_1, \varepsilon_2) + \lambda J(\varepsilon_1, \varepsilon_2). \qquad (17.86)$$

The condition for $K(\varepsilon_1, \varepsilon_2)$ to be an extremum is now written as

$$\left[\frac{\partial K}{\partial \varepsilon_1}\right]_{\substack{\varepsilon_1=0 \\ \varepsilon_2=0}} = \left[\frac{\partial K}{\partial \varepsilon_2}\right]_{\substack{\varepsilon_1=0 \\ \varepsilon_2=0}} = 0. \qquad (17.87)$$

In integral form this becomes

$$K\left(\varepsilon_1, \varepsilon_2\right) = \int_{x_A}^{x_B} h\left(x, \bar{y}, \bar{y}'\right) dx, \tag{17.88}$$

where the h function is defined as

$$h = f + \lambda g. \tag{17.89}$$

Differentiating with respect to these parameters and integrating by parts and using the boundary conditions we get

$$\left[\frac{\partial K}{\partial \varepsilon_j}\right] = \int_{x_A}^{x_B} \left[\frac{\partial h}{\partial \bar{y}} - \frac{d}{dx}\frac{\partial h}{\partial \bar{y}'}\right] \frac{\partial \bar{y}}{\partial \varepsilon_j} dx, \quad j = 1, 2. \tag{17.90}$$

Taking the variations as

$$\eta_j\left(x\right) = \left(\frac{\partial \bar{y}}{\partial \varepsilon_j}\right)_{\substack{\varepsilon_1 = 0 \\ \varepsilon_2 = 0}} \tag{17.91}$$

and using Equation (17.87) we write

$$\int_{x_A}^{x_B} \left[\frac{\partial h}{\partial y} - \frac{d}{dx}\frac{\partial h}{\partial y'}\right] \eta_j\left(x\right) dx = 0, \quad j = 1, 2. \tag{17.92}$$

Because the variations η_j are arbitrary, we set the quantity inside the square brackets to zero and obtain the differential equation

$$\frac{\partial h}{\partial y} - \frac{d}{dx}\frac{\partial h}{\partial y'} = 0. \tag{17.93}$$

Solutions of this differential equation contain two integration constants and a Lagrange undetermined multiplier λ. The two integration constants come from the boundary conditions [Eq. (17.82)], and λ comes from the constraint that fixes the value of J, thus completing the solution of the problem.

Another way to reach this conclusion is to consider the variation of the two functionals (17.80) and (17.81) as

$$\delta I = \int \frac{\delta f}{\delta y} \delta y \, dx$$

and

$$\delta J = \int \frac{\delta g}{\delta y} \delta y \, dx.$$

We now require that for all δy that makes $\delta J = 0$, δI should also vanish. This is possible if and only if

$$\frac{\delta f}{\delta y} \quad \text{and} \quad \frac{\delta g}{\delta y}$$

are constants independent of x, that is,

$$\left(\frac{\delta f}{\delta y}\right) \Big/ \left(\frac{\delta g}{\delta y}\right) = -\lambda (\text{const.}).$$

This is naturally equivalent to extremizing the functional

$$\int (f + \lambda g) dx$$

with respect to arbitrary variations δy.

When we have m constraints like $J_1, \ldots, J_m$, the above method is easily generalized by taking h as

$$h = f + \sum_{i=1}^{m} \lambda_i g_i \tag{17.94}$$

with m Lagrange undetermined multipliers. Constraining integrals now become

$$J_i = \int_{x_A}^{x_B} g_i (x, y, y') \, dx = \text{const.}, \quad i = 1, 2, \ldots, m. \tag{17.95}$$

If we also have n dependent variables, we have a system of n Euler equations given as

$$\frac{\partial h}{\partial y_i} - \frac{d}{dx} \frac{\partial h}{\partial y_i'} = 0, \quad i = 1, \ldots, n, \tag{17.96}$$

where h is given by Equation (17.94).

Example 17.5. *Isoperimetric problems:* Let us find the maximum area that can be enclosed by a closed curve of fixed perimeter L on a plane. We can define a curve on a plane in terms of a parameter t by giving its $x(t)$ and $y(t)$ functions. Now the enclosed area is

$$A = \frac{1}{2} \int_{t_A}^{t_B} (xy' - x'y) \, dt, \tag{17.97}$$

and the fixed perimeter condition is expressed as

$$L = \int_A^B ds = \int_{t_A}^{t_B} \sqrt{x'^2 + y'^2} \, dt, \tag{17.98}$$

where the prime denotes differentiation with respect to the independent variable t, and x and y are the two dependent variables. Our only constraint is given by Equation (17.98); thus we have a single Lagrange undetermined multiplier and the h function is written as

$$h = \frac{1}{2} (xy' - x'y) + \lambda \sqrt{x'^2 + y'^2}. \tag{17.99}$$

Writing the Euler equation for $x(t)$ we get

$$\frac{\partial h}{\partial x} - \frac{d}{dt}\frac{\partial h}{\partial x'} = 0,$$

$$\frac{1}{2}y' - \frac{d}{dt}(-\frac{1}{2}y + \lambda\frac{x'}{\sqrt{x'^2 + y'^2}}) = 0,$$

$$y' - \frac{d}{dt}(\lambda\frac{x'}{\sqrt{x'^2 + y'^2}}) = 0 \qquad (17.100)$$

and similarly for $y(t)$:

$$x' + \frac{d}{dt}\left(\lambda\frac{y'}{\sqrt{x'^2 + y'^2}}\right) = 0. \qquad (17.101)$$

The first integral of this system of equations [Eqs. (17.100) and (17.101)] can easily be obtained as

$$y - \lambda\frac{x'}{\sqrt{x'^2 + y'^2}} = y_0 \qquad (17.102)$$

and

$$x + \lambda\frac{y'}{\sqrt{x'^2 + y'^2}} = x_0. \qquad (17.103)$$

Solutions of these are given as

$$y - y_0 = \lambda\frac{x'}{\sqrt{x'^2 + y'^2}} \qquad (17.104)$$

and

$$x - x_0 = -\lambda\frac{y'}{\sqrt{x'^2 + y'^2}}, \qquad (17.105)$$

which can be combined to obtain the equation of the closed curve as

$$(x - x_0)^2 + (y - y_0)^2 = \lambda^2. \qquad (17.106)$$

This is the equation of a circle with its center at (x_0, y_0) and radius λ. Because the circumference is L, we determine λ as

$$\lambda = \frac{L}{2\pi}. \qquad (17.107)$$

Example 17.6. *Shape of a freely hanging wire with fixed length:* We now find the shape of a wire with length L and fixed at both ends at (x_A, y_A) and (x_B, y_B). The potential energy of the wire is

$$I = \rho g \int_{x_A}^{x_B} y\,ds = \rho g \int_{x_A}^{x_B} y\sqrt{1 + y'^2}\,dx. \qquad (17.108)$$

Because we take its length as fixed, we take our constraint as

$$L = \int_{x_A}^{x_B} \sqrt{1 + y'^2} \, dx. \tag{17.109}$$

For simplicity we use a Lagrange undetermined multiplier defined as $\lambda = -\rho g y_0$ and write the h function as

$$h = \rho g \left(y - y_0\right) \sqrt{1 + y'^2}, \tag{17.110}$$

where g is the acceleration of gravity and ρ is the density of the wire. We change our dependent variable to

$$y \to \eta = y - y_0, \tag{17.111}$$

which changes our h function to

$$h = \rho g \eta(x) \sqrt{1 + \eta'^2}. \tag{17.112}$$

After we write the Euler equation we find the solution as

$$y = y_0 + b \cosh\left(\frac{x - x_0}{b}\right). \tag{17.113}$$

Using the fact that the length of the wire is L and the end points are at (x_A, y_A) and (x_B, y_B), we can determine the Lagrange multiplier y_0 and the other constants x_0 and b.

17.7 APPLICATION TO CLASSICAL MECHANICS

With the mathematical techniques developed in the previous sections, we can conveniently express a fairly large part of classical mechanics as a variational problem. If a classical system is described by the generalized coordinates $q_i(t)$, $i = 1, 2, \dots, n$ and has a potential $V(q_i, t)$, then its Lagrangian can be written as

$$L(q_i, \dot{q}_i, t) = T(q_i, \dot{q}_i) - V(q_i, t), \tag{17.114}$$

where T is the kinetic energy and a dot denotes differentiation with respect to time. We now show that the equations of motion follow from Hamilton's principle:

Hamilton's principle: As a system moves from some initial time t_1 to t_2, with prescribed initial values $q_i(t_1)$ and $q_i(t_2)$, the actual path followed by the system is the one that makes the integral

$$I = \int_{t_1}^{t_2} L(q_i, \dot{q}_i, t) dt \tag{17.115}$$

an extremum. I is called the action.

From the conclusions of Section 17.2 the desired path comes from the solutions of

$$\frac{\partial L}{\partial q_i} - \frac{d}{dt}\left(\frac{\partial L}{\partial \dot{q}_i}\right) = 0, \quad i = 1, 2, ..., n, \tag{17.116}$$

which are now called the Lagrange (or Euler-Lagrange) equations. They are n simultaneous second-order differential equations to be solved for $q_i(t)$, where the $2n$ arbitrary integration constants are determined from the initial conditions $q_i(t_1)$ and $q_i(t_2)$.

For a particle of mass m and moving in an arbitrary potential $V(x_1, x_2, x_3)$ the Lagrangian is written as

$$L = \frac{1}{2}m(\dot{x}_1^2 + \dot{x}_2^2 + \dot{x}_3^2) - V(x_1, x_2, x_3). \tag{17.117}$$

Lagrange equations now become

$$m\ddot{x}_i = -\frac{\partial V}{\partial x_i}, \quad i = 1, 2, 3, \tag{17.118}$$

which are nothing but Newton's equations of motion.

The main advantage of the Lagrangian formulation of classical mechanics is that it makes applications to many particle systems and continuous systems possible. It is also a must in making the transition to quantum mechanics and quantum field theories. For continuous systems we define a Lagrangian density $\mathcal{L}$ as

$$L = \int_V \mathcal{L} d^3 \vec{r}, \tag{17.119}$$

where V is the volume. Now, the action in Hamilton's principle becomes

$$I = \int_{t_1}^{t_2} L dt \tag{17.120}$$

$$= \int_{t_1}^{t_2} \left[\int_V \mathcal{L} d^3 \vec{r}\right] dt.$$

For a continuous time-dependent system with n independent fields, $\phi_i(\vec{r}, t)$, $i = 1, 2, ..., n$, the Lagrangian density is given as

$$\mathcal{L}(\phi_i, \phi_{it}, \phi_{ix}, \phi_{iy}, \phi_{iz}, \vec{r}, t), \tag{17.121}$$

where

$$\phi_{it} = \frac{\partial \phi_i}{\partial t}, \quad \phi_{ix} = \frac{\partial \phi_i}{\partial x}, \quad i = 1, 2, ..., n. \tag{17.122}$$

We can now use the conclusions of Section 17.4 to write the n Lagrange equations as

$$\frac{\partial \mathcal{L}}{\partial \phi_i} - \frac{\partial}{\partial t}\frac{\partial \mathcal{L}}{\partial \phi_{it}} - \frac{\partial}{\partial x}\frac{\partial \mathcal{L}}{\partial \phi_{ix}} - \frac{\partial}{\partial y}\frac{\partial \mathcal{L}}{\partial \phi_{iy}} - \frac{\partial}{\partial z}\frac{\partial \mathcal{L}}{\partial \phi_{iz}} = 0. \qquad (17.123)$$

For time-independent fields, $\phi_i(\vec{r})$, $i = 1, 2, ..., n$, the Lagrange equations become

$$\frac{\partial \mathcal{L}}{\partial \phi_i} - \sum_{j=1}^{3} \frac{\partial}{\partial x_j}\frac{\partial \mathcal{L}}{(\partial \phi_i/\partial x_j)} = 0, \quad i = 1, 2, ..., n. \qquad (17.124)$$

As an example, consider the Lagrange density

$$\mathcal{L} = \frac{1}{2}\vec{\nabla}\phi(\vec{r}) \cdot \vec{\nabla}\phi(\vec{r}) + m^2\phi(\vec{r})^2 \qquad (17.125)$$

$$= \frac{1}{2}\left[\left(\frac{\partial \phi}{\partial x}\right)^2 + \left(\frac{\partial \phi}{\partial y}\right)^2 + \left(\frac{\partial \phi}{\partial y}\right)^2\right] + m^2\phi^2,$$

where the corresponding Lagrange equation is the Laplace equation

$$\vec{\nabla}^2\phi(\vec{r}) - m^2\phi(\vec{r}) = 0. \qquad (17.126)$$

17.8   EIGENVALUE PROBLEM AND VARIATIONAL ANALYSIS

For the variational problems we have considered the end product was a differential equation to be solved for the desired function. We are now going to approach the problem from the other direction and ask the question: Given a differential equation, is it always possible to obtain it as the Euler equation of a variational integral such as

$$\delta J = \delta \int_a^b f \, dt = 0? \qquad (17.127)$$

When the differential equation is an equation of motion, then this question becomes: Can we drive it from a Lagrangian? This is a rather subtle point. Even though it is possible to write theories that do not follow from a variational principle, they eventually run into problems.

We have seen that solving the Laplace equation within a volume V is equivalent to extremizing the functional

$$I[\phi(\vec{r})] = \frac{1}{2}\int_V \left(\vec{\nabla}\phi\right)^2 d^3\vec{r}, \qquad (17.128)$$

with the appropriate boundary conditions. Another frequently encountered differential equation in science and engineering is the Sturm-Liouville equation

$$\frac{d}{dx}\left[p(x)\frac{du(x)}{dx}\right] - q(x)u(x) + \lambda\rho(x)u(x) = 0, \quad x \in [a, b]. \qquad (17.129)$$

It can be obtained by extremizing the functional

$$I[u(x)] = \int_a^b \left[pu'^2 + (q - \lambda\rho) u^2 \right] dx. \qquad (17.130)$$

However, because the eigenvalues λ are not known a priori, this form is not very useful. It is better to extremize

$$I[u(x)] = \int_a^b \left[pu'^2 + qu^2 \right] dx, \qquad (17.131)$$

subject to the constraint

$$J[u(x)] = \int_a^b \rho u^2 dx = \text{const}. \qquad (17.132)$$

In this formulation eigenvalues appear as the Lagrange multipliers. Note that the constraint [Eq. (17.132)] is also the normalization condition of $u(x)$; thus we can also extremize

$$K[u(x)] = \frac{I[u(x)]}{J[u(x)]}. \qquad (17.133)$$

If we multiply the Sturm-Liouville equation by $u(x)$ and then integrate by parts from a to b, we see that the extremums of $K[u(x)]$ correspond to the eigenvalues λ. In a Sturm-Liouville problem (Morse and Feshbach, Section 6.3)

i) There exists a minimum eigenvalue.

ii) $\lambda_n \to \infty$ as $n \to \infty$.

iii) To be precise, $\lambda_n \sim n^2$ as $n \to \infty$.

Thus the minimums of Equation (17.133) give the eigenvalues λ_n. In fact, from the first property the absolute minimum of K is the lowest eigenvalue λ_0. This is very useful in putting an upper bound to the lowest eigenvalue.

To estimate the lowest eigenvalue we choose a trial function $u(x)$ and expand in terms of the exact eigenfunctions, $u_i(x)$, which are not known:

$$u = u_0 + c_1 u_1 + c_2 u_2 + \cdots . \qquad (17.134)$$

Depending on how close our trial function is to the exact eigenfunction, the coefficients $c_1, c_2, \ldots$ will be small numbers. Before we evaluate $K[u(x)]$, let us substitute our trial function into Equation (17.131):

$$\int_a^b \left[p \left(u'_0 + c_1 u'_1 + c_2 u'_2 + \cdots \right)^2 + q \left(u_0 + c_1 u_1 + c_2 u_2 + \cdots \right)^2 \right] dx. \qquad (17.135)$$

Since the set $\{u_i\}$ is orthonormal, using the relations

$$\int_a^b \left[pu_i'^2 + qu_i^2 \right] dx = \lambda_i \qquad (17.136)$$

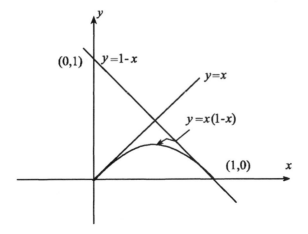

Fig. 17.4 $\sin(\pi x)$ could be approximated by $x(1-x)$

and

$$\int_a^b \left[pu_i'u_j' + qu_iu_j \right] dx = 0 \quad (i \neq j),$$

(17.137)

we can write

$$K\left[u\left(x\right)\right] = \frac{\int_a^b \left[pu'^2 + qu^2\right] dx}{\int_a^b \rho u^2 dx}$$

(17.138)

as

$$K\left[u\left(x\right)\right] \simeq \frac{\lambda_0 + c_1^2\lambda_1 + c_2^2\lambda_2 + \cdots}{1 + c_1^2 + c_2^2 + \cdots}.$$

(17.139)

Because $c_1, c_2, \ldots$ are small numbers, K gives us the approximate value of the lowest eigenvalue as

$$K \simeq \lambda_0 + c_1^2\left(\lambda_1 - \lambda_0\right) + c_2^2\left(\lambda_2 - \lambda_0\right) + \cdots .$$

(17.140)

What is significant here is that even though our trial function is good to the first order, our estimate of the lowest eigenvalue is good to the second order. This is also called the Hylleraas-Undheim theorem. Because the eigenvalues are monotonic increasing, this estimate is also an upper bound to the lowest eigenvalue.

Example 17.7. *How to estimate lowest eigenvalue:* Let us estimate the lowest eigenvalue of

$$\frac{d^2u}{dx^2} + \lambda u = 0,$$

(17.141)

with the boundary conditions $u(0) = 0$ and $u(1) = 0$. As shown in Figure 17.4 we can take our trial function as

$$u = x(1 - x). \tag{17.142}$$

This gives us

$$\lambda_0 \leq \frac{\int_0^1 u'^2 dx}{\int_0^1 u^2 dx} = \frac{\frac{1}{3}}{\frac{1}{30}} = 10. \tag{17.143}$$

This is already close to the exact eigenvalue π^2. For a better upper bound we can improve our trial function as

$$u = x(1 - x)(1 + c_1 x + \cdots) \tag{17.144}$$

and determine c_i by extremizing K. For this method to work our trial function

i) Must satisfy the boundary conditions.
ii) Should reflect the general features of the exact eigenfunction.
iii) Should be sufficiently simple to allow analytic calculations.

Example 17.8. *Vibrations of a drumhead:* We now consider the wave equation

$$\vec{\nabla}^2 u + k^2 u = 0, \quad k^2 = \frac{\omega^2}{c^2} \tag{17.145}$$

in two dimensions and in spherical polar coordinates. We take the radius as a and use $u(a) = 0$ as our boundary condition. This suggests the trial function

$$u = 1 - \frac{r}{a}. \tag{17.146}$$

Now the upper bound for the lowest eigenvalue k_0^2 is obtained from

$$k_0^2 \leq \frac{\int_0^a \int_0^{2\pi} \left(\vec{\nabla} u\right)^2 r dr d\theta}{\int_0^a \int_0^{2\pi} u^2 r dr d\theta} \tag{17.147}$$

as

$$k_0^2 \leq \frac{\pi}{\pi a^2 / 6} = \frac{6}{a^2}. \tag{17.148}$$

Compare this with the exact eigenvalue

$$k_0^2 = \frac{5.78}{a^2}.$$

Example 17.9. *Harmonic oscillator problem:* The Schrödinger equation can be driven from the functional

$$\frac{\Psi \cdot H\Psi}{\Psi \cdot \Psi}, \tag{17.149}$$

where $\Psi_1 \cdot \Psi_2$ means $\int_{-\infty}^{\infty} \Psi_1^* \Psi_2 dx$. For the harmonic oscillator problem the Schrödinger equation is written as

$$H\Psi = -\frac{d^2\Psi}{dx^2} + x^2\Psi = E\Psi, \quad x \in (-\infty, \infty). \tag{17.150}$$

For the lowest eigenvalue we take our trial function as

$$\Psi = (1 + \alpha x^2)e^{-x^2}. \tag{17.151}$$

For an upper bound this gives

$$E_0 \le \frac{\Psi \cdot H\Psi}{\Psi \cdot \Psi} = \frac{\dfrac{5}{4} - \dfrac{\alpha}{8} + \dfrac{43\alpha^2}{64}}{1 + \dfrac{\alpha}{2} + \dfrac{3\alpha^2}{16}}. \tag{17.152}$$

To find its minimum we solve

$$23\alpha^2 + 56\alpha - 48 = 0 \tag{17.153}$$

to find

$$\alpha = 0.6718. \tag{17.154}$$

Thus the upper bound to the lowest energy is obtained as

$$E_0 \le 1.034,$$

where the exact eigenvalue is 1.0. This method can also be used for the higher-order eigenvalues. However, one must make sure that the chosen trial function is orthogonal to the eigenfunctions corresponding to the lower eigenvalues.

17.9 RAYLEIGH-RITZ METHOD

In this method we aim to find an approximate solution to a differential equation satisfying certain boundary conditions. We first write the solution in terms of suitably chosen $\phi_n(x)$ functions as

$$y(x) \simeq \phi_0(x) + c_1\phi_1(x) + c_2\phi_2(x) + c_3\phi_3(x) + \cdots + c_n\phi_n(x), \tag{17.155}$$

where $c_1, c_2, ..., c_n$ are constants to be determined. $\phi_k(x)$ are chosen functions so that $y(x)$ satisfies the boundary conditions for any choice of the c values. In general $\phi_0(x)$ is chosen such that it satisfies the boundary conditions at the end points of our interval, and $\phi_1(x), \phi_2(x), ..., \phi_n(x)$ are chosen such that they vanish at the end points.

Example 17.10. *Loaded cable fixed between two points:* We consider a cable fixed between the points $(0,0)$ and (l, h). The cable carries a load along the y-axis distributed as

$$f(x) = -q_0 \frac{x}{l}. \tag{17.156}$$

To find the shape of this cable we have to solve the variational problem

$$\delta \int_0^l \left(\frac{1}{2} T_0 y'^2 + q_0 \frac{x}{l} y \right) dx = 0 \tag{17.157}$$

with the boundary conditions

$$y(0) = 0 \ \text{ and } \ y(l) = h, \tag{17.158}$$

where T_0 and q_0 are constants. Using the Rayleigh-Ritz method we choose $\phi_0(x)$ such that it satisfies the above boundary conditions and choose $\phi_1, ..., \phi_n$ such that they vanish at both end points:

$$\phi_0 = \frac{h}{l} x, \tag{17.159}$$

$$\phi_1 = x(x - l),$$

$$\phi_2 = x^2(x - l),$$

$$\vdots$$

$$\phi_n = x^n(x - l).$$

Now the approximate solution $y(x)$ becomes

$$y(x) \simeq \frac{h}{l} x + x(x - l) \left(c_1 + c_2 x + \cdots + c_n x^{n-1} \right). \tag{17.160}$$

For simplicity we choose $n = 1$ so that $y(x)$ becomes

$$y(x) \simeq \frac{h}{l} x + x(x - l) c_1. \tag{17.161}$$

Substituting this into the variational integral we get

$$\delta \int_0^l \left[\frac{1}{2} T_0 \left(\frac{h}{l} + (2x - l) c_1 \right)^2 + q_0 \frac{x}{l} \left(\frac{h}{l} x + x(x - l) c_1 \right) \right] dx = 0, \tag{17.162}$$

$$\delta \left[\frac{1}{2} T_0 \left(\frac{h^2}{l} + \frac{1}{3} l^3 c_1^2 \right) + q_0 \left(\frac{1}{3} h l - \frac{1}{12} c_1 l^3 \right) \right] = 0, \tag{17.163}$$

$$\frac{1}{3} T_0 l^3 c_1 \delta c_1 - \frac{1}{12} q_0 l^3 \delta c_1 = 0, \tag{17.164}$$

$$\frac{1}{3} T_0 l^3 \left(c_1 - \frac{q_0}{4 T_0} \right) \delta c_1 = 0. \tag{17.165}$$

Because the variation δc_1 is arbitrary, the quantity inside the brackets must vanish, thus giving

$$c_1 = \frac{q_0}{4 T_0} \tag{17.166}$$

and

$$y(x) \simeq \frac{h}{l} x + \frac{q_0}{4 T_0} x(x - l). \tag{17.167}$$

The Euler equation for the variational problem [Eq. (17.157)] can easily be written as

$$T_0 y'' - \frac{q_0 x}{l} = 0, \tag{17.168}$$

where the exact solution of this problem is

$$y(x) = \frac{h}{l} x + \frac{q_0}{6 T_0 l} x(x^2 - l^2). \tag{17.169}$$

As we will see in the next example, an equivalent approach is to start with the Euler equation (17.168), which results from the variational integral

$$\int_0^l \left[T_0 y'' - \frac{q_0 x}{l} \right] \delta y \, dx = 0. \tag{17.170}$$

Substituting Equation (17.161) into the above equation, we write

$$\int_0^l \left[2 c_1 T_0 - \frac{q_0 x}{l} \right] x(x - l) \delta c_1 \, dx = 0, \tag{17.171}$$

which after integration yields

$$\left[-\frac{1}{3} c_1 T_0 l^3 + \frac{1}{12} q_0 l^3 \right] \delta c_1 = 0. \tag{17.172}$$

Since δc_1 is arbitrary, we again obtain $c_1 = q_0 / 4 T_0$.

Example 17.11. *Rayleigh Ritz Method:* We now find the solution of the differential equation

$$\frac{d^2y}{dx^2} + xy = -x \tag{17.173}$$

with the boundary conditions $y(0) = 0$ and $y(1) = 0$ by using the Rayleigh-Ritz method. The variational problem corresponding to this differential equation can be written as

$$\int_0^1 (y'' + xy + x)\, \delta y dx = 0. \tag{17.174}$$

We take the approximate solution as

$$y(x) \simeq x(1-x)(c_1 + c_2 x + \cdots), \tag{17.175}$$

and substitute this in Equation (17.174) to obtain

$$\int_0^1 \left[\left(-2 + x^2 - x^3 \right) c_1 + \left(2 - 6x + x^3 - x^4 \right) c_2 + \cdots + x \right]$$
$$\times \left[\delta c_1 \left(x - x^2 \right) + \delta c_2 \left(x^2 - x^3 \right) + \cdots \right] dx = 0. \tag{17.176}$$

Solution with one term is given as

$$y^{(1)} = c_1 x(1-x), \quad c_1 = \frac{5}{19}, \tag{17.177}$$

while the solution with two terms is given as

$$y^{(2)} = c_1 x(1-x) + c_2 x^2 (1-x), \tag{17.178}$$

where

$$c_1 = 0.177,\ c_2 = 0.173\ . \tag{17.179}$$

Problems

17.1   For the variational problem

$$\delta \int_a^b F\left(x, y, y', \ldots, y^{(n)}\right) dx = 0,$$

show that the Euler equation is given as

$$F_y - \frac{d}{dx} F_{y'} + \frac{d^2}{dx^2} F_{y''} - \cdots + (-1)^n \frac{d^n}{dx^n} F_{y^{(n)}} = 0.$$

Assume that the variation at the end points is zero.

17.2   For the Sturm-Liouville system

$$y''(x) = -\lambda y(x), \quad y(0) = y(1) = 0,$$

find the approximate eigenvalues to first and second order.
    Compare your results with the exact eigenvalues.

17.3   Given the variational problem for the massive scalar field, with the potential $V(\vec{r})$, as

$$\delta \int \mathcal{L} d^3 \vec{r} \, dt = 0,$$

where

$$\mathcal{L}(\vec{r}, t) = \frac{1}{2}(\dot{\Phi}^2 - \left(\vec{\nabla}\Phi\right)^2 - m^2\Phi^2) - V(\vec{r}).$$

Find the equation of motion for

$$\Phi(\vec{r}, t).$$

17.4   Treat $\Psi(\vec{r}, t)$ and $\Psi^*(\vec{r}, t)$ as independent fields in the Lagrangian density

$$\mathcal{L} = \frac{\hbar^2}{2m} \vec{\nabla}\Psi \vec{\nabla}\Psi^* + V\Psi\Psi^* - \frac{i\hbar}{2}\left(\Psi^* \frac{\partial\Psi}{\partial t} - \Psi \frac{\partial\Psi^*}{\partial t}\right),$$

where

$$\int \mathcal{L} d^3 \vec{r} \, dt = 0.$$

Show that the corresponding Euler equations are the Schrödinger equations

$$H\Psi = (-\frac{\hbar^2}{2m} \vec{\nabla}^2 + V)\Psi = i\hbar \frac{\partial\Psi}{\partial t}$$

and

$$H\Psi^* = (-\frac{\hbar^2}{2m}\vec{\nabla}^2 + V)\Psi^* = -i\hbar\frac{\partial\Psi^*}{\partial t}.$$

17.5   Consider a cable fixed at the points $(0,0)$ and (l,h). It carries a load along the y-axis, distributed as

$$f(x) = -q\frac{x}{l}.$$

To find the shape of this cable we have to solve the variational problem

$$\delta \int_0^l \left(\frac{1}{2}T_0y'^2 + q_0\frac{x}{l}y\right) dx = 0$$

with the boundary conditions

$$y(0) = 0 \quad \text{and} \quad y(l) = h.$$

Find the shape of the wire accurate to second order.

Hint: See Example 17.10.

17.6   Show that the exact solution in Problem 17.5 is given as

$$y(x) = \frac{h}{l}x + \frac{q_0}{6T_0l}x(x^2 - l^2).$$

17.7   Find an upper bound for the lowest eigenvalue of the differential equation

$$\frac{d^2y}{dx^2} + \lambda xy = 0$$

with the boundary conditions

$$y(0) = y(1) = 0.$$

17.8   For a flexible elastic string, with constant tension and fixed at the end points:

$$y(0,t) = y(L,t) = 0,$$

Show that the Lagrangian density is given as

$$\mathcal{L} = \frac{1}{2}\rho(x)\left[\frac{\partial y(x,t)}{\partial t}\right]^2 - \frac{1}{2}\tau\left[\frac{\partial y(x,t)}{\partial x}\right]^2,$$

where ρ is the density and τ is the tension. Show that the Lagrange equation is

$$\frac{\partial^2 y(x,t)}{\partial x^2} - \frac{\rho}{\tau}\frac{\partial^2 y(x,t)}{\partial t^2} = 0.$$

17.9   For a given Lagrangian representing a system with n degrees of freedom, show that adding a total time derivative to the Lagrangian does not effect the equations of motion, that is, L and L' related by

$$L' = L + \frac{dF(q_1, q_2, ..., q_n, t)}{dt},$$

where F is an arbitrary function, have the same Lagrange equations.

17.10   For a given Lagrangian $L(q_i, \dot{q}_i, t)$, where $i = 1, 2, ...n$, show that

$$\frac{d}{dt}\left[\sum_{i=1}^{n} \dot{q}_i \frac{\partial L}{\partial \dot{q}_i} - L\right] + \frac{\partial L}{\partial t} = 0.$$

This means that if the Lagrangian does not depend on time explicitly, then the quantity, H, defined as

$$H(q_i, \frac{\partial L}{\partial \dot{q}_i}, t) = \sum_{i=1}^{n} \dot{q}_i \frac{\partial L}{\partial \dot{q}_i} - L$$

is conserved. Using Cartesian coordinates, interpret H.

17.11   **The brachistochrone problem**: Find the shape of the curve joining two points, along which a particle, initially at rest, falls freely under the influence of gravity from the higher point to the lower point in the least amount of time.

17.12   In an expanding flat universe the metric is given as

$$ds^2 = -dt^2 + a^2(t)(\ dx^2 + dy^2 + dz^2)$$

$$= -dt^2 + a^2(t)\delta_{ij}dx^i dx^j,$$

where $i = 1, 2, 3$, and $a(t)$ is the scale factor. Given this metric, consider the following variational integral for the geodesics:

$$\delta I = \frac{1}{2}\int \left[-\left(\frac{dt}{d\tau}\right)^2 + a^2(t)\delta_{ij}\frac{dx^i}{d\tau}\frac{dx^j}{d\tau}\right] d\tau,$$

where τ is the proper time. For the dependent variables $t(\tau)$ and $x^i(\tau)$ show that the Euler equations for the geodesics are:

$$\frac{d^2t}{d\tau^2} + a\dot{a}\delta_{ij}\frac{dx^i}{d\tau}\frac{dx^j}{d\tau} = 0$$

and

$$\frac{d^2x^i}{d\tau^2} + 2\frac{\dot{a}}{a}\frac{dt}{d\tau}\frac{dx^i}{d\tau} = 0,$$

where $a = da/dt$.

17.13   Using cylindrical coordinates, find the geodesics on a cone.

17.14   Write the Lagrangian and the Lagrange equations of motion for a double pendulum in uniform gravitational field.

17.15   Consider the following Lagrangian density for the massive scalar field in curved background spacetimes:

$$\mathcal{L}(x) = \frac{1}{2}[-\det g_{\alpha\beta}(x)]^{\frac{1}{2}} \left\{ g^{\mu\nu}(x)\partial_\mu \Phi(x)\partial_\nu \Phi(x) - [m^2 + \xi R(x)]\, \Phi^2(x) \right\},$$

where $\Phi(x)$ is the scalar field, m is the mass of the field quanta, and x stands for (x^0, x^1, x^2, x^3). Coupling between the scalar field and background geometry is represented by the term

$$\xi R(x)\Phi^2(x),$$

where ξ is called the coupling constant and $R(x)$ is the curvature (Ricci) scalar. The corresponding action is

$$S = \int \mathcal{L}(x)d^4x, \ (d^4x = dx^0 dx^1 dx^2 dx^3).$$

By setting the variation of the action with respect to $\Phi(x)$ to zero, show that the scalar field equation is given as

$$[\Box + m^2 + \xi R(x)]\,\Phi(x) = 0,$$

where $\Box = \partial_\mu \partial^\mu$ is the d'Alembert wave operator, ∂_μ stands for the covariant derivarive, and take the signature of the metric as $(+ - - -)$.

17.16   Find the extremals of the problem

$$\delta \int_{x_1}^{x_2} \left[a(x)y''^2 - p(x)y'^2 + q(x)y^2\right] dx = 0$$

subject to the constraint

$$\int_{x_1}^{x_2} r(x)y^2(x)dx = 1,$$

where $y(x_1), y'(x_1), y(x_2), y'(x_2)$ are prescribed.

18

INTEGRAL EQUATIONS

We have been rather successful with differential equations in representing physical processes. They are composed of the derivatives of an unknown function. Because derivatives are defined in terms of ratios of differences in the neighborhood of a point, differential equations are local. In mathematical physics there are also integral equations, where the unknown function appears under an integral sign. Because integral equations involve integrals of the unknown function over the entire space, they are global and in general much more difficult to solve.

An important property of differential equations is that to describe a physical problem completely, they must be supplemented with boundary conditions. Integral equations, on the other hand, constitute a complete description of a given problem, where extra conditions are neither needed nor could be imposed. Because the boundary conditions can be viewed as a convenient way of including global effects into a system, a connection between differential and integral equations is to be expected. In fact, under certain conditions integral and differential equations can be transformed into each other. Whether an integral or a differential equation is more suitable for expressing laws of nature is still an interesting problem, with some philosophical overtones that Einstein once investigated. Sometimes the integral equation formulation of a given problem may offer advantages over its differential equation description. At other times, as in some diffusion or transport phenomena, we may have no choice but to use integral equations.

In this chapter we discuss the basic properties of linear integral equations and introduce some techniques for obtaining their solutions. We also discuss the Hilbert-Schmidt theory, where an eigenvalue problem is defined in terms of linear integral operators.

18.1   CLASSIFICATION OF INTEGRAL EQUATIONS

Linear integral equations are classified under two general categories. Equations that can be written as

$$\alpha(x)y(x) = F(x) + \lambda \int_a^b \kappa(x,\xi)y(\xi)d\xi \qquad (18.1)$$

are called the **Fredholm equations**, where $\alpha(x)$, $F(x)$, and $\kappa(x,\xi)$ are known functions and $y(x)$ is the unknown function, while λ, a, and b are known constants. $\kappa(x,\xi)$ is called the kernel, which is closely related to Green's function. When the upper limit of the integral in a Fredholm equation is a variable, we have the **Volterra equation**:

$$\alpha(x)y(x) = F(x) + \lambda \int_a^x \kappa(x,\xi)y(\xi)d\xi. \qquad (18.2)$$

The Fredholm and Volterra equations also have the following kinds:

$$\begin{array}{ll} \alpha \neq 0 & \text{kind I} \\ \alpha = 1 & \text{kind II} \\ \alpha = \alpha(x) & \text{kind III} \end{array}$$

When $F(x)$ is zero, the integral equation is called **homogeneous**. Integral equations can also be defined in higher dimensions. In two dimensions we can write a linear integral equation as

$$\alpha(x,y)\omega(x,y) = F(x,y) + \lambda \int\int_R \kappa(x,y;\xi,\eta)\omega(\xi,\eta)d\xi d\eta. \qquad (18.3)$$

18.2   INTEGRAL AND DIFFERENTIAL EQUATIONS

Some integral equations can be obtained from differential equations. To see this connection we first derive a useful formula. We first consider the integral

$$I_n(x) = \int_a^x (x-\xi)^{n-1}f(\xi)d\xi, \qquad (18.4)$$

where $n > 0$ integer and a is a constant. Using

$$\frac{d}{dx}\int_{A(x)}^{B(x)} F(x,\xi)d\xi \qquad (18.5)$$

$$= \int_{A(x)}^{B(x)} \frac{\partial F(x,\xi)}{\partial x}d\xi + F(x,B(x))\frac{dB(x)}{dx} - F(x,A(x))\frac{dA(x)}{dx},$$

we take the derivative of $I_n(x)$ to write

$$\frac{dI_n}{dx} = (n-1)\int_a^x (x-\xi)^{n-2}f(\xi)d\xi + \left[(x-\xi)^{n-1}f(\xi)\right]_{\xi=x}. \qquad (18.6)$$

For $n > 1$ this gives

$$\frac{dI_n}{dx} = (n-1)I_{n-1}, \qquad (18.7)$$

and for $n = 1$ we have

$$\frac{dI_1}{dx} = f(x). \qquad (18.8)$$

Differentiating $I_n(x)$ k times gives

$$\frac{d^k I_n}{dx^k} = (n-1)(n-2)\cdots(n-k)I_{n-k}, \qquad (18.9)$$

which can be used to write

$$\frac{d^{n-1}I_n}{dx^{n-1}} = (n-1)!I_1(x) \qquad (18.10)$$

or

$$\frac{d^n I_n}{dx^n} = (n-1)!f(x). \qquad (18.11)$$

Since $I_n(a) = 0$ for $n \geq 1$ and from Equations (18.10) and (18.11), we see that $I_n(x)$ and all of its derivatives up to order $(n-1)$ are zero at $x = a$. This gives us

$$I_1(x) = \int_a^x f(x_1)dx_1, \qquad (18.12)$$

$$I_2(x) = \int_a^x I_1(x_2)dx_2 = \int_a^x \int_a^{x_2} f(x_1)dx_1 dx_2 \qquad (18.13)$$

and in general

$$I_n(x) = (n-1)! \int_a^x \int_a^{x_n} \cdots \int_a^{x_3} \int_a^{x_2} f(x_1)dx_1 dx_2 ... dx_{n-1} dx_n. \qquad (18.14)$$

Using the above equation we can now write the following useful formula, which is also known as the **Cauchy formula**:

$$\int_a^x \int_a^{x_n} \cdots \int_a^{x_3} \int_a^{x_2} f(x_1)dx_1 dx_2 ... dx_{n-1} dx_n$$
$$= \frac{1}{(n-1)!} \int_a^x (x-\xi)^{n-1}f(\xi)d\xi. \qquad (18.15)$$

18.3 HOW TO CONVERT SOME DIFFERENTIAL EQUATIONS INTO INTEGRAL EQUATIONS

We now consider the following second-order ordinary differential equation with variable coefficients:

$$\frac{d^2y}{dx^2} + A(x)\frac{dy}{dx} + B(x)y = f(x), \tag{18.16}$$

which is frequently encountered in physics and engineering applications. Let the boundary conditions be given as $y(a) = y_0$ and $y'(a) = y_0'$. Integrating this differential equation once gives us

$$y'(x) - y_0' = -\int_a^x A(x_1)y'(x_1)dx_1 - \int_a^x B(x_1)y(x_1)dx_1$$
$$+ \int_a^x f(x_1)dx_1. \tag{18.17}$$

We integrate the first term on the right-hand side by parts and then solve for $y'(x)$ to write

$$y'(x) = -A(x)y(x) - \int_a^x [B(x_1) - A'(x_1)]\, y(x_1)dx_1$$
$$+ \int_a^x f(x_1)dx_1 + A(a)y_0 + y_0'. \tag{18.18}$$

We integrate again to obtain

$$y(x) - y_0 = -\int_a^x A(x_1)y(x_1)dx_1$$
$$- \int_a^x \int_a^{x_2} [B(x_1) - A'(x_1)]\, y(x_1)dx_1 dx_2$$
$$+ \int_a^x \int_a^{x_2} f(x_1)dx_1 dx_2$$
$$+ [A(a)y_0 + y_0']\,(x - a). \tag{18.19}$$

Using the Cauchy formula [Eq. (18.15)] we can write this as

$$y(x) = -\int_a^x \{A(\xi) + (x - \xi)\, [B(\xi) - A'(\xi)]\}\, y(\xi)d\xi \tag{18.20}$$
$$+ \int_a^x (x - \xi)f(\xi)d\xi + [A(a)y_0 + y_0']\,(x - a) + y_0$$

or

$$y(x) = \int_a^x \kappa(x, \xi)y(\xi)d\xi + F(x), \tag{18.21}$$

where $\kappa(x, \xi)$ and $F(x)$ are given as

$$\kappa(x, \xi) = -(x - \xi)[B(\xi) - A'(\xi)] - A(\xi),$$

$$F(x) = \int_a^x (x - \xi) f(\xi) d\xi + [A(a)y_0 + y_0'](x - a) + y_0. \qquad (18.22)$$

This is an inhomogeneous Volterra equation of the second kind. This integral Equation (18.22) is equivalent to the differential equation (18.16) plus the boundary conditions $y(a) = y_0$ and $y'(a) = y_0'$.

Example 18.1. *Conversion of differential equations into integral equations:*
Using (18.22) we can convert the differential equation

$$\frac{d^2 y}{dx^2} + \lambda y = f(x) \qquad (18.23)$$

and the boundary conditions

$$y(0) = 1, \quad y'(0) = 0, \qquad (18.24)$$

into an integral equation as

$$y(x) = \lambda \int_0^x (\xi - x) y(\xi) d\xi - \int_0^x (\xi - x) f(\xi) d\xi + 1. \qquad (18.25)$$

Example 18.2. *Conversion of differential equations into integral equations:*
In the previous example we had a single-point boundary condition. We now consider the differential equation

$$\frac{d^2 y}{dx^2} + \lambda y = 0 \qquad (18.26)$$

with a two-point boundary condition

$$y(0) = 0 \quad \text{and} \quad y(l) = 0. \qquad (18.27)$$

Integrating Equation (18.26) between $(0, x)$ we get

$$\frac{dy}{dx} = -\lambda \int_0^x y(\xi) d\xi + C. \qquad (18.28)$$

C is an integration constant that is equal to $y'(0)$, which is not given. A second integration gives

$$y(x) = -\lambda \int_0^x (x - \xi) y(\xi) d\xi + Cx, \qquad (18.29)$$

where we have used the Cauchy formula [Eq. (18.15)] and the boundary condition at $x = 0$. We now use the remaining boundary condition, $y(l) = 0$, to determine C as

$$C = \frac{\lambda}{l} \int_0^l (l - \xi)y(\xi)d\xi. \tag{18.30}$$

Substituting this back in Equation (18.29) we write the result as

$$y(x) = -\lambda \int_0^x (x - \xi)y(\xi)d\xi + \frac{x\lambda}{l} \int_0^l (l - \xi)y(\xi)d\xi \tag{18.31}$$

or

$$y(x) = \lambda \int_0^x \frac{\xi}{l}(l - x)y(\xi)d\xi + \lambda \int_x^l \frac{x}{l}(l - \xi)y(\xi)d\xi. \tag{18.32}$$

This is a homogeneous Fredholm equation of the second kind:

$$y(x) = \lambda \int_0^l \kappa(x, \xi)y(\xi)d\xi, \tag{18.33}$$

where the kernel is given as

$$\kappa(x, \xi) = \begin{cases} \frac{\xi}{l}(l - x), & \xi < x, \\[2mm] \frac{x}{l}(l - \xi), & \xi > x. \end{cases} \tag{18.34}$$

18.4 HOW TO CONVERT SOME INTEGRAL EQUATIONS INTO DIFFERENTIAL EQUATIONS

Volterra equations can sometimes be converted into differential equations. Consider the following integral equation

$$y(x) = x^2 - 2 \int_0^x ty(t)dt. \tag{18.35}$$

We define $f(x)$ as

$$f(x) = \int_0^x ty(t)dt, \tag{18.36}$$

where the derivative of $f(x)$ is

$$\frac{df(x)}{dx} = xy(x). \tag{18.37}$$

Using $f(x)$ in Equation (18.35) we can also write

$$y(x) = x^2 - 2f(x), \qquad (18.38)$$

which when substituted back into Equation (18.37) gives a differential equation to be solved for $f(x)$:

$$\frac{df(x)}{dx} = x^3 - 2xf(x), \qquad (18.39)$$

the solution of which is

$$f(x) = \frac{1}{2}(Ce^{-x^2} + x^2 - 1).$$

Finally substituting this into Equation (18.38) gives us the solution for the integral equation as

$$y(x) = 1 - Ce^{-x^2}. \qquad (18.40)$$

Because an integral equation also contains the boundary conditions, constant of integration is found by substituting this solution [Eq. (18.40)] into the integral Equation (18.35) as $C = 1$.

We now consider the Volterra equation

$$y(x) = g(x) + \lambda \int_0^x e^{x-t} y(t)dt \qquad (18.41)$$

and differentiate it with respect to x as

$$y'(x) = g'(x) + \lambda y(x) + \lambda \int_0^x e^{x-t} y(t)dt, \qquad (18.42)$$

where we have used Equation (18.5). Eliminating the integral between these two formulas we obtain

$$y'(x) - (\lambda + 1)y(x) = g'(x) - g(x). \qquad (18.43)$$

The boundary condition to be imposed on this differential equation follows from integral equation (18.41) as $y(0) = g(0)$.

18.5 SOLUTION OF INTEGRAL EQUATIONS

Because the unknown function appears under an integral sign, integral equations are in general more difficult to solve than differential equations. However, there are also quite a few techniques that one can use in finding their solutions. In this section we introduce some of the most commonly used techniques.

18.5.1   Method of Successive Iterations: Neumann Series

Consider a Fredholm equation given as

$$f(x) = g(x) + \lambda \int_a^b K(x,t)f(t)dt. \qquad (18.44)$$

We start the Neumann sequence by taking the first term as

$$f_0(x) = g(x). \qquad (18.45)$$

Using this as the approximate solution of Equation (18.44) we write

$$f_1(x) = g(x) + \int_a^b K(x,t)f_0(t)dt. \qquad (18.46)$$

We keep iterating like this to construct the Neumann sequence as

$$f_0(x) = g(x) \qquad (18.47)$$

$$f_1(x) = g(x) + \lambda \int_a^b K(x,t)f_0(t)dt \qquad (18.48)$$

$$f_2(x) = g(x) + \lambda \int_a^b K(x,t)f_1(t)dt \qquad (18.49)$$

$$\vdots$$

$$f_{n+1}(x) = g(x) + \lambda \int_a^b K(x,t)f_n(t)dt \qquad (18.50)$$

$$\vdots$$

This gives us the Neumann series solution as

$$f(x) = g(x) + \lambda \int_a^b K(x,x')g(x')dx' + \lambda^2 \int_a^b dx' \int_a^b dx'' K(x,x')K(x',x'')g(x'') + \cdots \qquad (18.51)$$

If we take

$$\int_a^b \int_a^b |K(x,t)|^2 \, dxdt = B^2, \quad (B > 0) \qquad (18.52)$$

and if the inequality

$$\int_a^b |K(x,t)|^2 \, dt \leq C \qquad (18.53)$$

is true, where $|\lambda| < \dfrac{1}{B}$, and C is a constant the same for all x in the interval $[a,b]$, then the following sequence is uniformly convergent in the interval $[a,b]$:

$$\{f_i\} = f_0, f_1, f_2, \cdots, f_n, \cdots \rightarrow f(x). \qquad (18.54)$$

The limit of this sequence, that is, $f(x)$, is the solution of Equation (18.44) and it is unique.

Example 18.3. *Neumann sequence:* For the integral equation

$$f(x) = x^2 + \frac{1}{2}\int_{-1}^1 (t-x)f(t)dt \qquad (18.55)$$

we start the Neumann sequence by taking $f_0(x) = x^2$ and continue to write:

$$f_1(x) = x^2 + \frac{1}{2}\int_{-1}^1 (t-x)t^2 \, dt$$

$$= x^2 - \frac{x}{3},$$

$$f_2(x) = x^2 + \frac{1}{2}\int_{-1}^1 (t-x)(t^2 - \frac{t}{3})dt$$

$$= x^2 - \frac{x}{3} - \frac{1}{9}$$

$$f_3(x) = x^2 + \frac{1}{2}\int_{-1}^1 (t-x)(t^2 - \frac{t}{3} - \frac{1}{9})dt$$

$$= x^2 - \frac{2x}{9} - \frac{1}{9} \qquad (18.56)$$

$$\vdots$$

Obviously, in this case the solution is of the form

$$f(x) = x^2 + C_1 x + C_2. \qquad (18.57)$$

Substituting this [Eq. (18.57)] into Equation (18.55) and comparing the coefficients of equal powers of x we obtain $C_1 = -\frac{1}{4}$ and $C_2 = -\frac{1}{12}$; thus the exact solution in this case is given as

$$f(x) = x^2 - \frac{1}{4}x - \frac{1}{12}. \qquad (18.58)$$

18.5.2   Error Calculation in Neumann Series

By using the nth term of the Neumann sequence as our solution we will have committed ourselves to the error given by

$$|f(x) - f_n(x)| < D\sqrt{C}\frac{B^n|\lambda|^{n+1}}{1 - B|\lambda|} , \quad D = \sqrt{\int_a^b |g^2(x)| \, dx}. \qquad (18.59)$$

Example 18.4. *Error calculation in Neumann series:* For the integral equation

$$f(x) = 1 + \frac{1}{10}\int_0^1 K(x,t)f(t)dt, \qquad (18.60)$$

$$K(x,t) = \begin{cases} x & 0 \le x \le t \\ t & t \le x \le 1, \end{cases}$$

since

$$B = \frac{1}{\sqrt{6}}, \ C = \frac{1}{3}, \ D = 1, \ \lambda = 0.1, \qquad (18.61)$$

Equations (18.51–18.54) tell us that the Neumann sequence is convergent. Taking $f_0(x) = 1$, we find the first three terms as

$$f_0(x) = 1, \qquad (18.62)$$
$$f_1(x) = 1 + (1/10)x - (1/20)x^2,$$
$$f_2(x) = 1 + (31/300)x - (1/20)x^2 - (1/600)x^3 + (1/2400)x^4.$$

If we take the solution as

$$f(x) \simeq f_2(x), \qquad (18.63)$$

the error in the entire interval will be less than

$$1.\sqrt{\frac{1}{3}}\frac{(1/6)(0.1)^3}{[1 - (0.1/\sqrt{6})]} = 0.0001. \qquad (18.64)$$

18.5.3   Solution for the Case of Separable Kernels

When the kernel is given in the form

$$K(x,t) = \sum_{j=1}^n M_j(x)N_j(t), \quad n \text{ is a finite number}, \qquad (18.65)$$

it is called separable or degenerate. In such cases we can reduce the solution of an integral equation to the solution of a linear system of equations. Let us write a Fredholm equation with a separable kernel as

$$y(x) = f(x) + \lambda \sum_{j=1}^{n} M_j(x) \left[\int_a^b N_j(t) y(t) dt \right] . \tag{18.66}$$

If we define the quantity inside the square brackets as

$$\int_a^b N_j(t) y(t) dt = c_j, \tag{18.67}$$

Equation (18.66) becomes

$$y(x) = f(x) + \lambda \sum_{j=1}^{n} c_j M_j(x) . \tag{18.68}$$

After the coefficients c_j are evaluated, this will give us the solution $y(x)$. To find these constants we multiply Equation (18.68) with $N_i(x)$ and integrate to get

$$c_i = b_i + \lambda \sum_{j=1}^{n} a_{ij} c_j, \tag{18.69}$$

where

$$b_i = \int_a^b N_i(x) f(x) dx \tag{18.70}$$

and

$$a_{ij} = \int_a^b N_i(x) M_j(x) dx. \tag{18.71}$$

We now write Equation (18.69) as a matrix equation:

$$\mathbf{b} = \mathbf{c} - \lambda \mathbf{Ac}, \quad (\mathbf{A} = a_{ij}) \tag{18.72}$$

$$\mathbf{b} = (\mathbf{I} - \lambda \mathbf{A})\mathbf{c}. \tag{18.73}$$

This gives us a system of n linear equations to be solved for the n coefficients c_j as

$$(1 - \lambda a_{11})c_1 - \lambda a_{12}c_2 - \lambda a_{13}c_3 - \cdots - \lambda a_{1n}c_n = b_1$$
$$-\lambda a_{21}c_1 + (1 - \lambda a_{22})c_2 - \lambda a_{23}c_3 - \cdots - \lambda a_{2n}c_n = b_2$$
$$\vdots$$
$$-\lambda a_{n1}c_1 - \lambda a_{n2}c_2 - \lambda a_{n3}c_3 - \cdots + (1 - \lambda a_{nn})c_n = b_n . \tag{18.74}$$

When the Fredholm equation is homogeneous ($f(x) = 0$) all b_i are zero; thus for the solution to exist we must have

$$\det[\mathbf{I} - \lambda \mathbf{A}] = 0. \tag{18.75}$$

Solutions of this equation give the eigenvalues λ_i. Substituting these eigenvalues into Equation (18.74) we can solve for the values of c_i.

Example 18.5. *The case of separable kernels:* Consider the homogeneous Fredholm equation given as

$$y(x) = \lambda \int_{-1}^{1} (2t + x) y(t) dt, \tag{18.76}$$

where

$$M_1(x) = 1, \quad M_2(x) = x,$$
$$N_1(t) = 2t, \quad N_2(t) = 1 \tag{18.77}$$

and with $\mathbf{A}$ written as

$$\mathbf{A} = \begin{bmatrix} 0 & 4/3 \\ 2 & 0 \end{bmatrix}. \tag{18.78}$$

Using Equation (18.75) we write

$$\det \begin{vmatrix} 1 & -\dfrac{4\lambda}{3} \\ -2\lambda & 1 \end{vmatrix} = 0, \tag{18.79}$$

to find the eigenvalues as

$$\lambda_{1,2} = \pm \frac{1}{2} \sqrt{\frac{3}{2}}. \tag{18.80}$$

Substituting these into Equation (18.74) we find two relations between the c_1 and the c_2 values as

$$c_1 \mp c_2 \sqrt{\frac{2}{3}} = 0. \tag{18.81}$$

As in the eigenvalue problems in linear algebra, we have only obtained the ratio, c_1/c_2, of these constants. Because Equation (18.76) is homogeneous, normalization is arbitrary. Choosing c_1 as one, we can write the solutions of Equation (18.76) as

$$y_1(x) = \frac{1}{2} \sqrt{\frac{3}{2}} \left(1 + \sqrt{\frac{3}{2}} x \right) \quad \text{for} \quad \lambda_1 = \frac{1}{2} \sqrt{\frac{3}{2}}, \tag{18.82}$$

$$y_2(x) = -\frac{1}{2} \sqrt{\frac{3}{2}} \left(1 - \sqrt{\frac{3}{2}} x \right) \quad \text{for} \quad \lambda_2 = -\frac{1}{2} \sqrt{\frac{3}{2}}. \tag{18.83}$$

When Equation (18.74) is inhomogeneous, the solution can still be found by using the techniques of linear algebra. We will come back to the subject of integral equations and eigenvalue problems shortly.

18.5.4 Solution of Integral Equations by Integral Transforms

Sometimes it may be possible to free the unknown function under the integral sign, thus making the solution possible.

18.5.4.1 **Fourier Transform Method:**   When the kernel is a function of $(x-t)$ and the range of the integral is from $-\infty$ to $+\infty$ we can use the Fourier transform method.

Example 18.6. *Fourier transform method:* Consider the integral equation

$$y(x) = \phi(x) + \lambda \int_{-\infty}^{\infty} K(x-t)y(t)dt. \qquad (18.84)$$

We take the Fourier transform of this equation to write

$$\widetilde{y}(k) = \widetilde{\phi}(k) + \lambda \widetilde{K}(k)\widetilde{y}(k), \qquad (18.85)$$

where tilde means the Fourier transform, which is defined as

$$\widetilde{y}(k) = \frac{1}{\sqrt{2\pi}} \int_{-\infty}^{\infty} y(x)e^{ikx}dx. \qquad (18.86)$$

In writing Equation (18.85) we have also used the convolution theorem:

$$f * g = \int_{-\infty}^{\infty} g(y)f(x-y)dy = \int_{-\infty}^{\infty} \widetilde{f}(k)\widetilde{g}(k)e^{-ikx}dk,$$

which indicates that the Fourier transform of the convolution, $f * g$, of two functions is the product of their Fourier transforms. We now solve (18.85) for $\widetilde{y}(k)$ to find

$$\widetilde{y}(k) = \frac{\widetilde{\phi}(k)}{1 - \lambda \widetilde{K}(k)}, \qquad (18.87)$$

which after taking the inverse transform will give us the solution in terms of a definite integral:

$$y(x) = \frac{1}{\sqrt{2\pi}} \int_{-\infty}^{\infty} \frac{\widetilde{\phi}(k)e^{-ikx}dk}{1 - \lambda \widetilde{K}(k)}. \qquad (18.88)$$

18.5.4.2 **Laplace Transform Method:**   The Laplace transform method is useful when the kernel is a function of $(x - t)$ and the range of the integral is from 0 to x. For example, consider the integral equation

$$y(x) = 1 + \int_0^x y(u) \sin(x - u) du. \tag{18.89}$$

We take the Laplace transform of this equation to write

$$\mathcal{L}\left[y(x)\right] = \mathcal{L}\left[1\right] + \mathcal{L}\left[\int_0^x y(u) \sin(x - u) du\right]. \tag{18.90}$$

After using the convolution theorem:

$$F(s)G(s) = \mathcal{L}\left[\int_0^x f(u)g(x - u) du\right], \tag{18.91}$$

where $F(s)$ and $G(s)$ indicate the Laplace transforms of $f(x)$ and $g(x)$, respectively, we obtain the Laplace transform of the solution as

$$Y(s) = \frac{1}{s} + \frac{Y(s)}{s^2 + 1}, \tag{18.92}$$

$$Y(s) = \frac{1 + s^2}{s^3}. \tag{18.93}$$

Taking the inverse Laplace transform, we obtain the solution:

$$y(x) = 1 + \frac{x^2}{2}. \tag{18.94}$$

18.6   INTEGRAL EQUATIONS AND EIGENVALUE PROBLEMS (HILBERT-SCHMIDT THEORY)

In the Sturm-Liouville theory we have defined eigenvalue problems using linear differential operators. We are now going to introduce the Hilbert-Schmidt theory, where an eigenvalue problem is defined in terms of linear integral operators.

18.6.1   Eigenvalues Are Real for Hermitian Operators

Using the Fredholm equation of the second kind, we can define an eigenvalue problem as

$$y(x) = \lambda \int_a^b K(x, t)y(t) dt. \tag{18.95}$$

For the eigenvalue λ_i we write

$$y_i(x) = \lambda_i \int_a^b K(x,t)y_i(t)dt, \tag{18.96}$$

where $y_i(t)$ denotes the corresponding eigenfunction. Similarly, we write Equation (18.95) for another eigenvalue λ_j and take its complex conjugate as

$$y_j^*(x) = \lambda_j^* \int_a^b K^*(x,t)y_j^*(t)dt. \tag{18.97}$$

Multiplying Equation (18.96) by $\lambda_j^* y_j^*(x)$ and Equation (18.97) by $\lambda_i y_i(x)$, and integrating over x in the interval $[a,b]$ we obtain two equations

$$\lambda_j^* \int_a^b y_j^*(x)y_i(x)dx = \lambda_i \lambda_j^* \int_a^b \int_a^b K(x,t)y_j^*(x)y_i(t)dtdx \tag{18.98}$$

and

$$\lambda_i \int_a^b y_j^*(x)y_i(x)dx = \lambda_i \lambda_j^* \int_a^b \int_a^b K^*(x,t)y_j^*(t)y_i(x)dtdx. \tag{18.99}$$

If the kernel satisfies the relation

$$K^*(x,t) = K(t,x), \tag{18.100}$$

Equation (18.99) becomes

$$\lambda_i \int_a^b y_j^*(x)y_i(x)dx = \lambda_i \lambda_j^* \int_a^b \int_a^b K(t,x)y_j^*(t)y_i(x)dtdx. \tag{18.101}$$

Subtracting Equations (18.98) and (18.101) we obtain

$$(\lambda_j^* - \lambda_i) \int_a^b y_j^*(x)y_i(x)dx = 0. \tag{18.102}$$

Kernels satisfying relation (18.100) are called Hermitian. For $i = j$ Equation (18.102) becomes

$$(\lambda_i^* - \lambda_i) \int_a^b |y_i(x)|^2 \, dx = 0. \tag{18.103}$$

Since $\int_a^b |y_i(x)|^2 \, dx \neq 0$, Hermitian operators have real eigenvalues.

18.6.2   Orthogonality of Eigenfunctions

Using the fact that eigenvalues are real, for $i \neq j$ Equation (18.102) becomes

$$(\lambda_j - \lambda_i) \int_a^b y_j^*(x)y_i(x)dx = 0. \qquad (18.104)$$

For distinct (nondegenerate) eigenvalues this gives

$$\int_a^b y_j^*(x)y_i(x)dx = 0, \quad (\lambda_j \neq \lambda_i). \qquad (18.105)$$

This means that the eigenfunctions for the distinct eigenvalues are orthogonal. In the case of degenerate eigenvalues, using the Gram-Schmidt orthogonalization method we can always choose the eigenvectors as orthogonal. Thus we can write

$$\int_a^b y_j^*(x)y_i(x)dx = 0 , \quad (i \neq j). \qquad (18.106)$$

Summary: For a linear integral operator

$$\pounds = \int_a^b dt K(x,t), \qquad (18.107)$$

we can define an eigenvalue problem as

$$y_i(x) = \lambda_i \int_a^b K(x,t)y_i(t)dt. \qquad (18.108)$$

For Hermitian kernels satisfying $K^*(x,t) = K(t,x)$, eigenvalues are real and the eigenfunctions are orthogonal; hence after a suitable normalization we can write:

$$\int_a^b y_i^*(x)y_j(x)dx = \delta_{ij}. \qquad (18.109)$$

18.6.3   Completeness of the Eigenfunction Set

Proof of the completeness of the eigenfunction set is rather technical for our purposes and can be found in Courant and Hilbert (chapter 3, vol. 1, p. 136). We simply quote the following theorem:

Expansion theorem: Every continuous function $F(x)$, which can be represented as the integral transform of a piecewise continuous function $G(x)$ and with respect to the real and symmetric kernel $K(x,x')$ as

$$F(x) = \int K(x,x')G(x')dx', \qquad (18.110)$$

can be expanded in a series in the eigenfunctions of $K(x, x')$; this series converges uniformly and absolutely.

This conclusion is also true for Hermitian kernels. We can now write

$$F(x) = \sum_{m=0}^{\infty} a_m y_m(x), \tag{18.111}$$

where the coefficients a_m are found by using the orthogonality relation as

$$
\begin{aligned}
\int_a^b F(x) y_m^*(x)\, dx &= \sum_n \int_a^b a_n y_n(x)\, y_m^*(x)\, dx, \\
&= \sum_n a_n \int_a^b y_n(x) y_m^*(x)\, dx, \\
&= \sum_n a_n \delta_{nm}, \\
&= a_m.
\end{aligned}
\tag{18.112}
$$

Substituting these coefficients back into Equation (18.111) we get

$$F(x) = \sum_{m=0}^{\infty} \int_a^b F(x') y_m^*(x') y_m(x)\, dx', \tag{18.113}$$

$$= \int_a^b F(x') \left[\sum_{m=0}^{\infty} y_m^*(x') y_m(x) \right] dx'. \tag{18.114}$$

This gives us a formal expression for the completeness of $\{y_m(x)\}$ as

$$\sum_{m=0}^{\infty} y_m^*(x') y_m(x) = \delta(x' - x). \tag{18.115}$$

Keep in mind that in general $\{y_i(x)\}$ do not form a complete set. Not just any function, but only the functions that can be generated by the integral transform [Eq. (18.110)] can be expanded in terms of them.

Let us now assume that a given Hermitian kernel can be expanded in terms of the eigenfunction set $\{y_i(x)\}$ as

$$K(x, x') = \sum_i c_i(x) y_i(x'), \tag{18.116}$$

where the expansion coefficients c_i carry the x dependence. From Equation (18.112) $c_i(x)$ are written as

$$c_i(x) = \int K(x, x') y_i^*(x') dx', \tag{18.117}$$

which after multiplying by λ_i becomes

$$\lambda_i c_i(x) = \lambda_i \int K(x, x') y_i^*(x') dx'. \tag{18.118}$$

We now take the Hermitian conjugate of the eigenvalue equation

$$y_i(x) = \lambda_i \int K(x, x')y_i(x')dx', \tag{18.119}$$

to write

$$y_i^*(x) = \lambda_i \int y_i^*(x')K^*(x', x)dx' \tag{18.120}$$

$$= \lambda_i \int y_i^*(x')K(x, x')dx' \tag{18.121}$$

$$= \lambda_i \int K(x, x')y_i^*(x')dx'. \tag{18.122}$$

We now substitute Equation (18.122) into Equation (18.118) and solve for $c_i(x)$:

$$c_i(x) = \frac{y_i^*(x)}{\lambda_i}. \tag{18.123}$$

Finally, substituting Equation (18.123) into Equation (18.116) we obtain an elegant expression for the Hermitian kernels in terms of the eigenfunctions as

$$K(x, x') = \sum_i \frac{y_i^*(x)y_i(x')}{\lambda_i}. \tag{18.124}$$

18.7 EIGENVALUE PROBLEM FOR THE NON-HERMITIAN KERNELS

In most of the important cases a non-Hermitian kernel in Equation (18.95) can be written as

$$y_i(x) = \lambda_i \int_a^b \left[\overline{K}(x, t)w(t)\right] y_i(t)dt, \tag{18.125}$$

where $\overline{K}(x, t)$ satisfies the relation

$$\overline{K}(x, t) = \overline{K}^*(t, x). \tag{18.126}$$

We multiply Equation (18.125) by $\sqrt{w(x)}$ and define

$$\sqrt{w(x)}y(x) = \psi(x), \tag{18.127}$$

to write

$$\psi_i(x) = \lambda_i \int_a^b \left[\overline{K}(x, t)\sqrt{w(x)w(t)}\right] \psi_i(t)dt. \tag{18.128}$$

Now the kernel, $\overline{K}(x,t)\sqrt{w(x)w(t)}$, in this equation is Hermitian and the eigenfunctions, $\psi_i(x)$, are orthogonal with respect to the weight function $w(x)$ as

$$\int_a^b w(x)\psi_i^*(x)\psi_i(x)dx = \delta_{ij}. \tag{18.129}$$

Problems

18.1   Find the solution of the integral equation

$$y(t) = 1 + \int_0^t y(u)\sin(t - u)du.$$

Check your answer by substituting into the above integral equation.

18.2   Show that the following differential equation and boundary conditions:

$$y''(x) - y(x) = 0, \ y(0) = 0 \text{ and } y'(0) = 1,$$

are equivalent to the integral equation

$$y(x) = x + \int_0^x (x - x')y(x')dx'.$$

18.3   Write the following differential equation and boundary conditions as an integral equation:

$$y''(x) - y(x) = 0,$$
$$y(1) = 0 \text{ and } y(-1) = 1.$$

18.4   Using the Neumann series method solve the integral equation

$$y(x) = x + \int_0^x (x' - x)y(x')dx.$$

18.5   For the following integral equation find the eigenvalues and the eigenfunctions:

$$y(x) = \lambda \int_0^{2\pi} \cos(x - x')y(x')dx'.$$

18.6   To show that the solution of the integral equation

$$y(x) = 1 + \lambda^2 \int_0^x (x - x')y(x')dx'$$

is given as

$$y(x) = \cosh \lambda x.$$

a) First convert the integral equation into a differential equation and then solve.

b) Solve by using Neumann series.

c) Solve by using the integral transform method.

18.7 By using different methods of your choice find the solution of the integral equation

$$y(x) = x + \lambda \int_0^1 xx'y(x')dx'.$$

Answer: $y(x) = 3x/(3 - \lambda)$.

18.8 Consider the damped harmonic oscillator problem, where the equation of motion is given as

$$\frac{d^2x(t)}{dt^2} + 2\varepsilon \frac{dx(t)}{dt} + \omega_0^2 x(t) = 0.$$

a) Using the boundary conditions $x(0) = x_0$ and $\dot{x}(0) = 0$ show that $x(t)$ satisfies the integral equation

$$x(t) = x_0 \cos \omega_0 t + \frac{2x_0 \varepsilon}{\omega_0} \sin \omega_0 t + 2\varepsilon \int_0^t x(t') \cos \omega_0 (t - t')dt'.$$

b) Iterate this equation several times and show that it agrees with the exact solution expanded to the appropriate order.

18.9 Obtain an integral equation for the anharmonic oscillator, where the equation of motion and the boundary conditios are given as

$$\frac{d^2x(t)}{dt^2} + \omega_0^2 x(t) = -bx^3(t),$$

$$x(0) = x_0 \text{ and } \dot{x}(0) = 0.$$

18.10 Consider the integral equation

$$y(x) = x + 2 \int_0^1 [x\theta(x' - x) + x'\theta(x - x')]y(x')dx'.$$

First show that a Neumann series solution exists and then find it.

18.11 Using the Neumann series method find the solution of

$$y(x) = x^2 + 6 \int_0^1 (x + t)y(t)dt.$$

19

GREEN'S FUNCTIONS

Green's functions are among the most versatile mathematical tools. They provide a powerful tool in solving differential equations. They are also very useful in transforming differential equations into integral equations, which are preferred in certain cases like the scattering problems. Propagator interpretation of Green's functions is also very useful in quantum field theory, and with their path integral representation they are the starting point of modern perturbation theory. In this chapter, we introduce the basic features of both the time-dependent and the time-independent Green's functions, which have found a wide range of applications in science and engineering.

19.1 TIME-INDEPENDENT GREEN'S FUNCTIONS

19.1.1 Green's Functions in One Dimension

We start with the differential equation

$$\mathcal{L}y(x) = \phi(x), \tag{19.1}$$

where $\mathcal{L}$ is the Sturm-Liouville operator

$$\mathcal{L} = \frac{d}{dx}\left(p(x)\frac{d}{dx}\right) + q(x), \tag{19.2}$$

567

with $p(x)$ and $q(x)$ as continuous functions defined in the interval $[a, b]$. Along with this differential equation we use the homogeneous boundary conditions

$$\alpha y(x) + \beta \frac{dy\,(x)}{dx}\bigg|_{x=a} = 0$$

and   (19.3)

$$\alpha y(x) + \beta \frac{dy\,(x)}{dx}\bigg|_{x=b} = 0,$$

where α and β are constants. Because $\phi(x)$ could also depend on the unknown function explicitly, we will also write it as

$$\phi(x, y(x)).$$

Note that even though the differential operator $\mathcal{L}$ is linear, the differential equation [Eq. (19.1)] could be nonlinear.

We now define a function $G(x, \xi)$, which for a given $\xi \in [a, b]$ reduces to $G_1(x)$ when $x < \xi$ and to $G_2(x)$ when $x > \xi$, and also has the following properties:

i) Both $G_1(x)$ and $G_2(x)$ satisfy

$$\mathcal{L}G\,(x) = 0, \tag{19.4}$$

in their intervals of definition, that is:

$$\mathcal{L}G_1\,(x) = 0, \qquad x < \xi,$$
$$\mathcal{L}G_2(x) = 0, \qquad x > \xi. \tag{19.5}$$

ii) $G_1(x)$ satisfies the boundary condition at $x = a$, and $G_2(x)$ satisfies the boundary condition at $x = b$.

iii) $G(x, \xi)$ is continuous at $x = \xi$:

$$G_2(\xi) = G_1(\xi). \tag{19.6}$$

iv) $G(x, \xi)$ is discontinuous by the amount $\dfrac{1}{p(\xi)}$ at $x = \xi$:

$$G_2'(\xi) - G_1'(\xi) = \frac{1}{p(\xi)}. \tag{19.7}$$

We also assume that $p(x)$ is finite in the interval (a, b); thus the discontinuity is of finite order.

We are now going to prove that if such a function can be found, then the problem defined by the differential equation plus the boundary conditions [Eqs. (19.1–19.3)] is equivalent to the equation

$$y(x) = \int_a^b G(x, \xi)\phi(\xi, y(\xi))d\xi, \tag{19.8}$$

where $G(x, \xi)$ is called the Green's function. If $\phi(x, y(\xi))$ does not depend on $y(x)$ explicitly, then finding the Green's function is tantamount to solving the problem. For the cases where $\phi(x, y(\xi))$ depends explicitly on $y(x)$, then Equation (19.8) becomes the integral equation version of the problem defined by the differential equation plus the homogeneous boundary conditions [Eqs. (19.1–19.3)]. Before we prove the equivalence of Equations (19.8) and (19.1–19.3), we show how a Green's function can be constructed. However, we first drive a useful result called Abel's formula.

19.1.2   Abel's Formula

Let $u(x)$ and $v(x)$ be two linearly independent solutions of $\mathcal{L}y(x) = 0$, so that we can write

$$\frac{d}{dx}\left(p(x)\frac{d}{dx}\right)u(x) + q(x)u(x) = 0$$

and

$$\frac{d}{dx}\left(p(x)\frac{d}{dx}\right)v(x) + q(x)v(x) = 0,$$

respectively. Multiplying the first equation by v and the second by u and then subtracting gives us

$$v(x)\frac{d}{dx}\left(p(x)\frac{d}{dx}\right)u(x) - u(x)\frac{d}{dx}\left(p(x)\frac{d}{dx}\right)v(x) = 0.$$

After expanding and rearranging, we can write this as

$$\frac{d}{dx}\left[p(x)\left(uv' - vu'\right)\right] = 0,$$

which implies

$$(uv' - vu') = \frac{A}{p(x)}, \qquad (19.9)$$

where A is a constant. This result is known as Abel's formula.

19.1.3   How to Construct a Green's Function

Let $y = u(x)$ be a nontrivial solution of $\mathcal{L}y = 0$ satisfying the boundary condition at $x = a$ and let $y = v(x)$ be another nontrivial solution of $\mathcal{L}y = 0$ satisfying the boundary condition at $x = b$. We now define a Green's function as

$$G(x, \xi) = \begin{cases} c_1 u(x), & x < \xi, \\ c_2 v(x), & x > \xi. \end{cases} \qquad (19.10)$$

This Green's function satisfies conditions (i) and (ii). For conditions (iii) and (iv) we require c_1 and c_2 to satisfy the equations

$$c_2 v(\xi) - c_1 u(\xi) = 0 \tag{19.11}$$

and

$$c_2 v'(\xi) - c_1 u'(\xi) = \frac{1}{p(\xi)}. \tag{19.12}$$

For a unique solution of these equations we have to satisfy the condition

$$W[u, v] = \begin{vmatrix} u(\xi) & v(\xi) \\ u'(\xi) & v'(\xi) \end{vmatrix} = u(\xi) v'(\xi) - v(\xi) u'(\xi) \neq 0, \tag{19.13}$$

where $W[u, v]$ is called the Wronskian of the solutions $u(x)$ and $v(x)$. When these solutions are linearly independent, $W[u, v]$ is different from zero and according to Abel's formula $W[u, v]$ is equal to $\dfrac{A}{p(\xi)}$, where A is a constant independent of ξ. Equations (19.11) and (19.12) can now be solved for c_1 and c_2 to yield

$$c_1 = \frac{v(\xi)}{A} \quad \text{and} \quad c_2 = \frac{u(\xi)}{A}. \tag{19.14}$$

Now the Green's function becomes

$$G(x, \xi) = \begin{cases} \dfrac{1}{A} u(x) v(\xi), & x < \xi, \\[2mm] \dfrac{1}{A} u(\xi) v(x), & x > \xi. \end{cases} \tag{19.15}$$

Evidently, this Green's function is symmetric and unique. We now show that the integral

$$y(x) = \int_a^b G(x, \xi) \phi(\xi) d\xi \tag{19.16}$$

is equivalent to the differential equation [Eq. (19.1)] plus the boundary conditions [Eq. (19.3)]. We first write equation Equation (19.16) explicitly as

$$y(x) = \frac{1}{A} \left[\int_a^x v(x) u(\xi) \phi(\xi) d\xi + \int_x^b v(\xi) u(x) \phi(\xi) d\xi \right] \tag{19.17}$$

and evaluate its first- and second-order derivatives:

$$y'(x) = \frac{1}{A} \left[\int_a^x v'(x) u(\xi) \phi(\xi) d\xi + \int_x^b v(\xi) u'(x) \phi(\xi) d\xi \right],$$

$$y''(x) = \frac{1}{A} \left[\int_a^x v''(x) u(\xi) \phi(\xi) d\xi + \int_x^b v(\xi) u''(x) \phi(\xi) d\xi \right]$$

$$+ \frac{1}{A} [v'(x) u(x) - v(x) u'(x)] \phi(x), \tag{19.18}$$

where we have used the formula

$$\frac{d}{dx} \int_{A(x)}^{B(x)} F(x,\xi)d\xi = \int_{A(x)}^{B(x)} \frac{\partial F(x,\xi)}{\partial x}d\xi + F(x,B(x))\frac{dB(x)}{dx} - F(x,A(x))\frac{dA(x)}{dx}.$$

Substituting these derivatives into

$$\mathcal{L}y(x) = p(x)y''(x) + p'(x)y'(x) + q(x)y(x), \qquad (19.19)$$

we get

$$\mathcal{L}y(x) = \frac{1}{A}\left\{ \int_a^x [\mathcal{L}v(x)]\,u(\xi)\,\phi(\xi)d\xi + \int_x^b v(\xi)[\mathcal{L}u(x)]\,\phi(\xi)d\xi \right\}$$

$$+ \frac{1}{A}\left[p(x)\frac{A}{p(x)}\phi(x) \right]. \qquad (19.20)$$

Since $u(x)$ and $v(x)$ satisfy

$$\mathcal{L}u(x) = 0 \quad \text{and} \quad \mathcal{L}v(x) = 0, \qquad (19.21)$$

respectively, we obtain

$$\mathcal{L}y(x) = \phi(x).$$

To see which boundary conditions $y(x)$ satisfies we write

$$\left\{ \begin{array}{l} y(a) = \dfrac{1}{A}u(a)\int_a^b v(\xi)\,\phi(\xi)d\xi \\[2mm] y'(a) = \dfrac{1}{A}u'(a)\int_a^b v(\xi)\,\phi(\xi)d\xi \end{array} \right\} \qquad (19.22)$$

and

$$\left\{ \begin{array}{l} y(b) = \dfrac{1}{A}v(b)\int_a^b u(\xi)\,\phi(\xi)d\xi \\[2mm] y'(b) = \dfrac{1}{A}v'(b)\int_a^b u(\xi)\,\phi(\xi)d\xi \end{array} \right\} \qquad (19.23)$$

It is easily seen that $y(x)$ satisfies the same boundary condition with $u(x)$ at $x = a$ and with $v(x)$ at $x = b$.

In some cases it is convenient to write $\phi(x)$ as

$$\phi(x) = \lambda r(x)y(x) + f(x). \qquad (19.24)$$

Thus Equation (19.1) becomes

$$\mathcal{L}y(x) - \lambda r(x)y(x) = f(x). \qquad (19.25)$$

With the homogeneous boundary conditions this is equivalent to the integral equation

$$y(x) = \lambda \int_a^b G(x,\xi)r(\xi)\,y(\xi)\,d\xi + \int_a^b G(x,\xi)f(\xi)\,d\xi. \qquad (19.26)$$

19.1.4   The Differential Equation That the Green's Function Satisfies

To find the differential equation that the Green's function satisfies, we operate on $y(x)$ in Equation (19.16) with $\mathcal{L}$ to write

$$\mathcal{L}y(x) = \int_a^b \mathcal{L}G(x, \xi)\phi(\xi)\,d\xi.$$

Because the operator $\mathcal{L}$ [Eq. (19.2)] acts only on x, we can write this as

$$\phi(x) = \int_a^b [\mathcal{L}G(x, \xi)]\,\phi(\xi)\,d\xi, \tag{19.27}$$

which is the defining equation for the Dirac-delta function $\delta(x - \xi)$. Hence we obtain the differential equation for the Green's function as

$$\mathcal{L}G(x, \xi) = \delta(x - \xi). \tag{19.28}$$

Along with the homogeneous boundary conditions

$$\alpha G(x, \xi) + \beta \frac{dG(x, \xi)}{dx}\bigg|_{x=a} = 0$$

and   (19.29)

$$\alpha G(x, \xi) + \beta \frac{dG(x, \xi)}{dx}\bigg|_{x=b} = 0,$$

Equation (19.28) is the defining equation for $G(x, \xi)$.

19.1.5   Single-Point Boundary Conditions

We have so far used the boundary conditions in Equation (19.3), which are also called the two-point boundary conditions. In mechanics we usually encounter single-point boundary conditions, where the position and the velocity are given at some initial time. We first write the Green's function satisfying the homogeneous single-point boundary conditions $G(x_0, x') = 0$ and $G'(x_0, x') = 0$ as

$$\begin{aligned}
G(x, x') &= c_1 y_1(x) + c_2 y_2(x), \quad x > x', \\
G(x, x') &= \qquad\qquad 0, \qquad\qquad x < x',
\end{aligned} \tag{19.30}$$

where $y_1(x)$ and $y_2(x)$ are two linearly independent solutions of

$$\mathcal{L}y(x) = 0.$$

Following the steps of the method used for two-point boundary conditions (see Problem 19.4), we can find the constants c_1 and c_2, and construct the Green's function as

$$G(x, x') = -\frac{y_1(x)y_2(x') - y_2(x)y_1(x')}{p(x')W[y_1(x'), y_2(x')]}\theta(x - x'), \tag{19.31}$$

where $W[y_1(x'), y_2(x')]$ is the Wronskian.

Now the differential equation

$$\mathcal{L}y(x) = \phi(x),$$

with the given single-point boundary conditions

$$y(x_0) = y_0 \quad \text{and} \quad y'(x_0) = y_0'$$

is equivalent to the integral equation

$$y(x) = C_1 y_1(x) + C_2 y_2(x) + \int_{x_0}^{x} G(x, x')\phi(x')dx'. \tag{19.32}$$

The first two terms come from the solutions of the homogeneous equation. Because the integral term and its derivative vanish at $x = x_0$, we use C_1 and C_2 to satisfy the single-point boundary conditions.

19.1.6 Green's Function for the Operator d^2/dx^2

The Helmholtz equation in one dimension is written as

$$\frac{d^2y}{dx^2} + k_0^2 y = 0, \tag{19.33}$$

where k_0 is a constant. Using the homogeneous boundary conditions

$$y(0) = 0 \quad \text{and} \quad y(L) = 0, \tag{19.34}$$

we integrate Equation (19.33) between $(0, x)$ to write

$$\frac{dy}{dx} = -k_0^2 \int_0^x y(\xi)d\xi + C, \tag{19.35}$$

where C is an integration constant corresponding to the unknown value of the derivative at $x = 0$. A second integration yields

$$y(x) = -k_0^2 \int_0^x (x - \xi)y(\xi)d\xi + Cx, \tag{19.36}$$

where we have used one of the boundary conditions, that is, $y(0) = 0$. Using the second boundary condition, we can now evaluate C as

$$C = \frac{k_0^2}{L} \int_0^L (L - \xi)y(\xi)d\xi. \tag{19.37}$$

This leads us to the following integral equation for $y(x)$:

$$y(x) = -k_0^2 \int_0^x (x - \xi) y(\xi) d\xi + \frac{xk_0^2}{L} \int_0^L (L - \xi) y(\xi) d\xi. \tag{19.38}$$

To identify the Green's function for the operator $\mathcal{L} = \dfrac{d^2}{dx^2}$, we rewrite this as

$$y(x) = -k_0^2 \int_0^x (x - \xi) y(\xi) d\xi + \frac{xk_0^2}{L} \int_0^x (L - \xi) y(\xi) d\xi$$
$$+ \frac{xk_0^2}{L} \int_x^L (L - \xi) y(\xi) d\xi, \tag{19.39}$$

$$= k_0^2 \int_0^x \frac{\xi}{L} (L - x) y(\xi) d\xi + k_0^2 \int_x^L \frac{x}{L} (L - \xi) y(\xi) d\xi \tag{19.40}$$

and compare with

$$y(x) = \int_0^L G(x, \xi) [-k_0^2 y(\xi)] d\xi. \tag{19.41}$$

This gives the Green's function for the $\mathcal{L} = \dfrac{d^2}{dx^2}$ operator as

$$G(x, \xi) = \begin{cases} -\dfrac{x}{L} (L - \xi), & x < \xi, \\[2mm] -\dfrac{\xi}{L} (L - x), & x > \xi. \end{cases} \tag{19.42}$$

Now the integral equation (19.41) is equivalent to the differential equation

$$\frac{d^2 y}{dx^2} = -k_0^2 y, \tag{19.43}$$

with the boundary conditions

$$y(0) = y(L) = 0. \tag{19.44}$$

As long as the boundary conditions remain the same we can use this Green's function to express the solution of the differential equation

$$\frac{d^2 y}{dx^2} = \phi(x, y) \tag{19.45}$$

as

$$y(x) = \int_0^L G(x, \xi) \phi(\xi, y(\xi)) d\xi. \tag{19.46}$$

For a different set of boundary conditions one must construct a new Green's function.

Example 19.1. *Green's function for the* $\pounds = \frac{d^2}{dx^2}$ *operator:* We have obtained the Green's function [Eq. (19.42)] for the operator $\pounds = d^2/dx^2$ with the boundary conditions $y(0) = y(L) = 0$. Transverse waves on a uniform string of fixed length L with both ends clamped rigidly are described by

$$\frac{d^2}{dx^2}y(x) + k_0^2 y(x) = f(x,y), \tag{19.47}$$

where $f(x, y)$ represents external forces acting on the string. Using the Green's function for the d^2/dx^2 operator we can convert this into an integral equation as

$$y(x) = -k_0^2 \int_0^L G(x,\xi)y\left(\xi\right) d\xi + \int_0^L G(x,\xi)f\left(\xi,y(\xi)\right) d\xi. \tag{19.48}$$

19.1.7   Green's Functions for Inhomogeneous Boundary Conditions

In the presence of inhomogeneous boundary conditions we can still use the Green's function obtained for homogeneous boundary conditions and modify the solution [Eq. (19.8)] as

$$y(x) = P(x) + \int_a^b G(x,\xi)\phi\left(\xi\right) d\xi, \tag{19.49}$$

where $y(x)$ now satisfies

$$\pounds y(x) = \phi(x) \tag{19.50}$$

with the inhomogeneous boundary conditions. Operating on Equation (19.49) with $\pounds$ and using the relation between the Green's functions and the Dirac-delta function [Eq. (19.28)], we obtain a differential equation to be solved for $P(x)$ as

$$\phi(x) = \pounds\left[P(x)\right] + \int_a^b \left[\pounds G(x,\xi)\right]\phi\left(\xi\right) d\xi, \tag{19.51}$$

$$\pounds P(x) + \phi(x) - \phi(x) = 0, \tag{19.52}$$

$$\pounds P(x) = 0. \tag{19.53}$$

Because the second term in Equation (19.49) satisfies the homogeneous boundary conditions, $P(x)$ must satisfy the inhomogeneous boundary conditions.

Existence of $P(x)$ is guaranteed by the existence of $G(x,\xi)$. The equivalence of this approach with our previous method can easily be seen by defining a new unknown function

$$\bar{y}(x) = y(x) - P(x), \tag{19.54}$$

which satisfies the homogeneous boundary conditions.

Example 19.2. *Inhomogeneous boundary conditions:* Equation of motion of a simple plane pendulum of length l is given as

$$\frac{d^2\theta(t)}{dt^2} = -\omega_0^2 \sin\theta, \quad \omega_0^2 = g/l, \tag{19.55}$$

where g is the acceleration of gravity and θ represents the angle from the equilibrium position. We use the inhomogeneous boundary conditions:

$$\theta(0) = 0 \quad \text{and} \quad \theta(t_1) = \theta_1. \tag{19.56}$$

We have already obtained the Green's function for the d^2/dx^2 operator for the homogeneous boundary conditions [Eq. (19.42)]. We now solve

$$\frac{d^2}{dt^2}P(t) = 0 \tag{19.57}$$

with the inhomogeneous boundary conditions

$$P(0) = 0 \quad \text{and} \quad P(t_1) = \theta_1, \tag{19.58}$$

to find

$$P(t) = \frac{\theta_1 t}{t_1}. \tag{19.59}$$

Because $\phi(t)$ is

$$\phi(t) = -\omega_0^2 \sin\theta(t), \tag{19.60}$$

we can write the differential equation [Eq. (19.55)] plus the inhomogeneous boundary conditions [Eq. (19.56)] as an integral equation:

$$\theta(t) = \frac{\theta_1 t}{t_1} + \omega_0^2 \left[\frac{(t_1 - t)}{t_1} \int_0^t \xi \sin\theta(\xi)d\xi + \frac{t}{t_1} \int_t^{t_1} (t_1 - \xi)\sin\theta(\xi)d\xi \right]. \tag{19.61}$$

Example 19.3. *Green's function:* We now consider the differential equation

$$x^2 \frac{d^2y}{dx^2} + x\frac{dy}{dx} + \left(k^2x^2 - 1\right)y = 0 \tag{19.62}$$

with the boundary conditions given as

$$y(0) = 0 \text{ and } y(L) = 0. \tag{19.63}$$

We write this differential equation in the form

$$\left[\frac{d}{dx} \left(x \frac{dy}{dx} \right) - \frac{y}{x} \right] = -k^2 x y(x) \tag{19.64}$$

and define the $\mathcal{L}$ operator as

$$\mathcal{L} = \frac{d}{dx} \left(x \frac{d}{dx} \right) - \frac{1}{x}, \tag{19.65}$$

where

$$p(x) = x, \quad q(x) = -\frac{1}{x}, \quad r(x) = x. \tag{19.66}$$

The general solution of

$$\mathcal{L} y = 0 \tag{19.67}$$

is given as

$$y = c_1 x + c_2 x^{-1}. \tag{19.68}$$

Using the first boundary condition

$$y(0) = 0, \tag{19.69}$$

we find $u(x)$ as

$$y(x) = u(x),$$
$$= x. \tag{19.70}$$

Similarly, using the second boundary condition

$$y(L) = 0, \tag{19.71}$$

we find $v(x)$ as

$$v(x) = \frac{L^2}{x} - x. \tag{19.72}$$

We now evaluate the Wronskian of the u and the v solutions as

$$W[u, v] = u(x) v'(x) - v(x) u'(x) \tag{19.73}$$
$$= -\frac{2L^2}{x}$$
$$= \frac{A}{p(x)} = \frac{A}{x}, \tag{19.74}$$

which determines A as $-2L^2$. Putting all these together we obtain the Green's function as

$$
G(x,\xi) = \begin{cases} -\dfrac{x}{2L^2\xi}\left(L^2 - \xi^2\right), & x < \xi, \\[3mm] -\dfrac{\xi}{2L^2 x}\left(L^2 - x^2\right), & x > \xi. \end{cases} \tag{19.75}
$$

Using this Green's function we can now write the integral equation

$$
y(x) = -k^2 \int_0^L G(x,\xi)\xi y(\xi)d\xi, \tag{19.76}
$$

which is equivalent to the differential equation plus the boundary conditions given in Equations (19.62) and (19.63).

Note that the differential equation in this example is the Bessel equation and the only useful solutions are those with the eigenvalues k_n satisfying the characteristic equation

$$
J_1(k_n L) = 0. \tag{19.77}
$$

In this case the solution is given as

$$
y(x) = C J_1(k_n x), \tag{19.78}
$$

where C is a constant. The same conclusion is valid for the integral equation (19.76).

Note that we could have arranged the differential equation (19.62) as

$$
\left[\frac{d}{dx}\left(x\frac{dy}{dx}\right)\right] = \left(\frac{y}{x} - k^2 xy\right), \tag{19.79}
$$

where the operator $\mathcal{L}$ is now defined as

$$
\mathcal{L} = \frac{d}{dx}\left(x\frac{d}{dx}\right). \tag{19.80}
$$

If the new $\mathcal{L}$ and the corresponding Green's function are compatible with the boundary conditions, then the final answer, $y(x, k_n)$, will be the same. In the above example, Green's function for the new operator [Eq. (19.80)] has a logarithmic singularity at the origin. We will explore these points in Problems 19.11 and 19.12. In physical applications form of the operator is usually dictated to us by the physics of the problem. For example, in quantum mechanics $\mathcal{L}$ represents physical properties with well-defined expressions with their eigenvalues corresponding to observables like angular momentum and energy.

19.1.8   Green's Functions and the Eigenvalue Problems

Consider the differential equation

$$\mathcal{L}y(x) = f(x), \quad x \in [a, b] \tag{19.81}$$

with the appropriate boundary conditions, where $\mathcal{L}$ is the Sturm-Liouville operator

$$\mathcal{L} = \frac{d}{dx}\left(p(x)\frac{d}{dx}\right) - q(x). \tag{19.82}$$

We have seen that the $\mathcal{L}$ operator has a complete set of eigenfunctions defined by the equation

$$\mathcal{L}\phi_n(x) = \lambda_n\phi_n(x), \tag{19.83}$$

where λ_n are the eigenvalues. Eigenfunctions $\phi_n(x)$ satisfy the orthogonality relation

$$\int \phi_n^*(x)\phi_m(x)dx = \delta_{nm} \tag{19.84}$$

and the completeness relation

$$\sum_n \phi_n^*(x)\phi_n(x')dx = \delta(x - x'). \tag{19.85}$$

In the interval $x \in [a, b]$, we can expand $y(x)$ and $f(x)$ in terms of the set $\{\phi_n(x)\}$ as

$$\left.\begin{array}{l} y(x) = \sum_n^\infty \alpha_n\phi_n(x) \\ f(x) = \sum_n^\infty \beta_n\phi_n(x) \end{array}\right\}, \tag{19.86}$$

where α_n and β_n are the expansion coefficients:

$$\left.\begin{array}{l} \alpha_n = \int_a^b \phi_n^*(x)y(x)dx \\ \beta_n = \int_a^b \phi_n^*(x)f(x)dx \end{array}\right\}. \tag{19.87}$$

Operating on $y(x)$ with $\mathcal{L}$ we get

$$\mathcal{L}y(x) = \mathcal{L}\sum_n^\infty \alpha_n\phi_n(x)$$

$$= \sum_n^\infty \alpha_n\mathcal{L}\phi_n(x). \tag{19.88}$$

Using Equation (19.88) with the eigenvalue equation [Eq. (19.83)] and the Equation (19.86) we can write

$$\mathcal{L}y(x) = f(x) \tag{19.89}$$

as

$$\sum_{n}^{\infty} [\alpha_n \lambda_n - \beta_n] \phi_n(x) = 0. \tag{19.90}$$

Because ϕ_n are linearly independent, the only way to satisfy this equation for all n is to set the expression inside the square brackets to zero, thus obtaining

$$\alpha_n = \frac{\beta_n}{\lambda_n}. \tag{19.91}$$

We use this in Equation (19.86) to write

$$y(x) = \sum_{n}^{\infty} \frac{\beta_n}{\lambda_n} \phi_n(x). \tag{19.92}$$

After substituting the β_n given in Equation (19.87) this becomes

$$y(x) = \int \sum_{n}^{\infty} \frac{\phi_n(x)\phi_n^*(x')}{\lambda_n} f(x')dx'. \tag{19.93}$$

Using the definition of the Green's function, that is,

$$y(x) = \int G(x, x')f(x')dx', \tag{19.94}$$

we obtain

$$G(x, x') = \sum_{n}^{\infty} \frac{\phi_n(x)\phi_n^*(x')}{\lambda_n}. \tag{19.95}$$

Usually we encounter differential equations given as

$$\mathcal{L}y(x) - \lambda y(x) = f(x), \tag{19.96}$$

where the Green's function for the operator $(\mathcal{L} - \lambda)$ can be written as

$$G(x, x') = \sum_{n}^{\infty} \frac{\phi_n(x)\phi_n^*(x')}{\lambda_n - \lambda}. \tag{19.97}$$

Note that in complex spaces Green's function is Hermitian:

$$G(x, x') = G^*(x', x).$$

Example 19.4. ***Eigenfunctions and the Green's function for*** $\mathcal{L} = \frac{d^2}{dx^2}$:
Let us reconsider the $\mathcal{L} = \frac{d^2}{dx^2}$ operator in the interval $x \in [0, L]$. The corresponding eigenvalue equation is

$$\frac{d^2\phi_n}{dx^2} = -k_n^2\phi_n. \tag{19.98}$$

Using the boundary conditions

$$\phi_n(0) = 0 \text{ and } \phi_n(L) = 0, \tag{19.99}$$

we find the eigenfunctions and the eigenvalues as

$$\left\{ \begin{array}{c} \phi_n(x) = \sqrt{\dfrac{2}{L}} \sin\left(k_n x\right), \\[2mm] -\lambda_n = k_n^2 = \dfrac{n^2\pi^2}{L^2}, \quad n = 1, 2, 3, \dots . \end{array} \right\} \tag{19.100}$$

We now construct the Green's function as

$$G(x, x') = \frac{2}{L} \sum_{n}^{\infty} \frac{\sin\left(\dfrac{n\pi}{L}x\right)\sin\left(\dfrac{n\pi}{L}x'\right)}{-n^2\pi^2/L^2}. \tag{19.101}$$

For the same operator, using the Green's function in Equation (19.42), we have seen that the inhomogeneous equation

$$\frac{d^2y}{dx^2} = F(x, y) \tag{19.102}$$

and the boundary conditions

$$y(0) = y(L) = 0$$

can be written as an integral equation:

$$y(x) = \int_0^x (x - x')\, F(x')dx' - \frac{x}{L} \int_0^L (L - x')\, F(x')dx'. \tag{19.103}$$

Using the step function $\theta(x - x')$ we can write this as

$$y(x) = \int_0^L (x - x')\,\theta(x - x')\, F(x')dx' - \frac{x}{L} \int_0^L (L - x')\, F(x')dx', \tag{19.104}$$

or

$$y(x) = \int_0^L \left[(x - x')\,\theta(x - x') - \frac{x}{L}(L - x')\right] F(x')dx'. \tag{19.105}$$

This also gives the Green's function for the $\mathcal{L} = d^2/dx^2$ operator as

$$G(x, x') = \left[(x - x') \theta (x - x') - \frac{x}{L} (L - x') \right]. \tag{19.106}$$

One can easily show that the Green's function given in Equation (19.101) is the generalized Fourier expansion of the Equation (19.106) in terms of the complete and orthonormal set [Eq. (19.100)].

19.1.9 Green's Function for the Helmholtz Equation in One Dimension

Let us now consider the inhomogeneous Helmholtz equation

$$\frac{d^2 y(x)}{dx^2} + k_0^2 y(x) = f(x), \tag{19.107}$$

with the boundary conditions $y(0) = 0$ and $y(L) = 0$. Using Equations (19.96) and (19.97) we can write the Green's function as

$$G(x, x') = \frac{2}{L} \sum_n \frac{\sin k_n x \sin k_n x'}{k_0^2 - k_n^2}. \tag{19.108}$$

Using this Green's function, solution of the inhomogeneous Helmholtz equation (19.107) is written as

$$y(x) = \int_0^L G(x, x') f(x') dx',$$

where $f(x)$ represents the driving force in wave motion. Note that in this case the operator is defined as

$$\mathcal{L} = \frac{d^2}{dx^2} + k_0^2.$$

Green's function for this operator can also be obtained by direct construction, that is, by determining the u and the v solutions in Equation (19.15) as

$$\sin k_0 x \quad \text{and} \quad \sin k_0 (x - L),$$

respectively. We can now obtain a closed expression for $G(x, x')$ as

$$G(x, x') = \begin{cases} \dfrac{\sin k_0 x \sin k_0 (x' - L)}{k_0 \sin k_0 L}, & x < x', \\[4mm] \dfrac{\sin k_0 x' \sin k_0 (x - L)}{k_0 \sin k_0 L}, & x > x'. \end{cases} \tag{19.109}$$

19.1.10 Green's Functions and the Dirac-Delta Function

Let us operate on the Green's function [Eq. (19.95)] with the $\mathcal{L}$ operator:

$$\mathcal{L}G(x, x') = \mathcal{L}\sum_{n}^{\infty} \frac{\phi_n(x)\phi_n^*(x')}{\lambda_n}. \tag{19.110}$$

Because $\mathcal{L}$ is a linear operator acting on the variable x, we can write

$$\mathcal{L}G(x, x') = \sum_{n}^{\infty} \frac{\phi_n^*(x')\left[\mathcal{L}\phi_n(x)\right]}{\lambda_n}. \tag{19.111}$$

Using the eigenvalue equation $\mathcal{L}\phi_n(x) = \lambda_n\phi_n(x)$ we obtain

$$\mathcal{L}G(x, x') = \sum_{n}^{\infty} \phi_n^*(x')\phi_n(x) \equiv I(x, x'). \tag{19.112}$$

For a given function $f(x)$ we write the integral

$$\int I(x, x')f(x')dx' = \sum_{n}^{\infty} \phi_n(x) \int \phi_n^*(x')f(x')dx'$$

$$= \sum_{n}^{\infty} \phi_n(x)\,(\phi_n, f)\,. \tag{19.113}$$

For a complete and orthonormal set the right-hand side is the generalized Fourier expansion of $f(x)$; thus we can write

$$\int I(x, x')f(x')dx' = f(x). \tag{19.114}$$

Hence $I(x, x')$ is nothing but the Dirac-delta function:

$$I(x, x') = \mathcal{L}G(x, x') = \delta\left(x - x'\right). \tag{19.115}$$

Summary: A differential equation

$$\mathcal{L}y(x) = f(x, y)$$

defined with the Sturm-Liouville operator $\mathcal{L}$ [Eq. (19.2)] and with the homogeneous boundary conditions [Eq. (19.3)] is equivalent to the integral equation

$$y(x) = \int G(x, x')f(x', y(x'))dx',$$

where $G(x, x')$ is the Green's function satisfying

$$\mathcal{L}G(x, x') = \delta\left(x - x'\right),$$

with the same boundary conditions.

19.1.11   Green's Function for the Helmholtz Equation for All Space—Continuum Limit

We now consider the operator $\mathcal{L} = \dfrac{d^2}{dx^2} + k_0^2$ in the continuum limit with

$$\frac{d^2 y}{dx^2} + k_0^2 y = f(x), \quad x \in (-\infty, \infty). \tag{19.116}$$

Because the eigenvalues are continuous we use the Fourier transforms of $y(x)$ and $f(x)$ as

$$f(x) = \frac{1}{\sqrt{2\pi}} \int_{-\infty}^{\infty} dk' g(k') e^{ik'x} \tag{19.117}$$

and

$$y(x) = \frac{1}{\sqrt{2\pi}} \int_{-\infty}^{\infty} dk' \eta(k') e^{ik'x}. \tag{19.118}$$

Their inverse Fourier transforms are

$$g(k) = \frac{1}{\sqrt{2\pi}} \int_{-\infty}^{\infty} dx' f(x') e^{-ikx'} \tag{19.119}$$

and

$$\eta(k) = \frac{1}{\sqrt{2\pi}} \int_{-\infty}^{\infty} dx' y(x') e^{-ikx'}. \tag{19.120}$$

Using these in Equation (19.116) we get

$$\frac{1}{\sqrt{2\pi}} \int_{-\infty}^{\infty} dk' \left\{ \left[-k'^2 + k_0^2 \right] \eta(k') - g(k') \right\} e^{ik'x} = 0, \tag{19.121}$$

which gives us

$$\eta(k') = -\frac{g(k')}{(k'^2 - k_0^2)}. \tag{19.122}$$

Substituting this in Equation (19.118) we obtain

$$y(x) = -\frac{1}{\sqrt{2\pi}} \int_{-\infty}^{\infty} dk' \frac{g(k')}{(k'^2 - k_0^2)} e^{ik'x}. \tag{19.123}$$

Writing $g(k')$ explicitly this becomes

$$y(x) = -\frac{1}{2\pi} \int_{-\infty}^{\infty} dx' f(x') \int_{-\infty}^{\infty} dk' \frac{e^{ik'(x-x')}}{(k' - k_0)(k' + k_0)}, \tag{19.124}$$

which allows us to define the Green's function as

$$y(x) = \int_{-\infty}^{\infty} dx' f(x') \, G(x, x'),$$

where

$$G(x, x') = -\frac{1}{2\pi} \int_{-\infty}^{\infty} dk' \frac{e^{ik'(x-x')}}{(k' - k_0)(k' + k_0)}. \qquad (19.125)$$

Using one of the representations of the Dirac-delta function:

$$\frac{1}{2\pi} \int_{-\infty}^{\infty} e^{ik(x-x')} dk = \delta(x - x'),$$

it is easy to see that $G(x, x')$ satisfies the equation

$$\mathcal{L} G(x, x') = \delta(x - x'). \qquad (19.126)$$

The integral in Equation (19.125) is undefined at $k' = \pm k_0$. However, we can use the Cauchy principal value:

$$P \int_{-\infty}^{\infty} dx \frac{f(x)}{(x - a)} = \pm i\pi f(a), \qquad (19.127)$$

to make it well defined. The $+$ or $-$ signs depend on whether the contour is closed in the upper or lower z-planes, respectively. There are also other ways to treat these singular points in the complex plane, thus giving us a collection of Green's functions each satisfying a different boundary condition, which we study in the following example.

Example 19.5. *Helmholtz equation in the continuum limit:* We now evaluate the Green's function given in Equation (19.125) by using the Cauchy principal value and the complex contour integral techniques.

Case I. Using the contours in Figures 19.1 and 19.2 we can evaluate the integral

$$G(x, x') = -\frac{1}{2\pi} P \int_{-\infty}^{\infty} dk' \frac{e^{ik'(x-x')}}{(k' - k_0)(k' + k_0)}. \qquad (19.128)$$

For $(x - x') > 0$ we use the contour in Figure 19.1 to find

$$G(x, x') = -\frac{i\pi}{2\pi} \left[\frac{e^{ik_0(x-x')}}{2k_0} - \frac{e^{-ik_0(x-x')}}{2k_0} \right]$$

$$= \frac{1}{2k_0} \sin k_0 (x - x'). \qquad (19.129)$$

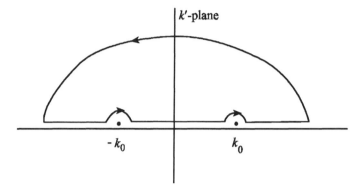

Fig. 19.1 Contour for Case I : $(x - x') > 0$

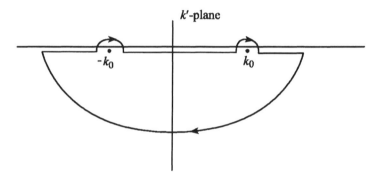

Fig. 19.2 Contour for Case I : $(x - x') < 0$

For $(x - x') < 0$ we use the contour in Figure 19.2 to find

$$G(x, x') = -\frac{1}{2k_0} \sin k_0 (x - x')$$

$$= \frac{1}{2k_0} \sin k_0 (x' - x). \qquad (19.130)$$

Note that for the $(x - x') < 0$ case the Cauchy principal value is $-i\pi f(a)$. In the following cases we add small imaginary pieces, $\pm i\varepsilon$, to the two roots, $+k_0$ and $-k_0$, of the denominator in Equation (19.125), thus moving them away from the real axis. We can now use the Cauchy integral theorems to evaluate the integral (19.125) and then obtain the Green's function in the limit $\varepsilon \to 0$.

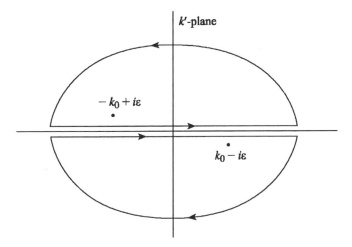

Fig. 19.3 Contours for Case II

Case II: Using the contours shown in Figure 19.3 we obtain the following Green's function:

For $(x - x') > 0$ we use the contour in the upper half complex k'-plane to find

$$G(x, x') = -\lim_{\varepsilon \to 0} \frac{1}{2\pi} \int_{-\infty}^{\infty} dk' \frac{e^{ik'(x-x')}}{(k' - k_0 + i\varepsilon)(k' + k_0 - i\varepsilon)}$$

$$= -\lim_{\varepsilon \to 0} \frac{2\pi i}{2\pi} \frac{e^{i(-k_0 + i\varepsilon)(x-x')}}{2(-k_0 + i\varepsilon)} \qquad (19.131)$$

$$= -\frac{e^{-ik_0(x-x')}}{2k_0 i}. \qquad (19.132)$$

For $(x - x') < 0$ we use the contour in the lower half-plane to get

$$G(x, x') = -\frac{e^{ik_0(x-x')}}{2k_0 i}. \qquad (19.133)$$

Note that there is an extra minus sign coming from the fact that the contour for the $(x - x') < 0$ case is clockwise; thus we obtain the Green's function as

$$G(x, x') = -\frac{e^{-ik_0(x-x')}}{2k_0 i} \theta(x - x') - \frac{e^{ik_0(x-x')}}{2k_0 i} \theta(x' - x). \qquad (19.134)$$

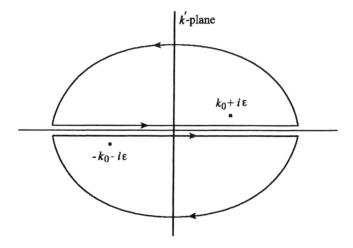

Fig. 19.4   Contours for Case III

Case III: Using the contours shown in Figure 19.4, Green's function is now given as the integral

$$G(x, x') = -\lim_{\varepsilon \to 0} \frac{1}{2\pi} \int_{-\infty}^{\infty} dk' \frac{e^{ik'(x-x')}}{(k' - k_0 - i\varepsilon)(k' + k_0 + i\varepsilon)}. \quad (19.135)$$

For $(x - x') > 0$ we use the upper contour in Figure 19.4 to find

$$G(x, x') = -\frac{1}{2\pi} 2\pi i \lim_{\varepsilon \to 0} \left[\frac{e^{i(k_0 + i\varepsilon)(x-x')}}{2(k_0 + i\varepsilon)} \right]$$

$$= \frac{e^{ik_0(x-x')}}{2k_0 i}, \quad (19.136)$$

while for $(x - x') < 0$ we use the lower contour to find

$$G(x, x') = -(-)\frac{2\pi i}{2\pi} \frac{e^{-ik_0(x-x')}}{-2k_0}$$

$$= \frac{e^{-ik_0(x-x')}}{2k_0 i}. \quad (19.137)$$

Combining these we write the Green's function as

$$G(x, x') = \frac{e^{ik_0(x-x')}}{2k_0 i} \theta(x - x') + \frac{e^{-ik_0(x-x')}}{2k_0 i} \theta(x' - x). \quad (19.138)$$

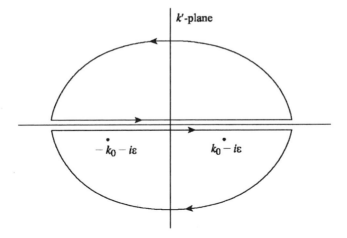

Fig. 19.5   Contours for Case IV

Case IV: Green's function for the contours in Figure 19.5:

For $(x - x') > 0$ we use the upper contour to find

$$G(x, x') = 0. \tag{19.139}$$

Similarly, for $(x - x') < 0$ we use the lower contour to obtain

$$
\begin{aligned}
G(x, x') &= -\frac{1}{2\pi}(-)2\pi i \lim_{\varepsilon \to 0} \left[\frac{e^{i(k_0 - i\varepsilon)(x - x')}}{2(k_0 - i\varepsilon)} + \frac{e^{i(-k_0 - i\varepsilon)(x - x')}}{2(-k_0 - i\varepsilon)} \right] \\
&= i \left[\frac{e^{ik_0(x - x')}}{2k_0} + \frac{e^{-ik_0(x - x')}}{2k_0} \right] \\
&= -\frac{\sin k_0 (x - x')}{k_0}.
\end{aligned} \tag{19.140}
$$

The combined result becomes

$$G(x, x') = -\frac{\sin k_0 (x - x')}{k_0} \theta (x' - x). \tag{19.141}$$

This Green's function is good for the boundary conditions given as

$$\lim_{x \to \infty} \left\{ \begin{array}{c} G(x, x') \to 0 \\ G'(x, x') \to 0 \end{array} \right\}. \tag{19.142}$$

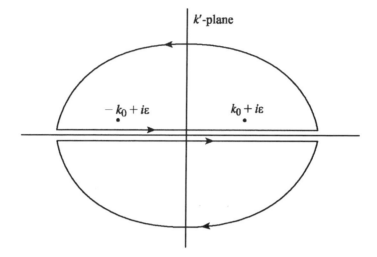

Fig. 19.6 Contours for Case V

Case V: Green's function using the contours in Figure 19.6:

For $(x - x') > 0$ we use the upper contour to find

$$G(x, x') = \frac{1}{i} \left[\frac{e^{ik_0(x - x')}}{2k_0} - \frac{e^{-ik_0(x - x')}}{2k_0} \right] \tag{19.143}$$

$$= \frac{\sin k_0 (x - x')}{k_0}.$$

Similarly, for $(x - x') < 0$ we use the lower contour to find

$$G(x, x') = 0.$$

The combined result becomes

$$G(x, x') = \frac{\sin k_0 (x - x')}{k_0} \theta (x - x'), \tag{19.144}$$

which is useful for the cases where

$$\lim_{x \to -\infty} \left\{ \begin{array}{c} G(x, x') \to 0 \\ G'(x, x') \to 0 \end{array} \right\}. \tag{19.145}$$

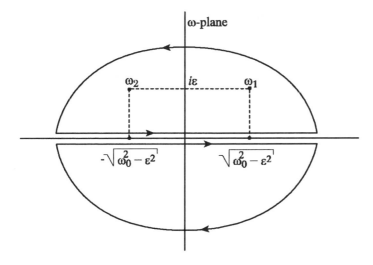

Fig. 19.7 Contours for the harmonic oscillator

Example 19.6. *Green's function for the harmonic oscillator:* For the damped driven harmonic oscillator the equation of motion is written as

$$\frac{d^2x}{dt^2} + 2\varepsilon\frac{dx}{dt} + \omega_0^2 x(t) = f(t), \qquad \varepsilon > 0. \tag{19.146}$$

In terms of a Green's function the solution can be written as

$$x(t) = C_1 x_1(t) + C_2 x_2(t) + \int_{-\infty}^{\infty} G(x, x') f(t') dt', \tag{19.147}$$

where $x_1(t)$ and $x_2(t)$ are the solutions of the homogeneous equation. Assuming that all the necessary Fourier transforms and their inverses exist, we take the Fourier transform of the equation of motion to write the Green's function as

$$G(t, t') = -\frac{1}{2\pi} \int_{-\infty}^{\infty} d\omega' \frac{e^{i\omega'(t-t')}}{(\omega'^2 - 2i\varepsilon\omega' - \omega_0^2)}. \tag{19.148}$$

Since the denominator has zeroes at

$$\omega'_{1,2} = \frac{2i\varepsilon \mp \sqrt{-4\varepsilon^2 + 4\omega_0^2}}{2}$$

$$= \mp\sqrt{\omega_0^2 - \varepsilon^2} + i\varepsilon, \tag{19.149}$$

we can write $G(t, t')$ as

$$G(t, t') = -\frac{1}{2\pi} \int_{-\infty}^{\infty} d\omega' \frac{e^{i\omega'(t-t')}}{(\omega' - \omega_1')(\omega' - \omega_2')}. \tag{19.150}$$

We can evaluate this integral by going to the complex ω-plane. For $(t-t') > 0$ we use the upper contour in Figure 19.7 to write the Green's function as

$$G(t, t') = -\frac{2\pi i}{2\pi} \left[\frac{e^{i\omega_1'(t-t')}}{(\omega_1' - \omega_2')} + \frac{e^{i\omega_2'(t-t')}}{(\omega_2' - \omega_1')} \right] \tag{19.151}$$

$$= \frac{1}{i} \left[\frac{e^{-\varepsilon(t-t')}}{2\sqrt{\omega_0^2 - \varepsilon^2}} \left(e^{i\sqrt{\omega_0^2 - \varepsilon^2}(t-t')} - e^{-i\sqrt{\omega_0^2 - \varepsilon^2}(t-t')} \right) \right]$$

$$= \frac{1}{i} \left[\frac{e^{-\varepsilon(t-t')}}{2\sqrt{\omega_0^2 - \varepsilon^2}} 2i \, \sin \left(\sqrt{\omega_0^2 - \varepsilon^2} \, (t - t') \right) \right].$$

For $(t - t') < 0$ we use the lower contour in Figure 19.7. Because there are no singularities inside the contour, Green's function is now given as

$$G(t, t') = 0.$$

Combining these results we write the Green's function as

$$G(t, t') = \left\{ \begin{array}{cc} \dfrac{e^{-\varepsilon(t-t')}}{\sqrt{\omega_0^2 - \varepsilon^2}} \sin \sqrt{\omega_0^2 - \varepsilon^2} \, (t - t'), & t - t' > 0 \\ 0, & t - t' < 0 \end{array} \right\} \tag{19.152}$$

or

$$G(t, t') = \frac{e^{-\varepsilon(t-t')}}{\sqrt{\omega_0^2 - \varepsilon^2}} \left[\sin \sqrt{\omega_0^2 - \varepsilon^2} \, (t - t') \right] \theta \, (t - t'). \tag{19.153}$$

It is easy to check that this Green's function satisfies the equation

$$\left[\frac{d^2}{dt^2} + 2\varepsilon \frac{d}{dt} + \omega_0^2 \right] G(t, t') = \delta \, (t - t'). \tag{19.154}$$

Example 19.7. *Damped driven harmonic oscillator:* In the previous example let us take the driving force as

$$f(t) = F_0 e^{-\alpha t}, \tag{19.155}$$

where α is a constant. For sinosoidal driving forces we could take α as $i\omega_1$, where ω_1 is the frequency of the driving force. If we start the system with the initial conditions $x(0) = \dot{x}(0) = 0$, C_1 and C_2 in Equation (19.147) is zero, hence the solution will be written as

$$x(t) = F_0 e^{-\varepsilon t} \int_0^t \frac{dt'}{\sqrt{\omega_0^2 - \varepsilon^2}} \sin\left[\sqrt{\omega_0^2 - \varepsilon^2}\,(t - t')\right] e^{(\varepsilon - \alpha)t'}$$

$$= \frac{F_0}{\sqrt{\omega_0^2 - \varepsilon^2}} \frac{\sin\left[\sqrt{\omega_0^2 - \varepsilon^2}\,t - \eta\right]}{\sqrt{\omega_0^2 + \alpha^2 - 2\alpha\varepsilon}} e^{-\varepsilon t} + \frac{F_0}{\omega_0^2 + \alpha^2 - 2\alpha\varepsilon} e^{-\alpha t},$$

where we have defined

$$\tan\eta = \frac{\sqrt{\omega_0^2 - \varepsilon^2}}{(\alpha - \varepsilon)}. \tag{19.156}$$

One can easily check that $x(t)$ satisfies the differential equation

$$\frac{d^2 x(t)}{dt^2} + 2\varepsilon \frac{dx(t)}{dt} + \omega_0^2 x(t) = F_0 e^{-\alpha t}. \tag{19.157}$$

For weak damping the solution reduces to

$$x(t) = \frac{F_0}{\omega_0} \frac{\sin[\omega_0 t - \eta]}{\sqrt{\omega_0^2 + \alpha^2}} + \frac{F_0}{\omega_0^2 + \alpha^2} e^{-\alpha t}. \tag{19.158}$$

As expected, in the $t \to \infty$ limit this becomes

$$x(t) = \frac{F_0}{\omega_0} \frac{\sin[\omega_0 t - \eta]}{\sqrt{\omega_0^2 + \alpha^2}}.$$

19.1.12   Green's Function for the Helmholtz Equation in Three Dimensions

The Helmholtz equation in three dimensions is given as

$$\left(\vec{\nabla}^2 + k_0^2\right) \psi(\vec{r}) = F(\vec{r}). \tag{19.159}$$

We now look for a Green's function satisfying

$$\left(\vec{\nabla}^2 + k_0^2\right) G(\vec{r}, \vec{r}\,') = \delta(\vec{r} - \vec{r}\,'). \tag{19.160}$$

We multiply the first equation by $G(\vec{r}, \vec{r}\,')$ and the second by $\psi(\vec{r})$ and then subtract, and integrate the result over the volume V to get

$$-\psi(\vec{r}\,') = \iiint_V \left[G(\vec{r}, \vec{r}\,')\vec{\nabla}^2 \psi(\vec{r}) - \psi(\vec{r})\vec{\nabla}^2 G(\vec{r}, \vec{r}\,')\right] d^3\vec{r}$$

$$- \iiint_V F(\vec{r})G(\vec{r}, \vec{r}\,')d^3\vec{r}. \tag{19.161}$$

Using Green's theorem

$$\iiint_V \left(F \vec{\nabla}^2 G - G \vec{\nabla}^2 F \right) d^3\vec{r} = \iint_S \left(F \vec{\nabla} G - G \vec{\nabla} F \right) \cdot \hat{\mathbf{n}} ds, \quad (19.162)$$

where S is a closed surface enclosing the volume V with the outward unit normal $\hat{\mathbf{n}}$, we obtain

$$\psi(\vec{r}') = \iiint_V F(\vec{r})G(\vec{r}, \vec{r}')d^3\vec{r} \qquad\qquad (19.163)$$
$$+ \iint_S \left[\psi(\vec{r})\vec{\nabla} G(\vec{r}, \vec{r}') - G(\vec{r}, \vec{r}')\vec{\nabla}\psi(\vec{r}) \right] \cdot \hat{\mathbf{n}} ds.$$

Interchanging the primed and the unprimed variables and assuming that the Green's function is symmetric in anticipation of the corresponding boundary conditions to be imposed later, we obtain the following remarkable formula:

$$\psi(\vec{r}) = \iiint_{V'} G(\vec{r}, \vec{r}')F(\vec{r}')d^3\vec{r}' \qquad\qquad (19.164)$$
$$+ \iint_{S'} \left[\vec{\nabla}'G(\vec{r}, \vec{r}')\psi(\vec{r}') \right] \cdot \hat{\mathbf{n}} ds' - \iint_{S'} G(\vec{r}, \vec{r}')\vec{\nabla}'\psi(\vec{r}') \cdot \hat{\mathbf{n}} ds'.$$

Boundary conditions:
The most frequently used boundary conditions are:
i) Dirichlet boundary conditions, where G is zero on the boundary.
ii) Neumann boundary conditions, where the normal gradient of G on the surface is zero:

$$\vec{\nabla} G \cdot \hat{\mathbf{n}} \Big|_{\text{boundary}} = 0.$$

iii) General boundary conditions:

$$\vec{\nabla} G + \vec{v}(\vec{r}')G \Big|_{\text{boundary}} = 0,$$

where $\vec{v}(\vec{r}')$ is a function of the boundary point $\vec{r}'$.
For any one of these cases, the Green's function is symmetric and the surface term in the above equation vanishes, thus giving

$$\psi(\vec{r}) = \iiint_V G(\vec{r}, \vec{r}')F(\vec{r}')d^3\vec{r}'. \qquad\qquad (19.165)$$

19.1.13 Green's Functions in Three Dimensions with a Discrete Spectrum

Consider the inhomogeneous equation

$$\mathbf{H}\psi(\vec{r}) = F(\vec{r}), \qquad\qquad (19.166)$$

where $\mathbf{H}$ is a linear differential operator. $\mathbf{H}$ has a complete set of orthonormal eigenfunctions, $\{\phi_\lambda(\vec{r})\}$, which are determined by the eigenvalue equation

$$\mathbf{H}\phi_\lambda(\vec{r}) = \lambda\phi_\lambda(\vec{r}),$$

where λ stands for the eigenvalues and the eigenfunctions satisfy the homogeneous boundary conditions given in the previous section. We need a Green's function satisfying the equation

$$\mathbf{H}G(\vec{r}, \vec{r}') = \delta(\vec{r} - \vec{r}'). \tag{19.167}$$

Expanding $\Psi(\vec{r})$ and $F(\vec{r})$ in terms of this complete set of eigenfunctions we write

$$\left\{ \begin{array}{l} \Psi(\vec{r}) = \sum_\lambda a_\lambda\phi_\lambda(\vec{r}) \\ F(\vec{r}) = \sum_\lambda c_\lambda\phi_\lambda(\vec{r}) \end{array} \right\}, \tag{19.168}$$

where the expansion coefficients are

$$\left\{ \begin{array}{l} a_\lambda = \iiint_V \phi_\lambda^*(\vec{r})\Psi(\vec{r})d^3\vec{r} \\ c_\lambda = \iiint_V \phi_\lambda^*(\vec{r})F(\vec{r})d^3\vec{r} \end{array} \right\}. \tag{19.169}$$

Substituting these into Equation (19.166) we obtain

$$a_\lambda = \frac{c_\lambda}{\lambda}. \tag{19.170}$$

Using this a_λ and the explicit form of c_λ in Equation (19.169) we can write $\Psi(\vec{r})$ as

$$\Psi(\vec{r}) = \iiint_V \left[\sum_\lambda \frac{\phi_\lambda(\vec{r})\phi_\lambda^*(\vec{r}')}{\lambda} \right] F(\vec{r}')d^3\vec{r}'. \tag{19.171}$$

This gives the Green's function as

$$G(\vec{r}, \vec{r}') = \sum_\lambda \frac{\phi_\lambda(\vec{r})\phi_\lambda^*(\vec{r}')}{\lambda}. \tag{19.172}$$

This Green's function can easily be generalized to the equation

$$(\mathbf{H} - \lambda_0)\,\Psi(\vec{r}) = F(\vec{r}), \tag{19.173}$$

for the operator $(\mathbf{H} - \lambda_0)$ as

$$G(\vec{r}, \vec{r}') = \sum_\lambda \frac{\phi_\lambda(\vec{r})\phi_\lambda^*(\vec{r}')}{\lambda - \lambda_0}. \tag{19.174}$$

We now find the Green's function for the three-dimensional Helmholtz equation

$$\left(\overrightarrow{\nabla}^2 + k_0^2\right)\psi(\overrightarrow{r}) = F(\overrightarrow{r})$$

in a rectangular region bounded by six planes:

$$\left\{ \begin{array}{ll} x = 0, & x = a \\ y = 0, & y = b \\ z = 0, & z = c \end{array} \right\}$$

and with the homogeneous Dirichlet boundary conditions. The corresponding eigenvalue equation is

$$\overrightarrow{\nabla}^2 \phi_{lmn}(\overrightarrow{r}) + k_{lmn}^2 \phi_{lmn}(\overrightarrow{r}) = 0. \tag{19.175}$$

The normalized eigenfunctions are easily obtained as

$$\phi_{lmn}(\overrightarrow{r}) = \frac{8}{abc} \sin\left(\frac{l\pi x}{a}\right) \sin\left(\frac{m\pi y}{b}\right) \sin\left(\frac{n\pi z}{c}\right), \tag{19.176}$$

where the eigenvalues are

$$k_{lmn}^2 = \frac{l^2\pi^2}{a^2} + \frac{m^2\pi^2}{b^2} + \frac{n^2\pi^2}{c^2}, \quad (l, m, n = \text{positive integer}). \tag{19.177}$$

Using these eigenfunctions [Eq. (19.176)] we can now write the Green's function as

$$G(\overrightarrow{r}, \overrightarrow{r}') = \sum_{lmn} \frac{\phi_{lmn}(\overrightarrow{r})\phi_{lmn}^*(\overrightarrow{r}')}{k_0^2 - k_{lmn}^2}. \tag{19.178}$$

19.1.14   Green's Function for the Laplace Operator Inside a Sphere

Green's function for the Laplace operator $\overrightarrow{\nabla}^2$ satisfies

$$\overrightarrow{\nabla}^2 G(\overrightarrow{r}, \overrightarrow{r}') = \delta(\overrightarrow{r}, \overrightarrow{r}'). \tag{19.179}$$

Using spherical polar coordinates this can be written as

$$\overrightarrow{\nabla}^2 G(\overrightarrow{r}, \overrightarrow{r}') = \frac{\delta(r - r')}{r'^2} \delta(\cos\theta - \cos\theta')\delta(\phi - \phi') \tag{19.180}$$

$$= \frac{\delta(r - r')}{r'^2} \sum_{l=0}^{\infty} \sum_{m=-l}^{m=l} Y_l^{*m}(\theta', \phi')Y_l^m(\theta, \phi), \tag{19.181}$$

where we have used the completeness relation of the spherical harmonics. For the Green's function inside a sphere, we use the boundary conditions

$$G(0, \overrightarrow{r}') = \text{finite}, \tag{19.182}$$

$$G(a, \overrightarrow{r}') = 0. \tag{19.183}$$

In spherical polar coordinates we can separate the angular part and write the Green's function as

$$G(\overrightarrow{r}, \overrightarrow{r}') = \sum_{l=0}^{\infty} \sum_{m=-l}^{m=l} g_l(r, r') Y_l^{*m}(\theta', \phi') Y_l^m(\theta, \phi). \tag{19.184}$$

We now substitute Equation (19.184) into Equation (19.181) to find the differential equation that $g_l(r, r')$ satisfies as

$$\frac{1}{r} \frac{d^2}{dr^2} [r g_l(r, r')] - \frac{l(l+1)}{r^2} g_l(r, r') = \frac{1}{r'^2} \delta(r - r'). \tag{19.185}$$

A general solution of the homogeneous equation

$$\frac{1}{r} \frac{d^2}{dr^2} [r g_l(r, r')] - \frac{l(l+1)}{r^2} g_l(r, r') = 0 \tag{19.186}$$

can be obtained by trying a solution as

$$c_0 r^l + c_1 r^{-(l+1)}. \tag{19.187}$$

We can now construct the radial part of the Green's function for the inside of a sphere by finding the appropriate u and the v solutions as

$$g_l(r, r') = \frac{r^l r'^l}{(2l+1)a^{2l+1}} \begin{cases} \left[1 - (\frac{a}{r'})^{2l+1} \right], & r < r', \\[2mm] \left[1 - (\frac{a}{r})^{2l+1} \right], & r > r'. \end{cases} \tag{19.188}$$

Now the complete Green's function can be written by substituting this result into Equation (19.184).

19.1.15 Green's Functions for the Helmholtz Equation for All Space—Poisson and Schrödinger Equations

We now consider the operator

$$\mathbf{H}_0 = \overrightarrow{\nabla}^2 + \lambda \tag{19.189}$$

in the continuum limit. Using this operator we can write the following differential equation:

$$\mathbf{H}_0 \Psi(\overrightarrow{r}) = F(\overrightarrow{r}). \tag{19.190}$$

Let us assume that the Fourier transforms $\widehat{\Psi}(\overrightarrow{k})$ and $\widehat{F}(\overrightarrow{k})$ of $\Psi(\overrightarrow{r})$ and $F(\overrightarrow{r})$ exists:

$$\widehat{F}(\overrightarrow{k}) = \frac{1}{(2\pi)^{3/2}} \iiint_V e^{-i\overrightarrow{k}\cdot\overrightarrow{r}} F(\overrightarrow{r}) d^3\overrightarrow{r}, \tag{19.191}$$

$$\widehat{\Psi}(\overrightarrow{k}) = \frac{1}{(2\pi)^{3/2}} \iiint_V e^{-i\overrightarrow{k}\cdot\overrightarrow{r}} \Psi(\overrightarrow{r}) d^3\overrightarrow{r}. \tag{19.192}$$

Taking the Fourier transform of Equation (19.190) we get

$$\frac{1}{(2\pi)^{3/2}} \iiint_V e^{-i\overrightarrow{k}\cdot\overrightarrow{r}} \overrightarrow{\nabla}^2 \Psi(\overrightarrow{r}) d^3\overrightarrow{r} + \lambda\widehat{\Psi}(\overrightarrow{k}) = \widehat{F}(\overrightarrow{k}). \tag{19.193}$$

Using the Green's theorem [Eq. (19.162)] we can write the first term in Equation (19.193) as

$$\frac{1}{(2\pi)^{3/2}} \iiint_V e^{-i\overrightarrow{k}\cdot\overrightarrow{r}} \overrightarrow{\nabla}^2 \Psi(\overrightarrow{r}) d^3\overrightarrow{r} \tag{19.194}$$

$$= \frac{1}{(2\pi)^{3/2}} \iiint_V \Psi(\overrightarrow{r}) \overrightarrow{\nabla}^2 e^{-i\overrightarrow{k}\cdot\overrightarrow{r}} d^3\overrightarrow{r}$$

$$+ \frac{1}{(2\pi)^{3/2}} \iint_S \left(e^{-i\overrightarrow{k}\cdot\overrightarrow{r}} \overrightarrow{\nabla}\Psi(\overrightarrow{r}) - \Psi(\overrightarrow{r})\overrightarrow{\nabla}e^{-i\overrightarrow{k}\cdot\overrightarrow{r}} \right) \cdot \widehat{\mathbf{n}} ds,$$

where S is a surface with an outward unit normal $\widehat{\mathbf{n}}$ enclosing the volume V. We now take our region of integration as a sphere of radius R and consider the limit $R \to \infty$. In this limit the surface term becomes

$$\frac{1}{(2\pi)^{3/2}} \lim_{R\to\infty} \iint_S \left(e^{-i\overrightarrow{k}\cdot\overrightarrow{r}} \overrightarrow{\nabla}\Psi(\overrightarrow{r}) - \Psi(\overrightarrow{r})\overrightarrow{\nabla}e^{-i\overrightarrow{k}\cdot\overrightarrow{r}} \right) \cdot \widehat{\mathbf{n}} ds \tag{19.195}$$

$$= \frac{1}{(2\pi)^{3/2}} \lim_{R\to\infty} R^2 \left\{ \iint_S \left[e^{-i\overrightarrow{k}\cdot\overrightarrow{r}} \frac{d\Psi}{dr} - \Psi \frac{d\left(e^{-i\overrightarrow{k}\cdot\overrightarrow{r}}\right)}{dr} \right] d\Omega \right\}_{r=R},$$

where $\widehat{\mathbf{n}} = \widehat{\mathbf{e}}_r$ and $d\Omega = \sin\theta d\theta d\phi$. If the function $\Psi(\overrightarrow{r})$ goes to zero sufficiently rapidly as $|\overrightarrow{r}| \to \infty$, that is, when $\Psi(\overrightarrow{r})$ goes to zero faster than $\frac{1}{r}$, then the surface term vanishes, thus leaving us with

$$\frac{1}{(2\pi)^{3/2}} \iiint_V e^{-i\overrightarrow{k}\cdot\overrightarrow{r}} \overrightarrow{\nabla}^2 \Psi(\overrightarrow{r}) d^3\overrightarrow{r} = -k^2 \widehat{\Psi}(\overrightarrow{k}) \tag{19.196}$$

in Equation (19.194). Consequently, Equation (19.193) becomes

$$\widehat{\Psi}(\overrightarrow{k}) = \frac{\widehat{F}(\overrightarrow{k})}{(-k^2 + \lambda)}. \tag{19.197}$$

In this equation we have to treat the cases $\lambda > 0$ and $\lambda \leq 0$ separately.

Case 1: $\lambda \leq 0$:

In this case we can write $\lambda = -\kappa^2$; thus the denominator $(k^2 + \kappa^2)$ in

$$\widehat{\Psi}(\overrightarrow{k}) = -\frac{\widehat{F}(\overrightarrow{k})}{k^2 + \kappa^2} \qquad (19.198)$$

never vanishes. Taking the inverse Fourier transform of this, we write the general solution of Equation (19.190) as

$$\Psi(\overrightarrow{r}) = \xi(\overrightarrow{r}) - \frac{1}{(2\pi)^{3/2}} \iiint \frac{\widehat{F}(\overrightarrow{k})}{k^2 + \kappa^2} e^{i\overrightarrow{k}\cdot\overrightarrow{r}} d^3\overrightarrow{k}, \qquad (19.199)$$

where $\xi(\overrightarrow{r})$ denotes the solution of the homogeneous equation

$$\mathbf{H}_0\xi(\overrightarrow{r}) = \left(\overrightarrow{\nabla}^2 - \kappa^2\right)\xi(\overrightarrow{r}) = 0. \qquad (19.200)$$

Defining a Green's function $G(\overrightarrow{r}, \overrightarrow{r}')$ as

$$G(\overrightarrow{r}, \overrightarrow{r}') = -\frac{1}{(2\pi)^3} \iiint \frac{e^{i\overrightarrow{k}\cdot(\overrightarrow{r} - \overrightarrow{r}')}}{k^2 + \kappa^2} d^3\overrightarrow{k}, \qquad (19.201)$$

we can express the general solution of Equation (19.190) as

$$\Psi(\overrightarrow{r}) = \xi(\overrightarrow{r}) + \iiint_V G(\overrightarrow{r}, \overrightarrow{r}')F(\overrightarrow{r}')d^3\overrightarrow{r}, \qquad (19.202)$$

The integral in the Green's function can be evaluated by using complex contour integral techniques. Taking the $\overrightarrow{k}$ vector as

$$\overrightarrow{k} = k\widehat{\mathbf{r}}, \quad \left(\widehat{\mathbf{r}} = \frac{\overrightarrow{r} - \overrightarrow{r}'}{|\overrightarrow{r} - \overrightarrow{r}'|}\right), \qquad (19.203)$$

we write

$$I = \iiint \frac{e^{i\overrightarrow{k}\cdot(\overrightarrow{r} - \overrightarrow{r}')}}{k^2 + \kappa^2} d^3\overrightarrow{k}, \qquad (19.204)$$

where $d^3\overrightarrow{k} = k^2\sin\theta dk d\theta d\phi$. We can take the ϕ and θ integrals immediately, thus obtaining

$$I = 2\pi \int_0^\infty \frac{kdk}{k^2 + \kappa^2} \cdot \left[\frac{e^{ik|\overrightarrow{r} - \overrightarrow{r}'|} - e^{-ik|\overrightarrow{r} - \overrightarrow{r}'|}}{i|\overrightarrow{r} - \overrightarrow{r}'|}\right]$$

$$= \frac{2\pi}{i|\overrightarrow{r} - \overrightarrow{r}'|} \int_{-\infty}^\infty dk \frac{ke^{ik|\overrightarrow{r} - \overrightarrow{r}'|}}{k^2 + \kappa^2}. \qquad (19.205)$$

Using Jordan's lemma (Section 13.7) we can show that the integral over the circle in the upper half complex k-plane goes to zero as the radius goes to infinity; thus we obtain I as

$$I = \frac{2\pi}{i\,|\overrightarrow{r} - \overrightarrow{r}'|} 2\pi i \sum_{k>0} \text{residues of} \left\{ \frac{k e^{ik|\overrightarrow{r} - \overrightarrow{r}'|}}{k^2 + \kappa^2} \right\} \tag{19.206}$$

$$= \frac{4\pi^2}{|\overrightarrow{r} - \overrightarrow{r}'|} \cdot \frac{i\kappa e^{-\kappa|\overrightarrow{r} - \overrightarrow{r}'|}}{2i\kappa} \tag{19.207}$$

$$= \frac{2\pi^2}{|\overrightarrow{r} - \overrightarrow{r}'|} e^{-\kappa|\overrightarrow{r} - \overrightarrow{r}'|}. \tag{19.208}$$

Using this in (19.201) we obtain the Green's function as

$$G(\overrightarrow{r}, \overrightarrow{r}') = -\frac{1}{4\pi} \frac{e^{-\kappa|\overrightarrow{r} - \overrightarrow{r}'|}}{|\overrightarrow{r} - \overrightarrow{r}'|}. \tag{19.209}$$

To complete the solution [Eq. (19.202)] we also need $\xi(\overrightarrow{r})$, which is easily obtained as

$$\xi(\overrightarrow{r}) = C_0 e^{\pm \kappa_1 x} e^{\pm \kappa_2 y} e^{\pm \kappa_3 z}, \tag{19.210}$$

$$\kappa^2 = \kappa_1^2 + \kappa_2^2 + \kappa_3^2.$$

Because this solution diverges for $|r| \to \infty$, for a finite solution everywhere we set $C_0 = 0$ and write the general solution as

$$\Psi(\overrightarrow{r}) = -\frac{1}{4\pi} \iiint_V \frac{e^{-\kappa|\overrightarrow{r} - \overrightarrow{r}'|}}{|\overrightarrow{r} - \overrightarrow{r}'|} F(\overrightarrow{r}') d^3 \overrightarrow{r}'. \tag{19.211}$$

In this solution if $F(\overrightarrow{r}')$ goes to zero sufficiently rapidly as $|r'| \to \infty$ or if $F(\overrightarrow{r}')$ is zero beyond some $|r'| = r_0$, we see that for large r, $\Psi(\overrightarrow{r})$ decreases exponentially as

$$\Psi(\overrightarrow{r}) \to C \frac{e^{-\kappa r}}{r}. \tag{19.212}$$

This is consistent with the neglect of the surface term in our derivation in Equation (19.195).

Example 19.8. *Green's function for the Poisson equation:* Using the above Green's function [Eq. (19.209)] with $\kappa = 0$, we can now convert the Poisson equation

$$\overrightarrow{\nabla}^2 \phi(\overrightarrow{r}) = -4\pi \rho(\overrightarrow{r}) \tag{19.213}$$

into an integral equation. In this case $\lambda = 0$; thus the solution is given as

$$\phi(\overrightarrow{r}) = -4\pi \iiint_V G(\overrightarrow{r}, \overrightarrow{r}') \rho(\overrightarrow{r}') d^3 \overrightarrow{r}', \tag{19.214}$$

where

$$G(\vec{r}, \vec{r}') = \frac{1}{4\pi} \frac{1}{|\vec{r} - \vec{r}'|}. \tag{19.215}$$

Example 19.9. *Green's function for the Schrödinger equation*$-E < 0$:
Another application for Green's function [Eq. (19.209)] is the time-independent Schrödinger equation:

$$\left(\vec{\nabla}^2 + \frac{2mE}{\hbar^2}\right) \Psi(\vec{r}) = \frac{2m}{\hbar^2} V(\vec{r}) \Psi(\vec{r}). \tag{19.216}$$

For central potentials and bound states $(E < 0)$ κ^2 is given as

$$\kappa^2 = -\frac{2m|E|}{\hbar^2}. \tag{19.217}$$

Thus the solution of Equation (19.216) can be written as

$$\Psi(\vec{r}) = -\frac{m}{2\pi\hbar^2} \iiint_V \frac{e^{-\kappa|\vec{r} - \vec{r}'|}}{|\vec{r} - \vec{r}'|} V(\vec{r}') \Psi(\vec{r}') d^3 \vec{r}'. \tag{19.218}$$

This is also the integral equation version of the time-independent Shrödinger equation for bound states.

Case II: $\lambda > 0$:

In this case the denominator in the definition [Eq. (19.197)] of $\widehat{\Psi}(\vec{k})$ has zeroes at $k = \pm\sqrt{\lambda}$. To eliminate this problem we add a small imaginary piece $i\varepsilon$ to the λ values as

$$\lambda = (q \pm i\varepsilon), \quad \varepsilon > 0 . \tag{19.219}$$

Substituting this in Equation (19.197) we get

$$\widehat{\Psi}_\pm(\vec{k}) = -\frac{\widehat{F}(\vec{k})}{k^2 - (q \pm i\varepsilon)^2}, \tag{19.220}$$

which is now well defined everywhere on the real k-axis. Taking the inverse Fourier transform of this we get

$$\Psi(\vec{r}) = \xi(\vec{r}) + \iiint_V G_\pm(\vec{r}, \vec{r}') F(\vec{r}') d^3 \vec{r}',$$

$$G_\pm(\vec{r}, \vec{r}') = -\frac{1}{(2\pi)^3} \iiint \frac{e^{i\vec{k} \cdot (\vec{r} - \vec{r}')}}{k^2 - (q \pm i\varepsilon)^2} d^3 \vec{k}. \tag{19.221}$$

We can now evaluate this integral in the complex k-plane using the complex contour integral theorems and take the limit as $\varepsilon \rightarrow 0$ to obtain the final result. Because our integrand has poles at

$$k = (q \pm i\varepsilon),$$

we use the Cauchy integral theorem. However, as before, we first take the θ and ϕ integrals to write

$$
G_{\pm}(\vec{r},\vec{r}') = -\frac{1}{8\pi^2 i \, |\vec{r} - \vec{r}'|} \int_{-\infty}^{\infty} k \, dk \left[\frac{e^{ik|\vec{r} - \vec{r}'|}}{(k - q \mp i\varepsilon)(k + q \pm i\varepsilon)} \right.
$$
$$
\left. - \frac{e^{-ik|\vec{r} - \vec{r}'|}}{(k - q \mp i\varepsilon)(k + q \pm i\varepsilon)} \right]. \tag{19.222}
$$

For the first integral we close the contour in the upper half complex k-plane and get

$$
\int_{-\infty}^{\infty} k \, dk \frac{e^{ik|\vec{r} - \vec{r}'|}}{(k - q \mp i\varepsilon)(k + q \pm i\varepsilon)} = \pi i e^{\pm iq|\vec{r} - \vec{r}'| - \varepsilon|\vec{r} - \vec{r}'|}. \tag{19.223}
$$

Similarly, for the second integral we close our contour in the lower half complex k-plane to get

$$
\int_{-\infty}^{\infty} k \, dk \frac{e^{-ik|\vec{r} - \vec{r}'|}}{(k - q \mp i\varepsilon)(k + q \pm i\varepsilon)} = -\pi i e^{\pm iq|\vec{r} - \vec{r}'| - \varepsilon|\vec{r} - \vec{r}'|}. \tag{19.224}
$$

Combining these we obtain the Green's function

$$
G_{\pm}(\vec{r},\vec{r}') = -\frac{1}{4\pi} \frac{e^{\pm iq|\vec{r} - \vec{r}'|}}{|\vec{r} - \vec{r}'|} e^{-\varepsilon|\vec{r} - \vec{r}'|} \tag{19.225}
$$

and the solution as

$$
\Psi_{\pm}(\vec{r}) = \xi(\vec{r}) - \frac{1}{4\pi} \iiint_V \frac{e^{i(\pm q + i\varepsilon)|\vec{r} - \vec{r}'|}}{|\vec{r} - \vec{r}'|} F(\vec{r}') d^3\vec{r}'. \tag{19.226}
$$

The choice of the $\pm$ sign is very important. In the limit as $|\vec{r}| \to \infty$ this solution behaves as

$$
\Psi_+(\vec{r}) \to \xi(\vec{r}) + C \frac{e^{iqr}}{r} \tag{19.227}
$$

or

$$
\Psi_-(\vec{r}) \to \xi(\vec{r}) + C \frac{e^{-iqr}}{r}, \tag{19.228}
$$

where C is a constant independent of r, but it could depend on θ and ϕ. The $\pm$ signs physically correspond to incoming and outgoing waves. We now look at the solutions of the homogeneous equation:

$$
\left(\vec{\nabla}^2 + q^2 \right) \xi(\vec{r}) = 0, \tag{19.229}
$$

which are now given as plane waves, $e^{i\vec{q}\cdot\vec{r}}$; thus the general solution becomes

$$\Psi_{\pm}(\vec{r}) = \frac{A}{(2\pi)^{3/2}}e^{i\vec{q}\cdot\vec{r}} - \frac{1}{4\pi}\iiint_V \frac{e^{i(\pm q + i\varepsilon)\left|\vec{r}-\vec{r}'\right|}}{\left|\vec{r}-\vec{r}'\right|}F(\vec{r}')d^3\vec{r}'.$$

(19.230)

The constant A and the direction of the $\vec{q}$ vector come from the initial conditions.

Example 19.10. *Green's function for the Schrödinger equation*—$E \geq 0$:
An important application of the $\lambda > 0$ case is the Schrödinger equation for the scattering problems, that is, for states with $E \geq 0$. Using the Green's function we have found [Eq. (19.225)] we can write the Schrödinger equation,

$$\left(\vec{\nabla}^2 + \frac{2mE}{\hbar^2}\right)\Psi(\vec{r}) = \frac{2m}{\hbar^2}V(\vec{r})\Psi(\vec{r}),$$

(19.231)

as an integral equation for the scattering states as

$$\Psi_{\pm}(\vec{r}) = \frac{A}{(2\pi)^{3/2}}e^{i\vec{q}_i\cdot\vec{r}} - \frac{m}{2\pi\hbar^2}\iiint_V \frac{e^{\pm iq_i\left|\vec{r}-\vec{r}'\right|}}{\left|\vec{r}-\vec{r}'\right|}V(\vec{r}')\Psi_{\pm}(\vec{r}')d^3\vec{r}'.$$

(19.232)

The magnitude of $\vec{q}_i$ is given as

$$q_i = \sqrt{\frac{2mE}{\hbar^2}}.$$

(19.233)

Equation (19.232) is known as the **Lipmann-Schwinger equation.** For bound state problems it is easier to work with the differential equation version of the Schödinger equation, hence it is preferred. However, for the scattering problems, the Lipmann-Schwinger equation is the starting point of modern quantum mechanics. Note that we have written the result free of ε in anticipation that the $\varepsilon \to 0$ limit will not cause any problems.

19.1.16   General Boundary Conditions and Applications to Electrostatics

In the problems we have discussed so far the Green's function and the solution were required to satisfy the same homogeneous boundary conditions (Section 19.1.12). However, in electrostatics we usually deal with cases in which we are interested in finding the potential of a charge distribution in the presence of conducting surfaces held at constant potentials. The question we now ask is: Can we still use the Green's function found from

$$\mathcal{L}G(\vec{r}, \vec{r}\,') = \delta(\vec{r} - \vec{r}\,') \tag{19.234}$$

with the homogeneous boundary conditions? To answer this question we start with a general second-order linear operator of the form

$$\mathcal{L} = \vec{\nabla} \cdot [p(\vec{r})\vec{\nabla}] + q(r), \tag{19.235}$$

which covers a wide range of interesting cases. The corresponding inhomogeneous differential equation is now given as

$$\mathcal{L}\Phi(\vec{r}) = F(\vec{r}), \tag{19.236}$$

where the solution $\Phi(\vec{r})$ is required to satisfy more complex boundary conditions than the usual homogeneous boundary conditions that the Green's function is required to satisfy. Let us first multiply Equation (19.236) with $G(\vec{r}, \vec{r}\,')$ and Equation (19.234) with $\Phi(\vec{r})$, and then subtract and integrate the result over V to write

$$\Phi(\vec{r}\,') = \iiint_V F(\vec{r})G(\vec{r}, \vec{r}\,')d^3\vec{r} \tag{19.237}$$

$$+ \iiint_V [\Phi(\vec{r})\mathcal{L}G(\vec{r}, \vec{r}\,') - G(\vec{r}, \vec{r}\,')\mathcal{L}\Phi(\vec{r})]\, d^3\vec{r}.$$

We now write $\mathcal{L}$ explicitly and use the following property of the $\vec{\nabla}$ operator:

$$\vec{\nabla} \cdot [f(\vec{r})\vec{v}(\vec{r})] = \vec{\nabla}f(\vec{r}) \cdot \vec{v}(\vec{r}) + f(\vec{r})\vec{\nabla} \cdot \vec{v}(\vec{r}),$$

to write

$$\Phi(\vec{r}\,') = \iiint_V F(\vec{r})G(\vec{r}, \vec{r}\,')d^3\vec{r}$$

$$+ \iiint_V \vec{\nabla} \cdot \left[p(\vec{r})\Phi(\vec{r})\vec{\nabla}G(\vec{r}, \vec{r}\,') - G(\vec{r}, \vec{r}\,')p(\vec{r})\vec{\nabla}\Phi(\vec{r})\right] d^3\vec{r}.$$

Using the fact that for homogeneous boundary conditions the Green's function is symmetric we interchange $\vec{r}\,'$ and $\vec{r}$:

$$\Phi(\vec{r}) = \iiint_V F(\vec{r}\,')G(\vec{r}, \vec{r}\,')d^3\vec{r}\,'$$

$$+ \iiint_V \vec{\nabla}' \cdot \left[p(\vec{r}\,')\Phi(\vec{r}\,')\vec{\nabla}'G(\vec{r}, \vec{r}\,') - G(\vec{r}, \vec{r}\,')p(\vec{r}\,')\vec{\nabla}'\Phi(\vec{r}\,')\right] d^3\vec{r}$$

We finally use the Gauss theorem to write

$$\Phi(\vec{r}) = \iiint_V F(\vec{r}\,')G(\vec{r}, \vec{r}\,')d^3\vec{r}\,' \tag{19.238}$$

$$+ \iint_S p(\vec{r}\,') \left[\Phi(\vec{r}\,')\vec{\nabla}'G(\vec{r}, \vec{r}\,') - G(\vec{r}, \vec{r}\,')\vec{\nabla}'\Phi(\vec{r}\,')\right] \cdot \hat{n}ds',$$

where $\hat{\mathbf{n}}$ is the outward unit normal to the surface S bounding the volume V. If we impose the same homogeneous boundary conditions on $\Phi(\vec{r})$ and $G(\vec{r},\vec{r}')$, the surface term vanishes and we reach the conclusions of Section 19.1.12.

In general, in order to evaluate the surface integral we have to know the function $\Phi(\vec{r})$ and its normal derivative on the surface. As boundary conditions we can fix the value of $\Phi(\vec{r})$, its normal derivative, or even their linear combination on the surface S, but not $\Phi(\vec{r})$ and its normal derivative at the same time. In practice, this difficulty is circumvented by choosing the Green's function such that it vanishes on the surface. In such cases the solution becomes

$$\Phi(\vec{r}) = \iiint_V F(\vec{r}')G(\vec{r},\vec{r}')d^3\vec{r}' + \iint_S \left[p(\vec{r}')\Phi(\vec{r}')\vec{\nabla}'G(\vec{r},\vec{r}') \right] \cdot \hat{\mathbf{n}}ds'.$$
$$(19.239)$$

As an example, consider electrostatics problems where we have

$$\left\{ \begin{array}{c} F(\vec{r}) = -4\pi\rho(\vec{r}), \\[2mm] p(\vec{r}) = 1, \\[2mm] q(\vec{r}) = 0. \end{array} \right\} \qquad (19.240)$$

The potential inside a region bounded by a conducting surface held at constant potential V_0 is now given as

$$\Phi(\vec{r}) = -\int_V 4\pi\rho(\vec{r})G(\vec{r},\vec{r}')d^3\vec{r}' + V_0 \oint_S \vec{\nabla}'G(\vec{r},\vec{r}') \cdot \hat{\mathbf{n}}ds',$$
$$(19.241)$$

where $G(\vec{r},\vec{r}')$ comes from the solution of Equation (19.234) subject to the (homogeneous) boundary condition, which requires it to vanish on the surface. The geometry of the surface bounding the volume V could in principle be rather complicated, and $\Phi(\vec{r}')$ in the surface integral does not have to be a constant.

Similarly, if we fix the value of the normal derivative $\vec{\nabla}\Phi(\vec{r}) \cdot \hat{\mathbf{n}}$ on the surface, then we use a Green's function with a normal derivative vanishing on the surface. Now the solution becomes

$$\Phi(\vec{r}) = \iiint_V F(\vec{r}')G(\vec{r},\vec{r}')d^3\vec{r}' - \iint_S p(\vec{r}') \left[G(\vec{r},\vec{r}')\vec{\nabla}'\Phi(\vec{r}') \right] \cdot \hat{\mathbf{n}}ds'.$$
$$(19.242)$$

19.2   TIME-DEPENDENT GREEN'S FUNCTIONS

19.2.1   Green's Functions with First-Order Time Dependence

We now consider differential equations, which could be written as

$$H\Psi(\overrightarrow{r},\tau) + \frac{\partial \Psi(\overrightarrow{r},\tau)}{\partial \tau} = 0, \qquad (19.243)$$

where τ is a timelike variable, and H is a linear differential operator independent of τ, which also has a complete set of orthonormal eigenfunctions. In applications we frequently encounter differential equations of this type. For example, the heat transfer equation is given as

$$\overrightarrow{\nabla}^2 T(\overrightarrow{r},t) = \frac{c}{k}\frac{\partial T(\overrightarrow{r},t)}{\partial t}, \qquad (19.244)$$

where $T(\overrightarrow{r},t)$ is the temperature, c is the specific heat per unit volume, and k is conductivity. Comparing this with Equation (19.243) we see that

$$\left\{ \begin{array}{c} H = -\overrightarrow{\nabla}^2 \\ \\ \tau = \dfrac{kt}{c} \end{array} \right\}.$$

Another example for the first-order time-dependent equations is the Schrödinger equation:

$$H\Psi(\overrightarrow{r},t) = i\hbar\frac{\partial \Psi(\overrightarrow{r},t)}{\partial t}, \qquad (19.245)$$

where H is the Hamiltonian operator. For a particle moving under the influence of a central potential $V(\overrightarrow{r})$, H is given as

$$H = -\frac{\hbar^2}{2m}\overrightarrow{\nabla}^2 + V(\overrightarrow{r}). \qquad (19.246)$$

Hence in Equation (19.243)

$$\left\{ \begin{array}{c} H = -\overrightarrow{\nabla}^2 + \dfrac{2m}{\hbar^2}V(\overrightarrow{r}) \\ \\ \tau = \left(\dfrac{i\hbar}{2m}\right)t \end{array} \right\}.$$

The diffusion equation is given as

$$\overrightarrow{\nabla}^2\rho = \frac{1}{a^2}\frac{\partial \rho}{\partial t}, \qquad (19.247)$$

where $\rho(r, t)$ is the density (or concentration) and a is the diffusion coefficient. In this case,

$$\left\{ \begin{array}{c} H = -\vec{\nabla}^2 \\ \\ \tau = a^2 t \end{array} \right\}.$$

Since H has a complete and orthonormal set of eigenfunctions, we can write the corresponding eigenvalue equation as

$$H\phi_m = \lambda_m \phi_m, \tag{19.248}$$

where λ_m are the eigenvalues and ϕ_m are the eigenfunctions. We now write the solution of Equation (19.243) as

$$\Psi(\vec{r},\tau) = \sum_m A_m(\tau)\phi_m(\vec{r}), \tag{19.249}$$

where the time dependence is carried in the expansion coefficients $A_m(\tau)$. Operating on $\Psi(\vec{r},\tau)$ with $\mathbf{H}$ and remembering that $\mathbf{H}$ is independent of τ, we obtain

$$\begin{aligned} H\Psi = H &\left[\sum_m A_m(\tau)\phi_m(\vec{r}) \right] \\ &= \sum_m A_m(\tau)H\phi_m(\vec{r}) \\ &= \sum_m \lambda_m A_m(\tau)\phi_m(\vec{r}). \end{aligned} \tag{19.250}$$

Using Equation (19.250) and the time derivative of Equation (19.249) in Equation (19.243) we get

$$\sum_m \left[\lambda_m A_m(\tau) + \frac{dA_m(\tau)}{d\tau} \right] \phi_m(\vec{r}) = 0. \tag{19.251}$$

Because $\{\phi_m\}$ is a set of linearly independent functions, this equation cannot be satisfied unless all the coefficients of ϕ_m vanish simultaneously, that is,

$$\frac{dA_m(\tau)}{d\tau} + \lambda_m A_m(\tau) = 0 \tag{19.252}$$

for all m. Solution of this differential equation can be written immediately as

$$A_m(\tau) = A_m(0)e^{-\lambda_m \tau}, \tag{19.253}$$

thus giving $\Psi(\vec{r},\tau)$ as

$$\Psi(\vec{r},\tau) = \sum_m A_m(0)\phi_m(\vec{r})e^{-\lambda_m \tau}. \tag{19.254}$$

To complete the solution we need an initial condition. Assuming that the solution at $\tau = 0$ is given as $\Psi(\overrightarrow{r},0)$, we write

$$\Psi(\overrightarrow{r},0) = \sum_m A_m(0)\phi_m(\overrightarrow{r}). \tag{19.255}$$

Because the eigenfunctions satisfy the orthogonality relation

$$\iiint_V \phi_m^*(\overrightarrow{r})\phi_n(\overrightarrow{r})d^3\overrightarrow{r} = \delta_{mn} \tag{19.256}$$

and the completeness relation

$$\sum_m \phi_m^*(\overrightarrow{r}')\phi_n(\overrightarrow{r}) = \delta(\overrightarrow{r} - \overrightarrow{r}'), \tag{19.257}$$

we can solve Equation (19.254) for $A_m(0)$ as

$$A_m(0) = \iiint_V \phi_m^*(\overrightarrow{r}')\Psi(\overrightarrow{r}',0)d^3\overrightarrow{r}'. \tag{19.258}$$

Substituting these $A_m(0)$ functions back into Equation (19.254) we obtain

$$\Psi(\overrightarrow{r},\tau) = \sum_m e^{-\lambda_m\tau}\phi_m(\overrightarrow{r})\iiint_V \phi_m^*(\overrightarrow{r}')\Psi(\overrightarrow{r}',0)d^3\overrightarrow{r}'. \tag{19.259}$$

Rearranging this expression as

$$\Psi(\overrightarrow{r},\tau) = \iiint_V G_1(\overrightarrow{r},\overrightarrow{r}',\tau)\Psi(\overrightarrow{r}',0)d^3\overrightarrow{r}', \tag{19.260}$$

we obtain a function

$$G_1(\overrightarrow{r},\overrightarrow{r}',\tau) = \sum_m e^{-\lambda_m\tau}\phi_m(\overrightarrow{r})\phi_m^*(\overrightarrow{r}'), \tag{19.261}$$

where the subscript 1 denotes the fact that we have first-order time dependence. Note that $G_1(\overrightarrow{r},\overrightarrow{r}',\tau)$ satisfies the relation

$$G_1(\overrightarrow{r},\overrightarrow{r}',0) = \sum_m \phi_m(\overrightarrow{r})\phi_m^*(\overrightarrow{r}') \tag{19.262}$$

$$= \delta^3(\overrightarrow{r}-\overrightarrow{r}')$$

and the differential equation

$$\left(H + \frac{\partial}{\partial\tau}\right)G_1(\overrightarrow{r},\overrightarrow{r}',\tau) = 0. \tag{19.263}$$

Because $G_1(\overrightarrow{r}, \overrightarrow{r}', \tau)$ does not satisfy the basic equation for Green's functions, that is,

$$\left(H + \frac{\partial}{\partial \tau}\right) G(\overrightarrow{r}, \overrightarrow{r}', \tau) = \delta^3\left(\overrightarrow{r} - \overrightarrow{r}'\right) \delta(\tau), \tag{19.264}$$

it is not yet the Green's function for this problem. However, as we shall see shortly, it is very closely related to it.

Note that if we take the initial condition as

$$\Psi(\overrightarrow{r}, 0) = \delta^3\left(\overrightarrow{r} - \overrightarrow{r}_0\right), \tag{19.265}$$

which is called the **point source initial condition**, G_1 becomes the solution of the differential equation (19.243), that is,

$$\Psi(\overrightarrow{r}, \tau) = G_1(\overrightarrow{r}, \overrightarrow{r_0}, \tau), \quad \tau \geq 0 \ . \tag{19.266}$$

19.2.2 Propagators

Because our choice of initial time as $\tau' = 0$ was arbitrary, for a general initial time τ', $\Psi(\overrightarrow{r}, \tau)$ and the G_1 functions become

$$\Psi(\overrightarrow{r}, \tau) = \iiint G_1(\overrightarrow{r}, \overrightarrow{r}', \tau, \tau')\Psi(\overrightarrow{r}', \tau')d^3\overrightarrow{r}', \tag{19.267}$$

$$G_1(\overrightarrow{r}, \overrightarrow{r}', \tau, \tau') = \sum_m e^{-\lambda_m\left(\tau - \tau'\right)}\phi_m(\overrightarrow{r})\phi_m^*(\overrightarrow{r}'). \tag{19.268}$$

From Equation (19.267) it is seen that, given the solution at $(\overrightarrow{r}', \tau')$ as $\Psi(\overrightarrow{r}', \tau')$, we can find the solution at a later time, $\Psi(\overrightarrow{r}, \tau > \tau')$, by using $G_1(\overrightarrow{r}, \overrightarrow{r}', \tau, \tau')$. It is for this reason that $G_1(\overrightarrow{r}, \overrightarrow{r}', \tau, \tau')$ is also called the **propagator**. In quantum field theory and perturbation calculations, propagator interpretation of G_1 is very useful in the interpretation of Feynman diagrams.

19.2.3 Compounding Propagators

Given a solution at τ_0, let us propagate it first to $\tau_1 > \tau_0$ and then to $\tau_2 > \tau_1$ as (from now on we use $\int d^3\overrightarrow{r}$ instead of $\iiint d^3\overrightarrow{r}$)

$$\Psi(\overrightarrow{r}, \tau_1) = \int G_1(\overrightarrow{r}, \overrightarrow{r}'', \tau_1, \tau_0)\Psi(\overrightarrow{r}'', \tau_0)d^3\overrightarrow{r}'', \tag{19.269}$$

$$\Psi(\overrightarrow{r}, \tau_2) = \int G_1(\overrightarrow{r}, \overrightarrow{r}', \tau_2, \tau_1)\Psi(\overrightarrow{r}', \tau_1)d^3\overrightarrow{r}'. \tag{19.270}$$

Using Equation (19.269) we can write the second equation as

$$\Psi(\overrightarrow{r},\tau_2) = \int \int G_1(\overrightarrow{r},\overrightarrow{r}',\tau_2,\tau_1)G_1(\overrightarrow{r}',\overrightarrow{r}'',\tau_1,\tau_0)\Psi(\overrightarrow{r}'',\tau_0)d^3\overrightarrow{r}'d^3\overrightarrow{r}''.$$

$$(19.271)$$

Using the definition of propagators [Eq. (19.268)] we can also write this as

$$\int G_1(\overrightarrow{r},\overrightarrow{r}',\tau_2,\tau_1)G_1(\overrightarrow{r}',\overrightarrow{r}'',\tau_1,\tau_0)d^3\overrightarrow{r}' \qquad\qquad (19.272)$$

$$= \int \sum_m e^{-\lambda_m(\tau_2-\tau_1)}\phi_m(\overrightarrow{r})\phi_m^*(\overrightarrow{r}')\sum_n e^{-\lambda_n(\tau_1-\tau_0)}\phi_n(\overrightarrow{r}')\phi_n^*(\overrightarrow{r}'')d^3\overrightarrow{r}'$$

$$= \sum_m e^{-\lambda_m(\tau_2-\tau_1)}\phi_m(\overrightarrow{r})\left[\int \phi_m^*(\overrightarrow{r}')\phi_n(\overrightarrow{r}')d^3\overrightarrow{r}'\right]\sum_n e^{-\lambda_n(\tau_1-\tau_0)}\phi_n^*(\overrightarrow{r}'').$$

Using the orthogonality relation

$$\int \phi_m^*(\overrightarrow{r}')\phi_n(\overrightarrow{r}')d^3\overrightarrow{r}' = \delta_{nm},$$

Equation (19.272) becomes

$$\int G_1(\overrightarrow{r},\overrightarrow{r}',\tau_2,\tau_1)G_1(\overrightarrow{r}',\overrightarrow{r}'',\tau_1,\tau_0)d^3\overrightarrow{r}'$$

$$= \sum_m e^{-\lambda_m(\tau_2-\tau_1)}\phi_m(\overrightarrow{r})\phi_m^*(\overrightarrow{r}'')e^{-\lambda_m(\tau_1-\tau_0)}$$

$$= \sum_m \phi_m(\overrightarrow{r})\phi_m^*(\overrightarrow{r}'')e^{-\lambda_m(\tau_2-\tau_0)} \qquad\qquad (19.273)$$

$$= G_1(\overrightarrow{r},\overrightarrow{r}'',\tau_2,\tau_0). \qquad\qquad (19.274)$$

Using this in Equation (19.271) we obtain the propagator, $G_1(\overrightarrow{r},\overrightarrow{r}'',\tau_2,\tau_0)$, that takes us from τ_0 to τ_2 in a single step in terms of the propagators, that take us from τ_0 to τ_1 and from τ_1 to τ_2, respectively, as

$$\Psi(\overrightarrow{r},\tau_2) = \int G_1(\overrightarrow{r},\overrightarrow{r}'',\tau_2,\tau_0)\Psi(\overrightarrow{r}'',\tau_0)d^3\overrightarrow{r}'', \qquad (19.275)$$

$$G_1(\overrightarrow{r},\overrightarrow{r}'',\tau_2,\tau_0) = \int G_1(\overrightarrow{r},\overrightarrow{r}',\tau_2,\tau_1)G_1(\overrightarrow{r}',\overrightarrow{r}'',\tau_1,\tau_0)d^3\overrightarrow{r}',$$

19.2.4 Propagator for the Diffusion Equation with Periodic Boundary Conditions

As an important example of the first-order time-dependent equations, we now consider the diffusion or heat transfer equations, which are both in the form

$$\overrightarrow{\nabla}^2\Psi(\overrightarrow{x},\tau) = \frac{\partial\Psi(\overrightarrow{x},\tau)}{\partial\tau}. \qquad\qquad (19.276)$$

To simplify the problem we consider only one dimension with

$$-\frac{L}{2} \leq x \leq \frac{L}{2}$$

and use the periodic boundary conditions:

$$\Psi(-\frac{L}{2},\tau) = \Psi(\frac{L}{2},\tau). \tag{19.277}$$

Because the H operator for this problem is

$$H = -\frac{d^2}{dx^2}, \tag{19.278}$$

we easily write the eigenvalues and the eigenfunctions as

$$-\frac{d^2\phi_m}{dx^2} = \lambda_m\phi_m, \tag{19.279}$$

$$\phi_m(x) = \frac{1}{\sqrt{L}}e^{i\sqrt{\lambda_m}x},$$

$$\lambda_m = \left(\frac{2\pi m}{L}\right)^2, \qquad m = \pm \text{ integer.}$$

If we define

$$k_m = \frac{2\pi m}{L}, \tag{19.280}$$

we obtain $G_1(x,x',\tau)$ as

$$G_1(x,x',\tau) = \sum_{m=-\infty}^{\infty} \frac{1}{L}e^{ik_m(x-x')}e^{-k_m^2\tau}. \tag{19.281}$$

19.2.5 Propagator for the Diffusion Equation in the Continuum Limit

We now consider the continuum limit of the propagator [Eq. (19.281)]. Because the difference of two neighboring eigenvalues is

$$\triangle k_m = \frac{2\pi}{L}, \tag{19.282}$$

we can write $G_1(x,x',\tau)$ as

$$G_1(x,x',\tau) = \frac{1}{2\pi} \sum_{m=-\infty}^{\infty} \triangle k_m e^{ik_m(x-x')}e^{-k_m^2\tau}. \tag{19.283}$$

In the continuum limit, $L \to \infty$, the difference between two neighboring eigenvalues becomes infinitesimally small; thus we may replace the summation with an integral as

$$\lim_{L\to\infty} \sum_m \triangle k_m f(k_m) \to \int f(k)dk. \tag{19.284}$$

This gives us the propagator as

$$G_1(x, x', \tau) = \frac{1}{2\pi} \int_{-\infty}^{\infty} dk e^{ik(x-x')} e^{-k^2\tau}. \tag{19.285}$$

Completing the square:

$$ik(x - x') - k^2\tau = -\tau \left(k - \frac{i(x - x')}{2\tau} \right)^2 - \frac{(x - x')^2}{4\tau} \tag{19.286}$$

and defining

$$\delta = \frac{(x - x')}{2\tau},$$

we can write $G_1(x, x', \tau)$ as

$$G_1(x, x', \tau) = \frac{1}{2\pi} e^{-\frac{(x-x')^2}{4\tau}} \int_{-\infty}^{\infty} dk e^{-\tau(k-i\delta)^2}. \tag{19.287}$$

This integral can be taken easily, thus giving us the propagator as

$$G_1(x, x', \tau) = \frac{1}{\sqrt{4\pi\tau}} e^{-\frac{(x-x')^2}{4\tau}}. \tag{19.288}$$

Note that G_1 is symmetric with respect to x and x'. In the limit as $\tau \to 0$ it becomes

$$\lim_{\tau \to 0} G_1(x, x', \tau) = \lim_{\tau \to 0} \frac{1}{\sqrt{4\pi\tau}} e^{-\frac{(x-x')^2}{4\tau}} \tag{19.289}$$
$$= I(x, x'),$$

which is one of the definitions of the Dirac-delta function; hence

$$I(x, x') = \delta(x - x'). \tag{19.290}$$

Plotting Equation (19.288) we see that it is a Gaussian (Fig. 19.8).

Because the area under a Gaussian is constant, that is,

$$\int_{-\infty}^{\infty} G_1(x, x')dx = \frac{1}{\sqrt{4\pi\tau}} \int_{-\infty}^{\infty} e^{-\frac{(x-x')^2}{4\tau}} dx = 1, \tag{19.291}$$

the total amount of the diffusing material is conserved. Using $G_1(x, x', \tau)$ and given the initial concentration $\Psi(x', 0)$, we can find the concentration at subsequent times as

$$\Psi(x, \tau) = \int_{-\infty}^{\infty} G_1(x, x', \tau)\Psi(x', 0)dx'$$
$$= \frac{1}{\sqrt{4\pi\tau}} \int_{-\infty}^{\infty} e^{-\frac{(x-x')^2}{4\tau}} \Psi(x', 0)dx'. \tag{19.292}$$

Note that our solution satisfies the relation

$$\int_{-\infty}^{\infty} \Psi(x, \tau)dx = \int_{-\infty}^{\infty} \Psi(x', 0)dx'. \tag{19.293}$$

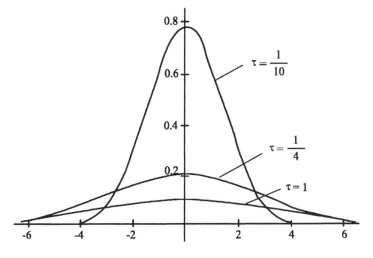

Fig. 19.8 Gaussian

19.2.6   Green's Functions in the Presence of Sources or Interactions

First-order time-dependent equations frequently appear with an inhomogeneous term:

$$H\Psi(\overrightarrow{r},\tau) + \frac{\partial \Psi(\overrightarrow{r},\tau)}{\partial \tau} = F(\overrightarrow{r},\tau), \qquad (19.294)$$

where $F(\overrightarrow{r},\tau)$ represents sources or interactions in the system; thus we need a Green's function which allows us to express the solution as

$$\Psi(\overrightarrow{r},\tau) = \Psi_0(\overrightarrow{r},\tau) + \int G(\overrightarrow{r}, \overrightarrow{r}',\tau,\tau')F(\overrightarrow{r}',\tau')d^3\overrightarrow{r}'d\tau', \qquad (19.295)$$

where $\Psi_0(\overrightarrow{r},\tau)$ represents the solution of the homogeneous part of Equation (19.294). We have seen that the propagator $G_1(\overrightarrow{r}, \overrightarrow{r}',\tau,\tau')$ satisfies the equation

$$\left(H + \frac{\partial}{\partial \tau}\right) G_1(\overrightarrow{r}, \overrightarrow{r}',\tau,\tau') = 0. \qquad (19.296)$$

However, the Green's function we need in Equation (19.295) satisfies

$$\left(H + \frac{\partial}{\partial \tau}\right) G(\overrightarrow{r}, \overrightarrow{r}',\tau,\tau') = \delta^3(\overrightarrow{r} - \overrightarrow{r}')\delta(\tau-\tau'). \qquad (19.297)$$

It is clear that even though $G_1(\overrightarrow{r}, \overrightarrow{r}',\tau,\tau')$ is not the Green's function, it is closely related to it. After all, except for the point $\overrightarrow{r} = \overrightarrow{r}'$ it satisfies the differential equation (19.297). Considering that $G_1(\overrightarrow{r}, \overrightarrow{r}',\tau,\tau')$ satisfies the relation

$$\lim_{\tau \to \tau'} G_1(\overrightarrow{r}, \overrightarrow{r}',\tau,\tau') = \delta^3(\overrightarrow{r} - \overrightarrow{r}'), \qquad (19.298)$$

we can expect to satisfy Equation (19.297) by introducing a discontinuity at $\tau = \tau'$. Let us start with

$$G(\overrightarrow{r}, \overrightarrow{r}',\tau,\tau') = G_1(\overrightarrow{r}, \overrightarrow{r}',\tau,\tau')\theta(\tau-\tau'), \qquad (19.299)$$

so that

$$\left\{ \begin{array}{ll} G = G_1, & \tau > \tau', \\[2mm] G = 0, & \tau < \tau'. \end{array} \right\} \qquad (19.300)$$

Substituting this in Equation (19.297), we get

$$\left(H + \frac{\partial}{\partial \tau}\right) G(\overrightarrow{r}, \overrightarrow{r}',\tau,\tau') \qquad (19.301)$$

$$= HG_1(\overrightarrow{r}, \overrightarrow{r}',\tau,\tau')\theta(\tau-\tau') + \frac{\partial}{\partial \tau}\left[G_1(\overrightarrow{r}, \overrightarrow{r}',\tau,\tau')\theta(\tau-\tau')\right],$$

$$= \theta(\tau-\tau')HG_1(\overrightarrow{r}, \overrightarrow{r}',\tau,\tau') + \theta(\tau-\tau')\frac{\partial}{\partial \tau}G_1(\overrightarrow{r}, \overrightarrow{r}',\tau,\tau')$$

$$+ G_1(\overrightarrow{r}, \overrightarrow{r}',\tau,\tau')\frac{\partial}{\partial \tau}\theta(\tau-\tau'),$$

$$= \theta(\tau-\tau')\left(H + \frac{\partial}{\partial \tau}\right) G_1(\overrightarrow{r}, \overrightarrow{r}',\tau,\tau') + G_1(\overrightarrow{r}, \overrightarrow{r}',\tau,\tau')\delta(\tau-\tau').$$

We have used the relation

$$\frac{d}{d\tau}\theta(\tau-\tau') = \delta(\tau-\tau'). \qquad (19.302)$$

Considering the fact that G_1 satisfies Equation (19.296), we obtain

$$\left(H + \frac{\partial}{\partial \tau}\right) G(\overrightarrow{r}, \overrightarrow{r}',\tau,\tau') = G_1(\overrightarrow{r}, \overrightarrow{r}',\tau,\tau')\delta(\tau-\tau'). \qquad (19.303)$$

Because the Dirac-delta function is zero except at $\tau=\tau'$, we only need the value of G_1 at $\tau = \tau'$, which is equal to $\delta^3(\overrightarrow{r} - \overrightarrow{r}')$; thus we can write Equation (19.303) as

$$\left(H + \frac{\partial}{\partial \tau}\right) G(\overrightarrow{r}, \overrightarrow{r}',\tau,\tau') = \delta^3(\overrightarrow{r} - \overrightarrow{r}')\delta(\tau-\tau'). \qquad (19.304)$$

From here we see that the Green's function for Equation (19.294) is

$$G(\overrightarrow{r}, \overrightarrow{r}',\tau,\tau') = G_1(\overrightarrow{r}, \overrightarrow{r}',\tau,\tau')\theta(\tau-\tau') \qquad (19.305)$$

and the general solution of Equation (19.294) is now written as

$$\Psi(\vec{r},\tau) = \Psi_0(\vec{r},\tau) + \int d^3\vec{r}' \int d\tau' G(\vec{r}, \vec{r}',\tau,\tau') F(\vec{r}',\tau')$$

$$= \Psi_0(\vec{r},\tau) + \int d^3\vec{r}' \int_{-\infty}^{\tau} d\tau' G_1(\vec{r}, \vec{r}',\tau,\tau') F(\vec{r}',\tau'). \quad (19.306)$$

19.2.7 Green's Function for the Schrödinger Equation for Free Particles

To write the Green's function for the Schrödinger equation for a free particle, we can use the similarity between the Schrödinger and the diffusion equations. Making the replacement $\tau \to \dfrac{i\hbar t}{2m}$ in Equation (19.288) gives us the propagator for a free particle as

$$G_1(x,x',t) = \frac{1}{\sqrt{2\pi i\hbar t/m}} e^{-m(x-x')^2/2i\hbar t}. \quad (19.307)$$

Now the solution of the Schrödinger equation

$$-\frac{\hbar^2}{2m}\frac{\partial^2}{\partial x^2}\Psi(x,t) = i\hbar\frac{\partial}{\partial t}\Psi(x,t), \quad (19.308)$$

with the initial condition $\Psi(x',0)$, can be written as

$$\Psi(x,t) = \int G_1(x,x',t)\Psi(x',0)dx'. \quad (19.309)$$

19.2.8 Green's Function for the Schrödinger Equation in the Presence of Interactions

When a particle is moving under the influence of a potential $V(x)$, the Schrödinger equation becomes

$$-\frac{\partial^2}{\partial x^2}\Psi(x,t) + \frac{2m}{i\hbar}\frac{\partial}{\partial t}\Psi(x,t) = -\frac{2m}{\hbar^2}V(x)\Psi(x,t), \quad (19.310)$$

For an arbitrary initial time t', Green's function is given as

$$G(x,x',t,t') = G_1(x,x',t,t')\theta(t-t') \quad (19.311)$$

and the solution becomes

$$\Psi(x,t) = \Psi_0(x,t) - \frac{i}{\hbar}\int dx' \int_{-\infty}^{t} dt' G(x,x',t,t')\, V(x')\Psi(x',t'). \quad (19.312)$$

19.2.9 Second-Order Time-Dependent Green's Functions

Most of the frequently encountered time-dependent equations with second-order time dependence can be written as

$$\left[H + \frac{\partial^2}{\partial \tau^2}\right] \Psi(\overrightarrow{r}, \tau) = 0, \tag{19.313}$$

where τ is a timelike variable and H is a linear differential operator independent of τ. We again assume that H has a complete set of orthonormal eigenfunctions satisfying

$$H\phi_n(\overrightarrow{r}) = \lambda_n \phi_n(\overrightarrow{r}), \tag{19.314}$$

where λ_n are the eigenvalues. We expand $\Psi(\overrightarrow{r}, \tau)$ in terms of this complete and orthonormal set as

$$\Psi(\overrightarrow{r}, \tau) = \sum_n A_n(\tau) \phi_n(\overrightarrow{r}), \tag{19.315}$$

where the coefficients $A_n(\tau)$ carry the τ dependence. Substituting this in Equation (19.313) we obtain

$$\sum_n \left[\ddot{A}_n(\tau) + \lambda_n A_n(\tau)\right] \phi_n(\overrightarrow{r}) = 0. \tag{19.316}$$

Because ϕ_n are linearly independent, we set the quantity inside the square brackets to zero to obtain the differential equation that the coefficients $A_n(\tau)$ satisfy as

$$\ddot{A}_n(\tau) + \lambda_n A_n(\tau) = 0. \tag{19.317}$$

The solution of this equation can be written immediately as

$$A_n(\tau) = a_n e^{i\sqrt{\lambda_n}\tau} + b_n e^{-i\sqrt{\lambda_n}\tau}. \tag{19.318}$$

Substituting these coefficients into Equation (19.315), we write $\Psi(\overrightarrow{r}, \tau)$ as

$$\Psi(\overrightarrow{r}, \tau) = \sum_n \left[a_n e^{i\sqrt{\lambda_n}\tau} + b_n e^{-i\sqrt{\lambda_n}\tau}\right] \phi_n(\overrightarrow{r}), \tag{19.319}$$

where the integration constants a_n and b_n are now to be determined from the initial conditions. Assuming that $\Psi(\overrightarrow{r}, 0)$ and $\dot{\Psi}(\overrightarrow{r}, 0)$ are given, we write

$$\Psi(\overrightarrow{r}, 0) = \sum_n [a_n + b_n] \phi_n(\overrightarrow{r}), \tag{19.320}$$

and

$$\dot{\Psi}(\overrightarrow{r}, 0) = i \sum_n \sqrt{\lambda_n} [a_n - b_n] \phi_n(\overrightarrow{r}). \tag{19.321}$$

Using the orthogonality relation [Eq. (19.256)] of $\phi_n(\vec{r})$ we obtain two relations between a_n and b_n as

$$[a_n + b_n] = \int \phi_n^*(\vec{r}')\Psi(\vec{r}',0)d^3\vec{r}' \tag{19.322}$$

and

$$[a_n - b_n] = \frac{-i}{\sqrt{\lambda_n}} \int \phi_n^*(\vec{r}')\dot{\Psi}(\vec{r}',0)d^3\vec{r}'. \tag{19.323}$$

Solving these for a_n and b_n we obtain the coefficients as

$$a_n = \frac{1}{2}\left[\int \phi_n^*(\vec{r}')\Psi(\vec{r}',0)d^3\vec{r}' \right. \tag{19.324}$$
$$\left. +\frac{1}{i\sqrt{\lambda_n}} \int \phi_n^*(\vec{r}')\dot{\Psi}(\vec{r}',0)d^3\vec{r}'\right]$$

and

$$b_n = \frac{1}{2}\left[\int \phi_n^*(\vec{r}')\Psi(\vec{r}',0)d^3\vec{r}' \right. \tag{19.325}$$
$$\left. -\frac{1}{i\sqrt{\lambda_n}} \int \phi_n^*(\vec{r}')\dot{\Psi}(\vec{r}',0)d^3\vec{r}'\right].$$

Substituting these back into $\Psi(\vec{r},\tau)$ gives us

$$\Psi(\vec{r},\tau) = \int \sum_n \cos\left(\sqrt{\lambda_n}\tau\right)\phi_n(\vec{r})\phi_n^*(\vec{r}')\Psi(\vec{r}',0)d^3\vec{r}' \tag{19.326}$$
$$+ \int \sum_n \sin\left(\sqrt{\lambda_n}\tau\right)\frac{1}{\sqrt{\lambda_n}}\phi_n(\vec{r})\phi_n^*(\vec{r}')\dot{\Psi}(\vec{r}',0)d^3\vec{r}'.$$

We write this as

$$\Psi(\vec{r},\tau) = \int G_2(\vec{r}, \vec{r}',\tau)\Psi(\vec{r}',0)d^3\vec{r}' + \int \tilde{G}_2(\vec{r}, \vec{r}',\tau)\dot{\Psi}(\vec{r}',0)d^3\vec{r}' \tag{19.327}$$

and define two functions G_2 and $\tilde{G}_2$ as

$$G_2(\vec{r}, \vec{r}',\tau) = \sum_n \cos\left(\sqrt{\lambda_n}\tau\right)\phi_n(\vec{r})\phi_n^*(\vec{r}') \tag{19.328}$$

and

$$\tilde{G}_2(\vec{r}, \vec{r}',\tau) = \sum_n \frac{\sin\left(\sqrt{\lambda_n}\tau\right)}{\sqrt{\lambda_n}}\phi_n(\vec{r})\phi_n^*(\vec{r}'). \tag{19.329}$$

Among these functions G_2 acts on $\Psi(\vec{r}',0)$ and $\tilde{G}_2$ acts on $\dot{\Psi}(\vec{r}',0)$. They both satisfy homogeneous equation

$$\left[H + \frac{\partial^2}{\partial\tau^2}\right] \left\{ \begin{array}{c} G_2(\vec{r}, \vec{r}',\tau) \\ \\ \tilde{G}_2(\vec{r}, \vec{r}',\tau) \end{array} \right\} = 0. \qquad (19.330)$$

Thus $\Psi(\vec{r},\tau)$ is a solution of the differential equation (19.313). Note G_2 and $\tilde{G}_2$ are related to each other by

$$G_2(\vec{r}, \vec{r}',\tau) = \frac{d}{d\tau}\tilde{G}_2(\vec{r}, \vec{r}',\tau). \qquad (19.331)$$

Hence we can obtain $G_2(\vec{r}, \vec{r}',\tau)$ from $\tilde{G}_2(\vec{r}, \vec{r}',\tau)$ by differentiation with respect to τ. Using Equation (19.328) and the completeness relation we can write

$$G_2(\vec{r}, \vec{r}',0) = \sum_n \phi_n(\vec{r})\phi_n^*(\vec{r}') = \delta^3(\vec{r} - \vec{r}'). \qquad (19.332)$$

Using the completeness relation (19.257) in Equation (19.326), one can easily check that $\Psi(\vec{r}, \tau)$ satisfies the initial conditions.

For an arbitrary initial time τ', $\Psi(\vec{r},\tau)$, $\tilde{G}_2$ and G_2 are written as

$$\Psi(\vec{r},\tau) = \int G_2(\vec{r}, \vec{r}',\tau, \tau')\Psi(\vec{r}',\tau')d^3\vec{r}'$$

$$+ \int \tilde{G}_2(\vec{r}, \vec{r}',\tau, \tau')\dot{\Psi}(\vec{r}',\tau')d^3\vec{r}', \qquad (19.333)$$

$$\tilde{G}_2(\vec{r}, \vec{r}',\tau, \tau') = \sum_n \frac{\sin\left[\sqrt{\lambda_n}\,(\tau - \tau')\right]}{\sqrt{\lambda_n}}\phi_n(\vec{r})\phi_n^*(\vec{r}'),$$

and

$$G_2(\vec{r}, \vec{r}',\tau, \tau') = \frac{d}{dt}\tilde{G}_2(\vec{r}, \vec{r}',\tau, \tau') \qquad (19.334)$$

$$= \sum_n \cos\left[\sqrt{\lambda_n}\,(\tau - \tau')\right]\phi_n(\vec{r})\phi_n^*(\vec{r}').$$

19.2.10   Propagators for the Scalar Wave Equation

An important example of the second-order time-dependent equations is the scalar wave equation:

$$\Box\Psi(\vec{r},t) = 0, \qquad (19.335)$$

where the wave (d'Alembert) operator is defined as

$$\Box = -\overrightarrow{\nabla}^2 + \frac{1}{c^2}\frac{\partial^2}{\partial t^2}.$$

Comparing with Equation (19.313) we have

$$\left\{ \begin{array}{c} H = -\overrightarrow{\nabla}^2 \\ \\ \tau = ct \end{array} \right\}.$$

Considering a rectangular region with the dimensions (L_1, L_2, L_3) and using periodic boundary conditions, eigenfunctions and the eigenvalues of the H operator are written as

$$H\phi_{n_1,n_2,n_3} = \lambda_{n_1,n_2,n_3}\phi_{n_1,n_2,n_3},$$

$$\phi_{n_1,n_2,n_3} = \frac{1}{\sqrt{L_1 L_2 L_3}}e^{ik_x x}e^{ik_y y}e^{ik_z z} \qquad (19.336)$$

and

$$k_x = \frac{2\pi n_1}{L_1}, \quad k_y = \frac{2\pi n_2}{L_2} \quad k_z = \frac{2\pi n_3}{L_3}; \quad n_i = \pm\text{integer and} \neq 0. \quad (19.337)$$

Eigenvalues satisfy the relation

$$\lambda_{n_1,n_2,n_3} = k_x^2 + k_y^2 + k_z^2. \qquad (19.338)$$

Using these eigenfunctions we can construct $\tilde{G}_2(\overrightarrow{r}, \overrightarrow{r}',\tau)$ as

$$\tilde{G}_2(\overrightarrow{r}, \overrightarrow{r}',\tau) \qquad (19.339)$$

$$= \frac{1}{L_1 L_2 L_3}\sum_{n_1,n_2,n_3}^{\infty} \frac{\sin\left(\sqrt{k_x^2 + k_y^2 + k_z^2}\,\tau\right)}{\sqrt{k_x^2 + k_y^2 + k_z^2}}e^{ik_x(x-x')}e^{ik_y(y-y')}e^{ik_z(z-z')}.$$

We now consider the continuum limit, where we make the replacements

$$\lim_{L_1\to\infty}\frac{1}{L_1}\sum_{n_1=-\infty}^{\infty} \to \frac{1}{2\pi}\int_{-\infty}^{\infty}dk_x \qquad (19.340)$$

$$\lim_{L_2\to\infty}\frac{1}{L_2}\sum_{n_2=-\infty}^{\infty} \to \frac{1}{2\pi}\int_{-\infty}^{\infty}dk_y \qquad (19.341)$$

$$\lim_{L_3\to\infty}\frac{1}{L_3}\sum_{n_3=-\infty}^{\infty} \to \frac{1}{2\pi}\int_{-\infty}^{\infty}dk_z. \qquad (19.342)$$

Thus $\tilde{G}_2(\vec{r}, \vec{r}', \tau)$ becomes

$$\tilde{G}_2(\vec{r}, \vec{r}', \tau) = \frac{1}{(2\pi)^3} \int_{-\infty}^{\infty} dk_x \int_{-\infty}^{\infty} dk_y \int_{-\infty}^{\infty} dk_z \frac{\sin k\tau}{k} e^{i\vec{k}\cdot\vec{\rho}}, \quad (19.343)$$

where

$$\vec{\rho} = (\vec{r} - \vec{r}'). \quad (19.344)$$

Defining a wave vector $\vec{k} = (k_x, k_y, k_z)$, and using polar coordinates we can write

$$\tilde{G}_2(\vec{r}, \vec{r}', \tau) = \frac{1}{(2\pi)^3} \int_0^\infty dk k^2 \frac{\sin k\tau}{k} \int_0^{2\pi} d\phi_k \int_{-1}^1 d(\cos\theta_k) e^{i\vec{k}\cdot\vec{\rho}}. \quad (19.345)$$

Choosing the direction of the $\vec{\rho}$ vector along the z-axis we write

$$\vec{k} \cdot \vec{\rho} = k\rho\cos\theta_k \quad (19.346)$$

and define x as

$$x = \cos\theta_k. \quad (19.347)$$

After taking the θ_k and ϕ_k integrals $\tilde{G}_2(\vec{r}, \vec{r}', \tau)$ becomes

$$\begin{aligned}
\tilde{G}_2(\vec{r}, \vec{r}', \tau) &= \frac{2\pi}{(2\pi)^3} \int_0^\infty dk k^2 \frac{\sin k\tau}{k} \int_{-1}^1 d(\cos\theta_k) e^{ik\rho\cos\theta_k} \\
&= \frac{1}{2\pi^2\rho} \int_0^\infty dk \sin k\tau \cdot \sin k\rho \\
&= \frac{1}{4\pi^2\rho} \int_{-\infty}^\infty dk \sin k\tau \cdot \sin k\rho \qquad (19.348) \\
&= \frac{1}{8\pi^2\rho} \int_{-\infty}^\infty dk[\cos k(\rho - \tau) - \cos k(\rho + \tau)].
\end{aligned}$$

Using one of the definitions of the Dirac-delta function, that is,

$$\delta(x - x') = \frac{1}{2\pi} \int_{-\infty}^\infty dk e^{ik(x-x')} = \frac{1}{2\pi} \int_{-\infty}^\infty dk \cos k(x - x'), \quad (19.349)$$

we can write $\tilde{G}_2(\vec{r}, \vec{r}', \tau)$ as

$$\tilde{G}_2(\vec{r}, \vec{r}', \tau) = \frac{1}{4\pi\rho}[\delta(\rho - \tau) - \delta(\rho + \tau)]. \quad (19.350)$$

Going back to our original variables, $\tilde{G}_2(\vec{r}, \vec{r}', t)$ becomes

$$\tilde{G}_2(\overrightarrow{r}, \overrightarrow{r}',t) = \frac{1}{4\pi|\overrightarrow{r} - \overrightarrow{r}'|} [\delta(|\overrightarrow{r} - \overrightarrow{r}'| - ct) - \delta(|\overrightarrow{r} - \overrightarrow{r}'| + ct)].$$

$$(19.351)$$

We write this for an arbitrary initial time t' to obtain the final form of the propagator as

$$\tilde{G}_2(\overrightarrow{r}, \overrightarrow{r}',t,t') \qquad (19.352)$$

$$= \frac{1}{4\pi|\overrightarrow{r} - \overrightarrow{r}'|} [\delta(|\overrightarrow{r} - \overrightarrow{r}'| - c(t - t')) - \delta(|\overrightarrow{r} - \overrightarrow{r}'| + c(t - t'))].$$

19.2.11   Advanced and Retarded Green's Functions

In the presence of a source, $\rho(\overrightarrow{r}, \tau)$, Equation (19.313) becomes

$$H\Psi(\overrightarrow{r}, \tau) + \frac{\partial^2 \Psi(\overrightarrow{r}, \tau)}{\partial \tau^2} = \rho(\overrightarrow{r}, \tau). \qquad (19.353)$$

To solve this equation we need a Green's function satisfying the equation

$$(H + \frac{\partial^2}{\partial \tau^2}) G(\overrightarrow{r}, \overrightarrow{r}', \tau, \tau') = \delta^3(\overrightarrow{r} - \overrightarrow{r}')\delta(\tau - \tau'). \qquad (19.354)$$

However, the propagators $G_2(\overrightarrow{r}, \overrightarrow{r}', \tau, \tau')$ and $\tilde{G}_2(\overrightarrow{r}, \overrightarrow{r}', \tau, \tau')$ both satisfy

$$\left[H + \frac{\partial^2}{\partial \tau^2}\right] \begin{pmatrix} G_2 \\ \tilde{G}_2 \end{pmatrix} = 0. \qquad (19.355)$$

Guided by our experience in G_1, to find the Green's function we start by introducing a discontinuity in either G_2 or $\tilde{G}_2$ as

$$G_R(\overrightarrow{r}, \overrightarrow{r}', \tau, \tau') = G_\zeta(\overrightarrow{r}, \overrightarrow{r}', \tau, \tau')\theta(\tau - \tau'). \qquad (19.356)$$

G_ζ stands for G_2 or $\tilde{G}_2$, while the subscript R will be explained later. Operating on $G_R(\overrightarrow{r}, \overrightarrow{r}', \tau, \tau')$ with $\left[H + \frac{\partial^2}{\partial \tau^2}\right]$ we get

$$\left[H + \frac{\partial^2}{\partial \tau^2}\right] G_R(\overrightarrow{r}, \overrightarrow{r}', \tau, \tau') \qquad (19.357)$$

$$= \theta(\tau - \tau') HG_\zeta + \frac{\partial^2}{\partial \tau^2} [G_\zeta(\overrightarrow{r}, \overrightarrow{r}', \tau, \tau')\theta(\tau - \tau')]$$

$$= \theta(\tau - \tau') \left[H + \frac{\partial^2}{\partial \tau^2}\right] G_\zeta + 2\left[\frac{\partial}{\partial \tau}\theta(\tau - \tau')\right] \frac{\partial}{\partial \tau}G_\zeta$$

$$+ G_\zeta \frac{\partial^2}{\partial \tau^2}\theta(\tau - \tau').$$

Since $G_2(\overrightarrow{r}, \overrightarrow{r}',\tau, \tau')$ and $\tilde{G}_2(\overrightarrow{r}, \overrightarrow{r}',\tau, \tau')$ both satisfy

$$\left[H + \frac{\partial^2}{\partial \tau^2}\right] G_\zeta = 0, \tag{19.358}$$

this becomes

$$\left[H + \frac{\partial^2}{\partial \tau^2}\right] G_R = 2 \left[\frac{\partial}{\partial \tau} \theta\left(\tau - \tau'\right)\right] \frac{\partial}{\partial \tau} G_\zeta + G_\zeta \frac{\partial^2}{\partial \tau^2} \theta\left(\tau - \tau'\right). \tag{19.359}$$

Using the fact that the derivative of a step function is a Dirac-delta function,

$$\frac{\partial}{\partial \tau} \theta\left(\tau - \tau'\right) = \delta\left(\tau - \tau'\right), \tag{19.360}$$

we can write

$$\left[H + \frac{\partial^2}{\partial \tau^2}\right] G_R(\overrightarrow{r}, \overrightarrow{r}',\tau, \tau') \tag{19.361}$$

$$= 2\delta\left(\tau - \tau'\right) \frac{\partial}{\partial \tau} G_\zeta(\overrightarrow{r}, \overrightarrow{r}',\tau, \tau') + \left[\frac{\partial}{\partial \tau} \delta\left(\tau - \tau'\right)\right] G_\zeta(\overrightarrow{r}, \overrightarrow{r}',\tau, \tau').$$

Using the following properties of the Dirac-delta function :

$$\delta\left(\tau - \tau'\right) \frac{\partial}{\partial \tau} G_\zeta(\overrightarrow{r}, \overrightarrow{r}',\tau, \tau') \tag{19.362}$$

$$= \left[\frac{\partial}{\partial \tau} G_\zeta(\overrightarrow{r}, \overrightarrow{r}',\tau, \tau')\right]_{\tau = \tau'} \delta\left(\tau - \tau'\right),$$

and

$$\left[\frac{\partial}{\partial \tau} \delta\left(\tau - \tau'\right)\right] G_\zeta(\overrightarrow{r}, \overrightarrow{r}',\tau, \tau') \tag{19.363}$$

$$= -\left[\frac{\partial}{\partial \tau} G_\zeta(\overrightarrow{r}, \overrightarrow{r}',\tau, \tau')\right]_{\tau = \tau'} \delta\left(\tau - \tau'\right)$$

we can write Equation (19.361) as

$$\left[H + \frac{\partial^2}{\partial \tau^2}\right] G_R(\overrightarrow{r}, \overrightarrow{r}',\tau, \tau') = \delta\left(\tau - \tau'\right) \left[\frac{\partial}{\partial \tau} G_\zeta(\overrightarrow{r}, \overrightarrow{r}',\tau, \tau')\right]_{\tau = \tau'}. \tag{19.364}$$

If we take G_2 as G_ζ, the right-hand side becomes zero; thus it is not useful for our purposes. However, taking $\tilde{G}_2$ we find

$$\left[H + \frac{\partial^2}{\partial \tau^2}\right] G_R(\overrightarrow{r}, \overrightarrow{r}',\tau, \tau') = \delta\left(\tau - \tau'\right) \delta^3\left(\overrightarrow{r} - \overrightarrow{r}'\right), \tag{19.365}$$

which means that the Green's function we need is

$$G_R(\vec{r}, \vec{r}',\tau, \tau') = \tilde{G}_2(\vec{r}, \vec{r}',\tau, \tau')\theta\left(\tau - \tau'\right). \tag{19.366}$$

The general solution can now be expressed as

$$\Psi_R(\vec{r},\tau) = \Psi_0(\vec{r},\tau) + \int d^3\vec{r}' \int_{-\infty}^{\tau} d\tau' \tilde{G}_2(\vec{r}, \vec{r}',\tau, \tau')\rho(\vec{r}',\tau'). \tag{19.367}$$

There is also another choice for the Green's function, which is given as

$$G_A(\vec{r}, \vec{r}',\tau, \tau') = -\tilde{G}_2(\vec{r}, \vec{r}',\tau, \tau')\theta\left(\tau' - \tau\right). \tag{19.368}$$

Following similar steps:

$$\left[H + \frac{\partial^2}{\partial\tau^2}\right] G_A$$

$$= -\theta\left(\tau' - \tau\right) H\tilde{G}_2 - \frac{\partial^2}{\partial\tau^2}\left[\tilde{G}_2(\vec{r}, \vec{r}',\tau, \tau')\theta\left(\tau' - \tau\right)\right]$$

$$= -\theta\left(\tau' - \tau\right)\left[H + \frac{\partial^2}{\partial\tau^2}\right]\tilde{G}_2 - 2\frac{\partial}{\partial\tau}\theta\left(\tau' - \tau\right)\frac{\partial}{\partial\tau}\tilde{G}_2 - \tilde{G}_2\frac{\partial^2}{\partial\tau^2}\theta\left(\tau' - \tau\right)$$

$$= -\left[\frac{\partial}{\partial\tau}\tilde{G}_2\right]\frac{\partial}{\partial\tau}\theta\left(\tau' - \tau\right) = \left[\frac{\partial}{\partial\tau}\tilde{G}_2\right]\frac{\partial}{\partial\tau'}\theta\left(\tau' - \tau\right)$$

$$= \left[\frac{\partial}{\partial\tau}\tilde{G}_2(\vec{r}, \vec{r}',\tau, \tau')\right]\delta\left(\tau' - \tau\right)$$

$$= \left[\frac{\partial}{\partial\tau}\tilde{G}_2(\vec{r}, \vec{r}',\tau, \tau')\right]_{\tau'=\tau}\delta\left(\tau' - \tau\right),$$

we see that $G_A(\vec{r}, \vec{r}',\tau, \tau')$ also satisfies the defining equation for the Green's function as

$$\left[H + \frac{\partial^2}{\partial\tau^2}\right] G_A(\vec{r}, \vec{r}',\tau, \tau') = \delta\left(\tau - \tau'\right)\delta^3\left(\vec{r} - \vec{r}'\right). \tag{19.369}$$

Now the general solution of Equation (19.353) can be written as

$$\Psi_A(\vec{r},\tau) = \Psi_0(\vec{r},\tau) - \int d^3\vec{r}' \int_{\tau}^{\infty} d\tau' \tilde{G}_2(\vec{r}, \vec{r}',\tau, \tau')\rho(\vec{r}',\tau'). \tag{19.370}$$

From Equation (19.367) it is seen the solution $\Psi_R(\vec{r},\tau)$ is determined by the past behavior of the source, that is, with source times $\tau' < \tau$, while $\Psi_A(\vec{r},\tau)$ is determined by the behavior of the source in the future, that is, with source times $\tau' > \tau$. We borrowed the subscripts from relativity, where R and A stand for the "retarded" and the "advanced" solutions, respectively. These terms acquire their true meaning with the relativistic wave equation discussed in the next section.

19.2.12   Advanced and Retarded Green's Functions for the Scalar Wave Equation

In the presence of sources or sinks the scalar wave equation is given as

$$\vec{\nabla}^2 \Psi(\vec{r},t) - \frac{1}{c^2} \frac{\partial^2 \Psi(\vec{r},t)}{\partial t^2} = \rho(\vec{r},t). \qquad (19.371)$$

We have already found the propagator $\tilde{G}_2$ for the scalar wave equation as

$$\tilde{G}_2(\vec{r}, \vec{r}',t,t') \qquad (19.372)$$

$$= \frac{1}{4\pi |\vec{r} - \vec{r}'|} [\delta (|\vec{r} - \vec{r}'| - c(t - t')) - \delta (|\vec{r} - \vec{r}'| + c(t - t'))].$$

Using Equation (19.366) we now write the Green's function for $t > t'$ as

$$G_R (\vec{r}, \vec{r}',t,t') = \frac{[\delta (|\vec{r} - \vec{r}'| - c(t - t')) - \delta (|\vec{r} - \vec{r}'| + c(t - t'))]}{4\pi |\vec{r} - \vec{r}'|} \theta(t - t').$$

For $t < t'$ the Green's function is

$$G_R = 0. \qquad (19.373)$$

For $t > t'$ the argument of the second Dirac-delta function never vanishes; thus the Green's function becomes

$$G_R(\vec{r}, \vec{r}',t,t') = \frac{1}{4\pi |\vec{r} - \vec{r}'|} \delta [|\vec{r} - \vec{r}'| - c(t - t')]. \qquad (19.374)$$

Now the general solution with this Green's function is expressed as

$$\Psi_R(\vec{r},t) = \Psi_0(\vec{r},t) + \frac{1}{4\pi} \int d^3 \vec{r}' \int_{-\infty}^{\infty} dt' \frac{\delta [|\vec{r} - \vec{r}'| - c(t - t')]}{|\vec{r} - \vec{r}'|} \rho(\vec{r}',t'), \qquad (19.375)$$

where $\Psi_0(\vec{r},\tau)$ is the solution of the homogeneous equation. Taking the t' integral we find

$$\Psi_R(\vec{r},t) = \Psi_0(\vec{r},t) + \frac{1}{4\pi} \int d^3 \vec{r}' \frac{[\rho(\vec{r}',t')]_R}{|\vec{r} - \vec{r}'|}, \qquad (19.376)$$

where

$$[\rho(\vec{r}',t')]_R \qquad (19.377)$$

means that the solution Ψ_R at $(\vec{r},t)$ is found by using the values of the source $\rho(\vec{r}',t')$ evaluated at the retarded times:

$$t' = t - \frac{|\vec{r} - \vec{r}'|}{c}. \qquad (19.378)$$

We show the source at the retarded times as

$$[\rho(\overrightarrow{r}', t')]_R = \rho(\overrightarrow{r}', t - \frac{|\overrightarrow{r} - \overrightarrow{r}'|}{c}), \qquad (19.379)$$

and the solution found by using $[\rho(\overrightarrow{r}', t')]_R$ is shown as $\Psi_R(\overrightarrow{r}, t)$. The physical interpretation of this solution is that whatever happens at the source point shows its effect at the field point later by the amount of time that signals (light) take to travel from the source to the field point. In other words, causes precede their effects.

Retarded solutions are of basic importance in electrodynamics, where the scalar potential $\Phi(\overrightarrow{r}, t)$, and the vector potential $\overrightarrow{A}(\overrightarrow{r}, t)$ satisfy the equations

$$\left[\overrightarrow{\nabla}^2 - \frac{1}{c^2}\frac{\partial^2}{\partial t^2}\right]\Phi(\overrightarrow{r}, t) = -4\pi\rho(\overrightarrow{r}, t), \qquad (19.380)$$

$$\left[\overrightarrow{\nabla}^2 - \frac{1}{c^2}\frac{\partial^2}{\partial t^2}\right]\overrightarrow{A}(\overrightarrow{r}, t) = -\frac{4\pi}{c}\overrightarrow{J}(\overrightarrow{r}, t), \qquad (19.381)$$

where $\rho(\overrightarrow{r}, t)$ and $\overrightarrow{J}(\overrightarrow{r}, t)$ stand for the charge and the current densities, respectively.

In search of a Green's function for Equation (19.371) we have added a discontinuity to $\tilde{G}_2$ as $\tilde{G}_2(\overrightarrow{r}, \overrightarrow{r}', \tau, \tau')\theta(\tau - \tau')$. However, there is also another alternative, where we take

$$G_A(\overrightarrow{r}, \overrightarrow{r}', \tau, \tau') = -\tilde{G}_2(\overrightarrow{r}, \overrightarrow{r}', \tau, \tau')\theta(\tau' - \tau). \qquad (19.382)$$

Solution of the wave equation with this Green's function is now given as

$$\Psi_A(\overrightarrow{r}, t) = \Psi_0(\overrightarrow{r}, t) + \frac{1}{4\pi}\int d^3\overrightarrow{r}' \frac{[\rho(\overrightarrow{r}', t')]_A}{|\overrightarrow{r} - \overrightarrow{r}'|}. \qquad (19.383)$$

In this solution A stands for advanced times, that is, $t' = t + |\overrightarrow{r} - \overrightarrow{r}'|/c$. In other words, whatever "happens" at the source point shows its effect at the field point before its happening by the amount of time $|\overrightarrow{r} - \overrightarrow{r}'|/c$, which is again equal to the amount of time that light takes to travel from the source to the field point. In summary, in advanced solutions effects precede their causes.

We conclude this section by saying that the wave equation (19.371) is covariant with c standing for the speed of light; hence the two solutions $\Psi_R(\overrightarrow{r}, t)$, and $\Psi_A(\overrightarrow{r}, t)$ are both legitimate solutions of the relativistic wave equation. Thus the general solution is in principle their linear combination:

$$\Psi(\overrightarrow{r}, t) = c_1\Psi_A(\overrightarrow{r}, t) + c_2\Psi_R(\overrightarrow{r}, t). \qquad (19.384)$$

However, Because we have no evidence of a case where causes precede their effects, as a boundary condition we set c_2 to zero, and take the retarded solution as the physically meaningful solution. This is also called the **principle of causality**.

Problems

19.1   Given the Bessel equation

$$\frac{d}{dx}\left[x\frac{dy}{dx}\right] + \left(kx - \frac{m^2}{x}\right)y(x) = 0$$

and its general solution

$$y(x) = A_0 J_m(x) + B_0 N_m(x),$$

find the Green's function satisfying the boundary conditions

$$y(0) = 0 \text{ and } y'(a) = 0.$$

19.2   For the operator $\mathcal{L} = d^2/dx^2$ and with the boundary conditions $y(0) = y(L) = 0$ we have found the Green's function as

$$G(x, x') = \left[(x - x')\theta(x - x') - \frac{x}{L}(L - x')\right].$$

Show that the trigonometric Fourier expansion of this is

$$G(x, x') = -\frac{2}{L}\sum_n \frac{\sin k_n x \sin k_n x'}{k_n^2}.$$

19.3   Show that the Green's function for the differential operator

$$\mathcal{L} = \frac{d^2}{dx^2} + k_0^2,$$

with the boundary conditions

$$y(0) = 0 \text{ and } y(L) = 0.$$

is given as

$$G(x, x') = \frac{1}{k_0 \sin k_0 L}\left\{\begin{array}{ll} \sin k_0 x \sin k_0(x' - L) & x < x' \\ \sin k_0 x' \sin k_0(x - L) & x > x' \end{array}\right\}.$$

Show that this is equivalent to the eigenvalue expansion

$$G(x, x') = \frac{2}{L}\sum_{n=1}^{\infty} \frac{\sin\dfrac{n\pi x}{L}\sin\dfrac{n\pi x'}{L}}{k_0^2 - (n\pi/L)^2}.$$

19.4   Single-point boundary condition: Consider the differential equation $\mathcal{L}y(x) = \phi(x)$, where $\mathcal{L}$ is the Sturm-Liouville operator

$$\mathcal{L} = \frac{d}{dx}\left(p(x)\frac{d}{dx}\right) + q(x).$$

Construct the Green's function satisfying the single-point boundary conditions $y(x_0) = y_0$ and $y'(x_0) = y_0'$.

Hint: First write the Green's function as

$$G(x, x') = Ay_1(x) + By_2(x), \quad x > x',$$

$$G(x, x') = Cy_1(x) + Dy_2(x), \quad x < x',$$

where $y_1(x)$ and $y_2(x)$ are two linearly independent solutions of $\mathcal{L}y(x) = 0$. Because the Green's function is continuous at $x = x'$ and its derivative has a discontinuity of magnitude $1/p(x)$ at $x = x'$, find the constants A, B, C, and D, thus obtaining the Green's function as

$$G(x, x') = Cy_1(x) + Dy_2(x) - \frac{[y_1(x)y_2(x') - y_2(x)y_1(x')]}{p(x')W[y_1(x'), y_2(x')]}, \quad x > x',$$
$$G(x, x') = Cy_1(x) + Dy_2(x), \quad x < x',$$

where $W[y_1(x), y_2(x)]$ is the Wronskian defined as $W[y_1, y_2] = y_1y_2' - y_2y_1'$. Now impose the single-point boundary conditions

$$G(x_0, x') = 0 \text{ and } G'(x_0, x') = 0$$

to show that $C = D = 0$. Finally show that the differential equation

$$\mathcal{L}y(x) = \phi(x)$$

with the single-point boundary conditions $y(x_0) = y_0$ and $y'(x_0) = y_0'$ is equivalent to the integral equation

$$y(x) = C_1y_1(x) + C_2y_2(x) + \int_{x_0}^{x} G(x, x')\phi(x')dx'.$$

19.5   Consider the differential operator

$$\mathcal{L} = \frac{d^2}{dt^2} + \omega_0^2.$$

For the single-point boundary conditions

$$x(0) = x_0 \text{ and } \dot{x}(0) = 0.$$

Show that the Green's function is given as

$$G(t, t') = \frac{\sin\omega_0(t - t')}{\omega_0}\theta(t - t')$$

and write the solution for $\ddot{x}(t) + \omega_0^2 x^2(t) = F(t)$.

19.6   Find a Green's function for the Sturm-Liouville operator

$$\mathcal{L} = a_3(x)\frac{d^3}{dx^3} + a_2(x)\frac{d^2}{dx^2} + a_1(x)\frac{d}{dx} + a_0(x),$$

satisfying the boundary conditions

$$G(x, x')|_{x=a} = 0, \qquad \frac{dG(x, x')}{dx}\Big|_{x=a} = 0, \qquad \frac{d^2G(x, x')}{dx^2}\Big|_{x=a} = 0,$$

in the interval $[a, b]$.

19.7   Find the Green's function for the differential equation

$$\frac{d^4y}{dx^4} = \phi(x, y),$$

with the boundary conditions

$$y(0) = y'(0) = y(1) = y'(1) = 0.$$

19.8   For the scalar wave equation

$$\vec{\nabla}^2 \Psi(\vec{r}, t) - \frac{1}{c^2}\frac{\partial^2 \Psi(\vec{r}, t)}{\partial t^2} = \rho(\vec{r}, t),$$

take the Green's function as

$$G_A(\vec{r}, \vec{r}', \tau, \tau') = -\tilde{G}_2(\vec{r}, \vec{r}', \tau, \tau')\theta(\tau' - \tau)$$

and show that the solution is given as

$$\Psi_A(\vec{r}, t) = \Psi_0(\vec{r}, t) + \frac{1}{4\pi}\int d^3\vec{r}' \frac{[\rho(\vec{r}', t')]_A}{|\vec{r} - \vec{r}'|}.$$

What does $[\rho(\vec{r}', t')]_A$ stand for? Discuss your answer. (Read Chapter 28 of *The Feynman Lectures on Physics*.)

19.9   Consider the partial differential equation

$$\left(\frac{\partial^2}{\partial x^2} - \frac{\partial}{\partial t}\right) y(x, t) = 0$$

with the boundary conditions

$$y(0, t) = 0, \quad y(L, t) = y_0.$$

If $y(x, 0)$ represents the initial solution, find the solution at subsequent times.

19.10   Using the Green's function technique solve the differential equation

$$\mathcal{L}y(x) = -\lambda xy(x), \quad x \in [0, L],$$

where

$$\mathcal{L}y(x) = \left[\frac{d}{dx}(x\frac{d}{dx}) - \frac{n^2}{x}\right]y(x), \quad n = \text{constant},$$

with the boundary conditions

$$y(0) = 0 \text{ and } y(L) = 0.$$

What is the solution of $\mathcal{L}y = -\lambda x^n$ with the above boundary conditions?

19.11   Find the Green's function for the problem

$$\mathcal{L}y(x) = F(x), \quad x \in [0, L],$$

where

$$\mathcal{L} = \frac{d}{dx}(x\frac{d}{dx}).$$

Use the boundary conditions

$$y(0) = \text{finite and } y(L) = 0.$$

Write Green's theorem [Eq. (19.162)] in one dimension. Does the surface term in (19.164) vanish?

19.12   Given the differential equation

$$y''(t) - 3y'(t) + 2y(t) = 2e^{-t}$$

and the boundary conditions $y(0) = 2$, $y'(0) = -1$.

i) Defining the operator in (19.1) as $\mathcal{L} = \dfrac{d^2}{dx^2} - 3\dfrac{d}{dx} + 2$ find the solution by using the Green's function method.

ii) Confirm your answer by solving the above problem using the Laplace transform technique.

iii) Using a different definition for $\mathcal{L}$ show that you get the same answer.

19.13   Consider the wave equation

$$\frac{\partial^2 y}{\partial x^2} - \frac{1}{c^2}\frac{\partial^2 y}{\partial t^2} = F(x, t) \text{ with } y(0, t) = y(L, t) = 0.$$

Find the Green's functions satisfying

$$\frac{\partial^2 G}{\partial x^2} - \frac{1}{c^2}\frac{\partial^2 G}{\partial t^2} = \delta(x - x')\delta(t - t')$$

and the initial conditions:

i)

$$y(x,0) = y_0(x), \quad \frac{\partial y(x,0)}{\partial t} = 0$$

ii)

$$y(x,0) = 0, \quad \frac{\partial y(x,0)}{\partial t} = v_0(x)$$

19.14   Consider the partial differential equation

$$\vec{\nabla}^2 \Psi(\vec{r}) = F(\vec{r}).$$

Show that the Green's function for the inside of a sphere satisfying the boundary conditions that $G(\vec{r}, \vec{r}')$ be finite at the origin and zero on the surface $r = a$ is given as

$$G(\vec{r}, \vec{r}') = \sum_{l=0}^{\infty} \sum_{m=-l}^{m=l} g_l(r,r') Y_l^{*m}(\theta', \phi') Y_l^m(\theta, \phi),$$

where

$$g_l(r,r') = \frac{r^l r'^l}{(2l+1)a^{2l+1}} \begin{cases} \left[1 - \left(\frac{a}{r'}\right)^{2l+1}\right] & r < r' \\[2mm] \left[1 - \left(\frac{a}{r}\right)^{2l+1}\right] & r > r' \end{cases}.$$

19.15   Consider the Helmholtz equation,

$$\vec{\nabla}^2 \Psi(\vec{r}) + k_0^2 \Psi(\vec{r}) = F(\vec{r}),$$

for the forced oscillations of a two-dimensional circular membrane (drumhead) with radius a, and with the boundary conditions

$$\Psi(0) = \text{finite, and } \Psi(a) = 0.$$

Show that the Green's function obeying

$$\vec{\nabla}^2 G(\vec{r}, \vec{r}') + k_0^2 G(\vec{r}, \vec{r}') = \delta(\vec{r} - \vec{r}')$$

is given as

$$G(\vec{r}, \vec{r}') = \sum_{m=0}^{\infty} \cos m(\theta - \theta') \times$$
$$\begin{cases} \dfrac{J_m(ka)N_m(kr') - N_m(ka)J_m(kr')}{2\epsilon_m J_m(ka)} J_m(kr), & r < r', \\[4mm] \dfrac{J_m(ka)N_m(kr) - N_m(ka)J_m(kr)}{2\epsilon_m J_m(ka)} J_m(kr'), & r > r', \end{cases}$$

where

$$
\epsilon_m =
\begin{cases}
2: & m = 0 \\
1: & m = 1, 2, 3, ...
\end{cases}
$$

Hint: use

$$
\delta(\vec{r} - \vec{r}') = \frac{\delta(r - r')}{r}\delta(\theta - \theta'),
$$

$$
= \frac{\delta(r - r')}{r}\frac{1}{2\pi}\sum_{m=-\infty}^{m=\infty} e^{im(\theta - \theta')}
$$

and separate the Green's function as

$$
G(\vec{r}, \vec{r}') = \frac{1}{2\pi}\sum_{m=-\infty}^{m=\infty} g_m(r, r')e^{im(\theta - \theta')}.
$$

One also needs the identity

$$
J_m(r)N_m'(r) - J_m'(r)N_m(r) = \frac{2}{\pi r},
$$

and ϵ_m is introduced when we combined the $\pm m$ terms to get $\cos m(\theta - \theta')$.

19.16   In the previous forced drumhead problem (19.15), first find the appropriate eigenfunctions and then show that the Green's function can also be written as

$$
G(\vec{r}, \vec{r}') = \sum_{m=0}^{\infty}\sum_{n=1}^{\infty} \frac{N_{mn}^2 J_m(k_{mn}r)J_m(k_{mn}r')\cos m(\theta - \theta')}{k^2 - k_{mn}^2},
$$

where the normalization constant N_{mn} is given as

$$
N_{mn} = \left[\sqrt{\frac{\pi\epsilon_m}{2}}aJ_m'(k_{mn}a)\right]^{-1}.
$$

Compare the two results.

19.17   Consider the differential equation

$$
\mathcal{L}\Phi(\vec{r}) = F(\vec{r})
$$

with the operator

$$
\mathcal{L} = \vec{\nabla}\cdot[p(\vec{r})\vec{\nabla}] + q(r).
$$

Show that the solution

$$
\Phi(\vec{r}') = \iiint_V F(\vec{r})G(\vec{r}, \vec{r}')d^3\vec{r}
$$

$$
+ \iiint_V [\Phi(\vec{r})\mathcal{L}G(\vec{r}, \vec{r}') - G(\vec{r}, \vec{r}')\mathcal{L}\Phi(\vec{r}')]d^3\vec{r}
$$

can be expressed as

$$\Phi(\vec{r}) = \int_V F(\vec{r}')G(\vec{r},\vec{r}')d^3\vec{r}'$$
$$+ \oint_S p(\vec{r}') \left[\Phi(\vec{r}')\vec{\nabla}G(\vec{r},\vec{r}') - G(\vec{r},\vec{r}')\vec{\nabla}\Phi(\vec{r}') \right] \cdot \hat{n}ds',$$

where $\hat{n}$ is the outward normal to the surface S bounding V.

19.18   Find the Green's function $G(\rho,\phi,\rho',\phi')$ for the two-dimensional Helmholtz equation

$$\left[\vec{\nabla}^2 + \kappa^2 \right] \Psi(\rho,\phi) = 0$$

for the full region outside a cylindrical surface $\rho = a$, which is appropriate for the following boundary conditions:

   i) Ψ is specified everywhere on $\rho = a$.

   ii) As $\rho \to \infty$, $\Psi \to \dfrac{e^{i\kappa\rho}}{\sqrt{\rho}} f(\phi)$ (outgoing cylindrical wave). Note that Ψ is independent of z.

19.19   Find the Green's function for the three-dimensional Helmholtz equation

$$\left[\vec{\nabla}^2 + \kappa^2 \right] \Psi(\vec{r}) = 0$$

for the region bounded by two spheres of radii a and b $(a > b)$ and which is appropriate for the boundary condition where $\Psi(\vec{r})$ is specified on the spheres of radius $r = a$ and $r = b$.

19.20   Find the Green's function for the operator

$$\mathcal{L} = \frac{d}{dx}\left(x\frac{d}{dx}\right) - \frac{n^2}{x}, \quad n = \text{integer},$$

with the boundary conditions $y(0) = 0$ and $y(L) = y_L$.

19.21   In Example 19.2 show that the solution for small oscillations is

$$\theta = \theta_1 \frac{\sin \omega_0 t}{\sin \omega_0 t_1}.$$

Show that this result satisfies the integral equation (19.61) in the small oscillations limit.

20

GREEN'S FUNCTIONS
and PATH INTEGRALS

In 1827 Brown investigates the random motions of pollen suspended in water under a microscope. The irregular movements of the pollen particles are due to their random collisions with the water molecules. Later it becomes clear that many small objects interacting randomly with their environment behave the same way. Today this motion is known as Brownian motion and forms the prototype of many different phenomena in diffusion, colloid chemistry, polymer physics, quantum mechanics, and finance. During the years 1920–1930 Wiener approaches Brownian motion in terms of path integrals. This opens up a whole new avenue in the study of many classical systems. In 1948 Feynman gives a new formulation of quantum mechanics in terms of path integrals. In addition to the existing Schrödinger and Heisenberg formulations, this new approach not only makes the connection between quantum and classical physics clearer, but also leads to many interesting applications in field theory. In this Chapter we introduce the basic features of this technique, which has many interesting existing applications and tremendous potential for future uses.

20.1 BROWNIAN MOTION AND THE DIFFUSION PROBLEM

Starting with the principle of conservation of matter, we can write the diffusion equation as

$$\frac{\partial \rho(\overrightarrow{r}, t)}{\partial t} = D \overrightarrow{\nabla}^2 \rho(\overrightarrow{r}, t), \tag{20.1}$$

where $\rho(\vec{r}, t)$ is the density of the diffusing material and D is the diffusion constant, which depends on the characteristics of the medium. Because the diffusion process is also many particles undergoing Brownian motion at the same time, division of $\rho(\vec{r}, t)$ by the total number of particles gives the probability, $w(\vec{r}, t)$, of finding a particle at $\vec{r}$ and t as

$$w(\vec{r}, t) = \frac{1}{N}\rho(\vec{r}, t). \tag{20.2}$$

Naturally, $w(\vec{r}, t)$ also satisfies the diffusion equation:

$$\frac{\partial w(\vec{r}, t)}{\partial t} = D\vec{\nabla}^2 w(\vec{r}, t). \tag{20.3}$$

For a particle starting its motion from $\vec{r} = 0$, we have to solve Equation (20.3) with the initial condition

$$\lim_{t \to 0} w(\vec{r}, t) \to \delta(\vec{r}). \tag{20.4}$$

In one dimension we write Equation (20.3) as

$$\frac{\partial w(x, t)}{\partial t} = D\frac{\partial^2 w(x, t)}{\partial x^2} \tag{20.5}$$

and by using the Fourier transform technique we can obtain its solution as

$$w(x, t) = \frac{1}{\sqrt{4\pi Dt}}\exp\left\{-\frac{x^2}{4Dt}\right\}. \tag{20.6}$$

Note that, consistent with the probability interpretation, $w(x, t)$ is always positive. Because it is certain that the particle is somewhere in the interval $(-\infty, \infty)$, $w(x, t)$ also satisfies the normalization condition

$$\int_{-\infty}^{\infty} dx w(x, t) = \int_{-\infty}^{\infty} dx \frac{1}{\sqrt{4\pi Dt}}\exp\left\{-\frac{x^2}{4Dt}\right\}$$
$$= 1. \tag{20.7}$$

For a particle starting its motion from an arbitrary point, (x_0, t_0), we write the probability distribution as

$$W(x, t, x_0, t_0) = \frac{1}{\sqrt{4\pi D(t - t_0)}}\exp\left\{-\frac{(x - x_0)^2}{4D(t - t_0)}\right\}, \tag{20.8}$$

where $W(x, t, x_0, t_0)$ is the solution of

$$\frac{\partial W(x, t, x_0, t_0)}{\partial t} = D\frac{\partial^2 W(x, t, x_0, t_0)}{\partial x^2} \tag{20.9}$$

satisfying the initial condition

$$\lim_{t \to t_0} W(x, t, x_0, t_0) \to \delta(x - x_0) \tag{20.10}$$

and the normalization condition

$$\int_{-\infty}^{\infty} dx W(x, t, x_0, t_0) = 1. \tag{20.11}$$

From our discussion of Green's functions in Chapter 19 we recall that $W(x, t, x_0, t_0)$ is also the propagator of the operator

$$\pounds = \frac{\partial}{\partial t} - D \frac{\partial^2}{\partial x^2}. \tag{20.12}$$

Thus, given the probability at some initial point and time, $w(x_0, t_0)$, we can find the probability at subsequent times, $w(x, t)$, by using $W(x, t, x_0, t_0)$ as

$$w(x, t) = \int_{-\infty}^{\infty} dx_0 W(x, t, x_0, t_0) w(x_0, t_0), \qquad t > t_0. \tag{20.13}$$

Combination of propagators gives us the **Einstein-Smoluchowski-Kolmogorov-Chapman (ESKC) equation**:

$$W(x, t, x_0, t_0) = \int_{-\infty}^{\infty} dx' W(x, t, x', t') W(x', t', x_0, t_0), \ \ t > t' > t_0. \tag{20.14}$$

The significance of this equation is that it gives the causal connection of events in Brownian motion as in the Huygens-Fresnel equation.

20.2 WIENER PATH INTEGRAL APPROACH TO BROWNIAN MOTION

In Equation (20.13) we have seen how to find the probability of finding a particle at (x, t) from the probability at (x_0, t_0) by using the propagator $W(x, t, x_0, t_0)$. We now divide the interval between t_0 and t into $N + 1$ equal segments:

$$\begin{aligned} \Delta t_i &= t_i - t_{i-1} \\ &= \frac{t - t_0}{N + 1}, \end{aligned} \tag{20.15}$$

which is covered by the particle in N steps. The propagator of each step is given as

$$W(x_i, t_i, x_{i-1}, t_{i-1}) = \frac{1}{\sqrt{4\pi D(t_i - t_{i-1})}} \exp\left\{ -\frac{(x_i - x_{i-1})^2}{4D(t_i - t_{i-1})} \right\}. \tag{20.16}$$

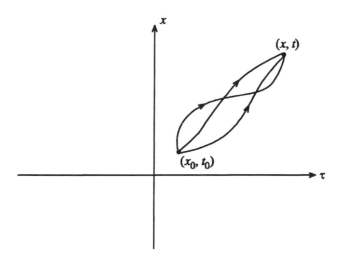

Fig. 20.1 Paths $C[x_0, t_0; x, t]$ for the pinned Wiener measure

Assuming that each step is taken independently, we combine propagators N times by using the ESKC relation to get the propagator that takes us from (x_0, t_0) to (x, t) in a single step as

$$W(x, t, x_0, t_0) = \int \cdots \int \exp\left\{ -\sum_{i=1}^{N+1} \frac{(x_i - x_{i-1})^2}{4D(t_i - t_{i-1})} \right\} \prod_{i=1}^{N} \frac{dx_i}{\sqrt{4\pi D(t_i - t_{i-1})}}.$$

(20.17)

This equation is valid for $N > 0$. Assuming that it is also valid in the limit as $N \to \infty$, that, is as $\Delta t_i \to 0$, we write

$$W(x, t, x_0, t_0) = \qquad\qquad\qquad\qquad (20.18)$$

$$\lim_{N \to \infty, \, \Delta t_i \to 0} \int \cdots \int \exp\left\{ -\frac{1}{4D} \sum_{i=1}^{N+1} \left(\frac{x_i - x_{i-1}}{t_i - t_{i-1}} \right)^2 \Delta t_i \right\} \prod_{i=1}^{N} \frac{dx_i(\tau)}{\sqrt{4\pi D \Delta t_i}},$$

$$W(x, t, x_0, t_0) = \int \cdots \int \exp\left\{ -\frac{1}{4D} \int_{t_0}^{t} \dot{x}^2(\tau) d\tau \right\} \prod_{\tau=t_0}^{t} \frac{dx(\tau)}{\sqrt{4\pi D d\tau}}. \quad (20.19)$$

Here, τ is a time parameter (Fig. 20.1) introduced to parametrize the paths as $x(\tau)$. We can also write $W(x, t, x_0, t_0)$ in short as

$$W(x, t, x_0, t_0) = \check{N} \int \exp\left\{ -\frac{1}{4D} \int_{t_0}^{t} \dot{x}^2(\tau) d\tau \right\} \check{D}x(\tau), \qquad (20.20)$$

where $\check{N}$ is a normalization constant and $\check{D}x(\tau)$ indicates that the integral should be taken over all paths starting from (x_0, t_0) and end at (x, t). This expression can also be written as

$$W(x, t, x_0, t_0) = \int_{C[x_0, t_0; x, t]} d_w x(\tau), \tag{20.21}$$

where $d_w x(\tau)$ is called the **Wiener measure**. Because $d_w x(\tau)$ is the measure for all paths starting from (x_0, t_0) and ending at (x, t), it is called the **pinned (conditional) Wiener measure** (Fig. 20.1).

Summary: For a particle starting its motion from (x_0, t_0), the propagator $W(x, t, x_0, t_0)$ is given as

$$W(x, t, x_0, t_0) = \frac{1}{\sqrt{4\pi D(t - t_0)}} \exp\left\{-\frac{(x - x_0)^2}{4D(t - t_0)}\right\}. \tag{20.22}$$

This satisfies the differential equation

$$\frac{\partial W(x, t, x_0, t_0)}{\partial t} = D\frac{\partial^2}{\partial x^2} W(x, t, x_0, t_0) \tag{20.23}$$

with the initial condition $\lim_{t \to t_0} W(x, t, x_0, t_0) \to \delta(x - x_0)$.

In terms of the Wiener path integral the propagator $W(x, t, x_0, t_0)$ is also expressed as

$$W(x, t, x_0, t_0) = \int_{C[x_0, t_0; x, t]} d_w x(\tau). \tag{20.24}$$

The measure of this integral is

$$d_w x(\tau) = \exp\left\{-\frac{1}{4D}\int_{t_0}^{t} \dot{x}^2(\tau)d\tau\right\}\prod_{i=1}^{N}\frac{dx_i}{\sqrt{4\pi D d\tau}}. \tag{20.25}$$

Because the integral is taken over all continuous paths from (x_0, t_0) to (x, t), which are shown as $C[x_0, t_0; x, t]$, this measure is also called the pinned Wiener measure (Fig. 20.1).

For a particle starting from (x_0, t_0) the probability of finding it in the interval Δx at time t is given by

$$\Delta x \int_{C[x_0, t_0; t]} d_w x(\tau). \tag{20.26}$$

In this integral, because the position of the particle at time t is not fixed, $d_w x(\tau)$ is called the **unpinned (or unconditional) Wiener measure**. At

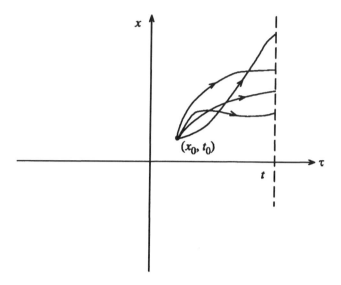

Fig. 20.2 Paths $C[x_0, t_0; t]$ for the unpinned Wiener measure

time t, because it is certain that the particle is somewhere in the interval $x \in [-\infty, \infty]$, we write (Fig. 20.2)

$$\int_{C[x_0, t_0; t]} d_w x(\tau) = \int_{-\infty}^{\infty} dx \int_{C[x_0, t_0; x, t]} d_w x(\tau) \qquad (20.27)$$
$$= 1.$$

The average of a functional, $F[x(t)]$, found over all paths $C[x_0, t_0; t]$ at time t is given by the formula

$$\langle F[x(t)] \rangle_C = \int_{C[x_0, t_0; t]} d_w x(\tau) F[x(\tau)] \qquad (20.28)$$
$$= \int_{-\infty}^{\infty} dx \int_{C[x_0, t_0; x, t]} d_w x(\tau) F[x(\tau)].$$

In terms of the Wiener measure we can express the ESKC relation as

$$\int_{C[x_0, t_0; x, t]} d_w x(\tau) = \int_{-\infty}^{\infty} dx' \int_{C[x_0, t_0; x', t']} d_w x(\tau) \int_{C[x', t'; x, t]} d_w x(\tau). \quad (20.29)$$

20.3   THE FEYNMAN-KAC FORMULA AND THE PERTURBATIVE SOLUTION OF THE BLOCH EQUATION

We have seen that the propagator of the diffusion equation,

$$\frac{\partial w(x,t)}{\partial t} - D\frac{\partial^2 w(x,t)}{\partial x^2} = 0, \tag{20.30}$$

can be expressed as a path integral [Eq. (20.24)]. However, when we have a closed expression as in Equation (20.22), it is not clear what advantage this new representation has. In this section we study the diffusion equation in the presence of interactions, where the advantages of the path integral approach begin to appear. In the presence of a potential $V(x)$, the diffusion equation can be written as

$$\frac{\partial w(x,t)}{\partial t} - D\frac{\partial^2 w(x,t)}{\partial x^2} = -V(x,t)w(x,t). \tag{20.31}$$

We now need a Green's function, W_D, that satisfies the inhomogeneous equation

$$\frac{\partial W_D(x,t,x',t')}{\partial t} - D\frac{\partial^2 W_D(x,t,x',t')}{\partial x^2} = \delta(x-x')\delta(t-t'), \tag{20.32}$$

so that we can express the general solution of (20.31) as

$$w(x,t) = w_0(x,t) - \int\int W_D(x,t,x',t')V(x',t')w(x',t')dx'dt', \tag{20.33}$$

where $w_0(x,t)$ is the solution of the homogeneous part of Equation (20.31), that is, Equation (20.5). We can construct $W_D(x,t,x',t')$ by using the propagator, $W(x,t,x',t')$, that satisfies the homogeneous equation (Chapter 19)

$$\frac{\partial W(x,t,x',t')}{\partial t} - D\frac{\partial^2 W(x,t,x',t')}{\partial x^2} = 0, \tag{20.34}$$

as

$$W_D(x,t,x',t') = W(x,t,x',t')\theta(t-t'). \tag{20.35}$$

Because the unknown function also appears under the integral sign, Equation (20.33) is still not the solution, that is, it is just the integral equation version of Equation (20.31). On the other hand, $W_B(x,t,x',t')$, which satisfies

$$\frac{\partial W_B(x,t,x',t')}{\partial t} - D\frac{\partial^2 W_B(x,t,x',t')}{\partial x^2} = -V(x,t)W_B(x,t,x',t'), \tag{20.36}$$

is given as

$$W_B(x,t,x_0,t_0) = W_D(x,t,x_0,t_0) \tag{20.37}$$

$$- \int_{-\infty}^{\infty} dx' \int_{-\infty}^{\infty} dt' W_D(x,t,x',t')V(x',t')W_B(x',t',x_0,t_0).$$

The first term on the right-hand side is the solution of the homogeneous equation [Eq. (20.34)], which is W. However, because $t > t_0$ we could also write it as W_D.

A very useful formula called the **Feynman-Kac formula** (theorem) is given as

$$W_B(x, t, x_0, 0) = \int_{C[x_0, 0; x, t]} d_w x(\tau) \exp \left\{ - \int_0^t d\tau V[x(\tau), \tau] \right\}. \qquad (20.38)$$

This is a solution of Equation (20.36), which is also known as the **Bloch equation,** with the initial condition

$$\lim_{t \to t'} W_B(x, t, x', t') = \delta(x - x'). \qquad (20.39)$$

The Feynman-Kac theorem constitutes a very important step in the development of path integrals. We leave its proof to the next section and continue by writing the path integral in Equation (20.38) as a Riemann sum:

$$W_B(x, t, x_0, 0) = \lim_{N \to \infty, \varepsilon \to 0} (4\pi D\varepsilon)^{-(N+1)/2} \int_{-\infty}^{\infty} dx_1 \int_{-\infty}^{\infty} dx_2 \cdots \int_{-\infty}^{\infty} dx_N$$

$$\times \exp \left\{ -\frac{1}{4D\varepsilon} \sum_{j=1}^{N+1} (x_j - x_{j-1})^2 - \varepsilon \sum_{j=1}^{N} V(x_j, t_j) \right\}. \qquad (20.40)$$

We have taken

$$\varepsilon = t_i - t_{i-1}$$
$$= \frac{t - t_0}{N+1}. \qquad (20.41)$$

The first exponential factor in Equation (2.40) is the solution [Eq. (2.18)] of the homogeneous equation. After expanding the second exponential factor as

$$\exp \left\{ -\varepsilon \sum_{j=1}^{N} V(x_j, t_j) \right\} \qquad (20.42)$$

$$= 1 - \varepsilon \sum_{j=1}^{N} V(x_j, t_j) + \frac{1}{2} \varepsilon^2 \sum_{j=1}^{N} \sum_{k=1}^{N} V(x_j, t_j) V(x_k, t_k) - \cdots,$$

we integrate over the intermediate x variables and rearrange to obtain

$$W_B(x, t, x_0, t_0) = W(x, t, x_0, t_0) \qquad (20.43)$$

$$-\varepsilon \sum_{j=1}^{N} \int_{-\infty}^{\infty} dx_j W(x, t, x_j, t_j) V(x_j, t_j) W(x_j, t_j, x_0, t_0)$$

$$+\frac{1}{2!} \varepsilon^2 \sum_{j=1}^{N} \sum_{k=1}^{N} \int_{-\infty}^{\infty} dx_j \int_{-\infty}^{\infty} dx_k W(x, t, x_j, t_j) V(x_j, t_j) W(x_j, t_j, x_k, t_k)$$

$$\times V(x_k, t_k) W(x_k, t_k, x_0, t_0) + \cdots.$$

In the limit as $\varepsilon \to 0$ we make the replacement $\varepsilon \sum_j \to \int_{t_0}^{t} dt_j$. We also suppress the factors of factorials, $(1/n!)$, because they are multiplied by ε^n, which also goes to zero as $\varepsilon \to 0$. Besides, because times are ordered in Equation (20.43) as

$$t_0 < t_1 < t_2 < \cdots < t,$$

we can replace W with W_D in the above equation and write W_B as

$$W_B(x, t, x_0, t_0) = W_D(x, t, x_0, t_0) \tag{20.44}$$

$$- \int_{-\infty}^{\infty} dx' \int_{t_0}^{t} dt' W_D(x, t, x', t') V(x', t') W_D(x', t', x_0, t_0)$$

$$+ \int_{-\infty}^{\infty} dx' \int_{t_0}^{t} dt' \int_{-\infty}^{\infty} dx'' \int_{t_0'}^{t'} dt'' W_D(x, t, x', t') V(x', t') W_D(x', t', x'', t'')$$

$$\times V(x'', t'') W_D(x'', t'', x_0, t_0) + \cdots .$$

Now $W_B(x, t, x_0, t_0)$ no longer appears on the right-hand side of this equation. Thus it is the perturbative solution of Equation (20.37) by the iteration method. Note that $W_B(x, t, x_0, t_0)$ satisfies the initial condition given in Equation (20.39).

20.4 DERIVATION OF THE FEYNMAN-KAC FORMULA

We now show that the Feynman-Kac formula,

$$W_B(x, t, x_0, 0) = \int_{C[x_0, 0; x, t]} d_w x(\tau) \exp\left\{ -\int_0^t d\tau V[x(\tau), \tau] \right\}, \tag{20.45}$$

is identical to the iterative solution to all orders of the following integral equation:

$$W_B(x, t, x_0, t_0) = W_D(x, t, x_0, t_0) \tag{20.46}$$

$$- \int_{-\infty}^{\infty} dx' \int_0^t dt' W_D(x, t, x', t') V(x', t') W_B(x', t', x_0, t_0),$$

which is equivalent to the differential equation

$$\frac{\partial W_B(x, t, x', t')}{\partial t} - D \frac{\partial^2 W_B(x, t, x', t')}{\partial x^2} = -V(x, t) W_B(x, t, x', t') \tag{20.47}$$

with the initial condition given in Equation (20.39).

We first show that the Feynman-Kac formula satisfies the ESKC [Eq. (20.14)] relation. Note that we write $V[x(\tau)]$ instead of $V[x(\tau), \tau]$ when there is no room for confusion:

$$\int_{-\infty}^{\infty} dx_s W_B(x, t, x_s, t_s) W_B(x_s, t_s, x_0, 0) \tag{20.48}$$

$$= \int_{-\infty}^{\infty} dx_s \int_{C[x_0,0;x_s,t_s]} d_w x(\tau) \exp\left\{-\int_0^s d\tau V[x(\tau)]\right\}$$

$$\times \int_{C[x_s,t_s;x,t]} d_w x(\tau') \exp\left\{-\int_{t_s}^t d\tau' V[x(\tau')]\right\}.$$

In this equation x_s denotes the position at t_s and x denotes the position at t.

Because $C[x_0, 0; x_\tau, t_\tau; x, t]$ denotes all paths starting from $(x_0, 0)$, passing through (x_τ, t_τ) and then ending up at (x, t), we can write the right hand-side of the above equation as

$$\int_{-\infty}^{\infty} dx_s \int_{[x_0,0;x_s,t_s;x,t]} d_w x(\tau) \exp\left\{-\int_0^{t_s} d\tau V[x(\tau)]\right\} \tag{20.49}$$

$$= \int_{C[x_0,0;x,t]} d_w x(\tau) \exp\left\{-\int_0^t d\tau V[x(\tau)]\right\} \tag{20.50}$$

$$= W_B(x, t, x_0, 0). \tag{20.51}$$

From here, we see that the Feynman-Kac formula satisfies the ESKC relation as

$$\int_{-\infty}^{\infty} dx_s W_B(x, t, x_s, t_s) W_B(x_s, t_s, x_0, 0) = W_B(x, t, x_0, 0). \tag{20.52}$$

With the help of Equations (20.21) and (20.22), we see that the Feynman-Kac formula satisfies the initial condition

$$\lim_{t \to 0} W_B(x, t, x_0, 0) \to \delta(x - x_0) \tag{20.53}$$

and the functional in the Feynman-Kac formula satisfies the equality

$$\exp\left\{-\int_0^t d\tau V[x(\tau)]\right\} \tag{20.54}$$

$$= 1 - \int_0^t d\tau \left(V[x(\tau)] \exp\left\{-\int_0^\tau ds V[x(s)]\right\}\right).$$

We can easily show that this is true by taking the derivative of both sides. Because this equality holds for all continuous paths $x(s)$, we take the integral of both sides over the paths $C[x_0, 0; x, t]$ via the Wiener measure to get

$$\int_{C[x_0,0;x,t]} d_w x(\tau) \exp\left\{-\int_0^t d\tau V[x(\tau)]\right\} \tag{20.55}$$

$$= \int_{C[x_0,0;x,t]} d_w x(\tau)$$

$$- \int_{C[x_0,0;x,t]} d_w x(\tau) \int_0^t d\tau \left(V[x(\tau)] \exp\left\{-\int_0^\tau ds V[x(s)]\right\}\right).$$

The first term on the right-hand side is the solution of the homogeneous part of Equation (20.36). Also, for $t > 0$, we can write $W_D(x_0, 0, x, t)$ instead of $W(x_0, 0, x, t)$. Because the integral in the second term involves exponentially decaying terms, it converges. Thus we interchange the order of the integrals to write

$$
\int_{C[x_0, 0; x, t]} d_w x(s) \int_0^t ds \left(V[x(s)] \exp\left\{ -\int_0^s d\tau V[x(\tau)] \right\} \right)
$$

$$
= \int_0^t ds \int_{C[x_0, 0; x, t]} d_w x(s) \left[V[x(s)] \exp\left\{ -\int_0^s d\tau V[x(\tau)] \right\} \right],
$$

$$
= \int_0^t ds \int_{-\infty}^{\infty} dx_s \int_{C[x_0, 0; x_s, t_s; x, t]} d_w x(s) \left[V[x(s)] \exp\left\{ -\int_0^s d\tau V[x(\tau)] \right\} \right],
$$

$$
= \int_0^t ds \int_{-\infty}^{\infty} dx_s V[x(s)] \int_{C[x_0, 0; x_s, t_s]} d_w x(\tau) \left[\exp\left\{ -\int_0^s d\tau V[x(\tau)] \right\} \right] \times \int_{C[x_s, t_s; x, t]} d_w x(\tau),
$$

$$
= \int_0^t ds \int_{-\infty}^{\infty} dx_s V[x(s)] W_B(x_s, t_s, x_0, 0) W_D(x, t, x_s, t_s), \qquad (20.56)
$$

where we have used the ESKC relation. We now substitute this result into Equation (20.55) and use Equation (20.45) to write

$$
W_B(x, t, x_0, 0) = \int_{C[x_0, 0; x, t]} d_w x(\tau) \exp\left\{ -\int_0^t d\tau V[x(\tau), \tau] \right\}, \qquad (20.57)
$$

$$
= W_D(x, t, x_0, 0) \qquad (20.58)
$$

$$
- \int_{-\infty}^{\infty} dx' \int_0^t dt' W_D(x, t, x', t') V(x', t') W_B(x', t', x_0, 0),
$$

thus proving the Feynman-Kac formula. Generalization to arbitrary initial time t_0 is obvious.

20.5   INTERPRETATION OF $V(x)$ IN THE BLOCH EQUATION

We have seen that the solution of the Bloch equation

$$
\frac{\partial W_B(x, t, x_0, t_0)}{\partial t} - D \frac{\partial^2 W_B(x, t, x_0, t_0)}{\partial x^2} = -V(x, t) W_B(x, t, x_0, t_0), \qquad (20.59)
$$

with the initial condition

$$
W_B(x, t, x_0, t_0)|_{t=t_0} = \delta(x - x_0), \qquad (20.60)
$$

is given by the Feynman-Kac formula

$$
W_B(x, t; x_0, t_0) = \int_{C[x_0, t_0; x, t]} d_w x(\tau) \exp\left\{ -\int_{t_0}^t V[x(\tau), \tau] d\tau \right\}. \qquad (20.61)
$$

In these equations, even though $V(x)$ is not exactly a potential, it is closely related to the external forces acting on the system.

In fluid mechanics the probability distribution of a particle undergoing Brownian motion and under the influence of an external force satisfies the differential equation

$$\frac{\partial W(x,t;x_0,t_0)}{\partial t} - D\frac{\partial^2 W(x,t;x_0,t_0)}{\partial x^2} = -\frac{1}{\eta}\frac{\partial}{\partial x}\left[F(x)W(x,t;x_0,t_0)\right],$$

$$(20.62)$$

where η is the friction coefficient in the drag force, which is proportional to the velocity. In Equation (20.62), if we try a solution of the form

$$W(x,t;x_0,t_0) = \exp\left\{\frac{1}{2\eta D}\int_{x_0}^x dxF(x)\right\}\widetilde{W}(x,t;x_0,t_0),\qquad (20.63)$$

we obtain a differential equation to be solved for $\widetilde{W}(x,t;x_0,t_0)$:

$$\frac{\partial\widetilde{W}(x,t;x_0,t_0)}{\partial t} - D\frac{\partial^2\widetilde{W}(x,t;x_0,t_0)}{\partial x^2} = V(x)\widetilde{W}(x,t;x_0,t_0),\qquad (20.64)$$

where we have defined $V(x)$ as

$$V(x) = \frac{1}{4\eta^2 D}F^2(x) + \frac{1}{2\eta}\frac{dF(x)}{dx}.\qquad (20.65)$$

Using the Feynman-Kac formula as the solution of Equation (20.64), we can write the solution of Equation (20.62) as

$$W(x,t;x_0,t_0) = \exp\left\{\frac{1}{2\eta D}\int_{x_0}^x dxF(x)\right\}\int_{C[x_0,t_0;x,t]} d_w x(\tau)\exp\left\{-\int_{t_0}^t V[x(\tau)]d\tau\right\}.$$

$$(20.66)$$

Using the Wiener measure, Equation (20.25), we write this equation as

$$W(x,t;x_0,t_0) = \int_{C[x_0,t_0;x,t]}\prod_{\tau=t_0}^t\frac{dx(\tau)}{\sqrt{4\pi D d\tau}}\qquad (20.67)$$

$$\times\exp\left\{\frac{1}{2\eta D}\int_{x_0}^x dxF(x) - \frac{1}{4D}\int_{t_0}^t d\tau\dot{x}^2(\tau) - \int_{t_0}^t d\tau V[x(\tau)]\right\}.$$

Finally, using the equality

$$\int_{x_0}^x dxF(x) = \int_{t_0}^t d\tau\dot{x}F(x),\qquad (20.68)$$

this becomes

$$W(x, t; x_0, t_0) = \int_{C[x_0, t_0; x, t]} \prod_{\tau = t_0}^{t} \frac{dx(\tau)}{\sqrt{4\pi D d\tau}} \tag{20.69}$$

$$\times \exp\left\{\frac{1}{2\eta D}\int_{t_0}^{t} d\tau \dot{x} F(x) - \frac{1}{4D}\int_{t_0}^{t} d\tau \dot{x}^2(\tau) - \int_{t_0}^{t} d\tau V[x(\tau)]\right\}$$

$$= \int_{C[x_0, t_0; x, t]} \exp\left\{-\frac{1}{4D}\int_{t_0}^{t} d\tau L[x(\tau)]\right\} \prod_{\tau = t_0}^{t} \frac{dx(\tau)}{\sqrt{4\pi D d\tau}}. \tag{20.70}$$

In the last equation we have defined

$$L[x(\tau)] = \left(\dot{x} - \frac{F}{\eta}\right)^2 + 2\frac{D}{\eta}\frac{dF}{dx} \tag{20.71}$$

and used Equation (20.65).

As we see from here, $V(x)$ is not quite the potential, nor is $L[x(\tau)]$ the Lagrangian. In the limit as $D \to 0$ fluctuations in the Brownian motion disappear and the argument of the exponential function goes to infinity. Thus only the path satisfying the condition

$$\int_{t_0}^{t} d\tau \left(\dot{x} - \frac{F}{\eta}\right)^2 = 0 \tag{20.72}$$

or

$$\frac{dx}{d\tau} - \frac{F}{\eta} = 0 \tag{20.73}$$

contributes to the path integral in Equation (20.70). Comparing this with

$$m\ddot{x} = -\eta\dot{x} + F(x), \tag{20.74}$$

we see that it is the deterministic equation of motion of a particle with negligible mass, moving under the influence of an external force $F(x)$ and a friction force $-\eta\dot{x}$ (Pathria, p. 463).

When the diffusion constant differs from zero, the solution is given as the path integral

$$W(x, t, x_0, t_0) = \int_{C[x_0, t_0; x, t]} \exp\left\{-\frac{1}{4D}\int_{t_0}^{t} d\tau L[x(\tau)]\right\} \prod_{\tau = t_0}^{t} \frac{dx(\tau)}{\sqrt{4\pi D d\tau}}. \tag{20.75}$$

In this case all the continuous paths between (x_0, t_0) and (x, t) will contribute to the integral. It is seen from equation Equation (20.75) that each path contributes to the propagator $W(x, t, x_0, t_0)$ with the weight factor

$$\exp\left\{-\frac{1}{4D}\int_{t_0}^{t} d\tau L[x(\tau)]\right\}.$$

Naturally, the majority of the contribution comes from places where the paths with comparable weights cluster. These paths are the ones that make the functional in the exponential an extremum, that is,

$$\delta \int_{t_0}^{t} d\tau L[x(\tau)] = 0. \tag{20.76}$$

These paths are the solutions of the Euler-Lagrange equation:

$$\frac{\partial L}{\partial x} - \frac{d}{d\tau} \left[\frac{\partial L}{\partial(dx/d\tau)} \right] = 0. \tag{20.77}$$

At this point we remind the reader that $L[x(\tau)]$ is not quite the Lagrangian of the particle undergoing Brownian motion. It is interesting that $V(x)$ and $L[x(\tau)]$ gain their true meaning only when we consider applications of path integrals to quantum mechanics.

20.6   METHODS OF CALCULATING PATH INTEGRALS

We have obtained the propagator of

$$\frac{\partial w(x,t)}{\partial t} - D\frac{\partial^2 w(x,t)}{\partial x^2} = -V(x,t)w(x,t) \tag{20.78}$$

as

$$W(x,t;x_0,t_0) = \check{N} \int_{C[x_0,t_0;x,t]} \exp\left\{ -\frac{1}{4D}\int_{t_0}^{t}\dot{x}^2(\tau)d\tau - \int_{t_0}^{t}V[x(\tau)]d\tau \right\} \check{D}x(\tau). \tag{20.79}$$

In term of the Wiener measure this can also be written as

$$W(x,t;x_0,t_0) = \check{N} \int_{C[x_0,t_0;x,t]} \exp\left\{ -\int_{t_0}^{t}V[x(\tau)]d\tau \right\} d_w x(\tau), \tag{20.80}$$

where $d_w x(\tau)$ is defined as

$$d_w x(\tau) = \exp\left\{ -\frac{1}{4D}\int_{t_0}^{t}\dot{x}^2(\tau)d\tau \right\} \prod_{\tau=t_0}^{t} \frac{dx(\tau)}{\sqrt{4\pi Ddt}}. \tag{20.81}$$

The average of a functional $F[x(\tau)]$ over the paths $C[x_0,t_0;x,t]$ is defined as

$$\langle F[x(\tau)]\rangle_C = \int_{C[x_0,t_0;x,t]} F[x(\tau)]\exp\left\{ -\int_{t_0}^{t}V[x(\tau)]d\tau \right\} d_w x(\tau), \tag{20.82}$$

where $C[x_0,t_0;x,t]$ denotes all continuous paths starting from (x_0,t_0) and ending at (x,t). Before we discuss techniques of evaluating path integrals, we should talk about a technical problem that exists in Equation (20.80). In this expression, even though all the paths in $C[x_0, t_0; x, t]$ are continuous, because of the nature of the Brownian motion they zig zag. The average distance squared covered by a Brown particle is given as

$$\langle x^2 \rangle = \int_{-\infty}^{\infty} w(x, t) x^2 \, dx \propto t. \tag{20.83}$$

From here we find the average distance covered during time t as

$$\sqrt{\langle x^2 \rangle} \propto \sqrt{t}, \tag{20.84}$$

which gives the velocity of the particle at any point as

$$\lim_{t \to 0} \frac{\sqrt{t}}{t} \to \infty. \tag{20.85}$$

Thus $\dot{x}$ appearing in the propagator [Eq. (20.79)] is actually undefined for all t values. However, the integrals in Equations (20.80) and (20.81) are convergent for $V(x) \geq c$, where c is some constant. In this expression $W(x, t, x_0, t_0)$ is always positive and thus consistent with its probability interpretation and satisfies the ESKC relation [Eq. (20.14)], and the normalization condition

$$\int_{C[x_0, t_0; x, t]} d_w x(\tau) = 1. \tag{20.86}$$

In summary: If we look at the propagator [Eq. (20.80)] as a probability distribution, it is Equation (20.79) written as a path integral, evaluated over all Brown paths with a suitable weight factor depending on the potential $V(x)$.

The zig zag motion of the particles in Brownian motion is essential in the fluid exchange process of living cells. In fractal theory, paths of Brown particles are two-dimensional fractal curves. The possible connections between fractals, path integrals, and differintegrals are active areas of research.

20.6.1   Method of Time Slices

Let us evaluate the path integral of the functional $F[x(\tau)]$ with the Wiener measure. We slice a given path $x(\tau)$ into N equal time intervals and approximate the path in each slice with a straight line $l_N(\tau)$ as

$$l_N(t_i) = x(t_i) = x_i, \qquad i = 1, 2, 3, ..., N. \tag{20.87}$$

This means that for a given path, $x(\tau)$, and a small number ε we can always find a number $N = N(\varepsilon)$ independent of τ such that

$$|x(\tau) - l_N(\tau)| < \varepsilon \tag{20.88}$$

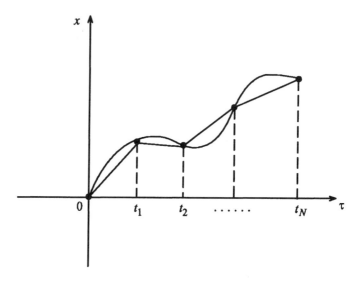

Fig. 20.3 Paths for the time slice method

is true. Under these conditions for smooth functionals (Fig. 20.3) the inequality

$$|F[x(\tau)] - F[l_N(\tau)]| < \delta(\varepsilon) \tag{20.89}$$

is satisfied such that the limit $\lim_{\varepsilon \to 0} \delta(\varepsilon) \to 0$ is true. Because all the information about $l_N(\tau)$ is contained in the set $x_1 = x(t_1), ..., x_N = x(t_N)$, we can also describe the functional $F[l_N(\tau)]$ by

$$F[l_N(\tau)] = F_N(x_1, x_2, ..., x_N), \tag{20.90}$$

which means that

$$\left| \int_{C[0,0;t]} d_w x(\tau) F[x(\tau)] - \int_{C[0,0;t]} d_w x(\tau) F_N(x_1, x_2, ..., x_N) \right|$$

$$\leq \int_{C[0,0;t]} d_w x(\tau) \, |F[x(\tau)] - F_N(x_1, x_2, \cdots, x_N)| \, ,$$

$$\leq \int_{C[0,0;t]} d_w x(\tau) \delta(\varepsilon),$$

$$\leq \delta(\varepsilon) \int_{C[0,0;t]} d_w x(\tau),$$

$$\leq \delta(\varepsilon). \tag{20.91}$$

Because for $N = 1, 2, 3, ...$, the function set $F_N(x_1, x_2, ..., x_N)$ forms a Cauchy set approaching $F[x(\tau)]$, for a suitably chosen N we can use the integral

$$\int_{C[0,0;t]} d_w x(\tau) F_N(x_1, x_2, ..., x_N) \tag{20.92}$$

$$= \int_{-\infty}^{\infty} \frac{dx_1}{\sqrt{4\pi Dt_1}} \cdots \frac{dx_N}{\sqrt{4\pi D(t_N - t_{N-1})}} F_N(x_1, x_2, ..., x_N)$$

$$\times \exp\left\{ -\frac{1}{4D} \sum_{i=1}^{N} \frac{(x_i - x_{i-1})^2}{t_i - t_{i-1}} \right\}$$

to evaluate the path integral

$$\int_{C[0,0;t]} d_w x(\tau) F[x(\tau)] \tag{20.93}$$

$$= \lim_{N \to \infty} \int_{-\infty}^{\infty} \frac{dx_1}{\sqrt{4\pi Dt_1}} \cdots \frac{dx_N}{\sqrt{4\pi D(t_N - t_{N-1})}} F_N(x_1, x_2, ..., x_N)$$

$$\times \exp\left\{ -\frac{1}{4D} \sum_{i=1}^{N} \frac{(x_i - x_{i-1})^2}{t_i - t_{i-1}} \right\}.$$

For a given ε the difference between the two approaches can always be kept less than a small number, $\delta(\varepsilon)$, by choosing a suitable $N(\varepsilon)$. In this approach a Wiener path integral $\int_{C[0,0;t]} d_w x(\tau) F[x(\tau)]$ will be converted into an N-dimensional integral [Eq. (20.92)].

20.6.2 Evaluating Path Integrals with the ESKC Relation

We introduce this method by evaluating the path integral of a functional $F[x(\tau)] = x(\tau)$, in the interval $[0, t]$ via the unpinned Wiener measure. Let τ be any time in the interval $[0, t]$. Using Equation (20.28) and the ESKC relation, we can write the path integral $\int_{C[x_0,0;t]} d_w x(\tau) x(\tau)$ as

$$\int_{C[x_0,0;t]} d_w x(\tau) x(\tau)$$

$$= \int_{-\infty}^{\infty} dx \int_{-\infty}^{\infty} dx_\tau x_\tau \int_{C[x_0,0;x_\tau,\tau]} d_w x(\tau) \int_{C[x_\tau,\tau;x,t]} d_w x(\tau),$$

$$= \int_{-\infty}^{\infty} dx_\tau x_\tau \int_{C[x_0,0;x_\tau,\tau]} d_w x(\tau) \int_{-\infty}^{\infty} dx \int_{C[x_\tau,\tau;x,t]} d_w x(\tau),$$

$$= \int_{-\infty}^{\infty} dx_\tau x_\tau \int_{C[x_0,0;x_\tau,\tau]} d_w x(\tau) \int_{C[x_\tau,\tau;t]} d_w x(\tau). \tag{20.94}$$

From Equation (20.27), the value of the last integral is one. Finally, using Equations (20.24) and (20.22), we obtain

$$
\int_{C[x_0,0;t]} d_w x(\tau) x(\tau)
$$

$$
= \int_{-\infty}^{\infty} dx_\tau x_\tau W(x_\tau, \tau, x_0, 0),
$$

$$
= \int_{-\infty}^{\infty} dx_\tau x_\tau \frac{1}{\sqrt{4\pi D\tau}} \exp\left\{ -\frac{(x_\tau - x_0)^2}{4D\tau} \right\},
$$

$$
= x_0. \tag{20.95}
$$

20.6.3   Path Integrals by the Method of Finite Elements

We now evaluate the path integral we have found above for the functional $F[x(\tau)] = x(\tau)$ by using the formula [Eq. (20.17)]:

$$
\int_{C[x_0,0;t]} d_w x(\tau) x(\tau) \tag{20.96}
$$

$$
= \int dx \int_{C[x_0,0;x,t]} d_w x(\tau) x(\tau) \tag{20.97}
$$

$$
= \int dx_{N+1} \int \cdots \int \prod_{i=1}^{N} \frac{dx_i}{\sqrt{4\pi D(t_i - t_{i-1})}} \exp\left\{ -\sum_{i=1}^{N+1} \frac{(x_i - x_{i-1})^2}{4D(t_i - t_{i-1})} \right\} x_{N+1},
$$

$$
= \int_{-\infty}^{\infty} \frac{dx_{N+1} x_{N+1}}{\sqrt{4\pi D\Delta t_N}} \int \cdots \int \prod_{i=1}^{N} \frac{dx_i}{\sqrt{4\pi D\Delta t_i}} \exp\left\{ -\sum_{i=1}^{N+1} \frac{(x_i - x_{i-1})^2}{4D(t_i - t_{i-1})} \right\},
$$

$$
= \int_{-\infty}^{\infty} dx \frac{x}{\sqrt{4\pi Dt}} \exp\left\{ -\frac{(x - x_0)^2}{4Dt} \right\}, \tag{20.98}
$$

$$
= x_0. \tag{20.99}
$$

In this calculation we have assumed that τ lies in the last time slice denoted by $N + 1$. Complicated functionals can be handled by Equation (20.92).

20.6.4   Path Integrals by the "Semiclassical" Method

We have seen that the propagator of the Bloch equation in the presence of a nonzero diffusion constant is given as

$$
W(x, t, x_0, t_0) = \int_{C[x_0,t_0;x,t]} \exp\left\{ -\frac{1}{4D} \int_{t_0}^{t} d\tau L[x(\tau)] \right\} \prod_{\tau=t_0}^{t} \frac{dx(\tau)}{\sqrt{4\pi D d\tau}}.
$$

$$
\tag{20.100}
$$

Naturally, the major contribution to this integral comes from the paths that satisfy the Euler-Lagrange equation

$$\frac{\partial L}{\partial x} - \frac{d}{d\tau}\left[\frac{\partial L}{\partial(dx/d\tau)}\right] = 0. \qquad (20.101)$$

We show these "classical" paths by $x_c(\tau)$. These paths also make the integral $\int L d\tau$ an extremum, that is,

$$\delta \int L d\tau = 0. \qquad (20.102)$$

However, we should also remember that in the Bloch equation $V(x)$ is not quite the potential and L is not the Lagrangian. Similarly, $\int L d\tau$ in Equation (20.102) is not the action, $S[x(\tau)]$, of classical physics. These expressions gain their conventional meanings only when we apply path integrals to the Schrödinger equation. It is for this reason that we have used the term "semi-classical".

When the diffusion constant is much smaller than the functional S, that is, $D/S << 1$, we write an approximate solution to Equation (20.100) as

$$W(x,t;x_0,t_0) \simeq \phi(t-t_0)\exp\left\{-\frac{1}{4D}\int_{t_0}^{t} d\tau L[x_c(\tau)]\right\}, \qquad (20.103)$$

where $\phi(t-t_0)$ is called the **fluctuation factor**. Even though methods of finding the fluctuation factor are beyond our scope (see Chaichian and Demichev), we give two examples for its appearance and evaluation.

Example 20.1. *Evaluation of* $\int_{C[x_0,0;x,t]} d_w x(\tau)$: To find the propagator $W(x,t,x_0,0)$ we write

$$W(x,t,x_0,0) = \int_{C[x_0,0;x,t]} \exp\left\{-\frac{1}{4D}\int_0^t d\tau \dot{x}^2\right\} \prod_{\tau=0}^{t} \frac{dx(\tau)}{\sqrt{4\pi D d\tau}} \qquad (20.104)$$

and the Euler-Lagrange equation

$$\ddot{x}_c(\tau) = 0, \; x_c(0) = x_0, \; x(t) = x, \qquad (20.105)$$

with the solution

$$x_c(\tau) = x_0 + \frac{\tau}{t}(x - x_0). \qquad (20.106)$$

We show the deviation from the classical path $x_c(\tau)$ as $\eta(\tau)$ so that we write $x(\tau) = x_c(\tau) + \eta(\tau)$. At the end points $\eta(\tau)$ satisfies (Fig. 20.4)

$$\eta(0) = \eta(t) = 0. \qquad (20.107)$$

In terms of $\eta(\tau)$, $W(x, t, x_0, 0)$ is given as

$$W(x, t, x_0, 0) = \exp\left\{-\frac{1}{4D}\int_0^t d\tau \dot{x}_c^2\right\} \tag{20.108}$$

$$\times \int_{C[0,0;0,t]} \exp\left\{-\frac{1}{4D}\int_0^t d\tau\left(\dot{\eta}^2 + 2\,\dot{x}_c\dot{\eta}\right)\right\}\prod_{\tau=0}^t \frac{d\eta(\tau)}{\sqrt{4\pi D d\tau}}.$$

We have to remember that the paths $x(\tau)$ do not have to satisfy the Euler-Lagrange equation. Because we can write

$$\int_0^t d\tau \dot{x}_c \dot{\eta} = \left.\frac{(x - x_0)}{t}\eta(\tau)\right|_0^t = 0,$$

Equation (20.108) is

$$W(x, t, x_0, 0) = \exp\left\{-\frac{1}{4D}\int_0^t d\tau \dot{x}_c^2\right\} \tag{20.109}$$

$$\times \int_{C[0,0;0,t]} \exp\left\{-\frac{1}{4D}\int_0^t d\tau \dot{\eta}^2\right\}\prod_{\tau=0}^t \frac{d\eta(\tau)}{\sqrt{4\pi D d\tau}}.$$

Because $\dot{x}_c = (x - x_0)/t$ is independent of τ, we can evaluate the factor in front of the integral on the right-hand side as

$$\exp\left\{-\frac{1}{4D}\int_0^t d\tau \dot{x}_c^2\right\} = \exp\left\{-\frac{1}{4D}\frac{(x-x_0)^2}{t}\right\}. \tag{20.110}$$

Because the integral

$$\int_{C[0,0;0,t]} \exp\left\{-\frac{1}{4D}\int_{t_0}^t d\tau \dot{\eta}^2\right\}\prod_{\tau=0}^t \frac{d\eta(\tau)}{\sqrt{4\pi D d\tau}} \tag{20.111}$$

only depends on t, we show it as $\phi(t)$ and write the propagator as

$$W(x, t, x_0, 0) = \phi(t)\exp\left\{-\frac{(x-x_0)^2}{4Dt}\right\}. \tag{20.112}$$

The probability density interpretation of the propagator gives us the condition

$$\int_{-\infty}^{\infty} dx W(x, t, x_0, 0) = 1, \tag{20.113}$$

which leads us to the $\phi(t)$ function as

$$\phi(t) = \frac{1}{\sqrt{4\pi Dt}}. \tag{20.114}$$

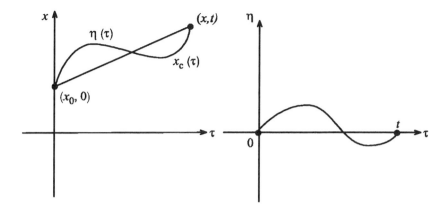

Fig. 20.4 Path and deviation in the "semiclassical" method

Finally the propagator is obtained as

$$W(x, t, x_0, 0) = \frac{1}{\sqrt{4\pi D t}} \exp\left\{ -\frac{(x - x_0)^2}{4Dt} \right\}. \qquad (20.115)$$

In this case the "semiclassical" method has given us the exact result. For more complicated cases we could use the method of time slices to find the factor $\phi(t - t_0)$. In this example we have also given an explicit derivation of Equation (20.22) for $t_0 = 0$, from Equation (20.24).

Example 20.2. *Evaluation of $\varphi(t)$ by the method of time slices:* Because our previous example is the prototype of many path integral applications, we also evaluate the integral

$$\phi(t) = \int_{C[0,0;0,t]} \exp\left\{ -\frac{1}{4D} \int_0^t d\tau \dot\eta^2 \right\} \prod_{\tau=0}^{t} \frac{d\eta(\tau)}{\sqrt{4\pi D d\tau}} \qquad (20.116)$$

by using the method of time slices. We divide the interval $[0, t]$ into $(N + 1)$ equal segments:

$$t_i - t_{i-1} = \varepsilon \qquad (20.117)$$

$$= \frac{t}{(N+1)} \quad i = 1, 2, ..., (N + 1).$$

Now the integral (20.116) becomes

$$\phi(t) = \lim_{N \to \infty, \varepsilon \to 0} \frac{1}{\left[\sqrt{4\pi D \varepsilon}\right]^{N+1}} \qquad (20.118)$$

$$\times \int d\eta_1 \int d\eta_2 \cdots \int d\eta_N \exp\left\{ -\frac{1}{4D\epsilon} \sum_{i=0}^{N} (\eta_{i+1} - \eta_i)^2 \right\}.$$

The argument of the exponential function (aside from a minus sign) is a quadratic of the form

$$\frac{1}{4D\varepsilon}\sum_{i=0}^{N}(\eta_{i+1}-\eta_i)^2 = \sum_{k=1}^{N}\sum_{l=1}^{N}\eta_k A_{kl}\eta_l. \tag{20.119}$$

We can write A_{kl} as an $N \times N$ matrix ($\eta_0 = \eta_{N+1} = 0$):

$$A = \frac{1}{4D\varepsilon}\begin{pmatrix} 2 & -1 & 0 & 0 & \cdots & \cdots & 0 \\ -1 & 2 & -1 & 0 & \cdots & \cdots & 0 \\ 0 & -1 & 2 & -1 & 0 & \cdots & 0 \\ \vdots & \vdots & \vdots & \vdots & \vdots & \vdots & \vdots \\ 0 & \cdots & 0 & -1 & 2 & -1 & 0 \\ 0 & \cdots & \cdots & 0 & -1 & 2 & -1 \\ 0 & \cdots & \cdots & \cdots & 0 & -1 & 2 \end{pmatrix}. \tag{20.120}$$

Using the techniques of linear algebra we can evaluate the integral

$$\int d\eta_1 \int d\eta_2 \cdots \int d\eta_N \exp\left\{-\sum_{k=1}^{N}\sum_{l=1}^{N}\eta_k A_{kl}\eta_l\right\} \tag{20.121}$$

as (Problem 20.7)

$$\int d\eta_1 \int d\eta_2 \cdots \int d\eta_N \exp\left\{-\sum_{k=1}^{N}\sum_{l=1}^{N}\eta_k A_{kl}\eta_l\right\}$$
$$= \frac{(\sqrt{4\pi D\varepsilon})^N}{\sqrt{\det A_N}}. \tag{20.122}$$

Using the last column of A, we find a recursion relation that $\det A_N$ satisfies:

$$\det A_N = 2\det A_{N-1} - \det A_{N-2}. \tag{20.123}$$

For the first two values of N, $\det A_N$ is found as

$$\det A_1 = 2 \tag{20.124}$$

and

$$\det A_2 = 3. \tag{20.125}$$

This can be generalized to $N-1$ as

$$\det A_{N-1} = N. \tag{20.126}$$

Using the recursion relation [Eq. (20.123)], this gives

$$\det A_N = N + 1, \tag{20.127}$$

which leads us to the $\phi(t)$ function

$$\phi(t) = \frac{(\sqrt{4\pi D\varepsilon})^N}{(\sqrt{4\pi D\varepsilon})^{N+1}\sqrt{N+1}}$$

$$= \frac{1}{\sqrt{4\pi Dt}}. \tag{20.128}$$

Another way to calculate the integral in Equation (20.116) is to evaluate the η integrals one by one using the formula

$$\int_{-\infty}^{\infty} d\eta \exp\left\{-a(\eta - \eta')^2 - b(\eta - \eta'')^2\right\}$$

$$= \left[\frac{\pi}{a+b}\right]^{1/2} \exp\left\{-\frac{ab}{a+b}(\eta' - \eta'')^2\right\}. \tag{20.129}$$

20.7 FEYNMAN PATH INTEGRAL FORMULATION OF QUANTUM MECHANICS

20.7.1 Schrödinger Equation for a Free Particle

We have seen that the propagator for a particle undergoing Brownian motion with its initial position at (x_0, t_0) is given as

$$W(x, t, x_0, t_0) = \frac{1}{\sqrt{4\pi D(t-t_0)}} \exp\left\{-\frac{(x-x_0)^2}{4D(t-t_0)}\right\}. \tag{20.130}$$

This satisfies the diffusion equation

$$\frac{\partial W(x, t, x_0, t_0)}{\partial t} = D\frac{\partial^2 W(x, t, x_0, t_0)}{\partial x^2} \tag{20.131}$$

with the initial condition $\lim_{t \to t_0} W(x, t, x_0, t_0) \to \delta(x - x_0)$. We have also seen that this propagator can also be written as a Wiener path integral:

$$W(x, t, x_0, t_0) = \int_{C[x_0, t_0; x, t]} d_w x(\tau). \tag{20.132}$$

In this integral $C[x_0, t_0; x, t]$ denotes all continuous paths starting from (x_0, t_0) and ending at (x, t), where $d_w x(\tau)$ is called the Wiener measure and is given as

$$d_w x(\tau) = \exp\left\{-\frac{1}{4D}\int_{t_0}^{t} \dot{x}^2(\tau)d\tau\right\} \prod_{i=1}^{N} \frac{dx_i}{\sqrt{4\pi D d\tau}}. \tag{20.133}$$

For a free particle of mass m Schrödinger's equation is given as

$$\frac{\partial \Psi(x,t)}{\partial t} = \frac{i\hbar}{2m} \frac{\partial^2 \Psi(x,t)}{\partial x^2}. \tag{20.134}$$

For this equation, propagator $K(x,t,x',t')$ satisfies the equation

$$\frac{\partial K(x,t,x',t')}{\partial t} = \frac{i\hbar}{2m} \frac{\partial^2 K(x,t,x',t')}{\partial x^2}. \tag{20.135}$$

Given the solution at (x',t'), we can find the solution at another point (x,t) by using this propagator as

$$\Psi(x,t) = \int K(x,t,x',t')\Psi(x',t')dx', \quad (t > t'). \tag{20.136}$$

Because the diffusion equation becomes the Schrödinger equation by the replacement

$$D \to \frac{i\hbar}{2m}, \tag{20.137}$$

we can immediately write the propagator of the Schrödinger equation by making the same replacement in Equation (20.130):

$$K(x,t,x',t') = \frac{1}{\sqrt{\dfrac{2\pi i\hbar}{m}(t-t_0)}} \exp\left\{-\frac{m(x-x_0)^2}{2i\hbar(t-t_0)}\right\}. \tag{20.138}$$

Even though this expression is mathematically correct, at this point we begin to encounter problems and differences between the two cases. For the diffusion phenomena, we have said that the solution of the diffusion equation gives the probability of finding a Brown particle at (x,t). Thus the propagator,

$$W(x,t,x_0,t_0) = \frac{1}{\sqrt{4\pi D(t-t_0)}} \exp\left\{-\frac{(x-x_0)^2}{4D(t-t_0)}\right\}, \tag{20.139}$$

is always positive and satisfies the normalization condition

$$\int_{-\infty}^{\infty} dx W(x,t,x_0,t_0) = 1. \tag{20.140}$$

For the Schrödinger equation the argument of the exponential function is proportional to i, which makes $K(x,t,x',t')$ oscillate violently; hence $K(x,t,x',t')$ cannot be normalized. This is not too surprising, because the solutions of the Schrödinger equation are the probability amplitudes, which are more fundamental, and thus carry more information than the probability density. In

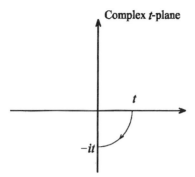

Fig. 20.5 Rotation by $-\frac{\pi}{2}$ in the complex-t plane

quantum mechanics probability density, $\rho(x,t)$, is obtained from the solutions of the Schrödinger equation by

$$\rho(x,t) = \Psi(x,t)\Psi^*(x,t) \tag{20.141}$$
$$= |\Psi(x,t)|^2,$$

where $\rho(x,t)$ is now positive definite and a Gaussian, which can be normalized.

Can we also write the propagator of the Schrödinger equation as a path integral? Making the $D \to \dfrac{i\hbar}{2m}$ replacement in Equation (20.132) we get

$$K(x,t,x',t') = \int_{C[x',t';x,t]} d_F x(\tau), \tag{20.142}$$

where

$$d_F x(\tau) = \exp\left\{\frac{i}{\hbar}\int_{t_0}^{t}\frac{1}{2}m\dot{x}^2(\tau)d\tau\right\}\prod_{i=1}^{N}\frac{dx_i}{\sqrt{\dfrac{2\pi i\hbar}{m}}\,d\tau}. \tag{20.143}$$

This definition was given first by Feynman, and $d_F x(\tau)$ is known as the Feynman measure. The problem in this definition is again the fact that the argument of the exponential, which is responsible for the convergence of the integral, is proportional to i, and thus the exponential factor oscillates. An elegant solution to this problem comes from noting that the Schrödinger equation is analytic in the lower half complex t-plane. Thus we make a rotation by $-\pi/2$ and write $-it$ instead of t in the Schrödinger equation (Fig. 20.5). This reduces the Schrödinger equation to the diffusion equation with the diffusion constant $D = \hbar/2m$. Now the path integral in Equation (20.142) can be taken as a Wiener path integral, and then going back to real time, we can obtain the propagator of the Schrödinger equation as Equation (20.138).

20.7.2 Schrödinger Equation in the Presence of Interactions

In the presence of interactions the Schrödinger equation is given as

$$\frac{\partial \Psi(x,t)}{\partial t} = \frac{i\hbar}{2m}\frac{\partial^2 \Psi(x,t)}{\partial x^2} - \frac{i}{\hbar}V(x)\Psi(x,t) \ . \tag{20.144}$$

Making the transformation $t \to -it$, we obtain the Bloch equation as

$$\frac{\partial \Psi(x,t)}{\partial t} = \frac{\hbar}{2m}\frac{\partial^2 \Psi(x,t)}{\partial x^2} - \frac{1}{\hbar}V(x)\Psi(x,t). \tag{20.145}$$

Using the Feynman-Kac theorem, we write the propagator and then transform back to real time to get

$$K(x,t,x_0,t_0) = \int_{C[x_0,t_0;x,t]} d_F x(\tau)\exp\left\{-\frac{i}{\hbar}\int_{t_0}^{t}d\tau V[x(\tau)]\right\} \tag{20.146}$$

or

$$K(x,t,x_0,t_0) = \int_{C[x_0,t_0;x,t]} \prod_{i=1}^{N} \frac{dx_i}{\sqrt{\frac{2i\hbar}{m}\pi d\tau}}\exp\left\{\frac{i}{\hbar}\int_{t_0}^{t}\left[\frac{1}{2}m\dot{x}^2(\tau)-V(x)\right]d\tau\right\}. \tag{20.147}$$

This propagator was first given by Feynman. Using this propagator we can write the solution of the Schrödinger equation as

$$\Psi(x,t) = \int K(x,t,x',t')\Psi(x',t')dx'. \tag{20.148}$$

Today the path integral formulation of quantum mechanics, after the Schrödinger and the Heisenberg formulations, has become the foundation of modern quantum mechanics. Writing the propagator [Eq. (20.147)] as

$$K(x,t,x_0,t_0) = \int_{C[x_0,t_0;x,t]} \prod_{i=1}^{N} \frac{dx_i}{\sqrt{\frac{2i\hbar}{m}\pi d\tau}}\exp\left\{\frac{i}{\hbar}S[x(\tau)]\right\}, \tag{20.149}$$

we see that, in contrast to the Bloch equation, $S[x(\tau)]$ in the propagator is the classical action:

$$S[x(\tau)] = \int_{t_0}^{t} L[x(\tau)]d\tau, \tag{20.150}$$

where $L[x(\tau)]$ is the classical Lagrangian:

$$L[x(\tau)] = \left[\frac{1}{2}m\dot{x}^2(\tau) - V(x)\right]. \tag{20.151}$$

In the Feynman propagator (20.149) we should note that $C[x_0, t_0; x, t]$ includes not only the paths that satisfy the Euler-Lagrange equation, but all the continuous paths that start from (x_0, t_0) and end at (x, t). In the classical limit, that is,

$$\hbar \to 0, \tag{20.152}$$

the exponential term

$$\exp\left\{\frac{i}{\hbar} S[x(\tau)]\right\} \tag{20.153}$$

in the propagator oscillates violently. In this case the major contribution to the integral comes from the paths with comparable weights bunched together. These paths are naturally the ones that make $S[x]$ an extremum, that is, the paths that satisfy the Euler-Lagrange equation

$$\frac{d}{dt}\left(\frac{\partial L}{\partial \dot{x}}\right) - \frac{\partial L}{\partial x} = 0. \tag{20.154}$$

In most cases this extremum is a minimum (Morse and Feshbach, p. 281).

As in the applications of path integrals to neural networks, sometimes a system can have more than one extremum. In such cases, a system could find itself in a local maximum or minimum. Is it then possible for such systems to reach the desired global minimum? If possible, how is this achieved and how long will it take? These are all very interesting questions and potential research topics, indicating that path integral formalism still has a long way to go.

20.8 FEYNMAN PHASE SPACE PATH INTEGRAL

The propagator of the Schrödinger equation expressed as a path integral,

$$K(x, t, x_0, t_0) = \int_{C[x_0, t_0; x, t]} \prod_{i=1}^{N} \frac{dx_i}{\sqrt{\frac{2\pi i \hbar}{m} d\tau}} \exp\left\{\frac{i}{\hbar}\int_{t_0}^{t}\left[\frac{1}{2}m\dot{x}^2(\tau) - V(x)\right] d\tau\right\}, \tag{20.155}$$

is useful if the Lagrangian can be expressed as $T - V$. However, as in the case of the Lagrangian of a free relativistic particle,

$$L(x) = -m_0 c^2 \sqrt{1 - \frac{\dot{x}^2}{c^2}}, \tag{20.156}$$

where the Lagrangian cannot be written as $T - V$, it is not much of a help. For this reason in 1951 Feynman introduced the phase space version of the path integral:

$$K(q'',t'',q',t') = \check{N} \int \exp\left\{\frac{i}{\hbar} \int_{t'}^{t''} dt \left[p\dot{q} - H(p,\dot{q})\right]\right\} \check{D}p\check{D}q. \quad (20.157)$$

This integral is to be taken over t, where $t \in [t',t'']$. $\check{D}q$ means that the integral is taken over the paths $q(t)$, fixed between $q''(t'') = q''$ and $q'(t') = q'$ and which make $S[x]$ an extremum. The integral over momentum p is taken over the same time interval but without any restrictions.

To bring this integral into a form that can be evaluated in practice, we introduce the phase space lattice by dividing the time interval $t \in [t',t'']$ into $N+1$ slices as

$$\varepsilon = \frac{t''-t'}{(N+1)}. \quad (20.158)$$

Now the propagator becomes

$$K(q'',t'',q',t')$$
$$= \lim_{\varepsilon \to 0} \int \cdots \int \exp\left\{\left(\frac{i}{\hbar}\right) \sum_{l=0}^{N} \left[p_{l+1/2}(q_{l+1} - q_l) - \varepsilon H\left(p_{l+1/2}, \frac{1}{2}(q_{l+1} + q_l)\right)\right]\right\}$$
$$\times \prod_{l=0}^{N} \frac{dp_{l+1/2}}{2\pi\hbar} \prod_{l=1}^{N} dq_l. \quad (20.159)$$

In this expression, except for the points at $q_{N+1} = q''$ and $q_0 = q'$, we have to integrate over all q and p. Because the Heisenberg uncertainty principle forbids us from determining the momentum and position simultaneously at the same point, we have taken the momentum values at the center of the time slices as $p_{l+1/2}$. In this equation one extra integral is taken over p. It is easily seen that this propagator satisfies the ESKC relation:

$$K(q''',t''',q',t') = \int K(q''',t''',q'',t'')K(q'',t'',q',t')dq''. \quad (20.160)$$

20.9   FEYNMAN PHASE SPACE PATH INTEGRAL IN THE PRESENCE OF QUADRATIC DEPENDENCE ON MOMENTUM

In phase space, the exponential function in the Feynman propagator [Eq. (20.159)] is written as

$$\exp\left\{\left(\frac{i}{\hbar}\right) \sum_{l=0}^{N} \left[p_{l+1/2}(q_{l+1} - q_l) - \varepsilon H\left(p_{l+1/2}, \frac{1}{2}(q_{l+1} + q_l)\right)\right]\right\}.$$

'hen the Hamiltonian has quadratic dependence on p as in

$$H(q,p) = \frac{p^2}{2m} + V(q,t), \tag{20.161}$$

is exponential function becomes

$$\exp\left\{\left(\frac{i}{\hbar}\right)\sum_{l=0}^{N}\varepsilon\left[\frac{p_{l+1/2}(q_{l+1}-q_l)}{\varepsilon} - \frac{p_{l+1/2}^2}{2m} - V\left(\frac{1}{2}(q_l + q_{l+1}), t_l\right)\right]\right\}. \tag{20.162}$$

ompleting the square in the expression inside the brackets we can write this

$$\exp\left\{\left(\frac{i}{\hbar}\right)\sum_{l=0}^{N}\varepsilon\right.$$
$$\times\left[-\frac{1}{2m}\left(p_{l+1/2} - \frac{(q_{l+1}-q_l)}{\varepsilon}m\right)^2 + \frac{m}{2}\left(\frac{q_{l+1}-q_l}{\varepsilon}\right)^2 - V\left(\frac{q_l + q_{l+1}}{2}, t_l\right)\right]\right\}. \tag{20.163}$$

ıbstituting this in Equation (20.159) and taking the momentum integral, we
d the propagator as

$$K(q'',t'',q',t') = \lim_{\substack{N\to\infty \\ \varepsilon\to 0}} \frac{1}{\sqrt{2\pi i\hbar\frac{\varepsilon}{m}}}\prod_{l=0}^{N}\int_{-\infty}^{\infty}\left[\frac{dq_l}{\sqrt{2\pi i\hbar\frac{\varepsilon}{m}}}\right]\exp\left\{\frac{i}{\hbar}S\right\}, \tag{20.164}$$

ıere S is given as

$$S = \sum_{l=0}^{N}\varepsilon\left[\frac{m}{2}\left(\frac{(q_{l+1}-q_l)}{\varepsilon}\right)^2 - V\left(\frac{1}{2}(q_l + q_{l+1}), t_l\right)\right]. \tag{20.165}$$

the continuum limit this becomes

$$S[q] = \lim_{\substack{N\to\infty \\ \varepsilon\to 0}}\sum_{l=0}^{N}\varepsilon\left[\frac{m}{2}\left(\frac{q_{l+1}-q_l}{\varepsilon}\right)^2 - V\left(\frac{1}{2}(q_l + q_{l+1}), t_l\right)\right]$$
$$= \int_{t'}^{t''} dt\, L[q,\dot{q},t], \tag{20.166}$$

ere

$$L[q,\dot{q},t] = \frac{1}{2}m\dot{q}^2 - V(q,t)$$

the classical action. In other words, the phase space path integral reduces
the standard Feynman path integral.

We can write the free particle propagator in terms of the phase space path integral as

$$K(x,t,x_0,t_0) = \check{N} \int_{C[x_0,t_0;x,t]} \check{D}p\check{D}x \exp\left\{\frac{i}{\hbar}\int_{t_0}^t d\tau\left[p\dot{x} - \frac{p^2}{2m}\right]\right\}. \quad (20.167)$$

After we take the momentum integral and after putting all the new constants coming into $\check{D}$, Equation (20.167) becomes

$$K(x,t,x_0,t_0) = \check{N} \int_{C[x_0,t_0;x,t]} \check{D}x \exp\left\{\frac{i}{\hbar}\int_{t_0}^t d\tau\left(\frac{1}{2}m\dot{x}^2\right)\right\}. \quad (20.168)$$

We can convert this into a Wiener path integral by the $t \to -it$ rotation, and after evaluating it, we return to real time to obtain the propagator as

$$K(x,t,x_0,t_0) = \frac{1}{\sqrt{2\pi i\hbar(t-t_0)/m}} \exp\frac{i}{\hbar}\frac{m(x-x_0)^2}{2(t-t_0)}. \quad (20.169)$$

We conclude by giving the following useful rules for path integrals with $N+1$ segments [Eq. (20.15)]:

For the pinned Wiener measure:

$$\int d_w x(\tau) = \frac{1}{(4\pi D\varepsilon)^{(N+1)/2}} \int_{-\infty}^{\infty} dx_1 \cdots \int_{-\infty}^{\infty} dx_N \exp\left\{-\frac{1}{4D\varepsilon}\sum_{i=1}^{N+1}(x_i - x_{i-1})^2\right\},$$

$$\int \prod_{\tau=0}^t \frac{dx(\tau)}{\sqrt{4\pi D d\tau}} = \frac{1}{(4\pi D\varepsilon)^{(N+1)/2}} \int_{-\infty}^{\infty} dx_1 \cdots \int_{-\infty}^{\infty} dx_N. \quad (20.170)$$

For the unpinned Wiener measure:

$$\int d_w x(\tau) = \frac{1}{(4\pi D\varepsilon)^{(N+1)/2}} \int_{-\infty}^{\infty} dx_1 \cdots \int_{-\infty}^{\infty} dx_{N+1} \exp\left\{-\frac{1}{4D\varepsilon}\sum_{i=1}^{N+1}(x_i - x_{i-1})^2\right\}.$$

$$\int \prod_{\tau=0}^t \frac{dx(\tau)}{\sqrt{4\pi D d\tau}} = \frac{1}{(4\pi D\varepsilon)^{(N+1)/2}} \int_{-\infty}^{\infty} dx_1 \cdots \int_{-\infty}^{\infty} dx_{N+1}. \quad (20.171)$$

Also,

$$\int_0^t d\tau\dot{x}^2(\tau) = \frac{1}{\varepsilon}\sum_{i=1}^{N+1}(x_i - x_{i-1})^2, \quad (20.172)$$

$$\int_0^t d\tau V(\tau) = \varepsilon\sum_{i=1}^{N+1} V\left(\frac{1}{2}(x_i + x_{i-1}), t_i\right) \overset{\text{or}}{=} \varepsilon\sum_{i=1}^{N+1} V(x_i, t_i). \quad (20.173)$$

Problems

20.1   Show that

$$W(x, t, x_0, t_0) = \frac{1}{\sqrt{4\pi D(t - t_0)}} \exp\left\{-\frac{(x - x_0)^2}{4D(t - t_0)}\right\}$$

satisfies the normalization condition

$$\int_{-\infty}^{\infty} dx W(x, t, x_0, t_0) = 1.$$

20.2   By differentiating both sides with respect to t show that the following equation is true:

$$\exp\left\{-\int_0^t d\tau V[x(\tau)]\right\}$$
$$= 1 - \int_0^t d\tau \left(V[x(\tau)] \exp\left\{-\int_0^\tau ds V[x(s)]\right\}\right).$$

20.3   Show that $V(x)$ in Equation (20.64):

$$\frac{\partial \widetilde{W}(x, t; x_0, t_0)}{\partial t} - D\frac{\partial^2 \widetilde{W}(x, t; x_0, t_0)}{\partial x^2} = V(x)\widetilde{W}(x, t; x_0, t_0),$$

is defined as

$$V(x) = \frac{1}{4\eta^2 D}F^2(x) + \frac{1}{2\eta}\frac{dF(x)}{dx}.$$

20.4   Show that the propagator

$$W(x, t, x_0, t_0) = \frac{1}{\sqrt{4\pi D(t - t_0)}} \exp\left\{-\frac{(x - x_0)^2}{4D(t - t_0)}\right\}$$

satisfies the ESKC relation [Eq. (20.14)].

20.5   Derive equation

$$W_B(x, t, x_0, t_0) = W(x, t, x_0, t_0)$$
$$-\varepsilon \sum_{j=1}^{N} \int_{-\infty}^{\infty} dx_j W(x, t, x_j, t_j)V(x_j, t_j)W(x_j, t_j, x_0, t_0)$$
$$+\frac{1}{2!}\varepsilon^2 \sum_{j=1}^{N}\sum_{k=1}^{N} \int_{-\infty}^{\infty} dx_j \int_{-\infty}^{\infty} dx_k W(x, t, x_j, t_j)V(x_j, t_j)W(x_j, t_j, x_k, t_k)$$
$$\times V(x_k, t_k)W(x_k, t_k, x_0, t_0) + \cdots .$$

given in Section 20.3.

20.6   Using the semiclassical method show that the result of the Wiener integral

$$W(x, t, x_0, t_0) = \int_{C[x_0, 0; x, t]} d_w x(\tau) \exp\left\{-k^2 \int_{t_0}^t d\tau \dot{x}^2\right\}$$

is given as

$$W(x, t, x_0, t_0) = \left[\frac{k}{2\pi\sqrt{D}\sinh(2k\sqrt{D}(t - t_0))}\right]^{(1/2)}$$

$$\times \exp\left\{-k\frac{(x^2 + x_0^2)\cosh(2k\sqrt{D}(t - t_0)) - 2x_0 x}{2\sqrt{D}\sinh(2k\sqrt{D}(t - t_0))}\right\}.$$

20.7   By diagonalizing the real symmetric matrix, A, show that

$$\int d\eta_1 \int d\eta_2 \cdots \int d\eta_N \exp\left\{-\sum_{k=1}^N \sum_{l=1}^N \eta_k A_{kl} \eta_l\right\} = \frac{(\sqrt{\pi})^N}{\sqrt{\det A}}.$$

20.8   Use the formula

$$\int_{-\infty}^{\infty} d\eta \exp\left\{-a(\eta - \eta')^2 - b(\eta - \eta'')^2\right\}$$

$$= \left[\frac{\pi}{a + b}\right]^{1/2} \exp\left\{-\frac{ab}{a + b}(\eta' - \eta'')^2\right\}$$

to evaluate the integral

$$\phi(t) = \int_{C[0,0;0,t]} \exp\left\{-\frac{1}{4D}\int_0^t d\tau \dot{\eta}^2\right\} \prod_{\tau=0}^t \frac{d\eta(\tau)}{\sqrt{4\pi D d\tau}}.$$

20.9   By taking the momentum integral in Equation (20.159) derive the propagator Equation (20.164):

$$K(q'', t'', q', t') = \lim_{\substack{N \to \infty \\ \varepsilon \to 0}} \frac{1}{\sqrt{2\pi i\hbar\frac{\varepsilon}{m}}} \prod_{l=1}^N \int_{-\infty}^{\infty} \left[\frac{dq_l}{\sqrt{2\pi i\hbar\frac{\varepsilon}{m}}}\right] \exp\left\{\frac{i}{\hbar}S\right\},$$

where S is given as

$$S = \sum_{l=0}^N \varepsilon\left[\frac{m}{2}\left(\frac{(q_{l+1} - q_l)}{\varepsilon}\right)^2 - V\left(\frac{1}{2}(q_l + q_{l+1}), t_l\right)\right].$$

References

. Akhiezer, N.I., *The Calculus of Variations*, Blaisdell, New York, 1962

. Arfken, G. B., and H. J. Weber, *Essential Mathematical Methods for Physicists*, Academic Press, 2003.

. Arfken, G. B., and H. J. Weber, *Mathematical Methods of Physics*, Academic Press, sixth edition, 2005.

. Artin, E., *The Gamma Function*, Holt, Rinehart and Winston, New York, 1964.

. Beiser, A., *Concepts of Modern Physics*, McGraw-Hill, sixth edition, 2002.

. Bell, W.W., *Special Functions for Scientists and Engineers*, Dover Publications, 2004.

. Bluman, W. B., and Kumei, S., *Symmetries and Differential Equations*, Springer Verlag, New York, 1989.

. Boas, M.L., *Mathematical Methods in the Physical Sciences*, Wiley, third edition, 2006.

. Bradbury, T.C., *Theoretical Mechanics*, Wiley, international edition, 1968.

. Bromwich, T.J.I., *Infinite Series*, Chelsea Publishing Company, 1991.

. Brown, J.W., and R.V. Churchill, *Complex Variables and Applications*, McGraw-Hill, New York, 1995.

. Butkov, E., *Mathematical Physics*, Addison-Wesley, New York, 1968.

. Byron, W. Jr., and R.W. Fuller, *Mathematics of Classical and Quantum Physics*, Dover Publications, New York, 1970.

. Chaichian, M., and A. Demichev, *Path Integrals in Physics, Volume I and II,* Institute of Physics Publishing, 2001.

. Churchill, R.V., *Fourier Series and Boundary Value Problems*, McGraw-Hill, New York, 1963.

. Courant, E., and D. Hilbert, *Methods of Mathematical Physics, Volume I and II*, Wiley, New York, 1991.

. Dennery, P., and A. Krzywicki, *Mathematics for Physics*, Dover Publications, New York, 1995.

. Doniach, S., and E.H. Sondheimer, *Green's Functions for Solid State Physics*, World Scientific, 1998.

. Dwight, H.B., *Tables of Integrals and Other Mathematical Data*, Prentice Hall, fourth edition, 1961.

. Erdelyi, A., *Asymptotic Expansions*, Dover Publications, New York, 1956.

. Erdelyi, A., Oberhettinger, M.W., and Tricomi. F.G., *Higher Transcendental Functions*, Krieger, vol. I, New York,1981.

. Feynman, R., R.B. Leighton, and M. Sands, *The Feynman Lectures on Physics,* Addison-Wesley, 1966.

. Feynman, R., and Hibbs, A.R., *Quantum Mechanics and Path Integrals*, McGraw-Hill,1965.

. Gantmacher, F.R., *The Theory of Matrices*, Chelsea Publishing Company, New York, 1960.

. Gluzman, S., and D. Sornette, Log Periodic Route to Fractal Functions, *Physical Review,* **E65**, 036142, (2002).

. Goldstein, H., C. Poole, and J. Safko, *Classical Mechanics*, Addison-Wesley, third edition, 2002.

. Hamermesh, M., *Group Theory and its Application to Physical Problems*, Addison-Wesley, 1962.

. Hartle, J.B., *An Introduction to Einstein's General Relativity*, Addison-Wesley, 2003.

. Hassani, S., *Mathematical Methods: for Students of Physics and Related Fields*, Springer Verlag, 2000.

. Hassani, S., *Mathematical Physics*, Springer Verlag, second edition, 2002.

. Hildebrand, F.B., *Methods of Applied Mathematics*, Dover Publications, second reprint edition, 1992.

. Hilfer, R., *Applications of Fractional Calculus*, World Scientific, 2000.

. Hydon, P. E., *Symmetry Methods for Differential Equations: A Beginner's Guide*, Cambridge, 2000.

. Ince, E.L., *Ordinary Differential Equations*, Dover Publications, New York, 1958.

. Infeld, L., and T.E. Hull, The Factorization Method, *Reviews of Modern Physics*, **23**, 21-68 (1951).

. Jackson, J.D., *Classical Electrodynamics*, Wiley, third edition, 1999.

. Jacobson, T.A., and R. Parentani, An Echo of Black Holes, *Scientific American*, December, 48–55 (2005).

. Kaplan, W., *Advanced Calculus*, Addison-Wesley, New York, 1973.

. Kleinert, H., *Path Integrals in Quantum Mechanics, Statistics, Polymer Physics and Financial Markets*, World Scientific, third edition, 2003.

. Lambrecht, A., The Casimir Effect: A Force from Nothing, *Physics World*, September, 29-32 (2002).

. Lebedev, N.N., *Special Functions and their Applications*, Prentice-Hall, 1965.

. Lebedev, N.N., I.P. Skolskaya, and Uflyand, *Problems of Mathematical Physics*, Prentice-Hall, Englewood Cliffs, 1965.

. Marion, J. B., *Classical Dynamics of Particles and Systems*, Academic Press, second edition, 1970.

. Mathews, J., and R.W. Walker, *Mathematical Methods of Physics*, Addison-Wesley, Marlo Park, second edition, 1970.

. McCollum, P.A., and B.F. Brown, *Laplace Transform Tables and Theorems*, Holt, Rinehart and Winston, New York, 1965.

. Milton, K.A., *The Casimir Effect*, World Scientific, 2001.

. Morse, P. M., and H. Feshbach, *Methods of Theoretical Physics*, McGraw-Hill, 1953.

. Oldham, B.K., and J. Spanier, *The Fractional Calculus*, Academic Press, 1974.

. Osler, T.J., Leibniz Rule for Fractional Derivatives and an Application to Infinite Series, *SIAM, Journal of Applied Math*ematics, **18**, 658-674 (1970).

. Osler, T.J., The Integral Analogue of the Leibniz Rule, *Mathematics of Computation*, **26**, 903-915 (1972).

. Pathria, R. K., *Statistical Mechanics*, Pergamon Press, 1984.

. Podlubny, I., *Fractional Differential Equations*, Academic Press, 1999.

. Rektorys, K., *Survey of Applicable Mathematics Volumes I and II*, Springer, second revised edition, 1994.

. Roach, G. F., *Green's Functions*, Cambridge University Press, second edition, 1982.

. Ross, S.L., *Differential Equations*, Wiley, New York, third edition, 1984.

. Samko, S.G., A.A. Kilbas, and O.I. Marichev, *Fractional Integrals and Derivatives*, Gordon and Breach Science Publishers, 1993.

. Schulman, L.S., Techniques and Applications of Path Integration, Dover Publications, 2005.

. Sokolov, I.M., J. Klafter, and A. Blumen, Fractional Kinetics, *Physics Today*, November 2002, pgs.48−54.

. Spiegel, M.R., *Advanced Mathematics for Engineers and Scientists: Schaum's Outline Series in Mathematics*, McGraw-Hill, 1971.

. Stephani, H., *Differential Equations-Their Solutions Using Symmetries*, Cambridge University Press, 1989.

. Szekerez, P., *A Course in Modern Mathematical Physics: Group, Hilbert Space and Differential Geometry*, Cambridge University Press, 2004.

. Titchmarsh, E.C., *The Theory of Functions*, Oxford University Press, New York, 1939.

. Wan, F.Y.M., *Introduction to the Calculus of Variations and its Applications*, ITP, 1995.

. Whittaker, E.T., and G.N. Watson, *A Course on Modern Analysis*, Cambridge University Press, New York, 1958.

. Wyld, H. W., *Mathematical Methods for Physics*, Addison-Wesley, New York, 1976.

Index